CLIMATE CHANGE AND OCEAN GOVERNANCE

Politics and Policy for Threatened Seas

Climate Change and Ocean Governance brings together authors from political science and cognate disciplines to examine the political and policy dimensions of climate change for our oceans. The environmental, social, and economic consequences of oceanic change present tremendous challenges for governments and other actors. New and innovative policies for governing oceans and seas – and for managing vital marine resources – have never been more important. Existing national and international institutions for marine governance that were created when oceanic conditions were relatively static may not be adequate for a future characterized by continuous oceanic change. Responses to oceanic change will result in winners and losers, and thus will involve politics in all its manifestations. This book reveals the unavoidable connections between climate change, the oceans, and questions of governance. It provides valuable lessons for researchers, policymakers, and activists concerned about governing oceanic change into the future.

PAUL G. HARRIS is a political scientist and the Chair Professor of Global and Environmental Studies at the Education University of Hong Kong. He is author and editor of many books on global environmental politics, policy, and ethics, including *Global Ethics and Climate Change, Second Edition* (2016, Edinburgh University Press), the *Routledge Handbook of Global Environmental Politics* (2016, Routledge), and *What's Wrong with Climate Politics and How to Fix It* (2013, Polity).

CLIMATE CHANGE AND OCEAN GOVERNANCE

Politics and Policy for Threatened Seas

Edited by

PAUL G. HARRIS
Education University of Hong Kong

CAMBRIDGE
UNIVERSITY PRESS

CAMBRIDGE
UNIVERSITY PRESS

University Printing House, Cambridge CB2 8BS, United Kingdom

One Liberty Plaza, 20th Floor, New York, NY 10006, USA

477 Williamstown Road, Port Melbourne, VIC 3207, Australia

314–321, 3rd Floor, Plot 3, Splendor Forum, Jasola District Centre, New Delhi – 110025, India

79 Anson Road, #06-04/06, Singapore 079906

Cambridge University Press is part of the University of Cambridge.

It furthers the University's mission by disseminating knowledge in the pursuit of education, learning, and research at the highest international levels of excellence.

www.cambridge.org
Information on this title: www.cambridge.org/9781108422482
DOI: 10.1017/9781108502238

© Cambridge University Press 2019

First published 2019

Printed and bound in Great Britain by Clays Ltd, Elcograf S.p.A.

A catalogue record for this publication is available from the British Library.

ISBN 978-1-108-42248-2 Hardback

Additional resources for this publication at www.cambridge.org/harris2019

Contents

Contributors

Justin Alger is a doctoral candidate in Political Science at the University of British Columbia.

Mohammad Al-Saidi is an assistant professor for Sustainable Development Policy at the Center for Sustainable Development, Qatar University.

Lisa Benjamin is an assistant professor in the LL.B. Program and co-founder of the Climate Change Initiative at the University of the Bahamas.

Thomas L. Brewer is a senior fellow at the International Center for Trade and Sustainable Development, and a visiting scholar at the MIT Center for Energy and Environmental Policy Research.

Dorothy J. Dankel is a researcher in the Department of Biological Sciences at the University of Bergen, Norway and a board member of the Nordic Marine Think Tank.

Pedro Fidelman is a senior research fellow in the Centre for Policy Futures at the University of Queensland, Australia.

David Glassom is a lecturer in the School of Life Sciences at the University of KwaZulu-Natal, South Africa.

Arne Harms is a lecturer at Uni Leipzig's Institute of Anthropology and senior researcher at the Collaborative Research Center on Processes of Spatialization under the Global Condition.

Paul G. Harris is the Chair Professor of Global and Environmental Studies at the Education University of Hong Kong.

Marcus Haward is a professor of Ocean and Antarctic Governance at the Institute for Marine and Antarctic Studies, University of Tasmania, Australia.

Benjamin Hofmann is a research associate and doctoral candidate at the Institute of Political Science, University of St.Gallen, Switzerland.

Yanasivan Kisten is a postdoctoral fellow in the Department of Zoology at Nelson Mandela University, South Africa.

Konstantia Koutouki is a professor of law at the faculty de droit, Université de Montréal, and executive director of the Nomomente Institute, Canada.

Anastasia Kuswardani is a researcher at the Agency for Marine and Fisheries Research and Human Resource Development, Ministry of Marine Affairs and Fisheries, Republic of Indonesia.

Judith van Leeuwen is an assistant professor in the Environmental Policy Group at Wageningen University, the Netherlands.

Reuben Makomere is a doctoral candidate in the Faculty of Law, University of Tasmania, Hobart, Australia.

Kennedy Liti Mbeva is a doctoral candidate in the Department of Political Science at the University of Melbourne, Australia.

Ian McGregor is a lecturer in Management at the UTS Business School, University of Technology, Sydney (UTS).

Elizabeth Mendenhall is an assistant professor in the Department of Marine Affairs at the University of Rhode Island.

Wendy E. Morrison is an analyst of Fishery Management Policy at the US National Marine Fisheries Service.

Kapil Narula is a senior research fellow at the University of Geneva and a former Indian naval officer.

John T. Oliver is the senior ocean policy advisor on the Emerging Policy Staff, US Coast Guard Headquarters, and an adjunct professor at the Georgetown Law Center, Washington, DC.

Joe-Mar Perez is a training specialist at the Office of Civil Defense, the Philippines, and a member of the ASEAN Emergency Response and Assessment Team.

Freedom-Kai Phillips is a research associate with the International Law Research Program at the Centre for International Governance Innovation, Canada.

Kamleshan Pillay is a climate finance researcher at the Center for International Climate and Environmental Research-Oslo (CICERO), Norway.

Achmad Poernomo is the senior research scientist and lecturer, Agency for Marine and Fisheries Research and Human Resource Development, Ministry of Marine Affairs and Fisheries, Republic of Indonesia.

Christina Reichert is a policy counsel of the Climate and Energy Program, Nicholas Institute for Environmental Policy Solutions, Duke University.

Ori Sharon is a Kenan Fellow in the Kenan Institute for Ethics, and a Moskowitz-Stern environmental doctoral candidate in the School of Law at Duke University.

Albertus Smit is an associate professor in the Department of Biodiversity and Conservation Biology at the University of the Western Cape, South Africa.

Peter Stoett is the Dean of Social Science and Humanities at the University of Ontario Institute of Technology, Canada.

Anastasia Telesetsky is a senior lecturer at the University of Auckland Law School, New Zealand.

Valerie Termini is the executive director of the California Fish and Game Commission.

Adelle Thomas is an assistant professor of Geography and co-founder of the Climate Change Initiative at the University of the Bahamas.

Rachel Tiller is a senior research scientist at SINTEF Ocean, Norway.

Andrew Tirrell is an assistant professor of Political Science and International Relations at the University of San Diego.

Steven M. Tucker is the Marine Protected Resources Program Manager at the US Coast Guard Headquarters, Washington, DC.

Noralene Uy is a lecturer at the Department of Environmental Science, Ateneo de Manila University.

Joanna Vince is a senior lecturer at the School of Social Sciences, University of Tasmania.

John Virdin is director of the Ocean and Coastal Policy Program, Nicholas Institute for Environmental Policy Solutions, Duke University.

Hilary Yerbury is an adjunct professor in the Faculty of Arts and Social Sciences, University of Technology, Sydney (UTS).

Acknowledgments

I am grateful to the contributors for sharing my interest in, and concerns about, climate change and ocean governance. Without their scholarship, this volume would not be possible. (A number of contributors have expressed acknowledgments in their own chapters.)

Matt Lloyd, Editor for Earth and Environmental Sciences, and Publishing Director for Science, Technology and Medicine (Americas), at Cambridge University Press, diligently guided the project through the complex review process, for which he has my genuine thanks. I am grateful to the anonymous reviewers commissioned by Cambridge University Press. They saw the scholarly merits of the project and provided valuable insights. The Press Syndicate of Cambridge University Press (the Press' editorial board) generously agreed to publish the volume, for which I am grateful. My thanks also go to other people at Cambridge University Press for their help in bringing this volume to readers, including Mariela Valdez-Cordero, editorial assistant; Esther Miguéliz Obanos, content manager for academic books in science, technology, and medicine; Ashwani Radjassegarane, project manager; Lisa McCoy, copyeditor; and other individuals and teams at the press who have contributed behind the scenes.

Finally, I give my special thanks to Keith, Mobie, and Susie for companionship and forbearance during this project and many similar projects in the past.

This volume is dedicated to the countless people around the world who are actively engaged in ocean governance in its myriad manifestations. If this volume assists their efforts to respond effectively to climate change at sea, it will be very worthwhile indeed.

Part I

Introduction

1

Climate Change at Sea
Interactions, Impacts, and Governance

PAUL G. HARRIS

Introduction

Climate change is the greatest challenge facing humanity. But the challenge would be far greater where it is not for the role played by oceans and seas. This is because the oceans have restrained global warming. They have absorbed the vast majority of the heat in global warming and are the largest environmental storehouse or "sink" for the most consequential greenhouse gas (GHG) – humanity's carbon dioxide (CO_2) pollution. Without these buffering services provided by the oceans, global warming, and other manifestations of climate change would be vastly worse – for the atmosphere, for terrestrial ecosystems, and for societies. However, in performing these services, ocean ecosystems themselves are being severely undermined. Indeed, some of the most profound effects of climate change will occur beneath oceans, across seas, and along coastlines. These effects are already being manifested in rising ocean temperatures. As ocean waters warm and land ice melts, sea levels are rising and vast areas are being transformed. What is more, as CO_2 pollution is absorbed by the oceans, seawater is becoming more acidic. Marine biodiversity and ecosystems are suffering the effects. These and many other changes create enormous difficulties for communities that depend upon marine and coastal ecosystems for sustenance, economic wellbeing, and ways of life.

The environmental, social, and economic consequences of *oceanic change* – the changes to seas and oceans arising from broader climate change – present tremendous challenges for governments and other actors. Existing national and international institutions for marine governance, which were created when oceanic conditions were relatively stable and humanity's exploitation of the oceans were much less than they are today, may not be adequate for a future characterized by continuous oceanic change. The impacts of climate change on oceans and seas will have political implications at all levels – local, national, international, and global. Responses to oceanic change will result in winners and losers. This will require politically difficult choices. New and innovative policies for governing oceans and seas, and for managing vital marine resources, have never been more important.

The objective of this book is to explore and understand possibilities for *ocean governance amidst climate change*. We conceive of governance broadly as "a social function centered on efforts to steer societies or human groups away from

3

collectively undesirable outcomes (e.g., the tragedy of the commons) and toward socially desirable outcomes (e.g., the maintenance of a benign climate system)" (Delmas and Young, 2009: 6). Governance is often performed by governments alone, but increasingly involves other actors ranging from international organizations to nongovernmental organizations (NGOs) and businesses. Sometimes those actors work completely independently of governments (Rosenau and Czempiel, 1992). To introduce the topic of climate change and ocean governance, some of the major connections between climate change and oceans (and seas) are briefly described in this chapter. The chapter then highlights some of the international responses to climate change, especially key agreements for climate governance. Finally the case studies presented in subsequent chapters are introduced and summarized.

Climate Change and Oceans: Interactions and Impacts

Science is central to understanding and responding to climate change. This volume is not about the science of climate change per se, so we leave a detailed examination of that to others (see, e.g., IPCC, 2013). However, each of the chapters that follows is necessarily informed by climate science. Similarly, while this volume is not about the science of oceanic change per se, nearly everything that is examined here is intimately connected to that science (see, e.g., IPCC, 2013: 361–484). Governance of the environment without science is impossible. This is not to say that other factors, such as values, preferences, capabilities, and political bargaining are not central to climate governance; rightly or wrongly, they can be more important than science. But to discount science in the governance process virtually assures that governance will fail. To be sure, there are still persons, including those in high office, who wish to deny the realities of climate science, presumably including what it tells us about the oceans. This denial of reality can have real-world consequences, as manifested in the policy decisions of the Trump Administration in the United States. At the time of this writing in 2018, it is working diligently to undermine American and global efforts to address climate change effectively, and indeed it is working equally diligently to shrink dramatically existing marine protected areas and to open up coastal and offshore areas for fossil fuel extraction (Friedman, 2018). These efforts will undermine, or at least delay, effective governance of climate change generally, and governance of oceanic change particularly. Ironically, such actions will make the actual science of climate change, and what it tells us about the future, all that more important.

Climate Change Science and Ocean Interactions

The science of climate change is not new. The idea that carbon emissions from human activities could warm Earth's atmosphere and lead to unnatural global warming was first hypothesized in the nineteenth century. However, it was only in the 1970s that the problem started to become prominent on an international agenda, and it took another decade before governments agreed to create a global organization to study it – the Intergovernmental Panel on Climate Change (IPCC). Since 1990, the IPCC has produced a number of reports examining the science of climate change and describing its causes, environmental impacts, and socioeconomic consequences (see, most recently, IPCC, 2013, 2014). As time has passed, the scientific understanding of climate change has increased markedly, as have the predictions of dire consequences. Indeed, the harmful consequences that were seen as relatively unlikely possibilities a quarter century ago are now viewed as virtual certainties. What is more, climate change has gone from being perceived as mostly a future problem to be avoided, to a problem that is being experienced today. As highlighted in a number of the chapters in this volume, the world is now getting a taste of things to come.

The primary causes and major consequences of climate change will be familiar to most readers of this volume. To recap, in simple terms, climate change involves long-term and large-scale anthropogenic change to Earth's climate system, including global warming – the "Greenhouse Effect" – and follow-on changes to other Earth systems. For our purposes, this includes unnatural changes in Earth's oceanic systems. Global warming and other manifestations of climate change are by-products of global CO_2 emissions produced during humanity's burning of fossil fuels – coal, oil, and natural gas – since the start of the Industrial Revolution more than two centuries ago, as well as emissions of other climate-effecting pollutants, often referred to as "greenhouse gases." These pollutants have a number of effects, among the most prominent for the oceans being ocean warming – a major aspect of *global warming* – and chemical reactions between CO_2 and seawater that lead to ocean acidification. These and other changes in oceans and seas associated with climate change are referred to here as oceanic change.

One cannot understate the importance of the oceans in the wider global *problematique* of climate change. Perhaps the most profound example of this is the role that the oceans play both as sinks for CO_2 pollution and for the global heat that results from that pollution. The oceans have absorbed at least one-third of the CO_2 emitted into the atmosphere since the Industrial Revolution (Khatiwala et al., 2013) and more than 90 percent of the heat resulting from humanity's CO_2 emissions (Wijjfels et al., 2016). However, the capacity of the oceans to absorb CO_2 pollution may decline and their ability to absorb the extra heat of global warming will have its limits. Indeed, it is possible that the oceans may switch from absorbing the heat of global warming to releasing that heat back into the atmosphere. This could happen on timescales that will have very real consequences for humanity (Tollefson, 2016).

In essence, the oceans have been a planetary dump for the much of the negative environmental externalities of global industrialization and modernization. If a stable global climate is what we value, then we owe an enormous debt of gratitude to the oceans. Without their role in mitigating climate change, and global warming in particular, environmental changes would be far greater and more rapid than they have been and will be in the future. However, much as every other environmental sink has its limits, the oceans will not be able to continue to buffer climate change forever. The impacts of climate change on the oceans themselves are profound. The consequences for societies will be – are already, in many places – equally profound.

Impacts of Oceanic Change

The effects of climate change at sea are probably as varied as the oceans themselves. Among the biggest effects are ocean warming, sea-level rise, and acidification. These effects have associated direct and indirect impacts that are of consequence to human societies. Ocean warming arises when oceans absorb the heat of global warming. Ocean warming is largely a function of ocean circulation among the sea surface and deep oceans, and among different regions via major currents. Worryingly, the ocean circulation that enables the transfer of heat from the atmosphere into the deep oceans may result in changes to those very circulation systems, with potentially paradoxical consequences, such as substantial cooling of Europe's climate (Struzik, 2017). Among the major consequences of ocean warming are mortality of coral reefs, typified by coral "bleaching" events (see Chapters 17 and 22). In combination with acidification and other environmental stressors such as river runoff containing agricultural pollution, warming seas have resulted in dramatic declines in corals in most areas where they traditionally thrive (Langlais et al., 2017). Just as dramatic are the consequences of ocean warming for the Arctic Ocean (see Chapters 12, 14, and 15). Warming there has resulted in radical reductions in the thickness and total area of ice coverage, with consequential major effects for Arctic marine ecosystems. What is more, reduced ice coverage is opening new areas of the Arctic to exploitation of oil, gas, and minerals, potentially exacerbating climate change still further. Frighteningly, warming in the Arctic could result in the release of seabed methane, substantially contributing to total global GHG emissions and creating a "positive feedback loop" to drive additional global warming (Wadhams, 2016).

A manifestation of oceanic change that is particularly evident to many people is that of sea-level rise (see especially Chapters 5, 8, and 18). As the oceans warm, they expand, resulting in some of the rise in sea levels that has already been experienced and will be much more so in the future. Furthermore, as oceans warm, they contribute to the melting of glaciers on land at the edge of the sea, thereby adding to sea-level rise. Rising seas are also a consequence of the melting of inland and mountain glaciers due to atmospheric warming. Estimates of sea-level rise are on the order of one meter or more by the end of this century, and potentially multiples of that in later centuries, with actual rises depending on location (see, for example, IPCC, 2013: 285–91; DeConto & Pollard, 2016). Roughly one-third of sea level rise is attributed to thermal expansion, with the remainder mostly the consequence of melting of glaciers and ice sheets on land (National Research Council, 2012: 33–53). Rising seas have myriad adverse consequences for coastlines and shallow seas (see the chapters in Part II). They inundate estuaries, which are vital nurseries for many fish and other marine species, and they can rise too quickly for coral reefs to adapt, in turn harming entire reef ecosystems.

Even if global warming were, by some cosmic intervention, to stop suddenly, the oceans and countless species that live within them would be threatened. That is because ocean acidification – the decrease in the pH of seawater resulting from absorption of CO_2 – is affecting seawater itself: the very chemistry of the oceans is undergoing quite a rapid change. Ocean acidification is contributing to coral bleaching and making survival difficult for other marine species dependent on calcification, such as many species of plankton and shellfish (Tynan, 2016). Ocean waters are changing in other ways as a consequence of climate change. For example, due to glacial runoff and intensified rain, salinity is undergoing significant change, affecting ocean ecosystems and circulation, thereby impacting the distribution of marine life and even weather phenomena (Balaguru et al., 2016; Lange and Marshall, 2017).

Taken together, oceanic change is already resulting in substantial deviations in marine species and habitats from longstanding historical norms. The impacts will be felt by people who live, often precariously, along coastlines, and by those who rely on vital resources from the sea. Rising seas are already major threats to some of the world's poorest countries, most obviously many vulnerable small-island states (see Chapters 8 and 18). Fish species are disappearing from their normal habitats, sometimes being lost altogether or migrating to new areas where regulatory protections are weaker (see the chapters in Part III). Other impacts include loss of inhabited areas along and near coasts, damage to infrastructure and loss of agricultural land to the sea, impacts from more powerful storms, and threats to coastal and high-seas fisheries (see the chapters in Parts II and III). Some island communities and entire nation-states face the existential threat of becoming uninhabitable within decades (Storlazzi, Elias, and Berkowitz, 2015). The chapters that follow explore these impacts in detail, in the process demonstrating the importance of climate change for the oceans and, in turn, the importance of oceanic change – and the importance of effective governance of it – for people and societies.

Climate Governance: Key Objectives and Agreements

Climate change has been on the global policy agenda for the better part of half a century. The climate change regime complex consists of formal international treaties, notably a framework convention and a protocol, associated nonbinding agreements, ongoing conference negotiations, and a variety of implementation mechanisms at regional, national, and local levels. Furthermore, the climate regime is intimately connected to, and arguably not complete without, agreements associated with other issues. For example, one major tool for reducing GHG pollution has been to use the Montreal Protocol on Substances that Deplete the Ozone Layer. The pollutants controlled by that agreement are powerful GHGs, so action in the context of the Montreal Protocol is effectively action on climate change. The climate regime also comprises commonly accepted overarching goals and standards. For example, it is now widely accepted by governments and most major industries that climate change demands action. Acceptance does not automatically translate into action, but it helps. This section briefly highlight some of the steps that have been taken internationally to craft the international regime for climate governance (see Harris, 2018, from which this section is adapted, for a more detailed description).

In response to concerns among scientists about the global implications of climate change, the United Nations Framework Convention on Climate Change (UNFCCC) was signed at the 1992 UN Conference on Environment and Development. The objective of the framework convention was the "stabilization of greenhouse gas concentrations in the atmosphere at a level that would prevent dangerous anthropogenic interference with the climate system" (UNFCCC, 1992: art. 2). The UNFCCC called on the world's developed states to reduce their emissions of GHGs to 1990 levels by 2000. That "soft" objective – there were no real penalties for noncompliance – was not achieved. However, the agreement of the UNFCCC was the start of a decades-long series of negotiations to find ways to formulate international objectives and rules for addressing climate change. The first UNFCCC "Conference of The Parties" (COP) was held in 1995, with COPs soon becoming annual (or nearly so) international meetings of diplomats and others to debate and negotiate implementation of the framework convention.

At the first COP, held in Berlin, diplomats agreed that the principle of "common but differentiated responsibilities" ought to guide responses to climate change. According to this principle, all of the world's states have common responsibility for climate change, but the developed states have more responsibility to do so. At the 1996 second COP, diplomats called for a legally binding protocol to the UNFCCC that would have specific targets and timetables for limiting GHG pollution coming from developed states. Toward that end, the Kyoto Protocol was agreed at the third COP in 1997. The protocol required developed states to reduce their collective GHG emissions by 5.2 percent below 1990 levels by 2012. The Kyoto Protocol was designed to provide flexibility in implementing its objectives. However, much as they failed to do what they promised in the UNFCCC, many developed states did not do what the Kyoto Protocol demanded. One contentious issue among all states was whether the use of carbon sinks, such as planting forests and making other land-use changes to remove GHGs from the atmosphere, should be counted alongside concrete reductions in greenhouse emissions (see Chapter 22). The effectiveness of such an approach is still subject to debate.

Subsequent international climate negotiations have been tortuous. They have resulted in incremental steps toward action on climate change, but in the process they have highlighted recurring differences among states about how best to achieve the fundamental objective of the UNFCCC to "prevent dangerous anthropogenic interference with the climate system." Many states have been unwilling to accept internationally mandated cuts in their GHG pollution. As with most other international collective action problems (not least those associated with governing the oceans), states that are required to take action frequently try to avoid doing so if there are significant financial or political costs involved (see Harris, 2013). At the seventeenth COP in 2011, diplomats affirmed an informal objective of limiting global warming to less than 2°C above the pre-industrial norm. At the same time, however, they admitted that twice that much global warming was likely without new national commitments to cut global emissions of GHGs far more aggressively. By the time of the twentieth COP in 2014, the UN Environment Program was making it clear that *urgent* action was necessary to limit GHG emissions, specifically *halving* them almost immediately and eliminating them completely later this century (UNEP, 2014). The "top-down" approach of dealing with climate change, whereby GHG limitations were decided internationally, was not working.

A shift in the approach to climate governance was agreed at the 2015 Paris COP. In the Paris Agreement on climate change, governments accepted that overall climate objectives could be agreed internationally, but that the commitments of each state – how much each country should be required to cut or otherwise limit its GHG emissions – would be determined nationally – by individual states themselves (United Nations, 2016). Perhaps as a consequence, unlike in previous international agreements on climate change, in the Paris Agreement developing states agreed to join developed ones in limiting their GHG emissions. As part of the Paris Agreement, all states accepted the common objective of limiting global warming to less than 2°C, and they acknowledged that it would be preferable to go further and aim for a target of 1.5°C. As part of the agreement, each state pledged to limit its national emissions in some way, although not necessarily to *reduce* them. The idea was that these pledges – formally known as "nationally determined contributions" – would become baselines for more action in the future (see Chapter 16). This approach garnered nearly universal participation. That said, this new "bottom-up" approach has so far resulted in pledges that will *not* achieve the objective of the UNFCCC to prevent "dangerous anthropogenic interference" in Earth's climate. Even if all of the Paris pledges were to be fully implemented, global warming would surpass 3°C (UNEP, 2016).

Because the international climate change agreements have been informed by science, the role of the oceans in governing climate change has been implicit from the start of negotiations. That said, for the most part the oceans have seldom played a major role in international policy making on climate change. After all, getting governments to agree to take climate change seriously, and specifically to actually limit and then cut their use of fossil fuels – to "decarbonize" their economies – is difficult enough. For most of them to also focus on the role of oceans is quite a lot to ask. However, recently this has started to change, as a number of the chapters in this volume point out. The oceans are now increasingly part of the official climate regime and a central feature of related governance initiatives, whether those be about the oceans per se – for example, fisheries agreements (see chapters in Part III) – or about other issues – for example, the UN's Sustainable Development Goals (see Chapters 16 and 19).

Governance of Threatened Seas: Case Studies

Scientific literature on the role of oceans and seas in climate change is now substantial. In contrast, the body of literature looking at the *governance* of oceanic change is relatively small. Through the chapters that follow, this volume aims to help address this shortfall in the literature. It brings together research findings from political science and cognate disciplines to examine the political and policy dimensions of climate change for the oceans. Collectively, the chapters give a snapshot of the current state of knowledge and portray a cross-section of research and analyses being conducted in this nascent and vital area of climate-related scholarship. All of the chapters make explicit connections between climate change, oceans (or seas), and questions of governance, particularly politics, policy formulation, and policy implementation at all levels, from the global to the local. Taken together, the chapters in this volume provide a comprehensive look at the state of climate change and ocean governance in its relative infancy.

Before the case studies begin, Part I of this volume continues with a chapter by Elizabeth Mendenhall. In Chapter 2, Mendenhall constructs some scaffolding that is useful for understanding subsequent chapters. She does this by surveying the most important international conventions and institutions for ocean governance. Mendenhall notes that the basic principles of ocean governance have evolved over many centuries. By the middle of the last century, increasing exploitation of the oceans had prompted unilateral claims by coastal states for control of resources well offshore. This increased the need for more effective rules for ocean governance. Toward that end, under the auspices of the United Nations, starting in the 1970s, the international community negotiated a framework for ocean governance: the UN Convention on the Law of the Sea (UNCLOS). The convention simultaneously nationalized, regionalized, and internationalized ocean governance (see the chapters in Part V). National zones of ownership and jurisdiction, and particularly exclusive economic zones, gave coastal states rights to the fish and other resources that could be found in a large proportion of the oceans (see Chapter 18). The Law of the Sea convention created new instruments for the resolution of maritime disputes, particularly those related to delimitation of national jurisdictions. The convention also contained several provisions to facilitate international cooperation, with one result being the emergence of regional bodies to manage fisheries and pollution (see the chapters in Part III). The UNCLOS reaffirmed the status of the high seas as a global commons area and designated the deep seabed as "common heritage" of humanity. As examined in subsequent chapters, especially those in Part V, while UNCLOS has established institutional mechanisms for governing the oceans in many ways, not all of these mechanisms have been successful. It is therefore of great importance to ask whether and how the current oceans regime can be deployed to govern the world's oceans as they grow warmer, higher, and more acidic (see, especially, Chapter 20).

Vulnerable Islands and Coasts

Our analyses begin with several chapters on the vulnerabilities of islands and coasts. The first case study, by Lisa Benjamin and Adelle Thomas, examines climate-related challenges for island states of the Caribbean. Small island developing states (SIDS) are among the areas most vulnerable to the impacts of climate change (see also Chapters 3, 6, 7, and 18). Rising sea levels, ocean acidification, coastal erosion, and loss of coral reefs are all highly likely to have significant ecological, economic, and societal impacts in SIDS. Climate change may ultimately threaten the territorial existence of some of them. In the Caribbean, coastlines and beaches are vital economic resources for national economies due to their close association with tourism. Coastal tourism contributes disproportionately to the economic development of countries in this region, often providing jobs for a high percentage of the working population. But the coasts of the Caribbean are particularly vulnerable to climate change. Reducing vulnerability through adaptation actions, such as through setting back coastal property development, would make some of the most valuable real estate unavailable to local economies. Benjamin and Thomas argue that such actions will restrict already-limited development options and thereby undermine economic development in Caribbean countries. Their chapter focuses especially on the contributions that coastlines make to the economies of Commonwealth Caribbean states and the detrimental impacts that sea-level rise may have on their political economy. An examination of the laws and policies on coastal development in The Bahamas is a case in point. Difficult political choices between economic survival and climate change adaptation will arise. The Bahamas case shows that policymakers in the Caribbean are often reluctant to implement adaptation measures that severely hamper options for near-term economic development along coastlines. This could have significant implications for long-term governance of climate change in the region, with governance implications for vulnerable SIDS in other parts of the world.

In Chapter 4, Mohammad Al-Saidi describes the many challenges that climate change presents to countries of the Persian/Arabian Gulf. Using the case of Qatar, Al-Saidi highlights the implications and policy responses among countries of the Gulf Cooperation Council (GCC). Not surprisingly, climate change will have significant impacts along the shorelines of GCC member states. Vital infrastructure, such as refineries, power plants, agricultural schemes, and desalination systems, will be negatively affected. Reduced rainfall, greater seasonal temperature variability, sea-level rise, and loss of agricultural production are some expected consequences of climate change in area, as are increased migration pressures. Vulnerable marine ecosystems of the Gulf are vital for food security, recreation, and cultural identity. Al-Saidi's chapter analyses climate-related challenges for developmental initiatives, coastal areas, and marine ecosystems. He identifies marine ecosystem vulnerabilities in the region and explores key drivers of that vulnerability, including coastal development ratios; dependency on vital services, such as those for water, energy, and food supplies; and the sensitivity of the ecosystems to climate change and their ability to recover from external shocks. Al-Saidi maps out the official stakeholders and marine policies related to climate change, identifies policy choices related to mitigation and adaptation, and highlights future policy-related risks. Using the case of coastal development and maritime policies of Qatar, his chapter presents examples from specific coastal industries in energy, desalination, real estate, and tourism.

His chapter identifies some of the internal societal pressures and external factors that influence policy, notes complexities arising from rivalries among states and the demand for economic development, and explores the potential for cooperation and joint action on coastal development amidst climate change.

Vulnerable islands and coasts must bear the brunt of some of the harshest impacts of climate change. More storms – or more intense storms – are bad enough, but when combined with other consequences of climate change, such as sea-level rise, the impacts on these areas can be devastating. Sometimes the impacts are acute, as when storms hit degraded shorelines; at other times they are chronic, as when shorelines erode, even in the best of times. This juxtaposition of acute and chronic impacts is highlighted in Chapter 5, by Arne Harms, which focuses on South Asia. As Harms makes abundantly clear, climate change will result in shifting weather patterns, groundwater depletion, and coastal erosion in South Asia. Along many of the region's low-lying and heavily populated shorelines, these impacts will add to an already long list of threats. The region's coastal populations have demonstrated a high degree of adaptation to recurring, but relatively infrequent, storms, and surges. However, they are facing greater difficulties in coping with normalized, climate change-related degradations. This is particularly the case with amplified coastal erosion due to sea-level rise, which has already displaced entire communities. This type of challenge is mirrored in humanitarian institutions, which are reasonably well-equipped to deal with traditional disasters, but not well-prepared to provide adequate assistance in situations of normalized degradation that are increasingly the consequence of climate change. Based on long-term fieldwork in coastal India, Harms argues that relatively effective measures taken by governments to safeguard populations from cyclones and storm surges have not been effective in accounting for chronic tidal incursions and coastal erosion. To explain this failure, Harms highlights the relative novelty of the massive scale of climate-caused coastal incursions and erosion. He looks at the framing of what counts as a disaster, and therefore, what mobilizes funds, technology, and labor toward needy populations. Harms' chapter shows that, because coastal degradation is becoming normalized, small-scale and only rarely deadly, it is not a priority for governmental and nongovernmental humanitarian institutions. To counter coastal erosion effectively would require considerably more funding, and indeed more political will. However, because development priorities across much of South Asia are not focused on rural coasts, neither adequate funds nor sufficient political will seem to be available.

In Chapter 6, Noralene Uy and Joe-Mar Perez also look at the implications of acute-but-recurring coastal impacts of climate change, specifically those associated with severe storms along coastal areas of the Philippines. The Philippines are composed of many islands, and thus, have long coastlines that are at risk to many ocean-related threats. Among the greatest acute threats are typhoons, which invariably strike the country several times each year due to its geographic location in the western Pacific, not far from where the majority of typhoons form over warm ocean waters. Increases in the frequency and severity of destructive typhoons, likely associated with climate change, have exacerbated other vulnerabilities of coastal areas in the Philippines. Uy and Perez's chapter examines many of the policy and institutional ramifications of strong typhoons. They review the institutional landscape on disaster risk reduction and management in the Philippines, as well as some of the country's policies for adaptation to climate change. They draw on the experience of Typhoon Haiyan, which struck the country in late 2013. They look at how plans for disaster risk reduction and management are translated (or not) from the level of the national government down to local governance in coastal areas. As they show, despite anticipating growing dangers from the sea that climate change is bringing, having plans in place does not guarantee effective responses when those dangers become real. The experience of the Philippines in preparing for the "new normal" of typhoons made worse by climate change highlights the need for careful adaptation governance in coastal areas. Uy and Perez's chapter describes some of the things that countries with vulnerable coastlines can learn from the case of the Philippines' response to typhoons. International governmental and NGOs concerned with humanitarian assistance can likewise draw lessons for delivering more effective support to coastal communities in developing countries.

Possibly no developing country is more vulnerable to the coastal impacts of climate change than is Indonesia. In Chapter 7, Achmad Poernomo and Anastasia Kuswardani describe Indonesia's vulnerabilities and related policy institutions and responses. As they point out, due to its geography, Indonesia is extremely reliant on ocean and coastal ecosystems. Threats from climate change faced by Indonesia include coastal erosion, coastal ecosystem degradation, sinking islands and cities, coral bleaching, potential loss of homes for coastal inhabitants, and loss of income from marine resources. Indonesia holds a significant share of the world's sea grass meadows and mangrove forests. These ecosystems are important for the mitigation of climate change due to their roles as natural sinks for atmospheric carbon (see also Chapter 22). Although many policies on climate change have been promulgated in Indonesia since the 1990s, those addressing ocean and coastal areas are surprisingly limited given the archipelagic nature of the country. What is more, there are quite extensive problems of related policy coordination between national and local governments, among ministries and agencies, and between presidential administrations. Poernomo and Kuswardani's chapter reviews the approaches of the Indonesian government to mitigate and adapt to the coastal and oceanic impacts. They give some suggestions for how policy might be improved in the future. Especially when combined with other chapters in Part II of this volume, their observations help us to better understand the challenges and opportunities for responding to climate change on vulnerable islands and coasts in other regions.

Our focus on vulnerable coasts and islands concludes with an analysis of narratives about rising sea levels among small island developing states. In Chapter 8, Ian McGregor and Hilary Yerbury look at such narratives at local, national, and international levels. The politics of climate change and rising tides include processes in which international NGOs, such as the Pacific Calling Partnership and the

Climate Action Network, collaborate with local NGOs. The most urgent climate-related concerns of local NGOs involve adaptation, vulnerability, and resilience. The politics of vulnerability is particularly clear in narratives around climate mitigation, especially when the intended audiences are foreign governments and citizens. Greenhouse gas emissions of SIDS are minimal, so the message coming from their governments and local NGOs is that developed countries should implement large and urgent emissions reductions to lessen sea-level rise and other adverse changes to oceans. Within the politics of resilience, however, the key audiences are primarily the populations of the island states themselves. These populations face a number of existential questions, such as whether they will be able to adapt to rising sea levels and increasing storm surges. In this respect, questions are about adaptation strategies that SIDS might implement to sustain their societies, and whether they can access the financial, technical, and other resources necessary to implement these strategies. Many local NGOs aim to educate climate change activists, supporters, and the global public about the risks faced by SIDS. McGregor and Yerbury's chapter shows how small-scale local actions to increase resilience, some of which are supported through NGOs, can present a perspective on climate change governance that contrasts with the dominant messages about vulnerability in international ocean politics.

Marine Fisheries and Pelagic Seas

Climate change will have direct and enormous impacts on fish in the sea. Rising ocean temperatures and increasing acidification of ocean waters are profoundly impacting the populations and ranges of commercially valuable fish species. These repercussions of climate change are creating both winners and losers in the short-term, but in the long-term the consequences are dire for the fishing industry as a whole. In Part III, we turn our attention to the difficult question of how to respond to these impacts, and specifically how to govern fisheries and protect their viability into the future. We begin in Chapter 9 by looking at what fishers themselves – especially small-scale fishers – say about climate change and official policies to manage impacted fisheries. In Chapter 9, Andrew Tirrell recounts what he learned by talking to fishers in the United States, New Zealand, and Norway. Tirrell considers the impacts of climate change in the context of the national fisheries management systems of these countries. He explores fishers' perceptions of the consequences of changing oceans for their industries. For each country case, Tirrell looks at the current political structures of fisheries management and considers how governance responses to climate change challenge local fishing economies. He shows that general levels of trust in government are closely related to whether, and how much, fishers trust their governments' efforts to respond to the fisheries-related consequences of climate change. Where historical trust in fisheries science and policies is relatively high, there is relatively high trust in related policies associated with climate change. Where that trust is low, fishers are highly skeptical of fisheries policies aimed at responding to climate change impacts at sea. Tirrell's chapter suggests some context-specific recommendations that have implications in other national or regional fisheries affected by climate change.

How might fisheries policies actually help fishers adapt to climate change? This question receives multiple answers from Wendy E. Morrison and Valerie Termini in Chapter 10. As Morrison and Termini point out, climate change will affect marine fisheries by altering ecosystem functions, fish abundance and productivity, distributions of fish populations, fish phenology, interactions of fisheries with nontarget species, bycatch rates and levels, and habitat use and availability. How should fisheries managers prepare for and respond to these changes? Morrison and Termini present a range of options currently being discussed in the scientific literature. In general, management approaches can be either proactive, thereby planning for future climate change at sea, or they can be reactive and respond to change after it has occurred. Proactive management alternatives can be implemented to increase resilience of fish stocks, species, and ecosystems, as well as local communities and businesses. Given the large uncertainties surrounding the effects of climate change, two potential governance approaches involve efforts to reduce uncertainty through research, and to devise management options that will be robust despite uncertainty. Morrison and Termini argue that policies seeking to increase management flexibility and provide incentives to the fishing industry to attempt new approaches, while simultaneously preserving genetic diversity of fished populations, should prove to be beneficial. Ideally, the advantages, disadvantages, and trade-offs associated with various management options should be evaluated to determine the best approach, or mix of approaches, given an understanding of likely environmental changes in the future. New approaches will have to emerge as fisheries management across the globe grapples with climate change.

Growing awareness about the effects of ocean warming and acidification has resulted in governments and environmental groups pursuing increasingly large marine protected areas (MPAs). These areas are the subject of Chapter 11, by Justin Alger. As Alger points out, eighteen of the world's nineteen largest MPAs have been established since 2006. Each of these areas exceeds 200,000 square kilometers. This increase in the scale of protected ocean spaces has mostly occurred in very remote regions. The diffuse risks of climate change are as salient in these regions as they are in coastal areas, but the remoteness of these risks makes them much less visible. Scientific studies have shown that large no-take MPAs increase the resilience of marine ecosystems to ocean warming and acidification. Consequently, creating MPAs in remote locations is a policy option for governments that are looking for ways to increase the resilience of their own seas and the living resources within them. However, protections for these remote areas has led to a disconnect between the greatest threat in them, namely climate change, and the targets of the regulations imposed on them, which is typically the fishing industry. Domestic fishing industries have tended to strongly resist the creation of MPAs. According to Alger, these industries perceive themselves to be under siege as governments prohibit fishing in remote areas that industry claims are already being sustainably managed. Drawing upon his own fieldwork and interviews with scores of stakeholders across five large MPAs, Alger argues that the opposition of fishing industries is disproportionate to the impact that MPAs have on their industries. The source of this disproportionality is a change in marine conservation priorities. Governments target MPAs, despite fishing industry opposition, because doing so represents a path of least resistance to addressing some of the impacts of climate change on the oceans.

As climate change brings on warming seas, marine species will migrate into new areas. This presents challenges to existing institutions for the governance of fisheries. A case in point is the unique fisheries management area around the Arctic Ocean archipelago of Svalbard, the subject of Chapter 12. In this chapter, Rachel Tiller and Dorothy Dankel ask whether the Svalbard Fisheries Protection Zone will be able to cope with the perturbations that arise when new species move into the zone due to oceanic change. Projections suggest that a warming climate will increase over-all marine species richness and abundance in the Arctic, resulting in increased catch potential around Svalbard, while also pushing fishers out of customary fishing grounds farther to the south. Tiller and Dankel describe the Svalbard Fisheries Protection Zone around the archipelago as a "contested management area." It has been managed by Norway for nearly a century. Other states have acquiesced to this management in practice because Norway has recognized their historical access to fishing grounds. However, as new species arrive, history may no longer be a reliable guide for determining which states should have full access to the zone's marine resources. Chapter 12 explores the implications that climate change may have for this fisheries' zone specifically, and on cooperation in the changing Arctic Ocean more generally (on the Arctic, see also Chapters 14 and 15.) To do this, Tiller and Dankel draw on the experiences of other international environmental regimes. They conclude that the adaptation of other regimes to environmental perturbations may not come as easily in the Svalbard case. As new species of fish move northward due to ocean warming, conflict over them is likely to increase.

Changing Polar Seas

In Part IV, our focus turns from the governance of fisheries amidst climate change to governance of the polar oceans, which both influence global climate and are being profoundly affected by it. In Chapter 13, Marcus Haward describes what he calls the "regime complex" for governing seas surrounding Antarctica. Haward highlights the significance of the Southern Ocean for Earth's climate system gener-ally, and global ocean circulation (currents) more specifically. He notes some of the challenges that climate change presents for this ocean region, including biophysical change and shifts in the ranges of marine species. Haward's chapter outlines the key institutions for governance of the Southern Ocean under the Antarctic Treaty System, particularly the Convention on the Conservation of Antarctic Marine Living Resources and its decision-making body, the Commission for the

Conservation of Antarctic Marine Living Resources. He looks at instruments relevant to addressing the impacts of climate change, and which of them are contributing to an evolving Southern Ocean regime complex. Haward reveals how climate change heightens the salience of ongoing research into the resilience of Antarctic regimes. In particular, he shows that there is growing cooperation between institutions created decades ago to manage the Southern Ocean and institutions that have been created to address climate change in particular. The regime complex for the Antarctic region is starting to overlap and interact with the regime complex for climate change.

In Chapter 14, we shift our attention from the polar seas of the South to those of the North. In this chapter, Benjamin Hofmann analyses international regulatory responses to growing environmental threats from increasing maritime industrial activities in a warming Arctic Ocean. Climate change accelerates the melting of Arctic Ocean sea ice, thereby making the area much more accessible to shipping and facilities for offshore oil and gas production. These industrial activities can have negative external effects on this ecologically vulnerable region. Hofmann investigates how states have responded to such threats, assessing and comparing the stringency of international environmental regulation of maritime industries in the Arctic. Stringency is a product of a regulation's formal "tightness" and its "ambition." "Tightness" refers to legality, precision, monitoring, and enforcement. "Ambition" includes changes in the scope and level of requirements imposed by regulations, as measured temporally and globally. Hofmann's chapter introduces a stringency database that encompasses regulations that partly or wholly covered the Arctic from 1950 to 2016. His chapter compares regulatory stringency across the shipping, oil, and gas industries; across regulators, such as the International Maritime Organization (IMO) and the Arctic Council; and across external effects and time periods. Hofmann shows that Arctic warming has been accompanied by increased regulatory activity to address the environmental impacts of maritime industries. However, the stringency of these regulations is found to vary considerably across regulatory bodies. Hofmann's findings can be used by policymakers to identify regulatory gaps and to create blueprints for more stringent regulations in a warming and threatened Arctic Ocean. His findings have important implications for governance in other ocean regions.

The profound environmental changes that are underway in the Arctic Ocean are manifested most dramatically by the melting of sea ice. In Chapter 15, Thomas Brewer reaffirms what is evident from the preceding chapter: the extent and speed of sea ice melt in the Arctic Ocean will soon result in increasing international maritime shipping across the region. With this increased presence of shipping in the Arctic will come an increase in the emissions of "black carbon," which is a highly potent climate change pollutant. Reports of the Arctic Council have focused attention on the problem of black carbon deposition in the region, but they have also noted that black carbon reaches the Arctic from other regions. Black carbon pollution also comes from aircraft flying over or near the Arctic. Existing institutions for Arctic governance are not able to impose regulatory measures to address this pollution. The IMO (see Chapter 2) has taken an interest in the problem of black carbon emissions from the diesel engines that power nearly all vessels, but it has yet to take action to address black carbon pollution. Similarly, in the case of aircraft, the International Civil Aviation Organization has not given the problem of black carbon any sustained attention, nor has it included black carbon in its plan for a carbon emissions offset program. Bearing in mind these weaknesses of existing forms of governance, in his chapter Brewer proposes that an Arctic Black Carbon Agreement be negotiated as a starting point for the development of a new regulatory framework. Such an approach, he predicts, could begin the process of more effectively responding to the impacts of black carbon pollution on the Arctic Ocean.

Institutions and Law for Ocean Governance

Building on analyses in earlier chapters, in Part V we focus on institutions and law for ocean governance amidst climate change. We begin in Chapter 16 with an analysis of "contested multilateralism" by Reuben Makomere and Kennedy Liti Mbeva. Makomere and Mbeva explore the process of "alignment" between the international regime for climate governance, exemplified in the UNFCCC and related international agreements, and the international regime for ocean governance, specifically the agreements and practices associated with the United Nations Convention on the Law of the Sea (see Chapter 2). Makomere and Mbeva argue that contemporary regimes for ocean and climate governance have not kept up with the growing complexity of changes underway in Earth's climate and ocean systems. This is not particularly surprising to students of international relations; it is emblematic of the broader challenge of gridlock and complex interdependence in other areas of global governance, including those related to the global environment. That said, Makomere and Mbeva see evidence of modest changes with respect to climate change and ocean governance. Their chapter examines pathways of alignment between the ocean and climate governance regimes. It analyses linkage politics as a catalyst for more coordinated climate-and-oceans governance, in the process examining how and why states have used mechanisms, such as the Paris Agreement's nationally determined contributions and the United Nations' Sustainable Development Goals, to push for greater alignment between the regimes for oceans and climate. Chapter 16 shows that coalitions of poor and small island states have worked together to highlight and promote key ocean-related issues. They have leveraged provisions of the climate regime in an effort to make the ocean regime more attuned to their needs.

In Chapter 17, Pedro Fidelman looks in detail at institutions for ocean governance amidst climate change, specifically an initiative that has the potential to enhance adaptive capacity. Fidelman identifies the role of power relationships among states in shaping institutions for climate change and ocean governance. The subject of his analysis is the Coral Triangle, a region of extraordinary ecological diversity encompassing seas around Malaysia, the Philippines, Indonesia, Timor Leste, Papua New Guinea, and the Solomon Islands. More than 120 million people in the region depend directly on coastal and marine resources for income, livelihood, and food security. As part of their efforts to address degradation of coastal and marine environments threated by climate change, and to use ocean natural resources more sustainably, governments of the region negotiated the Coral Triangle Initiative on Coral Reefs, Fisheries, and Food Security (CTI). The CTI is an example of a large-scale intervention to implement ecosystem-based management in the changing marine environment (see Chapter 11 for other examples). Governance institutions associated with those types of interventions can both facilitate and constrain the capacity of actors to adapt to climate change. Drawing upon empirical research, Fidelman examines how the CTI may enable such adaptive capacity. His chapter assesses the CTI in terms of its ability to encourage the involvement of a variety of actors, perspectives, and solutions; to enable actors to learn and improve governance institutions; to motivate stakeholders to self-organize, design, and reform their institutions; to mobilize leadership and resources for decision-making and implementation; and to support principles of fair governance. Fidelman concludes that the CTI has fostered collaboration among some stakeholders. However, broadening network relations and achieving effective collaboration remain a challenge. As the ocean impacts of climate change increase, it is increasingly important to cope with the relative power of the states (and other actors) involved in formulating and implementing institutional governance.

Shifting from international agreements and organizations for ocean governance amidst climate change, in Chapters 18–20 the focus turns to the institution of international law (see Chapter 2). In Chapter 18, Ori Sharon examines the vexing question of the future rights of small island states if – or, more likely, when – their territories encounter the existential threat of sea-level rise. Faced with the unthinkable reality of losing their territory and sovereignty to rising seas, governments of SIDS have been working to secure alternative territories for the resettlement of their populations. However, as developing states with marine-oriented economies, these countries have little to offer in exchange for territory. According to conventional interpretations, SIDS' most valuable national asset – sovereignty over resource-rich marine exclusive economic zones (EEZs) – will vanish, or at least be reduced, if these countries' territories are submerged. Territory is currently a prerequisite for enjoying the full rights of statehood, and it is also a prerequisite for an EEZ. However, Sharon questions this assumption. He argues that a disappearing EEZ does not necessarily entail the loss of associated rights. To identify whether a right is extinguished, one must first answer a series of questions pertaining to the nature of the right, the character of the right holder, the relationship that established the right, and the circumstances that might lead to the disappearance of that right. Sharon applies three ethical and political frameworks to address such questions: a rights-based approach that focuses on the origins of the basic human right to subsistence; a communitarian property theory of national resources; and a contractarian theory of international treaties. He shows how it may be possible to avoid the unjust outcome that a pre-climate change interpretation of territorial seas might otherwise prescribe. A key to such an outcome will be the equitable treatment of the peoples most affected.

The significance of equity is apparent again in Chapter 19, by Freedom-Kai Phillips and Konstantia Koutouki. Phillips and Koutouki argue that a number of apparently unrelated international treaties for governing water resources are highly germane to the governance of climate change. Their chapter introduces various measures under these treaties for managing and conserving international watercourses. It identifies potential synergies for adaptation to, and mitigation of, climate change. After highlighting significant legal provisions of climate change agreements, Phillips and Koutouki outline international legal norms relating to marine governance. Particular attention is given to related measures established under the Convention on the Law of the Non-Navigational Uses of International Watercourses (the New York Convention), the Convention on the Protection and Use of Transboundary Watercourses and International Lakes (the Helsinki Convention), the Convention on Wetlands of International Importance, especially as Waterfowl Habitat (the Ramsar Convention), and the Convention on Biological Diversity. Phillips and Koutouki point to mutually supportive mechanisms at the nexus of marine ecosystem governance and climate governance, including the interconnection of obligations under international law, the evolving application of the due diligence standard, and the broad intersections of the UN's Sustainable Development Goals. Use of "source-to-sea" governance, positive incentives for conservation and sustainability, equitable use of traditional knowledge, and effective governance of areas beyond national jurisdiction, are identified as a means by which marine ecosystem governance can be strengthened, in the process supporting climate change adaptation and mitigation. Phillips and Koutouki show that effective governance of marine ecosystems can help the world respond to the growing pressures of climate change.

The final chapter in Part V, Chapter 20, asks whether the legal framework of the UN Convention on the Law of the Sea (see Chapter 2) is sufficient to address the biological, chemical, and geographic impacts of climate change. To answer this question, Anastasia Telesetsky reviews UNCLOS obligations that are impacted by climate change. She starts by arguing that human-generated atmospheric GHGs are, in the lexicon of UNCLOS, "pollution of the marine environment." As such, GHG emissions violate UNCLOS-derived obligations of states to prevent, reduce, and control ocean pollution. Telesetsky then looks at fisheries management under UNCLOS and the need for that management to adapt to existing and anticipated changes in fishery migrations and habitat losses (see also chapters in Part III). Finally, Telesetsky's chapter examines UNCLOS precedents on establishing basepoints for measuring maritime jurisdictions. It evaluates the evidence for the formation of customary international law regarding the fixing of basepoints (see also Chapter 18). With these UNCLOS-derived obligations as the context, Telesetsky reflects on whether the right of coastal states to exploit fossil fuels on their continental shelves (see Chapter 24), which is a core part of the UNCLOS regime, will undermine efforts to mitigate ocean pollution, adapt fisheries management plans, and fix basepoints for states vulnerable to sea level rise. A question is whether the traditional provisions of UNCLOS – and indeed provisions of other extant regimes for governing seas and oceans – will be effective foundations for ocean governance amidst climate change.

Policies for Ocean Governance

In Part VI, we look in greater detail at particular areas of policymaking and policy implementation. This begins with an exploration in Chapter 21 of the links between the science, policy, and justice of plastic pollution of the oceans. Plastic pollution in the oceans is now an enormous problem. Perhaps surprisingly, this phenomenon has potentially important implications for climate governance. In Chapter 21, Peter Stoett and Joanna Vince ask whether there is a "plastic-climate nexus" that might assist us in better understanding the means for realizing ocean governance amidst climate change. The advent of massive amounts of plastic pollution in the oceans will both exacerbate climate change-related impacts at sea and further complicate political and policy responses to climate change. There is mounting concern that the density of marine plastic waste and associated toxicity will permanently damage the marine food web and reduce the ability of the oceans to absorb CO_2. Plastic debris can be a transport vector for pathogenic microbes and invasive species, thereby exacerbating climate-induced threats to marine life. From a wider governance perspective, the political similarity between the climate-related oceanic change and micro-plastic pollution may be very significant. Both of these environmental problems are, to a large extent, a consequence of the fossil-fuel industry, and in both cases small island states and other relatively disadvantaged coastal communities tend to suffer the most from them (see the chapters in Part II). Both problems will need significant industrial collaboration and normative development if solutions to them are to be found. Stoett and Vince reveal what those interested in ocean-and-climate change governance can learn from the micro-plastics issue. In particular, they show how the experience with plastic offers lessons for more effective and environmentally just ocean governance in the future.

In Chapter 22, we turn to questions of how biological systems can be protected for the benefit of both climate and oceans. Kamleshan Pillay, Yanasivan Kisten, Albertus Smit, and David Glassom explore whether lessons learned from efforts to protect global forests as carbon sinks can be applied to future efforts to protect coral reefs. As they point out, there is a growing need for innovative and well-structured climate policies that can contribute to mitigation of GHG pollution. Forest protection is already considered to be a vital policy for carbon sequestration in particular. Other biological systems have the potential to assist in these efforts. Coral reefs, which Pillay *et al.* refer to as the "forests of the sea," possess the highest biodiversity per unit area of any biome. Consequently, reefs hold the substantial potential to sequester carbon. Pillay *et al.* argue in favor of coral reefs being considered collectively as a "payment for ecosystem scheme" as part of efforts to realize the objectives of the Paris Agreement on climate change. Using forest-protection schemes as their reference point, they describe important attributes of a future coral reef protection mechanism (CRPM). Key to such a mechanism will be finance. Pillay *et al.* argue that developing a market-based mechanism might allow for a CRPM to generate finance that would make the mechanism more robust. Importantly, a successful CRPM would provide "co-benefits" to local communities, for example in the form of additional marine resources and jobs. Such benefits would encourage greater participation and "buy in," a lesson that arguably ought to be applied to other mechanisms for ocean governance in a future characterized by climate change.

Ships and shipping are synonymous with the oceans. But ships and shipping contribute to pollution, including pollution-causing climate change. In Chapter 23, Judith van Leeuwen points out that commercial shipping runs on highly polluting heavy fuel oil. In addition to contributing to climate change through CO_2 emissions, shipping emits sulfur, nitrogen oxide, particulates, and black carbon (the latter examined in Chapter 15). While most of these marine pollutants are regulated by international convention, CO_2 emissions from shipping have not been included in international climate change agreements. Instead, the IMO has in recent years developed related regulations for ships (see also Chapter 2). It has considered measures to make ships less polluting, for example by levying fees on fuel oil, using emissions cap-and-trade markets, and disseminating requirements for energy efficiency. In her chapter, van Leeuwen gives an overview of the way in which the climate change impact of shipping is regulated and reflects on the major bottlenecks that exist in moving forward internationally in this policy domain. She argues that there is a conflict of interest between developing maritime nations and European countries. There are few incentives worldwide for technical innovation to support a switch to more sustainable forms of energy in shipping. What is more, pressure from society is quite limited when it comes to shipping's climate change impact. To overcome these obstacles, new forms of governance are probably needed. These might include the use of market- or information-based measures that more effectively target ships' GHG emissions. Van Leeuwen shows how such new forms of governance might compensate for IMO's regulatory gaps, thereby helping to move toward decarbonizing a mode of transport that is vital to the global economy.

In Chapter 24, Christina Reichert and John Virdin look at technologies and related policies for extracting renewal energy from the oceans. They consider some of the implications of climate change for ocean-derived sources of energy. The offshore oil-and-gas sector has traditionally been virtually the only source of ocean energy for those living on land. Fossil fuels from the seabed remain very significant even as new sources of energy are developed. However, with growing emphasis on decarbonizing the world economy, as well as recognition of the adverse environmental effects of pollution from offshore oil and gas drilling, there is growing demand for a generation of renewable energy. The historical focus on offshore oil and gas extraction has shifted. Far more attention is being given to the development of technologies for extracting renewable energy from the sea, the worldwide potential of which is enormous. This potential has already resulted in the development of a number of prototype technologies, many of which are introduced in this chapter. However, as Reichert and Virdin point out, significant challenges to widespread commercialization of these technologies exist. These challenges include high investment costs, operational costs that sometimes compare unfavorably to traditional energy sources, and public opposition. Consequently, government assistance, for example in the form of support for research and development and guaranteed feed-in tariffs, will likely be needed if renewable ocean energy sources are to be fully exploited. Reichert and Virdin make a case for innovative governance that can overcoming some of the barriers to more widespread commercialization of renewable energy from the oceans.

While commercial ships are nearly always invisible to most of the world, being out of sight and seldom in the news, the same cannot be said of naval ships. Indeed, navies of the world will increasingly find themselves involved in operations that are affected by, or instigated by, the impacts of climate change. In Chapter 25, Kapil Narula tells us how navies are bracing for these impacts. He analyses the changing role of navies in response to the uncertainties that arise due to climate change. As Narula shows, as seen through the lens of national security, climate change will have both direct and indirect consequences for naval readiness. He examines these consequences and reveals how they have grown in salience within governments of many maritime countries. New threats from climate change range from damage to critical infrastructure, which could undermine operational capabilities of navies, to the reduced ability of navies to provide civilian aid during climate change-induced disasters. This contrasts with the likelihood – arguably the certainty – that climate change will increase the need for navies to undertake international humanitarian missions and provide disaster relief. This is expected to change the way in which ships are utilized and how military personnel are trained. In the long run, climate change could also lead to changing military strategies and force structures. One example of such a shift is the decrease in sea ice in the Arctic (see Chapters 14 and 15), which has already led the navies of China, the United States, and other countries to adopt new Arctic strategies. Using case studies from different countries, Narula proposes several policy responses to the impacts of climate change on navies. He argues that the world's navies can better brace themselves for climate change by taking timely action.

Increasingly, it will be the work of navies to respond to climate change at sea. But what might be done at sea to mitigate climate change proactively? One answer – an answer that some would say is radical – might be to "geoengineer" the oceans to bolster their ability to absorb the carbon pollution that is causing climate change. In Chapter 26, John T. Oliver and Steven M. Tucker describe one potential means for doing this: "fertilization" of large areas of the oceans. They propose that this could be done in areas where biological productivity is currently limited by consistently low nutrient levels. If the promise of ocean fertilization could be safely and efficiently realized, Oliver and Tucker argue, it might remove billions of tons of excess CO_2 from the atmosphere and from ocean waters. At the same time, they observe, it might increase the amount of biomass available to the marine food chain, ultimately reducing ocean acidification and increasing the amount of fish that are available for exploitation by humans. As Oliver and Tucker acknowledge, while any geoengineering proposal would have to be approached with extreme caution, it may be important to thoughtfully consider ocean fertilization as one of the responses to climate change. So far, discussion of ocean fertilization has been restrained, but Oliver and Tucker believe that policymakers and research scientists should do more to determine whether it is an environmentally safe and cost-effective policy response to climate change. Assuming that long-term safety and efficacy of this type of geoengineering can be clearly established, it might then be time to lay the groundwork for implementing the concept. If Oliver and Tucker are right, ocean governance amidst climate change may involve super-human attempts to bolster the role that the oceans have played already: to absorb human-generated pollution and heat.

Conclusion

The chapters that follow portray many of the problems of, and prospects for, climate change and ocean governance. One cannot overstate the importance of the oceans for climate change. Without the oceans, climate change would be far worse. Effective governance of the oceans may help the global community to both mitigate climate change and adapt to it. It is vital that the oceans become an integral part of the wider climate change regime, ranging from international agreements and principles to local policies and actions. In the real world, climate change and oceanic change are inseperable; they will have to be similarly inseperable in climate governance.

The case studies in this volume highlight the challenges, explore the policy responses, and provide suggestions for how to govern the oceans in what will be a very challenging future. To be sure, no single volume – nor dozens of them – can capture the full range of issues that are, and should be, considered in efforts to formulate and implement policies for effective governance of oceanic change. Nevertheless, the chapters that follow offer valuable information and insights that will have to be part of those efforts. They point to many lessons for the future. It is to those lessons that the final chapter turns. Therein, some of the common themes and conclusions of the preceding chapters are distilled. Some actual and potential strengths and weaknesses of ocean governance in the context of climate change are pinpointed. Likely and alternative future directions for ocean governance as the impacts of climate change inevitably become more pronounced are also discussed.

References

Balaguru, K., Foltz, G. R., Leung, L. R., and Emanuel, K. A. (2016). Global warming-induced upper-ocean freshening and the intensification of super typhoons. *Nature Communications*, **7**. Available at www.nature.com/articles/ncomms13670.

DeConto, R. M. and Pollard, D. (2016). Contribution of Antarctica to past and future sea-level rise. *Nature*, **531**, 591–7.

Delmas, M. A. and Young, O. R. (2009). Introduction: new perspectives on governance for sustainable development. In M. A. Delmas and O. R. Young, eds., *Governance for the Environment: New Perspectives*, Cambridge: Cambridge University Press, pp. 3–40.

Friedman, L. (2018) Trump moves to open nearly all offshore waters to drilling. *New York Times*, 4 January. Available at www.nytimes.com/2018/01/04/climate/trump-offshore-drilling.html.

Harris, P. G. (2013). *What's Wrong with Climate Politics and How to Fix It*, Cambridge: Polity.

Harris, P. G. (2018). Climate change: science, international cooperation and global environmental politics. In Gabriela Kütting and Kyle Herman, eds., *Global Environmental Politics: Concepts, Theories and Case Studies*, 2nd edn, London: Routledge, pp. 123–42.

Intergovernmental Panel on Climate Change (IPCC) (2013). *Climate Change 2013: The Physical Science Basis*, Cambridge: Cambridge University Press.

Intergovernmental Panel on Climate Change (IPCC) (2014). *Climate Change 2014: Synthesis Report*, Cambridge: Cambridge University Press.

Khatiwala, S., Tanhua, T., Fletcher, S. M., *et al.* (2013). Global ocean storage of anthropogenic carbon. *Biogeosciences*, **10**, 2169–91.

Lange, R. and Marshall, D. (2017). Ecologically relevant levels of multiple, common marine stressors suggest antagonistic effects. *Scientific Reports*, **7**. Available at www.nature.com/articles/s41598-017-06373-y.

Langlais, C. E., Lenton, A., Heron, S. F., *et al.* (2017). Coral bleaching pathways under the control of regional temperature variability. *Nature Climate Change*, **7**, 839–44.

National Research Council (2012). *Sea-Level Rise for the Coasts of California, Oregon, and Washington: Past, Present, and Future*, Washington, DC: National Academies Press.

Rosenau, J. N. and Czempiel, E. O. (eds.) (1992). *Governance without Government: Order and Change in World Politics*, Cambridge: Cambridge University Press.

Storlazzi, C. D., Elias, E. P. L., and Berkowitz, P. (2015). Many atolls may be uninhabitable within decades due to climate change. *Scientific Reports*, **5**. Available at www.nature.com/articles/srep14546.

Struzik, E. (2017). How a wayward Arctic current could cool the climate in Europe. *Yale Environment 360*, 11 December. Available at http://e360.yale.edu/features/how-a-wayward-arctic-current-could-cool-the-climate-in-europe.

Tollefson, J. (2016). How much longer can Antarctica's hostile ocean delay global warming? *Nature*, **539**, 346–8.

Tynan, E. (2016). Ocean acidification: emergence from pre-industrial conditions. *Nature Geoscience*, **9**(11), 804.

United Nations (2016). *Paris Agreement*. New York: United Nations. Available at http://unfccc.int/files/essential_background/convention/application/pdf/english_paris_agreement.pdf.

United Nations Environment Program (UNEP) (2014). *Emissions Gap Report 2014: A UNEP Synthesis Report*, Nairobi: UNEP.

United Nations Environment Program (UNEP) (2016). *Emissions Gap Report 2016: A UNEP Synthesis Report*, Nairobi, UNEP.

United Nations Framework Convention on Climate Change (UNFCCC) (1992). *United Nations Framework Convention on Climate Change*, New York: United Nations. Available at http://unfccc.int/resource/docs/convkp/conveng.pdf.

Wadhams, P. (2016). The global impacts of rapidly disappearing Arctic sea ice. Yale Environment 360, 26 September. Available at https://e360.yale.edu/features/as_arctic_ocean_ice_disappears_global_climate_impacts_intensify_wadhams.

Wijjfels, S., Roemmich, D., Monselesan, D., Church, J., and Gilson, J. (2016). Ocean temperatures chronical the ongoing warming of Earth. *Nature Climate Change*, **6**, 116–8.

2

The Ocean Governance Regime
International Conventions and Institutions

ELIZABETH MENDENHALL

Introduction

Like the ocean itself, the international institutions and organizations tasked with governing maritime activities are sprawling and complex. The contemporary ocean governance regime comprises the rules, norms, principles, and decision-making procedures designed to collectively manage the myriad users and multiple uses of the Earth's oceans. The regime itself has a rich and storied history, culminating in a flurry of regime-building activities in the second half of the twentieth century. In general, ideas about the collective governance of ocean space emerged alongside growth in human activities on and under the seas, and as the intensity of uses increased, so, too, did the calls for formulating international consensus about the status of marine space and maritime resources. This chapter reviews the historical development and contemporary status of the ocean governance regime. It lays the foundation for deeper analyses of climate change and ocean governance in subsequent chapters.

Customary International Law

The basic principles of contemporary ocean governance evolved out of centuries of state practice and jurisprudential debates, especially among the early modern Europeans, who elaborated and systematized customary international laws of the sea. Although Hugo Grotius is widely touted as the progenitor of the "freedom of the seas" principle, this concept has important antecedents in the ancient Indian Ocean trading system and Roman ideas about the status of the Mediterranean (Anand, 1983). A competing principle – the idea of a territorial sea that is owned and/or controlled by a coastal state – can be traced back to medieval claims over coastal European seas, and early modern declarations of dominion over broad swaths of the Atlantic by English and Iberian powers (Fulton, 1976). In his later works Grotius himself contributed to the idea that any area that could be physically controlled could be politically claimed, and this more limited notion of the territorial sea became embedded in state proclamations and practice starting in the eighteenth century. The contemporary ocean governance regime reflects a balance or compromise between these two competing principles: freedom of the seas and territorialization.

Another centuries-old customary international law of the sea is the practice of national flagging, whereby each ship must fly the national flag of the state under whose jurisdiction it falls. Although flags and other banners had been used as symbols of affiliation since ancient times, the practice of national registration originated in early modern Europe as part of taxation schemes related to maritime trade and was first codified by the British (Mansell, 2009). The "flag state" norm developed in part through efforts to control and eradicate piracy and privateering during the modern period. The national flag requirement aided in the identification of pirates, while defining the flag as a symbol of sovereign jurisdiction, but not sovereign violence, was a key part of the de-legitimization of privateering (Thomson, 1994).

Although these customary international laws are reflected in the contemporary ocean governance regime, the bulk of what we now call the "law of the sea" was negotiated and adopted by the international community in the last century. After World War II, increasing ocean use prompted a series of unilateral national claims and caused several visible environmental and ecological disasters. The need for more detailed and comprehensive rules became apparent, and under the auspices of the United Nations, the international community constructed a detailed and weighty architecture to manage the ocean. In particular, the United Nations Convention on the Law of the Sea (UNCLOS) serves as a framework or umbrella institution for ocean governance by empowering, coordinating and complementing other agreements related to maritime activities. This chapter will review this and other basic components of the contemporary ocean governance regime, focusing on the institutions (bodies of rules and norms), organizations (empowered agencies) and legal instruments tasked with achieving shared interests and solving collective problems in the vast ocean.

International Organizations

Several international agreements related to ocean governance were negotiated and implemented prior to UNCLOS, but most were eventually subsumed or replaced by it. Two important exceptions are the International Whaling Commission and the International Maritime Organization, which have evolved and grown alongside and complementary to the UNCLOS-centered ocean regime. These organizations are referred to indirectly in the UNCLOS text as "competent international organizations" through which the duties of state parties can be pursued and fulfilled.

International Whaling Commission

The International Whaling Commission (IWC) was established as a voluntary membership organization in 1946, with the goal of developing the commercial whaling industry while avoiding unsustainable whaling practices. In the first several decades of its operation, the IWC failed to prevent the serial collapse of whale populations. Its single global quota system (undivided by states) encouraged over-capitalization by whalers competing for a diminishing number of whales (Kalland and Moeran, 1992: 12). Specific quotas were difficult to set and enforce. The IWC depended on unreliable catch information provided by whalers themselves, and cetologists (zoologists who study whales and dolphins) had neither the data nor the consensus models required to make authoritative statements about the status of whale stocks (Peterson, 1992: 161).

In the 1970s, major shifts in the membership and institutional structure of the IWC changed this approach to the management of whaling. In 1972, the Stockholm Conference adopted Resolution 33, calling for a ten-year moratorium on commercial whaling. The United States presented this idea to the IWC, where it was rejected (Miyaoka, 2013: 31). Instead, in 1974 new procedures in the IWC Scientific Committee mandated the collection of more data and refinement of scientific models in order to strengthen the relationship between scientific information and decision making (Peterson, 1992: 164). During the late 1970s and early 1980s, a successful campaign by environmentalists and anti-whaling governments, especially that of the United States, encouraged more states to become IWC members. Although many of these states joined for domestic reasons, their presence tipped IWC decision making in favor of non-whaling states and anti-whaling interests (Stoett, 1997: 66). By 1982, the majority of IWC members had no involvement in whaling, thereby shifting the balance of opinion against whaling. The IWC voted to introduce a moratorium on commercial whaling starting in 1986.

The moratorium on commercial whaling persists today, although several whaling states have taken advantage of options to avoid compliance. The IWC rules allow violations in the case of formal objections. These have been filed by several states. The moratorium also contains an exception for whaling for the purposes of scientific research. This has been claimed by Japan, Iceland, and Norway, although the scientific merit of their whale kills is dubious. Another exception is aboriginal subsistence whaling, which takes place in Greenland, Russia, the United States and St. Vincent and the Grenadines. Despite these exceptions, the IWC moratorium on commercial whaling is understood to be durable, and a return to large-scale whaling appears to be inconceivable in contemporary society (Stoett, 1997: 77). The IWC remains the central institution for whaling issues, a status reaffirmed in Article 65 of the UNCLOS.

International Maritime Organization

The International Maritime Organization (IMO) is an inter-governmental organization and a specialized regulatory agency of the United Nations. In 1948, the fledgling United Nations drafted a convention to establish the Inter-Governmental Maritime Consultative Organization (IMCO), which entered into force in 1958. The goals outlined in the founding convention focused on promoting free access and non-discrimination in international shipping, with a secondary interest in maritime safety. The convention has been modified several times to clarify and extend the organization's purview and to alter its functions in line with changes in shipping technology and the interests of member states. In the late 1970s, amendments to the convention deleted the article that described IMCO functions as merely "consultative and advisory," added the prevention of marine pollution to the list of goals and changed the name of the IMCO to the "International Maritime Organization."

The IMO enjoys broad participation. In addition to 172 member states, 79 international non-governmental organizations have consultative status. The Assembly is the IMO's plenary body and its highest level of decision making, which includes all member states and meets every two years. The Council is the executive organ of the IMO, and its 32 member states manage ongoing business between Assembly sessions. Council members are chosen by the Assembly using specific criteria to ensure representation of states with significant interests in providing and utilizing international shipping services, and also geographical representation. Similar mechanisms to ensure appropriate representation of interested parties are found throughout the IMO institutional structure. Annual membership dues are calculated using a formula that emphasizes the tonnage of the registered merchant fleet. Many international agreements negotiated under the auspices of the IMO have a "double ratification threshold," such that a sufficient number of states representing a specific proportion of global registered shipping must ratify an agreement before it enters into force (DeSombre, 2006: 74).

The IMO has been described as "quasi-legislative" because it issues codes and recommendations to its members in addition to sponsoring and hosting intergovernmental negotiations and supporting implementation of resulting international conventions (Chircop, 2015: 429). Within the usual functioning of the organization, IMO member states drive the creation of guidelines, regulations and rules through a system of committees and sub-committees. All member states may become members of five main committees: Maritime Safety, Marine Environment Protection, Legal, Technical Cooperation, and Facilitation. A large number of sub-committees take on technical work, and within these, non-governmental observers advocate for specific interests and provide technical expertise (Chircop, 2015: 425). The ongoing process of updating maritime rules and regulations is facilitated by the IMO's use of a "tacit acceptance procedure" for amendments to many of its conventions (Biermann, 2014: 182). Under this procedure, an amendment automatically enters into force unless a specified number of parties to the original agreement object before a certain date.

The IMO plays a unique role in the ocean governance regime. Its policies shape the balance between the rights of coastal states and the "freedom of the seas" principle (Chircop, 2015: 418). For example, the IMO is responsible for evaluating and adopting ship routing schemes proposed by coastal states for the purpose of enhancing navigational safety and avoiding marine pollution. The organization also places conditions on the freedom of navigation. Regulations created by the IMO apply to all kinds of vessels at sea, including fishing boats and cruise ships, and IMO rules and standards cover all parts of a regulated ship's life cycle: design, construction, equipment, operation, and disposal. To improve at-sea monitoring, in 2000 the IMO made the use of Automatic Identification Systems mandatory for all ships of a certain size. Although these technical and operational issues have always been a core function of the IMO, the organization also promotes access to global shipping services.

One of the most controversial aspects of IMO governance is the so-called "flags of convenience" problem associated with ship registries. The IMO requires all ships to be registered in a country and to fly that country's flag as a signal of registry and jurisdiction. Around the time of the IMO's establishment, the practice of "open registries" became more prevalent. Open registry states allow ships owned and/or operated by nonnationals to register under their national flag. Such states often use ship registry as a source of domestic revenue, and attract ships registration with the promise of lax enforcement of maritime regulations. Due to the existence of these flags of convenience, the system of flag-state enforcement has been described as a weakness in IMO governance (Chircop 2015: 437).

Conventions on Marine Pollution

Conventions and agreements negotiated under the auspices of the IMO cover many topics related to maritime shipping, but those surrounding the issue of marine pollution have been especially influential in contemporary ocean governance. The right to pollute was an "implicit freedom of the high seas" for many centuries, but in the 1920s marine pollution from industrialized and transnational shipping networks began to arouse international concern (Caldwell, 1990: 294; Vogler, 2000: 57). Attempts in the 1950s and 1960s to regulate oil emissions in the open ocean were thwarted by insufficient monitoring and lack of infrastructure, and those attempts "had essentially no impact on improving the marine environment" (Mitchell, 1993: 245).

In the 1970s, two conventions negotiated through the IMO directly addressed the problems of vessel-source pollution and dumping at sea. The International Convention for the Prevention of Pollution from Ships (MARPOL) has been lauded for its innovative and effective requirements for reducing ship-based pollution, both operational and accidental. The first MARPOL agreement (1973) did not receive sufficient ratifications to enter into force due to the lobbying efforts of powerful shipping interests in maritime states (Chasek, Downie, and Brown, 2014: 24). After modifications to assuage the concerns of opponents, a new agreement, fused with the previous one to become MARPOL 73/78, entered into force in 1983 (DeSombre, 2006: 74).

MARPOL 73/78 introduced design requirements for oil tankers, including monitoring devices, separators (to reduce discharge) and segregated ballast tanks. The 1978 amendments to MARPOL added a requirement for washing out tanks with crude oil itself, instead of water. These changes facilitated new, less-polluting practices related to ballast exchange and tank cleaning. The MARPOL agreement also shifted responsibility from operators, who manage at-sea discharges, to owners, who purchase constructed ships. Because these provisions targeted the technology itself, instead of its operation, they shifted from the more challenging enforcement at sea to enforcement in port (Wonham, 1996). In 1997 the MARPOL conference of parties adopted a new protocol that limits emissions of air pollutants and the sulfur content of fuels. The treaty currently covers nearly 98 percent of registered global shipping, by weight (DeSombre, 2006: 74). The MARPOL agreement is generally regarded as a success, despite continued challenges with implementation and enforcement in the developing world (DeSombre, 2006: 75; Karim, 2010).

The 1972 Convention on the Prevention of Marine Pollution by Dumping of Wastes and Other Matter (commonly called the London Convention), also negotiated under the auspices of the IMO, created a "black list" of substances prohibited from dumping and a "gray list" of substances that could, under particular circumstances, be considered for dumping. The London Convention also mandated that state parties designate an authority to issue permits for all dumping and special permits for dumping of gray-list materials. The IMO is not empowered to monitor or enforce these rules, but the London Convention was the first agreement to authorize coastal states to enforce its provisions (Caldwell, 1990, 146; Chasek, Downie, and Brown, 2014, 24). Article 210 of UNCLOS enjoins member states to adopt national laws and regulations that are at least as effective as "global rules and standards" for marine dumping, which has typically been taken to mean the London Convention.

In 1996, a meeting of the parties to the London Convention adopted the London Protocol, which was intended to modify, update and eventually replace the London Convention. The London Protocol adopted the precautionary approach by prohibiting all dumping, except of those materials specially authorized by a formal list. Materials eligible for consideration – after assessment and licensing – include dredged material, manmade vessels and platforms, fish wastes, and sewage sludge. In addition to the positive listing approach, the London Protocol contains several other innovations when compared to the London Convention, including enhanced reporting requirements, a formalized dispute settlement procedure and a slightly broader definition of dumping. The London Protocol is also more adaptive and dynamic, and it has included amendments regarding marine geo-engineering (see Chapter 26) and carbon capture and sequestration in the seabed. Yet despite these advancements, the London Protocol did not succeed in replacing the London Convention, and it attracted relatively few state ratifications. The two conventions now exist alongside one another in an unusual informal arrangement described as "two treaties, one family." The governing bodies of each agreement typically hold joint meetings, although the majority of states in attendance have ratified only one of the treaties (Hong and Lee, 2015).

Although MARPOL and the London Convention and Protocol represent significant strides in the international regulation and restriction of marine pollution, they only target pollution from shipping and dumping. Around 70 percent of marine pollution comes from land-based activities (see Chapter 21), which indirectly or unintentionally deposit toxins and debris through wind, river outflows, coastal runoff and other pathways (Kirk 2015: 526). These sources of marine pollution have received little international attention in terms of regulation, in part because it was initially assumed that they would only have local impacts and could be effectively controlled by coastal states (Kirk 2015: 519).

The United Nations Convention on the Law of the Sea

The United Nations Convention on the Law of the Sea (see the chapters in Part V of this volume), also commonly referred to as the Law of the Sea Convention, is the centerpiece of contemporary ocean governance. This expansive institution emerged out of the third (and last) UN Conference on the Law of the Sea, which was tasked with addressing all matters pertaining to the oceans. In the decades following World War II, a series of unilateral and inconsistent declarations of ownership and jurisdiction over coastal ocean space had confused and complicated the application of customary international law. A new consensus was needed to resolve disputes and confront the phenomenon of "creeping jurisdiction." The first two Law of the Sea conferences, in 1958 and 1960, failed to resolve major disagreements about the width and nature of territorial seas and other jurisdiction zones, or to develop consensus and produce cooperation around issues related to high-seas fishing.

During the third UN Conference on the Law of the Sea, which took place from 1973 to 1982, representatives of over 150 states convened on 11 separate occasions to discuss the terms of UNCLOS. When the conference began, the law of the sea was in a chaotic state (Beesley, 1983: 183). Existing customary international law regarding the territorial sea had been called into question, and novel issues regarding resource use were emerging in the absence of a clear legal regime. As a result, the agenda for UNCLOS negotiations was very broad, including navigation, fishing, scientific research, seabed drilling and mining, the laying of seabed cables, marine pollution and territorial and jurisdiction claims. The basic goal was to produce a "package deal" treaty that would clarify and codify customary international law, establish new rules for emerging uses and ensure the sustainable and equitable use of ocean resources. Because of their scope, the UNCLOS negotiations produced diverse and shifting coalitions from issue to issue. Because of their length, the position of any given state could change with turnover in government administrations. And because of their broad international participation, negotiators and diplomats needed to balance a number of underlying divisions and antagonisms between North and South, East and West, and coastal, maritime and landlocked states. Despite these challenges, the conference managed to produce a Law of the Sea Convention covering all major issues, which would eventually win broad support from the international community.

Territory and Jurisdiction Zones

The territory and jurisdiction zones created by UNCLOS are a central feature of its legacy. These zones specify the duties and rights of various parties and determine who can legitimately access which resources. The seabed is divided into two types of zones: the continental shelf, which belongs to the nearest coastal state (up to 350 nautical miles) and "the Area" which is designated the "common heritage of mankind" and managed by the International Seabed Authority (see later) (Article 136). The water column is divided into four types of zones: the territorial sea, contiguous zone,

Exclusive Economic Zone (EEZ), and high seas (also called the "Area Beyond National Jurisdiction"). In general, the closer a zone is to the coastline, the more control a coastal state has. This section will briefly survey the basic features of the water column zones. Seabed zones will be reviewed in the following section, in conjunction with the instruments created to define and manage them.

The national jurisdiction zones created by UNCLOS are delineated with reference to the "baseline," which is normally drawn at the low-tide line as represented by official charts of the coastal state (Article 5). Special provisions exist for drawing baselines along coastlines with a high degree of topographical variation, and for archipelagic states made up of a large number of unevenly spaced islands (Article 7 and Article 47). The text of UNCLOS does not specify whether baselines shift in the case of dynamic coastlines and newly constructed or newly submerged islands (see Chapter 18 for a detailed discussion).

The territorial sea extends up to 12 nautical miles from the baseline (Article 3). It is defined as an extension of coastal state sovereignty, and has been recognized as customary international law even for nonparties to UNCLOS (Noyes, 2017: 94). Sovereignty over the territorial sea includes the seabed, the water column and the airspace above the territorial sea, but the exercise of sovereignty is limited. Coastal states have duties to provide information about hazards and regulations to nonnational ships, and they must allow all navigation that is "innocent passage." State parties disagree about whether military ships, nuclear-powered ships and ships carrying hazardous material should qualify as "innocent passage" (Noyes, 2017: 99). Another condition on sovereignty in the territorial sea applies to the situation of international straits, through which coastal states must allow "transit passage" (Article 38). Transit passage through international straits allows submarines to travel in "normal mode" (submerged) and permits the overflight of nonnational aircraft. Unlike "innocent passage" in the territorial sea, "transit passage" cannot be suspended by the coastal state.

Beyond the territorial sea lies the contiguous zone, which can be claimed up to 24 nautical miles from the baseline. In the contiguous zone, a coastal state can "exercise the control necessary" to prevent and punish the infringement of customs, fiscal, immigration or sanitary laws broken in the territorial sea or on state territory (Article 33). The coastal state is also empowered to protect objects of "archaeological and historical nature" within the contiguous zone (Article 303).

The 200-mile Exclusive Economic Zone is a novel and extremely significant contribution to the law of the sea. After the initial wave of major unilateral claims from Latin American states in the 1940s and 1950s, developing and newly independent states in Africa latched onto the concept of seaward extension of their territorial rights. In 1972 Kenya presented a working paper titled "Exclusive Economic Zone Concept" to an Asian-African Legislative Consultative Committee. It was this group that brought the idea to the UNCLOS negotiations. The EEZ concept was supported by developing coastal states as a way to protect their offshore resources from long-distance fishing by developed states (Scott, 2005: 33). The 200-mile EEZ gives states exclusive control over water column resources – most notably fisheries – for the purposes of exploitation, conservation, and management (Article 56) (see Chapter 18). EEZ jurisdiction is conditioned by the freedoms of navigation, overflight and the laying of submarine cables and pipelines by other states (Article 58). States are not obliged to demarcate their EEZs, and those that do may run into delimitation challenges because of overlap with the jurisdiction zones of other coastal states. The convention offers several options for dispute settlement, reviewed in the next section of this chapter.

Another area of potential disagreement in the creation of jurisdiction zones is the status of small islands, such as those found in the South China Sea. Islands that are "naturally formed" and never completely submerged can generate a territorial sea, contiguous zone, EEZ, and continental-shelf claim. In contrast, a mere rock that "cannot sustain human habitation or economic life" can only be used to generate a territorial sea and contiguous zone (Article 121). State practice regarding the distinction between rocks and islands is infrequent and inconsistent, such that there is no clear interpretation of UNCLOS and no coherent customary international law (Churchill, 2005: 106).

The "high seas" encompass all parts of the ocean that fall outside the other jurisdiction zones (Article 86). All states have rights to navigation, overflight, laying of cables and pipelines, construction of artificial installations, fishing, and scientific research, among other activities in the high seas (Article 87 provides a nonexhaustive list). The convention invalidates any sovereignty claims in the high seas, and it reserves high-seas areas for "peaceful purposes" (Article 88). Enforcement of international and national law on the high seas is the purview of "flag states." There is a limited "right of visit" for a warship if it suspects that a nonnational ship is engaged in piracy, slavery or unauthorized broadcasting, or if the ship is without a nationality (Article 91 and 110).

Instruments for Implementation

Several UNCLOS provisions associated with navigation, safety, and pollution obligate states to observe the rules of preexisting treaties and organizations. The convention also created three new instruments to facilitate implementation and resolve disputes among state parties: the Commission on the Limits of the Continental Shelf, the International Tribunal for the Law of the Sea and the International Seabed Authority. The features and functions of these institutions help make UNCLOS an evolving and adaptive institution.

The Commission on the Limits of the Continental Shelf (CLCS) is a technical organ established by Annex II of the convention. Its purpose is to make recommendations about the application of Article 76, which uses a complex and highly technical formula for the establishment of the outer limits of the continental shelf. Coastal states have the exclusive right to exploit the natural resources of their continental shelves (Article 77), so they have an incentive to maximize their continental-shelf claims. The 21 members of the CLCS are persons elected by states parties every five years, and those members must be experts in geology, geophysics or hydrography. Although the treaty does not precisely specify representational criteria for CLCS members, the parties have implemented an equitable geographical distribution scheme in the members' selection process. Commission members "serve in their personal capacities," and each member is funded by the state party that nominated that person for election (Annex II, Article 2). The convention also encourages the CLCS to cooperate and exchange scientific and technical information with other expert bodies, including the Intergovernmental Oceanographic Commission of UNESCO and the International Hydrographic Organization (Annex II, Article 3).

States seeking recommendations on the outer limits of their continental shelves must submit their claims and supporting scientific information to the commission within ten years after the convention enters into force for that state (Article 4). Since UNCLOS entered into force, the meetings of the states parties have effectively amended this provision to extend the deadline for submissions and to allow the submission of preliminary information to be considered and commented upon before the final submission (Churchill, 2017: 43).

In addition to the commission, UNCLOS provides multiple options for the settlement of disputes among its member states. These include adjudication by the International Court of Justice, two types of international arbitration and submission to the International Tribunal for the Law of the Sea as established in Annex VI (Article 287). Although states are obligated to resolve disputes via peaceful means, they are also permitted to exclude some issues from compulsory dispute settlement, including maritime boundary disputes and military activities (Article 298).

The International Tribunal for the Law of the Sea (ITLOS) has ruled on 25 cases and issued two advisory opinions since its instantiation. Like the CLCS, the tribunal is made up of 21 members with demonstrated competence in their subject matter (in this case, legal affairs) and representing an equitable geographical distribution and the "principal legal systems" of the world (Article 2). Tribunal members are nominated by state parties and elected to nine-year terms by secret ballot. A quorum of 11 members is required to constitute ITLOS, and decisions are made by a majority of those ruling on a case. The jurisdiction of the tribunal is broad, comprising "all disputes and all applications submitted to it in accordance with [UNCLOS]" (Article 21). It also has special jurisdiction over provisional measures while cases are pending and over situations requiring "prompt release" of seized vessels and crews (Articles 290 and 292). Its rulings are only binding between parties to a dispute, and they are not intended to set more general precedents (Annex VI, Article 33).

The creation of the International Seabed Authority (ISA) was the subject of prolonged contention during UNCLOS negotiations (Hollick, 1981: 287). The designation of the seabed beyond national jurisdiction as the "common heritage of mankind" – formally affirmed by UN General Assembly Resolution 2749 in 1970 – required the creation of a central institution to manage seabed resource development. The ISA is established within Part XI of UNCLOS, which covers all activities in the area. Its basic purpose is to "organize and control activities in the Area, particularly with a view to administering the resources of the Area" (Article 157). In effect, the ISA's role is to manage seabed mining in the Area Beyond National Jurisdiction.

The ISA is an autonomous institution that became fully operational in 1996. Its principal organs are an assembly, a council and a secretariat, all headquartered in Kingston, Jamaica. All state parties to UNCLOS are automatically members of the ISA, and all ISA members can designate one representative in the assembly. In addition to the adoption of general policies, the assembly elects 36 members that comprise the council. The members of the council must be elected from within five different categories: the largest consumers of minerals (four members), the largest investors in seabed mining (four members), the largest exporters of minerals (four members), developing countries with "special interests" such as large populations or land-locked status (six members) and whatever countries need to be placed on the council in order to achieve geographical representation (18 members) (Article 161).

For disputes related to seabed mining, ITLOS convenes a specific Seabed Disputes Chamber composed of 11 members of the tribunal, with a quorum threshold of 7 members. The Seabed Disputes Chamber has special jurisdiction over activities in the area, including disputes between state parties and the International Seabed Authority (Article 187). In its rulings, the chamber is empowered to apply "the rules, regulations, and procedures of the Authority" and "the terms of contracts concerning activities in the Area" (Annex VI, Article 38).

Implementing Agreements

Alterations to UNCLOS began before the convention entered into force in November 1994. Two "implementing agreements" were intended to rectify gaps and problems identified in UNCLOS (Harrison, 2011: 86). The first agreement – the Implementing Agreement on Part XI – emerged from four years of informal negotiations spearheaded by the UN Secretary General. There was major concern that UNCLOS would fail to achieve widespread acceptance – of the first 60 states to ratify UNCLOS, 58 were developing countries. The original provisions on seabed mining in Part XI were perceived as a central barrier to universal participation in UNCLOS, and the agreement aimed to win over nonparty industrialized countries who specifically objected to this section. The final implementing agreement amended Part XI of UNCLOS in order to strengthen the position of private investors and weaken the role of the enterprise. This included watering down provisions aimed at benefit sharing, including technology transfer and the taxation and redistribution of mining profits.

The July 1994 Implementation Agreement on Part XI was adopted by the UN General Assembly as a resolution and was combined with the original convention "to be interpreted and applied … as a single instrument" (A/RES/48/264). The agreement is therefore only open to those states that have already ratified UNCLOS, and any state that ratifies UNCLOS after the agreement was adopted must consent to be bound by both. In other words, states that ratified UNCLOS before July 1994 had to "opt in" to the implementing agreement, and states that ratified after July 1994 could not "opt out" of the agreement. Thus far, 146 states have ratified the implementing agreement in Part XI.

The second implementing agreement did not nullify, replace or amend any parts of UNCLOS, but rather clarified, elaborated, and modernized the provisions relating to fisheries management. The 1995 Agreement for the Implementation of the Provisions of the Convention Relating to the Conservation and Management of Straddling Fish Stocks and High Migratory Fish Stocks ("Fish Stocks Agreement") focused on the regulation of highly migratory fish stocks and those that straddled the national jurisdiction zones created by UNCLOS (see Chapter 11). (Unlike the Implementing Agreement on Part XI, the Fish Stocks Agreement is a freestanding treaty that can be ratified by nonmembers of UNCLOS. It entered into force in 2001, and currently has only 82 member states.) The original convention simply enjoined states that fished for straddling or migratory stocks to cooperate over their management (Articles 63 and 64). This requirement was seen as inadequate for effective governance, a fact that was formally acknowledged in the Agenda 21 document produced by the 1992 UN Conference on Environment and Development. In 1993, the UN General Assembly convened a UN Conference on Straddling Fish Stocks and Highly Migratory Stocks, which after three years of negotiations adopted the Fish Stocks Agreement by consensus in August 1995. The agreement was welcomed and promoted by UN General Assembly Resolution 50/24.

The Fish Stocks Agreement reaffirms the duty of states to cooperate in fisheries management and to use the best scientific evidence available (Article 119). It adds two principles of sustainable development to UNCLOS: the precautionary approach and the ecosystem approach. The Fish Stocks Agreement obligates states to consider uncertainty in scientific information about fish stock size and reproduction, and the impact of other marine activities on target and non-target species alike. It also requires states to generally "protect biodiversity in the marine environment" [Article 5(g)]. The Fish Stocks Agreement explicitly addresses the functions and features of regional fisheries management organizations (RFMOs) (described in detail in the next section; see also Chapters 9 and 10). One of the most controversial elements of the Fish Stocks Agreement restricts fishing in RFMO-governed areas, or fishing of RFMO-governed species, to those states that are RFMO members [Article 8(4)]. Member states are also empowered to engage in enforcement actions against nonmember states violating RFMO dictates (Articles 21 and 22). Some nonparties to the Fish Stocks Agreement argue that these provisions undermine the principles of freedom of the high seas and flag-state jurisdiction (Molenaar, 2011).

Upon the completion of the UNCLOS III conference, the president of the conference (Tommy Koh of Singapore) declared the creation of a "comprehensive constitution for the oceans which will stand the test of time." The description of UNCLOS as a "constitution" is appropriate given its comprehensive scope, widespread participation, hierarchical relationship to other institutions and embeddedness in the overall ocean governance regime (Churchill, 2017: 45). Although the convention certainly has its failings, there is currently no significant interest within the international community to replace or significantly revise UNCLOS.

Regional Fisheries Management Organizations

RFMOs are autonomous organizations formed by agreement between a group of members that self-regulate their exploitation either of a particular species or of all commercial species within a particular area. Some RFMOs only have an advisory function, but most have a management function. They are open-membership, and they only create legal obligations for their members. There are around 18 RFMOs (the exact number depends on the criteria used), and they have become the preferred vehicle for fulfilling UNCLOS obligations related to the conservation of living resources (Article 117 and 118) (Rayfuse 2015: 440). Although each RFMO is different, common management tools include data collection, dispute settlement and limitations on fishing technology, capacity and effort.

RFMOs are widely understood to have failed at their primary task: maintaining the sustainability of fisheries (Cullis-Suzuki and Pauly, 2010). Although the Fish Stocks Agreement requires the adoption of a precautionary and ecosystem-based approach, RFMOs remain deeply flawed and have only made negligible progress in adherence to these principles (Gilman, Passfield, and Nakamura, 2014). Fisheries management organizations are particularly subject to the problem of "regulatory capture," whereby regimes are controlled by vested interests seeking to justify existing practices (Gjerde et al., 2013; Barkin and DeSombre, 2013). Many flag states simply do not become members of the RFMOs that regulate the fisheries that their nationals exploit. But fishers from RFMO member states can easily register in nonmember states to avoid regulation (Barkin and DeSombre, 2013: 32). Although the Fish Stocks Agreement authorizes RFMO member states to enforce RFMO provisions against nonmember states, this only applies to situations where the nonmember of the RFMO is a member of the Fish Stocks Agreement (Molenaar, 2011: 205). In general, little or no effort is made to keep nonmember fishers out of an RFMO area. The problem of RFMO management is summarized succinctly by Samuel Barkin and Elizabeth DeSombre (2013: 9): "a common pool resource cannot be successfully protected by a sub-group of users."

Advances in fisheries management aim to redress the insufficiency of RFMOs. Efforts have focused on making ships more traceable at sea and more accountable in port. The 2009 Food and Agricultural Organization Port State Measures Agreement has the explicit goal of blocking the flow of IUU (illegal, unreported, and unregulated) fish into markets. It allows port states to deny entry to foreign boats suspected of illegal fishing and to require detailed documentation and inspection to ensure the legality of catches. Despite these efforts, the persistence of "flags of convenience" and "ports of convenience" make effective governance of global fisheries an extremely difficult task (see Part III of this volume).

Conclusion

The heavy and complex architecture of the contemporary ocean governance regime represents decades, even centuries, of investment in diplomacy and legalization. Although lauded for its comprehensiveness and universality, the UNCLOS-centered regime can be criticized for its lethargic response to emerging issues associated with climate change (see Chapter 20). In particular, jurisdictional boundaries tend to assume that the ocean will not change fundamentally. The law of baselines, from which the territorial sea and EEZ are calculated, does not account for the possibility of a dynamic coastline. Whether the baseline shifts with rising seas remains an open question (see Chapter 18). The "regions" of RFMOs reflect the spatial extent of fish populations and fishing practices, which may shift in response to warming seas (see Chapters 12–14). Even the most successful parts of the contemporary governance regime fail to address risks associated with climate change. The London Convention and MARPOL are narrowly focused on vessel-source pollution, without any provision for land-based or atmospheric sources of harmful emissions that cause acidification. In general, current environmental protection efforts aim to prevent over-exploitation by human users, with little attention to more diffuse threats to marine habitats like warming and acidification.

The breadth and depth of UNCLOS represents a substantial "sunk cost" for the international community. Because there is little international enthusiasm for replacement or substantial reform, collective problems must be addressed within the basic framework established by UNCLOS (see Chapter 16). Whether adjustments and augmentations to the existing regime will be sufficient to confront issues associated with climate change promises to be a key question for the twenty-first century.

References

Anand, R. P. (1983). *Origin and Development of the Law of the Sea: History of International Law Revisited*. Publications on Ocean Development, v. 7. The Hague; Boston: Hingham, MA: Martinus Nijhoff; [distributed in the U.S. by] Kluwer Boston.

Barkin, J. S. and DeSombre, E. R. (2013). *Saving Global Fisheries: Reducing Fishing Capacity to Promote Sustainability*. Cambridge, MA: The MIT Press.

Beesley, A. (1983). The Negotiating Strategy of UNCLOS III: Developing and Developed Countries as Partners – A Pattern for Future Multilateral International Conferences? *Law and Contemporary Problems*, **46**(2), 183–94.

Biermann, F. (2014). Earth System Governance: A Core Research Project of the International Human Dimensions Programme on Global Environmental Change. *Earth System Governance: World Politics in the Anthropocene*. Cambridge, MA: The MIT Press.

Caldwell, L. K. (1990). *International Environmental Policy: Emergence and Dimensions (Duke Press Policy Studies)*, 2nd edn. Durham, NC; London: Duke University Press.

Chasek, P. S., Downie, D. L. and Brown, J. W. (2014). *Global Environmental Politics Dilemmas in World Politics*, 6th edn. Boulder, CO: Westview Press, a member of the Perseus Books Group.

Chircop, A. (2015). The International Maritime Organization. In D. R. Rothwell, A. G. O. Elferink, K. N. Scott, and T. Stephens, eds., *The Oxford Handbook of the Law of the Sea*. New York: Oxford University Press, pp. 416–38.

Churchill, R. R. (2005). The Impact of State Practice on the Jurisdictional Framework Contained in the LOS Convention. In A. G. Oude Elferink and S. V. Shirley, eds., *Stability and Change in the Law of the Sea: The Role of the LOS Convention No. 24*. Leiden; Boston, MA: Martinus Nijhoff Publishers.

Churchill, R. R. (2017). The 1982 United Nations Convention on the Law of the Sea. In D. R. Rothwell, A. G. O. Elferink, K. N. Scott, and T. Stephens, eds., *The Oxford Handbook of the Law of the Sea*. New York: Oxford University Press, pp. 24–45.

Cullis-Suzuki, S. and Pauly, D. (2010). Failing the High Seas: A Global Evaluation of Regional Fisheries Management Organizations. *Marine Policy*, **34**(5), 1036–42. doi:10.1016/j.marpol.2010.03.002.

DeSombre, E. R. (2006). *Global Environmental Institutions (Global Institutions Series)*. London; New York: Routledge.

Fulton, T. W. (1976). *The Sovereignty of the Sea: An Historical Account of the Claims of England to the Dominion of the British Seas, and of the Evolution of the Territorial Waters, with Special Reference to the Rights of Fishing and the Naval Salute*. Millwood, NY: Kraus Reprint Co.

Gilman, E., Passfield, K. and Nakamura, K. (2014). Performance of Regional Fisheries Management Organizations: Ecosystem-Based Governance of Bycatch and Discards. *Fish and Fisheries*, **15**(2), 327–51. doi:10.1111/faf.12021.

Gjerde, K. M., Currie, D., Wowk, K. and Sack, K. (2013). Ocean in Peril: Reforming the Management of Global Ocean Living Resources in Areas beyond National Jurisdiction. *Marine Pollution Bulletin*, **74**(2), 540–51.

Harrison, J. (2011). *Making the Law of the Sea: A Study in the Development of International Law*. Cambridge; New York: Cambridge University Press.

Hollick, A. L. (1981). *U.S. Foreign Policy and the Law of the Sea*. Princeton, NJ: Princeton University Press.

Hong, G. H. and Lee, Y. J. (2015). Transitional Measures to Combine Two Global Ocean Dumping Treaties into a Single Treaty. *Marine Policy*, **55**, 47–56. doi:10.1016/j.marpol.2015.01.007.

Kalland, A. and Moeran, B. (1992). *Japanese Whaling: End of an Era? Scandinavian Institute for Asian Studies Monograph Series 61*. London: Curzon Press.

Karim, M. S. (2010). Implementation of the MARPOL Convention in Developing Countries. *Nordic Journal of International Law*, **79**(2), 303–37. doi:10.1163/157181010X12668401899110.

Kirk, E. A. (2015). Science and the International Regulation of Marine Pollution. In D. R. Rothwell, A. G. O. Elferink, K. N. Scott, and T. Stephens, eds., *The Oxford Handbook of the Law of the Sea*. New York: Oxford University Press, pp. 516–35.

Mansell, J. N. K. (2009). *Flag State Responsibility: Historical Development and Contemporary Issues*. Berlin: Springer.

Mitchell, R. (1993). Intentional Oil Pollution of the Oceans. In P. M. Haas, R. O. Keohane and M. A. Levy, eds., *Institutions for the Earth: Sources of Effective International Environmental Protection (Global Environmental Accords Series)*. Cambridge, MA: MIT Press.

Miyaoka, I. (2013). *Legitimacy in International Society: Japan's Reaction to Global Wildlife Preservation*. New York: Springer.

Molenaar, E. J. (2011). Non-Participation in the Fish Stocks Agreement: Status and Reasons. *The International Journal of Marine and Coastal Law*, **26**(2), 195–234. doi:10.1163/157180811X558956.

Noyes, J. E. (2017). The Territorial Sea and Contiguous Zone. In D. R. Rothwell, A.G.O. Elferink, K. N. Scott, and T. Stephens, eds., *The Oxford Handbook of the Law of the Sea*. New York: Oxford University Press, pp. 91–113.

Peterson, M. J. (1992). Whalers, Cetologists, Environmentalists, and the International Management of Whales. *International Organization*, **46**(1), 147–86.

Rayfuse, R. (2015). Regional Fisheries Management Organizations. In D. R. Rothwell, A.G.O. Elferink, K. N. Scott, and T. Stephens, eds., *The Oxford Handbook of the Law of the Sea*. New York: Oxford University Press, pp. 439–62.

Scott, S. V. (2005). The LOS Convention as a Constitutional Regime for the Oceans. In A. G. Oude Elferink and S. V. Shirley, eds., *Stability and Change in the Law of the Sea: The Role of the LOS Convention No. 24*. Leiden; Boston: Martinus Nijhoff Publishers.

Stoett, P. J. (1997). Cetapolitics: The IWC, Foreign Policies, and NGOs. In *The International Politics of Whaling*. Vancouver: UBC Press.

Thomson, J. E. (1994). *Mercenaries, Pirates, and Sovereigns: State-building and Extraterritorial Violence in Early Modern Europe*. Princeton, N.J.: Princeton University Press.

Vogler, J. (2000). *The Global Commons: Environmental and Technological Governance*. 2nd edn. Chichester, West Sussex, England; New York: Wiley.

Wonham, J. (1996). Some Recent Regulatory Developments in IMO for Which There Are Corresponding Requirements in the United Nations Convention on the Law of the Sea. A Challenge to Be Met by the States Parties? *Marine Policy*, **20**(5), 377–88. doi:10.1016/0308-597X(96)00030-9.

Part II
Vulnerable Islands and Coasts

3

Political Economy of Coastal Development
The Case of the Caribbean

LISA BENJAMIN AND ADELLE THOMAS

Introduction

Small-island developing states (SIDS) are among the areas most vulnerable to the impacts of climate change. Rising sea levels, ocean acidification, coastal erosion, loss of coral reefs, and other ecosystems are all highly likely to have significant ecological, economic, and societal impacts. Climate change may ultimately threaten the territorial existence of some of these states. Coastlines and beaches are valuable economic resources for SIDS in the Caribbean. Coastal tourism is often a significant driver for their economies, contributing greatly to their economic development and employing high percentages of their workforces. However, both coasts and coastal tourism are highly vulnerable to the impacts of climate change. Climate change affects destination attractiveness, deteriorates the natural resources that coastal tourism is dependent upon, and shifts global flows of tourists, resulting in reduced visitation to the region. Mitigation policies aimed at reducing greenhouse gas emissions associated with travel also threaten to negatively affect tourism industries in the Caribbean.

Reducing vulnerability through adaptation actions, such as coastal property setbacks, would remove the most valuable areas of these states from development. Such actions may severely restrict already limited development options and undermine economic growth. Although SIDS are highly vulnerable to climate change, coastal tourism is a significant economic driver. These countries may be reluctant to undertake the full suite of climate change adaptation options available to them in order to preserve coastal development options. Difficult political choices between economic survival and climate change adaptation will undoubtedly arise, leaving policy makers reluctant to implement adaptation measures that severely hamper near-term economic development along coastlines. This could have significant implications for long-term governance of climate change in the region.

This chapter examines the contributions that coastlines provide to the economies of Commonwealth Caribbean states and the detrimental impacts that sea-level rise may have on their political economy. First, the impacts of climate change on Caribbean SIDS are briefly outlined. Next, the importance of tourism to Caribbean SIDS, along with a synopsis of the industry's vulnerability to climate change, is provided. This is followed by an examination of explicit governance arrangements for coasts in the region. Finally, a review of the laws and policies on coastal development, with some emphasis on The Bahamas, illustrates the emergent dissonance between economic survival and adaptation to climate change.

Impacts of Climate Change on Caribbean SIDS

Climate change is projected to affect almost every facet of life in Caribbean SIDS, including biophysical variations, socioeconomic transformations, and changes to culture and senses of identity. Key drivers of climate change in the region include variations in atmospheric and oceanic temperatures, oceanic chemistry, precipitation, extreme events, and sea levels. These drivers have already resulted in observed impacts in Caribbean SIDS that are expected to intensify as global temperatures rise (Nurse et al., 2014: 1619). Sea-level rise is perhaps the most concerning impact for SIDS, particularly those with extensive low-lying coastal zones where the majority of residents, infrastructure, and economic activity are located (McGranahan et al., 2007: 24). Sea-level rise has already been identified as resulting in coastal erosion and the loss of beaches in several Caribbean SIDS (Cambers, 2009: 174). SIDS are also acknowledged as facing the largest increases in impacts as sea levels continue to rise, including complete submergence of low-lying areas (Nicholls and Cazenave, 2010: 1519). In addition to flooding of coastal land, sea-level rise may result in saltwater intrusion into groundwater (Werner and Simmons, 2009: 204), loss of coastal wetlands such as mangroves (Parkinson et al., 1994: 1084), degradation of ecosystems (Kirwan and Megonigal, 2013: 54), and loss of biodiversity (Courchamp et al., 2014: 127).

These biophysical changes will have resultant socioeconomic and cultural impacts. The loss of coastal land will likely result in migration and displacement of coastal populations, issues that Caribbean SIDS are currently ill equipped to address (Thomas and Benjamin, 2017a). Vital infrastructure such as hospitals, air and sea ports, road networks, and power generation plants are often located in coastal areas and face significant damages or complete loss (Lewsey et al., 2004: 407; Dasgupta et al., 2007: 11). Industries that are concentrated in coastal zones such as tourism and fisheries also face significant impacts due to sea-level rise, possibly resulting in loss of economic activity and changes to livelihoods (Allison et al., 2009: 15; Scott et al., 2012: 891). Noneconomic-related changes such as loss of culturally significant physical spaces and loss of identity associated with particular livelihoods or locations are also associated with the impacts of sea-level rise in Caribbean SIDS (Adger et al., 2013: 112–113).

Other impacts of climate change, including increased intensity of tropical storms and associated damages, higher temperatures resulting in more heat waves, and increased ocean acidity with resultant coral bleaching, paint a dire picture for Caribbean SIDS (Nurse et al., 2014: 1643). In addition to these significant economic, environmental, and social impacts, residents of SIDS may suffer from psychological impacts, including decreased mental health associated with being directly and indirectly affected by climate change (Doherty and Clayton, 2011: 265).

Tourism in the Caribbean: Vulnerability to Climate Change

Tourism and Caribbean SIDS

The Caribbean has been identified as one of the most tourism-dependent regions in the world (Thomas, 2015: 2). Tourism accounted for approximately 15 percent of total gross domestic product (GDP) and 13 percent of all employment for the region in 2016 (WTTC, 2017a: 5). These statistics positioned the Caribbean as having the highest relative contribution to GDP from tourism on a global scale, as well as the highest total contribution to employment, surpassing other frequently visited destinations such as Southeast Asia and Europe (WTTC, 2017a: 12). For some islands in the Caribbean, tourism is even more of an economic giant. For example, in 2016 tourism accounted for 45 percent of GDP in The Bahamas, 60 percent of GDP in Antigua and Barbuda, and a staggering 88 percent of GDP in Aruba (WTTC, 2017b). This economic reliance on tourism has not occurred by chance; it is a result of the post–World War II restructuring of newly independent Caribbean countries, which turned to tourism as a vehicle for economic growth, often to the detriment of other industries (Pattullo, 2005: 6–10).

Tourism in the Caribbean is often based on providing a "sun, sand and sea" experience. SIDS market their natural assets, such as warm climates, sandy beaches, and clear seas, to attract visitors from North America and Europe (Pattullo, 2005: 14). The natural beauty of the islands is a major component in luring the over 25 million visitors to the region annually (WTTC, 2017a: 5). Tourism facilities and attractions are generally concentrated in coastal areas, ensuring easy access to beaches and definable "tourism zones" (Weaver, 1993: 136–140). However, the expansion of tourism facilities along the coasts of islands has led to land-use change that often displaces residents, prohibits access to beaches, prices coastal real estate out of the reach of citizens, and places the industry at risk due to climate change (Thomas, 2016: 39).

Tourism's Vulnerability to Climate Change

On a global scale, climate change has been identified as the most significant challenge facing tourism industries (Scott et al., 2008: 175). Changes to the key drivers of climate change, resultant impacts on natural resources, and shifts in global patterns of tourist flows are all expected to transform industries, with particularly dire consequences for tourism-dependent regions such as the Caribbean (Thomas, 2012: 22–26). Climatic conditions have been shown to correspond to destination attractiveness (De Freitas, 2005: 9–11). Changing precipitation, temperature, wind, and so forth due to climate change may alter the attractiveness of various destinations and affect their suitability for tourism (e.g. Amelung and Viner, 2006; Lewis-Bynoe et al., 2009).

The Caribbean is a region with well-defined tourism seasonality; the peak season corresponds to winter in the Northern Hemisphere (Butler, 2014: 9). This is when temperatures in the region are appealing to Europeans and North Americans, and it is also well before hurricane season, which extends from June to November. Studies have shown that climate change will likely shorten the tourism season when climatic conditions are appealing to visitors and ultimately affect the temporal and financial patterns of tourist visitation (Uyarra et al., 2005: 16). As temperatures increase, hurricane seasons extend, and precipitation becomes more erratic, the once-inviting climate of the region will not be as appealing to northern visitors (Amelung et al., 2007: 295; Moore, 2010: 503).

Climate change will also have impacts on natural resources that are critical to Caribbean tourism. Warmer and more acidic oceans will result in bleaching of coral reefs, depleting an important attraction for scuba diving and snorkeling visitors (Uyarra et al., 2005: 11–12). Coastal erosion and the resultant loss of beaches decreases the appeal of islands to tourists (Scott et al., 2012: 887). Sea-level rise and coastal erosion also threaten the accommodations and activities often located either directly on, or in very close proximity to, coastlines. Adaptation strategies such as coastal protection structures important in stabilizing coastlines may be visually unappealing and may also prevent direct and easy access to beaches (Buzinde et al., 2010: 351).

These climatic and environmental changes are also associated with shifts in global tourism patterns. As climates around the world change, there will likely be a poleward shift in the location of climatically ideal tourist destinations (Amelung et al., 2007: 290–294). As the world warms, new destinations will become the locus of climates appealing to tourists. The tropics, currently the site of popular tourist regions such as the Caribbean and Southeast Asia, will no longer offer the ideal climates that attract tourists (Amelung et al., 2007: 292). Rather, more northerly destinations that are likely to be located in the same countries that the majority of international tourists originate from will have these conditions. Thus, there is expected to be a major increase in domestic tourism in North America and Europe with less international travel to regions currently reliant on visitors from these northerly regions (Scott et al., 2004: 116).

Another issue that increases the vulnerability of Caribbean tourism in an era of climate change is addressing its own contributions to climate change as an industry. Tourism and travel are major contributors to the greenhouse gases driving climate change, with emissions from air and sea transport, accommodations, and activities contributing approximately 5 percent of global carbon dioxide emissions (UNWTO, 2007: 3). Initiatives to mitigate the environmental impacts of long-haul travel have produced policies that have not been beneficial for the Caribbean (Blanc and Winchester, 2013: 360). For example, the United Kingdom's Air Passenger Duty required travelers to pay increased fees for emissions associated with their air travel, resulting in significant price changes in tickets and a decrease in travel to the Caribbean (Jamaica Observer, 2014). Policies such as these will likely affect tourist arrivals to destinations that are reliant on long-haul travelers, such as the Caribbean (Pentelow and Scott, 2011: 203).

Laws and Policies on Coastal Development

Coastal and marine resources are vital to the economic growth and social development of Caribbean states (Anderson, 2012: 317). Land tenure has always been a source of conflict in many countries, including along the coasts of SIDS (Cambers et al., 2003: 1). The small size of SIDS, as well as concentrations of industry and infrastructure along the coasts, exacerbate these conflicts. In addition, as set out earlier, coastal areas in SIDS are highly vulnerable to the impacts of climate change. Competing and unregulated use of limited spaces in these small states are degrading natural habitats. The SIDS Accelerated Modalities of Action Pathway has identified the compounding effect that climate change is having on existing vulnerabilities of SIDS (SIDS Accelerated Modalities of Action, 2014: para 15). It acknowledged that the insufficient levels of resources of SIDS have hampered their ability to create environments to implement previous sustainable development plans and programs, such as the Barbados Programme of Action and the Mauritius Strategy for the Further Implementation of the Programme of Action for the Sustainable Development of SIDS (SIDS Accelerated Modalities of Action, 2014: para 16).

Conflicts in Coastal Zones

In many Caribbean states, coastal areas up to the high-water mark are owned by the government and are therefore available for public use, subject to any coastal leases granted by the government. However, beach access remains a contentious issue in the region. In Jamaica, the Beach Control Act of 1956 has allowed the government to grant exclusive use of the foreshore and sea floor to licensees, often for tourism development. This has led to tensions as local beachgoers are excluded from beaches. In the 1980s, a movement called "windows to the sea" in Barbados was developed to ensure openings with views to the sea, as well as beach and coastal access for residents (Cambers et al., 2003: 8). Conflict erupted in St. Vincent and the Grenadines at the end of 2000 when residents of Canouan were denied access to Godahl Beach due to the construction of a large hotel complex (Cambers et al., 2003: 8). As recently as 2016, violence erupted in The Bahamas when The Cabbage Beach Vendors' Association tore down a fence that was erected to prevent them accessing a beach on Paradise Island, populated by large hotel chains (Wells, 2016). Coastal access and development have long been contentious issues in the region, with residents and tourists having to share increasingly shrinking coastal resources. These conflicts are only likely to increase as the impacts of climate change take hold. Management of coastal zone areas is important not only to secure economic and livelihood protection but also to manage social conflict within these states.

Regional Approaches to Coastal Zone Management

The management of coastal areas in the region through regulatory intervention has been patchy due to spatial, human, financial, and technical constraints. However, three main approaches to coastal protection can be identified. The first approach is the adoption of a dedicated coastal zone management regime with defined jurisdictional competence (UNEP/GPA, 2003: 6). The Coastal Zone Management Act of 1998 in Barbados and the Coastal Zone Management Act of 1998 in Belize established in legislation their own coastal zone management authority with defined coastal zones. Belize has now adopted the pursuit of sustainable coastal resource management as the "spine" of its marine protected areas initiative (Belize Coastal Zone Management Authority and Institute, 2010: 11). The second approach is to protect coastal zones through broad environmental legislation and to reserve specific institutional arrangements for coastal zones (UNEP/GPA, 2003: 6). The Natural Resources Conservation Authority Act of 1991 in Jamaica established the Natural Resources Conservation Authority (now the National Environment and Planning Agency) as the lead environmental agency for the country, and it included a specific coastal management zone administrative unit.

The third approach is a traditional, ad hoc, and sectoral one, which does not establish a dedicated management institution or legislation for the coastal region. This traditional approach is based on the ecological reasoning that a small-island nation cannot be separated into territorial and coastal regions (UNEP/GPA, 2003: 6). For example, Trinidad and Tobago has traditionally resisted the establishment of coastal zone legislation on the basis that the integrated coastal zone management concept was developed primarily for continental states, and small islands were not large enough to support a legislative differentiation between coastal and territorial management (UNEP/GPA, 2003: 36). Recently, Trinidad and Tobago appointed a multisectoral steering committee to reconsider its traditional approach to coastal zone management due to large economic, industrial, and population concentrations in coastal areas. This has led to unsustainable utilization of coastal resources and ocean spaces (Integrated Coastal Zone Management Steering Committee, 2014: 7).

Coastal zone management has become an increasing priority for these states. Challenges such as tourism development and agricultural and recreational activities, as well as growing urbanization, pollution, and loss of habitats are plaguing regulatory efforts (Anderson, 2012: 317). Threats to marine and coastal ecosystems are dramatically increasing in the region, and existing governance and institutional arrangements are not curbing the increasing deterioration of these economic and socially valuable areas (Innis, Braithwaite, and Rowe, 2008: 109). Challenges such as these are being exacerbated by the impacts of climate change. Sea-level rise will change existing coastal boundaries by elevating high-water marks (see Chapter 8). Even in Barbados, often considered a leader in the region for its coastal zone management, systemic problems have been identified that lead to poor management of the coastal zone (Anderson, 2012: 345).

Legal and Policy Approaches: The Case of The Bahamas

In The Bahamas, permits for foreign direct investment, including hotel development, are granted through the National Economic Council (a nonstatutory body comprising mainly a committee of members of the Cabinet). Although project proposals are often subject to an environmental review and the requirement of an environmental impact assessment (EIA), there is no detailed legislative requirement to submit an EIA. When submitted, EIAs are often reviewed by the Bahamas Environment Science and Technology (BEST) Commission, which sits under the Ministry of Environment and Housing. Neither the ministry nor the BEST Commission are established by statute. The approval process for developments is therefore nontransparent and does not require a legally prescribed public-participation process (Benjamin, 2014: 65).

The main piece of development legislation in The Bahamas is the Planning and Subdivision Act of 2010 (PSA 2010). Land under the act is defined as including land under water up to the limits of the territorial sea. The act establishes a Town Planning Committee, which must take into account a number of criteria when issuing approvals for developments. These include conservation of natural resources, preservation of environmentally sensitive land and wetlands, and flood control and encroachment on surrounding coastlines (PSA, 2010: Section 12[c]–[f] and [q]). The act also requires that land-use plans be developed for each island consistent with national land-use development policies (PSA, 2010: Section 16). Land-use plans are required to designate areas unsuitable for development for reasons such as flooding, erosion, subsidence, instability, or other hazards or environmental considerations (PSA, 2010: Section 16[2][c]). Land-use plans must also take into account amenities listed in the Second Schedule, which includes coastal zone protection (PSA, 2010: Part IV[18]). The act also requires the adoption of zoning by-laws for each island, prohibiting the erection of buildings or structures on lands subject to flooding, which are otherwise unstable or hazardous, or are subject to erosion (PSA, 2010: Section 23[1][d]). The act therefore develops a legislative and governance structure that can protect coastal resources and limit development in vulnerable areas. However, due to capacity constraints, political pressure, or both, the act is not enforced. A recent State of the Nation report (the precursor to the National Development Plan) notes that, due to lack of enforcement, the PSA 2010 has been rendered "inactive," constituting a "serious risk to the orderly regulation of national planning and development activities, as well as for overarching principles such as the rule of law" (NDP Secretariat, 2016: 35), demonstrating a critical lack of emphasis on the enforcement of planning legislation by the government.

Other pieces of legislation touch on coastal zone management, but not completely. The Disaster Preparedness and Response Act of 2006 is limited in that it focuses primarily on disaster response and relief, not on disaster management or risk identification. Additionally, the National Policy for the Adaptation to Climate Change of 2005 is already outdated; most of its policy directives have not been implemented. The Coast Protection Act of 1968 provides the government the authority to carry out coastal protection work, including the erection of coastal defenses, but the act contains no other development controls.

Coastal development in the country often involves dredging, which can be controversial. Although a permit for dredging is required, it does not necessarily have to contain details of the type or period of dredging or any remedial action. The only legislation regulating dredging activities is the PSA 2010. Controversial dredging of almost 1,000,000 cubic feet of seabed, including the destruction of coral reefs in Bimini Bay, illustrates the ecological and social sensitivities of the subject in the country (Benjamin, 2015: 129–133). In addition, very little monitoring of development takes place due to limited resources, as well due to the spread-out, archipelagic nature of the country. Travel between islands is expensive, and neither the Ministry of the Environment and Housing nor the BEST Commission have statutory footing to conduct monitoring of developments. The State of the Nation report notes the challenges the country is facing regarding coastal development. It states that significant threats to the coastal and marine areas are occurring due to inadequate management of the environment (NDP Secretariat, 2016: 55–56). The report continues that these risks include increased vulnerability to storm surge and sea-level rise as a result of direct coastal impacts such as dredging, removal of mangroves, and landfilling, as well as from indirect impacts such as pesticides and sewage discharge and runoff (NDP Secretariat, 2016: 55–56).

The governance framework for coastal development in The Bahamas remains fragmented and sectoral, with many unwritten policy directives governing activities in coastal zones. This provides the government with the flexibility to permit coastal development without taking into account the value of coastal resources that may be destroyed in the process. The focus, therefore, through the coastal development governance structure is firmly on near-term economic benefits and does not take into account the need to incorporate climate change adaptation activities within coastal development. The country has recently entered into a technical cooperation grant from the Inter-American Development Bank for the phased establishment of an integrated coastal zone management unit.

The Dichotomy between Coastal Development and Climate Change Adaptation

Two main trends can be identified as affecting coastal resources in the Caribbean: the major environmental impacts of tourism on coastal areas and the impacts from climate change (Jones and Phillips, 2008: 379). Despite these negative trends, the development of consistent and effective coastal regulation and governance has been identified as problematic within the region, as exemplified by The Bahamas. Poor development choices are exacerbating the vulnerability of coastal areas in the region to climate change (Mycoo and Donovan, 2017: 25). Poor land-use practices, outdated land registration systems, and weak enforcement of development plans and legislation (including environmental regulations), combined with profound governance deficits and lack of appropriate infrastructure, are all resulting in incomplete and inadequate coastal zone regulation (Mycoo and Donovan, 2017: 25).

Poor Coastal Development Policies and the "Coastal Squeeze"

Regular political interference in development decisions is rife in Caribbean small-island states due to the economic and political value of coastal areas. Development of coastal areas contributes to the tourism product in the form of both hotels and the sale of land. Because coastal development is one of the primary income generators for these states, this dichotomy has engendered conflicting approaches to the management of coastal resources. The lack of effective coastal zone regulation may therefore be both due to the capacity and spatial constraints of these small states but also due to purposeful policy directives of governments. Unwritten policy rules, lack of implementation, and understaffing of critical departments provide governments with the flexibility to allocate, sell, and develop coastal lands for near-term economic development, regardless of the environmental impacts of the development or the vulnerability of the area. Due to their small sizes, there is little spatial opportunity for development outside of coastal areas of small-island states. The value of coastal lands is therefore very high, creating socioeconomic conflict. This approach to coastal development may no longer be viable for these states as the impacts of climate change, including sea-level rise, increase. Many SIDS are experiencing this "coastal squeeze" wherein development pressures combined with climate change impacts, notably sea-level rise, are decreasing available coastal areas (Mycoo and Donovan, 2017: 61).

Climate Change Adaptation Options

Mycoo and Donovan (2017) have identified three main adaptation options available for SIDS in coastal regions. These are planned retreat using planning and development tools, accommodating by adjusting human use of the coastal zone to hazards, and coastal protection using soft or hard engineering projects (Mycoo and Donovan, 2017: 69). Planned retreat would include the use of coastal setbacks, establishing "no-build" zones, prohibiting reconstruction in areas damaged by extreme events, land acquisition and land-use restrictions, and reducing fiscal incentives for development in vulnerable coastal areas (Mycoo and Donovan, 2017: 79). The high value of coastal lands, however, is the very thing that interferes with the implementation of climate change adaptation options such as planned retreat. Implementing development controls such as coastal setbacks removes valuable areas of these countries from development. SIDS such as Barbados have already had to confront these difficult political realities. Due to strong political pressure and fears of lack of compensation for coastal landowners, Barbados has chosen to implement a "hold the line" adaptation option, using soft or hard engineering projects to protect and retain coastal properties (Mycoo and Donovan, 2017: 77). Although these projects may retain near-term economic benefits by protecting coastal areas, these protection projects are both expensive and contain within them residual risks (Mycoo and Donovan, 2017: 77). The impacts from climate change, including extreme events and sea-level rise, are becoming increasingly unpredictable. Engineering options adopted today may not be resilient to the climate of tomorrow.

Protection-only adaptation options may no longer be viable for some SIDS. Extreme events, such as hurricanes with increased storm surge and regular flooding, have already forced some states to implement ad hoc no-build zones in devastated areas. In The Bahamas, after Hurricane Joaquin in 2015, the Ministry of Environment and Housing chose to rebuild homes inland away from the coasts, and it adopted an informal, unwritten policy not to rebuild in coastal areas devastated by extreme events. It is unclear, however, whether permission to construct tourism developments in coastal areas affected by past hurricanes would be refused by the government. It is again evident that climate change is having increased impacts on the economies of these states.

Decreasing Resilience and Available Policy Space

As a percentage of GDP, the costs of sea-level rise are anticipated to be highest in SIDS by 2100 (Mycoo and Donovan, 2017: 1). Caribbean states are already experiencing some of the highest number of natural disasters in the world per square kilometer (Ruprah et al., 2014: 7). The direct economic impacts of these disasters are also higher for these small economies (Ruprah et al., 2014: 7). The impacts of these events are already diverting precious public resources toward reconstruction efforts and away from other developmental investments, including climate change adaptation efforts (Thomas and Benjamin, 2017a; Benjamin and Haynes, 2018). As Lyster (2015: 126) notes, climate change often leads to uncompensated damages in the developing world as governments have to absorb large welfare losses from public budgets. She also notes that after disasters, governments face further macroeconomic strains, including depleted tax bases, declining reserves and credit ratings, and difficulty in borrowing, leading these states to divert capital from other social programs to postdisaster efforts (Lyster, 2015: 140–141). Thomas and Benjamin (2017b: 7–8) have found that these states are unprepared for the impacts of loss and damage due to climate change, with very few states having concrete policies or data to address these increasing impacts. Thomas and Benjamin also found that SIDS are unprepared for climate-induced displacement or migration, with very few states having formal policies for planned retreat (Thomas and Benjamin, 2017a). These states, therefore, suffer from policy deficits in the areas of coastal regulation and impacts from climate change.

Sea-level rise is anticipated to exacerbate the vulnerability of these states to natural disasters due to increased storm surges and flooding events. The impacts of climate change on economic development may therefore begin to outweigh the contributions of tourism to these states. These events and anticipated impacts leave these states at increased risk and vulnerability, with decreased resilience and policy vacuums, while at the same time potentially removing existing economic opportunities, thus exacerbating their economic and social vulnerability. Alternative development options that are less vulnerable to climate change must therefore be developed. Despite the difficulties, SIDS do have policy opportunities for adaptation.

The "blue economy" provides one potential way forward to diversify economic development. The ocean resources of these states hold the potential to transform their productive bases, diversifying livelihoods and reducing poverty and food and energy insecurities (Rustomjee, 2016: 2). The implementation of ocean-energy projects (see Chapter 24), as well as access and benefit sharing for marine resources in the development of new pharmaceutical medicines, could constitute new and climate-resilient development options for Caribbean states (Rustomjee, 2016: 2). Diversified value chains could be developed from marine resources. These could provide important revenue streams to supplement declining revenues from tourism. However, due to capacity constraints and lack of financial resources, these economic opportunities have yet to be realized. In addition, as ocean and marine resources become further degraded, the impacts of climate change are circumscribing the ability of SIDS to pursue a blue-economy developmental trajectory (Rustomjee, 2016: 4).

Conclusion

SIDS are highly vulnerable to climate change and rely heavily on tourism and the natural resources that underpin the tourism product. The traditional economic trajectory of tourism as a development model, however, is putting perverse pressures on SIDS to not implement all climate change adaptation options available to them. Coastal retreat is economically problematic, as it removes valuable land from development and serves to further hamper economic development options. Caribbean SIDS suffer from policy and legislative deficits in their coastal zone regulation. Poor development practices and lack of enforcement of development plans result from capacity and spatial constraints. Such approaches may have been purposeful due to economic necessity. Much of this is exemplified in governance in The Bahamas. There, detailed planning legislation was passed in 2010, yet that legislation remains unenforced. Political interference in development decisions is still rife in the region. This is due to the high value placed on coastal areas under existing development patterns. Coastal tourism is also vulnerable to climate change, and the reluctance or inability of SIDS to regulate coastal activities is fueling social conflict, increasing the vulnerability of residents and degrading coastal resources. SIDS are therefore left with few, and ever decreasing, economic opportunities.

Coastal erosion and the increasing impacts of climate change on tourism developments will lead to increasing concerns over the viability and merits of remedial actions (Jones and Philips, 2008: 376). If states adopt the protection approach of Barbados, a pressing question will be who will pay for these investments. In already highly indebted states, protection approaches may not be economically feasible. With increasing impacts of climate change, adaptation approaches may not be durable. Risk assessments will have to be undertaken to determine the costs and benefits of adopting particular adaptation strategies along coastlines (Jones and Phillips, 2008: 381) (see Chapter 18).

In order for risk assessments to be effective, adequate and localized data are needed (Thomas and Benjamin, 2017b: 8–9). SIDS should adopt cohesive and integrated development trajectories that take into account not only coastal zone management but also climate change impacts, loss and damage due to climate change, and disaster-risk management. Considering the existing policy deficits in these states, adopting new and integrated development trajectories may be challenging. International financing to increase capacity within environmental and disaster-risk agencies, as well as finance departments or ministries, is therefore critical, as is funding to acquire and collate localized data on climate change impacts to enable SIDS to design the best possible future for their coastal resources, economies, and societies.

References

Adger, W. N., Barnett, J., Brown, K., Marshall, N., and O'Brien, K. (2013). Cultural dimensions of climate change impacts and adaptation. *Nature Climate Change*, 3(2), 112.

Allison, E. H., Perry, A. L., Badjeck, M. C., et al. (2009). Vulnerability of national economies to the impacts of climate change on fisheries. *Fish and Fisheries*, 10(2), 173–96.

Amelung, B., Nicholls, S., and Viner, D. (2007). Implications of global climate change for tourism flows and seasonality. *Journal of Travel Research*, 45(3), 285–96.

Amelung, B. and Viner, D. (2006). Mediterranean tourism: Exploring the future with the tourism climatic index. *Journal of Sustainable Tourism*, 14(4), 349–66.

Anderson, W. (2012). *Principles of Caribbean Environmental Law*, Washington, DC: Environmental Law Institute.

Belize Coastal Zone Management Authority and Institute (2010). The National Integrated Coastal Zone Management Strategy for Belize. Available at www .coastalzonebelize.org/wp-content/uploads/2010/04/national_integrated_CZM_ strategy.pdf.

Benjamin, L. (2014). Country Report: The Bahamas Access to Environmental Information, EIAs and Public Participation in Development Decisions. *IUCN Academy of Environmental Law eJournal*, 5, 54–65.

Benjamin, L. (2015). Country Report: The Bahamas The Problem of Unpermitted Development and Fragmented Environmental Laws. *IUCN Academy of Environmental Law eJournal*, 6, 127–35.

Benjamin, L. and Haynes, R. (2017). Climate change and human rights in the Commonwealth Caribbean: Case studies of The Bahamas and Trinidad and Tobago. In S. Jodoin, S. Duyck and A. Johl, eds., *Routledge Handbook of Human Rights and Climate Governance*. Routledge.

Blanc, É. and Winchester, N. (2013). The impact of the EU emissions trading system on air passenger arrivals in the Caribbean. *Journal of Travel Research*, 52(3), 353–63.

Butler, R. (2014). Addressing seasonality in tourism: The development of a proto-type. In *Conclusions and Recommendations Resulting from the Punta del Este Conference*, May 2014. UNWTO.

Buzinde, C. N., Manuel-Navarrete, D., Yoo, E. E., and Morais, D. (2010). Tourists' perceptions in a climate of change: Eroding destinations. *Annals of Tourism Research*, **37**(2), 333–54.

Cambers, G. (2009). Caribbean beach changes and climate change adaptation. *Aquatic Ecosystem Health and Management*, **12**(2), 168–76.

Cambers, G., Muehlig-Hofmann, A., and Troost, D. (2003). Coastal land tenure: a small-islands' perspective. Environment and development in coastal regions and in small islands. *CSI*. pp. 1–15. Available at http://portal.unesco.org/en/ev.php-URL_ID=13784&URL_DO=DO_TOPIC&URL_SECTION=201.html.

Coast Protection Act 1968, Chapter 204, Statute Law of the Commonwealth of The Bahamas.

Courchamp, F., Hoffmann, B. D., Russell, J. C., Leclerc, C., and Bellard, C. (2014). Climate change, sea-level rise, and conservation: Keeping island biodiversity afloat. *Trends in Ecology and Evolution*, **29**(3), 127–30.

Dasgupta, S., Laplante, B., Meisner, C., Wheeler, D., and Yan, J. (2007). *The impact of sea level rise on developing countries: A comparative analysis.* World Bank Policy Research Working Paper 4136, February 2007. Washington, DC, World Bank.

Doherty, T. J. and Clayton, S. (2011). The psychological impacts of global climate change. *American Psychologist*, **66**(4), 265.

De Freitas, C. R. (2005). The climate-tourism relationship and its relevance to climate change impact assessment. In C. M. Hall and J. Higham, eds., *Tourism, Recreation and Climate Change: International Perspectives*. Bristol, UK: Channelview Press, pp. 29–43.

Disaster Preparedness and Response Act 2006, Chapter 34A, Statute Law of the Commonwealth of The Bahamas.

Innis, L. V., Braithwaite, A., and Rowe, A. (2008). Governance, ecosystem monitoring and coastal engineering in some Caribbean small islands. In R. R. Krishnaumurthy, B. C. Glavovic, A. Kannen, et al., eds., *Integrated Coastal Zone Management*. Singapore, Chennai: Research Publishing, pp. 109–25.

Integrated Coastal Zone Management Steering Committee (2014). *Integrated Coastal Zone Management (ICZM) Policy Framework. Ministry of Environment and Water Resources.* Available at www.ima.gov.tt/home/images/docs/Ingrated_Coastal_Zone_Mment_Policy_Framework_Minister_April_2014.pdf.

Jamaica Observer (2014). Jamaica, Caribbean welcome UK Air Passenger Duty reforms. Available at www.jamaicaobserver.com/news/Jamaica--Caribbean-welcome-UK-Air-Passenger-Duty-reforms, accessed 15 August, 2017.

Jones, A. L. and Phillips, M. R. (2008). Tourism development in the coastal zone: Managing natural and cultural change. In R. R. Krishnaumurthy, B. C. Glavovic, A. Kannen, et al., eds., *Integrated Coastal Zone Management*. Singapore, Chennai: Research Publishing, pp. 375–89.

Kirwan, M. L. and Megonigal, J. P. (2013). Tidal wetland stability in the face of human impacts and sea-level rise. *Nature*, **504**(7478), 53.

Lewis-Bynoe, D., Howard, S., and Moore, W. (2009). *Climate Change and Tourism Features in the Caribbean.* Munich: Munich Personal RePEc Archive.

Lewsey, C., Cid, G. and Kruse, E. (2004). Assessing climate change impacts on coastal infrastructure in the Eastern Caribbean. *Marine Policy*, **28**(5), 393–409.

Lyster, R. (2015). A fossil fuel-funded climate disaster response fund under the Warsaw International Mechanism for Loss and Damage Associated with Climate Change Impacts. *Transnational Environmental Law*, **6**(1), 125–51.

McGranahan, G., Balk, D., and Anderson, B. (2007). The rising tide: Assessing the risks of climate change and human settlements in low elevation coastal zones. *Environment and Urbanization*, **19**(1), 17–37.

Moore, W. R. (2010). The impact of climate change on Caribbean tourism demand. *Current Issues in Tourism*, **13**(5), 495–505.

Mycoo, M. and Donovan, M. G. (2017). *A Blue Urban Agenda: Adapting to Climate Change in the Coastal Cities of Caribbean and Pacific Small Island Developing States.* New York and Washington, DC: IDB Monograph. http://dx.doi.org/10.18235/0000690.

NDP Secretariat. (2016). *State of the Nation Report: Vision 2040 National Development Plan of The Bahamas.* Available at www.vision2040bahamas.org/media/.../State_of_the_Nation_Summary_Report.pdf.

Nicholls, R. J. and Cazenave, A. (2010). Sea-level rise and its impact on coastal zones. *Science*, **328**(5985), 1517–20.

Nurse, L. A., McLean, R. F., Agard, J., et al. (2014). Small islands. In C. B. Field, V. R. Barros, D. J. Dokken, et al., eds., *Climate Change 2014: Impacts, Adaptation, and Vulnerability. Part A: Global and Sectoral Aspects. Contribution of Working Group II to the Fifth Assessment Report of the Intergovernmental Panel on Climate Change.* Cambridge, New York: Cambridge University Press, pp. 1613–54.

Parkinson, R. W., DeLaune, R. D., and White, J. R. (1994). Holocene sea-level rise and the fate of mangrove forests within the wider Caribbean region. *Journal of Coastal Research*, **10**(4), 1077–86.

Pattullo, P. (2005). *Last Resorts: The Cost of Tourism in the Caribbean.* New York: NYU Press.

Pentelow, L. and Scott, D. J. (2011). Aviation's inclusion in international climate policy regimes: Implications for the Caribbean tourism industry. *Journal of Air Transport Management*, **17**(3), 199–205.

Planning and Subdivision Act 2010 (No. 4 of 2010), Statute Law of the Commonwealth of The Bahamas.

Ruprah, I., Melgarejo, K., and Sierra, R. (2014). *Is There a Caribbean Sclerosis? Stagnating Economic Growth in the Caribbean*, Inter-American Development Bank Monograph; 178.

Rustomjee, C. (2016). *Developing the Blue Economy in Caribbean and Other Small Island States.* CIGI Policy Brief No. 75. Available at www.cigionline.org/sites/default/files/pb_no.75web.pdf.

Scott, D., Simpson, M. C., and Sim, R. (2012). The vulnerability of Caribbean coastal tourism to scenarios of climate change related sea level rise. *Journal of Sustainable Tourism*, **20**(6), 883–98.

Scott, D., Amelung, B., Becken, S., et al. (2008). *Climate Change and Tourism: Responding to Global Challenges.* Madrid: UNWTO, p. 269.

Scott, D., McBoyle, G., and Schwartzentruber, M. (2004). Climate change and the distribution of climatic resources for tourism in North America. *Climate Research*, **27**(2), 105–17.

SIDS Accelerated Modalities of Action (SAMOA) Pathway (2014). A/RES/69/15. www.sids2014.org/index.php?menu=1537.

Thomas, A. (2012). *An Integrated View: Multiple Stressors and Small Tourism Enterprises in The Bahamas.* New Brunswick, NJ: Rutgers University, p. 236.

Thomas, A. (2016). Tourism and land use change in Bahamas. *Caribbean Geography*, **21**, 24–44.

Thomas, A. and Benjamin, L. (2017a). Policies and mechanisms to address climate-induced migration and displacement in Pacific and Caribbean small island developing states. *International Journal of Climate Change Strategies and Management.*

Thomas, A. and Benjamin, L. (2017b). Management of loss and damage in small island developing states: Implications for a 1.5C or warmer world. *Regional Environmental Change*, 1–10. Available at http://DOI10.1007/s10113-017-1184-7.

Thomas, D. (2015) *The Caribbean Tourism Industry in the 21st Century: An Assessment*, KGLACC Working Paper No. 3/2015 Miami: Florida International University.

UNEP/GPA. (2003). Review of National Legislations Related to Coastal Zone Management in the English-Speaking Caribbean. Prepared by Dr. Winston Anderson. Available at www.cep.unep.org/publications-and-resources/databases/document-database/other/gpa-review-of-caribbean-coastal-zone-legislation.pdf/at_download/file.

UNWTO (2007). *Tourism and Climate Change: Confronting the Common Challenges.* UNWTO. Available at http://sdt.unwto.org/sites/all/files/docpdf/docuconfrontinge.pdf, accessed 15 August, 2017.

Uyarra, M. C., Cote, I. M., Gill, J. A., et al. (2005). Island-specific preferences of tourists for environmental features: Implications of climate change for tourism-dependent states. *Environmental Conservation*, **32**(1), 11–19.

Weaver, D. B. (1993). Model of urban tourism for small Caribbean islands. *Geographical Review*, **83**(2), 134–40.

Wells, R. (2016). Police and vendors clash over cabbage beach access. *The Tribune.* Available at www.tribune242.com/news/2016/mar/01/cabbage-beach-access-reopened-temporarily-after-di/.

Werner, A. D. and Simmons, C. T. (2009). Impact of sea level rise on sea water intrusion in coastal aquifers. *Groundwater*, **47**(2), 197–204.

WTTC (2017a). *Travel and Tourism Economic Impact 2017 Caribbean.* London: World Travel and Tourism Council, p. 24.

WTTC (2017b). *WTTC Data Gateway.* World Travel and Tourism Council. Available at www.wttc.org/datagateway/, accessed 11 August 2017.

4

Coastal Development and Climate Risk Reduction in the Persian/Arabian Gulf
The Case of Qatar

MOHAMMAD AL-SAIDI

Introduction

Climate change and climate variability represent serious regional challenges for the countries of the Gulf Cooperation Council (GCC). This highly arid region has experienced high socioeconomic growth in recent decades, mainly along the coasts. The impacts of climate change on coastal development are tangible. Vital coastal infrastructure of many of the GCC countries, such as refineries, power plants, agricultural schemes, and desalination systems, will be negatively affected. Reduced rainfall, greater seasonal temperature variability, sea-level rise, and loss of agricultural production are some expected consequences, as are increased migration pressures. Vulnerable marine ecosystems in the region are vital for food security, recreation, and cultural identity. Most of the heavily populated cities of the region are situated along coasts, and development projects and urban expansion through land reclamation are pushing coastlines further out to sea and increasing future climatic risks.

The study of the strategies and politics involved in development projects and initiatives related to climate risks of coastal areas and marine ecosystems in the GCC is important. Many stakeholders and marine policies related to climate change are still evolving. This chapter identifies policy choices related to mitigation and adaptation and explains policy gaps among GCC countries. It also outlines marine ecosystem vulnerabilities in the region and explores key drivers of those vulnerabilities, including coastal development ratios; dependency on vital services such as water, energy, and food supplies; and the sensitivity of the ecosystems to climate change and their ability to recover from external shocks. Using the example of coastal development and maritime politics of Qatar, the chapter provides examples from specific coastal industries in energy, desalination, real estate, and tourism. It highlights internal societal pressures, actors affecting policy development and state rivalry, and the potential for cooperation and joint action on coastal development policies.

Climate change consideration in coastal developments in the GCC regions are neither adequately studied nor incorporated into specific policies. The rapid pace of coastal development seeks to accommodate short-term to mid-term growth needs, while climate vulnerability issues are rarely addressed on project-based terms. Climate policies are seen as an overarching environmental issue that can be addressed in a future phase of "post-growth." The scattered national-level policy initiatives toward climate risk reduction need to evolve into a regional approach encompassing all Persian/Arabian Gulf countries. The waters of the Gulf can be considered a common resource vulnerable to degradation, increased salinity, and the loss of marine ecosystems. The example of the State of Qatar shows that economic transformation toward a knowledge-based economic model can help mobilize technological options to increase climate resilience and develop effective institutional arrangements. In light of recent international climate agreements and increasing evidence of climate impacts, the need and urgency for actions are increasing. These may lead to more targeted development policies for reducing climate-related risks.

Coastal Climate Vulnerabilities in the Gulf Region

The effects of climate change and increasing environmental variability on the GCC countries – Bahrain, Kuwait, Oman, Qatar, Saudi Arabia, and United Arab Emirates (UAE) – are concrete and alarming. Development projects and urban areas on coasts will be highly vulnerable to the predicted increases of water extremes, notably falls in precipitation, more frequent heatwaves, and sea-level rise (SLR), as in other regions of the world (IPCC, 2011). The wider region of the Middle East is considered to be a highly vulnerable region. The mean temperature is set to increase between 3°C and 5°C, while precipitation and runoff may decrease between 20 and 30 percent (IPCC, 2007; Milly et al., 2005; Elasha, 2010). Considering the minimal development of adaptive capacities of many countries in this region (Sowers et al., 2011), the effects on people, nature, and infrastructure from reduced rainfall, greater seasonal temperature variability, and SLR will be felt (Wodon et al., 2014).

The GCC region shows a significant trend of climate warming across coastal areas that can be directly related to urbanization and development projects. Temperature has increased in the region in recent decades (AlSarmi and Washington, 2011), an acknowledged fact by all Gulf countries in their national communications to the United Nations Framework Convention on Climate Change (UNFCCC). Single-country studies indicate consistent warming trends, for example, 0.06°C/year in Saudi Arabia (Almazroui et al., 2013) and 0.55°C/decade in Qatar (Cheng et al., 2015). Rainfall variability is also increasing, including in Qatar (Al-Mamoon and Rahman, 2017). In fact, rapid urbanization is often linked to the observed warming trend, as studies of Bahrain show (Elagib and Abdu, 2010; Elagib and Alvi, 2013). The phenomenon of the "urban heat island" (UHI) is present all across coastal areas and can be attributed to large construction projects, the decrease of green areas, and the ongoing land reclamation projects (Radhi and Sharples, 2013). Urbanization, development, and large energy-use footprints in the region are also leading to increased pollution affecting people and marine ecosystems (Farahat, 2016).

The increasing ratio of coastal development is leading to higher emissions, pollution, and increased climate vulnerability. Such a high ratio is fueled by socioeconomic growth and a high reliance on energy abundance to compensate for the lack of land and water resources. The GCC region is characterized by a high level of natural scarcity of freshwater and arable land. On average, less than 5 percent of the total land of GCC countries is suitable for agriculture, and all GCC countries exhibit high levels of water scarcity. Under such conditions, serious constraints on food security and economic growth are expected. However, the abundance of energy reserves (i.e., oil and gas) is compensating for water and land scarcities. The GCC region holds around 40 percent and 20 percent of the global oil and gas reserves, respectively (BP, 2016).

Much of the available fossil fuel in the GCC countries is exported, leading to economic prosperity and impressive socioeconomic growth. Between 2000 and 2015, the GCC economies grew by more than 5 percent annually, while population increased rapidly due to the influx of expatriate workers, now constituting 30 to 88 percent of the countries' total populations. These rapid growth trends; changing lifestyles; and the still-missing infrastructure and standards for sustainable mobility, energy efficiency, and resource reuse resulted in the GCC countries having some of largest per capita resource footprints in the world (i.e., per capita consumption of carbon, waste, water, energy, and calories). Further, energy abundance has facilitated urban development and is linked to coastal infrastructure. Energy is used locally in coastal desalination plants to provide for domestic water needs, industrial applications (e.g., cooling and air-conditioning, minerals production, refineries, and tourism), or even for agricultural production. At least 4 to 12 percent of total electricity use in different GCC countries is consumed by desalination plants (Siddiqi and Anadon, 2011). Through energy subsidies, water and energy services are provided at very low prices, often free of charge. Such subsidies account for over 8.5 percent of the region's gross domestic product (GDP) and 22 percent of government revenues (Meltzer et al., 2014). The availability of energy together with the high dependency on food imports and a relatively high socioeconomic stability may make the GCC region less vulnerable to current and future water supply problems than the big agricultural countries in the Middle East and North Africa (MENA) region (Al-Saidi et al., 2016).

Desalination capacity in the GCC region accounts for around 45 percent of the global capacity (Mezher et al., 2011) and relies heavily on fossil fuels. At the same time, there are ambitious targets for the development of renewable energies to increase domestic supply of electricity and also for the use in desalination or for pumping groundwater for irrigation. Much of these new renewable energy projects will be located on or close to coastal areas in the Gulf. For example, the world's largest solar photovoltaic desalination plant, located near the coastal city of Khafji in Saudi Arabia, will have a production capacity of 60,000 cubic meters per day.

New projects will ensure that water and energy remain intensely coupled and inter-dependent in the GCC region – and potentially vulnerable to climate change impacts. At the same time, such new developments are negatively affecting marine ecosystems and thus decreasing their ability to recover from external shocks associated with climate change. For example, the increased rate of seawater desalination and the high evaporation rate of Gulf waters are leading to increased salinity, including so-called hypersalinity conditions (Smith et al., 2007). At the same time, energy and power generation using steam plants is leading to thermal pollution and the release of pollutants that endanger sensitive marine ecosystems (Lattemann and Höpner, 2008).

Climate Policies and Coastal Sectors at Risk

Strategies and policies addressing climate change in the GCC region are evolving both on a national and international track. Although GCC countries have joined major international agreements like the 2015 Paris Agreement, the urge to act on climate change is recent. In light of the recent volatility and uncertainty in global markets for fossil fuels, GCC countries started to diversify and increase the share of the nonhydrocarbon parts of their economies (Callen et al., 2014). Climate change policies are being mainstreamed into diversification policies, as outlined in so-called "national visions." These national visions are aspirational documents outlining future investments and reform priorities. Common elements of the national visions are renewables, diversification, and the promotion of knowledge as a basis for economic growth. The growth agenda also includes a wide range of initiatives and mega-projects like national and regional rails; the development of sustainable agricultural options (e.g., aquaculture and hydroponics); the promotion of local innovations in hotspot technologies (e.g., membranes and desalination); or the encouragement of smart urban solutions for housing, work, and mobility.

Despite environmental sustainability and the reduction of emissions generally being understood as a residual of the larger goal of diversified growth, climate change is still considered a priority by policy makers in certain countries. In fact, not all GCC countries have had a favorable attitude toward responses to climate change. Saudi Arabia, for example, has traditionally been a skeptic of climate change and an obstructionist in climate negotiations (Depledge, 2008). However, its national vision for 2030, adopted in 2016, outlines specific targets for emissions reduction and renewables. Other GCC countries revaluated climate change in terms of stakeholders and policies. Both Oman and UAE have incorporated climate change responses into new ministries for environment and climate affairs in 2007 and 2016, respectively.

The UAE was one of the first countries in the region to launch large-scale investments in renewable energies. Being the host of the headquarters of the International Renewable Energy Agency (IRENA) and home to the carbon-neutral Masdar City as a global showcase of transformation and renewable energy (Reiche, 2010a),

UAE's engagement expands beyond the local scale. Similarly, the State of Qatar has had global ambitions in climate change diplomacy. For example, it hosted the conference of the parties of the UNFCCC in 2012 (the so-called COP 18), and it has provided international aid for countries adapting to climate change. In its National Vision 2030, Qatar aspires to have "a proactive and significant international role in assessing the impact of climate change and mitigating its negative impacts, especially on countries of the Gulf" (State of Qatar, 2008).

Alongside international climate diplomacy, renewables initiatives, and infrastructure mega-projects, GCC countries have also started to explore other mechanisms to reduce emissions – for example, carbon capture and storage, with networks and projects being implemented within national oil companies. However, major obstacles exist, including the absence of local regulation of markets and inadequate cross-border trade agreements (Tsai, 2013). In fact, there are major structural restrictions on the totality of climate policies in the GCC, including the lack of civil society, minimal taxation, domination of public stakeholders, and lack of data (Reiche, 2010b). Further, governmental patronage and lack of long-term sector strategies, as well as specific policies and legislation, are major problems for diversification (Hvidt, 2013). The development of renewable energy, for example, is expected to be jeopardized by the large energy subsidies and modest technology and research investments (Mondal et al., 2016).

The majority of response measures in the Intended National Determined Contributions (INDCs), which all GCC countries submitted as a part of the Paris Agreement, focus on hard infrastructure, energy efficiency in oil industries, and renewables. Coastal climate vulnerability features in all of the INDCs, but it is not always addressed specifically. All countries acknowledge climate threats like tropical cyclones, coastal erosion, SLR, and migration of marine species and sea birds. Concrete measures to address the marine environment and make use of the carbon sequestration potential of the seas (so-called "blue carbon"; see Chapter 26) are mentioned by UAE and Oman in terms of demonstration projects for mangrove cultivation and the increase of seagrass beds.

Coastal climate policies in the GCC region will have to go beyond this narrow project focus to address key economic sectors at risk. Several key categories of affected sectors will be discussed here: First, tourism, recreation, and fisheries depend on healthy marine ecosystems. Second, land reclamation can be directly affected by climate change. Third, supply infrastructure, especially energy and water production plants on the coasts, depend on the Gulf's waters and are thus highly vulnerable.

Marine ecosystems of the Gulf are vulnerable to climate change but also to anthropogenic factors like reclamation, dredging, industrial pollution, and environmental accidents (including, dramatically, an oil spill of around 11 million barrels during the 1991 Gulf War). Climatic stressors on marine ecosystems, which closely interact with anthropogenic factors in the Gulf, are well documented, including the temperature and sea-level rise mentioned earlier. Further, climate effects like ocean acidification and increased salinity due to rising temperatures could have disastrous effects on certain components of marine ecosystems. For example, Feary et al. (2013)

outlined two critical research areas regarding climate change impacts on the biology and ecology of the Gulf. First, rising temperature combined with ocean acidification will lead to bleaching events on reefs and a reduction in coral calcification, resulting in erosion and degradation at an unknown scale. Second, the extent of increased evaporation and salinity on species diversity needs to be investigated. The vulnerability of coral reefs to climatic extremes has been reviewed by Riegl and Purkis (2012), highlighting the special importance of the Gulf, which harbors species living under extreme thermal conditions. Thus, the Gulf is highly significant in terms of monitoring of local changes, as well as predicting future conditions of other water bodies under climate change scenarios.

The damage of marine ecosystems like coral reefs in the Gulf is not entirely linked to climate change. Sale et al. (2010) reviewed anthropogenic disturbances from waste, mismanagement, and economic expansion. In fact, human activities, such as recently increased land reclamation projects on the coasts, continue to negatively affect coastlines that are vulnerable to climate change. While reclamation is seen in national strategies as a viable way to secure land needs, evidence shows that it is leading to the alteration of natural processes related to salinity and water flow and damage to key biodiverse ecosystems (Fakhro, 2013). At the same time, countries intensively using land reclamation, like the UAE or Bahrain, which expanded its land area by 13 percent through reclamation, increase future climate risks. These countries will have to also ensure adequate protection of these reclaimed areas.

Another highly vulnerable sector is represented by the supply infrastructure for water and energy. In most GCC countries, desalination is the main source of potable water. Large desalination plants are constructed across the Gulf, supplying key cities in the region. Mega-plants on the coast use co-generation technology of both desalinated water and thermal power. For example, the Ras Al-Khair plant, which began operation in 2014, supplies water and electricity for most of the capital city of Riyad in Saudi Arabia. Similar large-scale co-production plants exist in other GCC countries, indicating an increased integration between water and energy supply systems and a heavy reliance on centralized supply infrastructure. In fact, the growing technological sophistication, intensification, and lack of decentralization can lead to growing risks from potential failures. The overall resilience of the water and energy supply infrastructure to external risks such as climatic extremes is little studied and not addressed properly in policy making. Concrete risk management strategies and vulnerability studies are important for the security of water supplies.

The Case of the State of Qatar

Overall Vulnerability, Climate Policies, and Diplomacy

The State of Qatar's ecological and human systems are prone to climate change, while its economic welfare and prosperity largely depend on the outcome of international climate change negotiations (State of Qatar, 2011). With more than 95 percent of the population living in coastal areas, SLR poses a serious threat to Qatar and could result in flooding of up to 18 percent of its land area if SLR reaches five meters (State of Qatar, 2011). Climate change is expected to negatively affect people's health, the health of terrestrial coastal and marine ecosystems, and state revenues. Alongside fighting these direct impacts, Qatar is also working on reducing its high-carbon footprint since it has, according to some estimates, the highest carbon emissions per capita in the world (UN Habitat, 2011). Therefore, Qatar proposes in its INDC a set of mitigation options that stem from its future vision of economic diversification, including the increase of energy efficiency, the promotion of renewables, and knowledge and education, as well as tourism, as sources of revenue (State of Qatar, 2015). Adaptation options focus on sustainable water and waste management, infrastructure, and mobility, as well as environmental awareness (State of Qatar, 2011). In this sense, Qatar has embarked on a number of initiatives to reduce its carbon footprint and its dependence on fossil fuels. Such initiatives include the promotion of solar power by installing ten gigawatts of capacity by 2030, increasing the certification of new projects according to Leadership in Energy and Environmental Design (LEED) criteria in large sites like the Education City and Lusail City, and proposing a research institute on climate impact and energy monitoring with a regional outreach (which has not yet been established).

Qatar's activism in the international arena for climate change was highlighted by hosting climate negotiations (COP18) in 2012, organizing international fora (e.g., the 2011 and 2013 Doha Carbon and Energy Forum), and promising climate aid to the least-developed countries. It seems, however, that such international activism slowed after COP18, although Qatar is still active in international negotiations and has ratified the Paris Agreement. One reason for the apparent decrease of interest in climate leadership might lie in the economic crisis since 2014, which is due to the decline of oil prices. In the past, Qatar established a National Committee for Climate Change and incorporated climate change into the Qatar National Vision 2030 (QNV2030) under the government's "fourth pillar" of environmental sustainability. Climate change featured prominently in Qatar's Second National Human Development Report of 2009 (State of Qatar, 2009) but not so much in the two subsequent reports under the themes of "Youth and Development" and "The Right to Development." Such national reports represent an effort to operationalize QNV2030 with the help of international organizations such as the United Nations Development Programme. In fact, until the recent reduction in oil prices, Qatar made several concrete initiatives, for example, joining the World Bank's Global Gas Flaring Reduction Partnership as the first Gulf country and initiating its first project under the Clean Development Mechanism (Al Shaheen Oil Field Gas Recovery and Utilization Project).

Nowadays, climate negotiators for Qatar stress the country's commitment to climate change but also the need for a careful and coordinated approach. In a recent interview the head of the Ministry of Development and Statistics Climate Change Department, established in 2014, emphasized the vulnerability of Qatar's oil industry to mitigation policies, the need for regional coordination, and the importance of technology and knowledge, as well as the alignment of climate goals to QNV2030 and economic diversification (Murthy, 2016).

Risk Reduction in Coastal Ecosystems

With 23 percent of the total coastline of the Gulf and 15 percent of its waters, Qatar harbors some of the richest marine ecosystems in the region and among the largest areas of seagrass, coral reefs, and mangroves surviving under extreme conditions of salinity and temperature (State of Qatar, 2011). Coastal ecosystems are threatened by ocean warming and acidification, leading to loss of corals, bird habitats, and biodiversity (State of Qatar, 2009). Temperature rise can also lead to the loss of large stocks of fish, thus threatening food security. For example, Al-Ansi et al. (2002) documented the loss of massive amounts of fish in the summer of 1996 and 1998 due to a temperature rise beyond 35 degrees. In fact, all coral reefs in Qatar are classified as "at risk" due to local anthropogenic and global climatic stressors (Burke et al., 2011). Other warming events were documented in 2002 and 2010. A study of these events showed that bleaching thresholds of Gulf corals are rising, indicating that coral reefs in the region are highly resilient and might be capable of adapting to future climate pressures (Riegl et al., 2012). Although this might be a good thing, one needs to study the combined effect of both anthropogenic pressures and the totality of climate stressors on these ecosystems. As Qatar's national communication to the UNFCC outlines, the physical, biological, and biochemical conditions of the oceans and coasts are expected to be affected by climate change, collectively leading to potentially profound impacts on the sustainability, productivity, and biodiversity of marine ecosystems (State of Qatar, 2011). According to Burt et al. (2012), reclamation of islands and manmade structures like pipelines or oil and gas platforms can provide important artificial reefs and hard-bottom breakwater habitats in the Gulf. However, these structures are rarely consciously designed to do so, and their impacts on natural reefs and other ecosystems, such as seagrass beds and mangroves, are not fully understood (Burt et al., 2012).

Protection, monitoring, and regulation are jointly needed to reduce climate risks to coastal infrastructure. The protection of the marine environment is officially a priority issue within the climate change policies of Qatar due to its unique importance for the cultural identity of the country and for the supply chains of water and food (State of Qatar, 2011). Qatar imports more than 90 percent of its food supplies, with only one-third of fish supplies produced locally. According to the government's National Food Security Plan of 2012, Qatar strives to increase local fish production to around 60 to 80 percent (State of Qatar, 2012). (This national plan had not been adopted as of 2017.) The 2017 political crisis between Qatar and other Gulf countries has led to the closing to Qatar of the borders, ports, and airspaces of Saudi Arabia, Bahrain, and the UAE, leading to the reorientation of Qatar's food-import strategies. Short-term turbulences like this could jeopardize Qatar's efforts to protect its marine environment, which is, in the long run, highly important for its future strategies for food security. At the same time, Qatar has started to establish both terrestrial and marine protected areas, adopt conservation laws (mainly for terrestrial land), and implement major environmental treaties to which it is a signatory (e.g., the 1992 Convention for Prevention of Marine Pollution by Dumping of Wastes and Other Matters). Around 22 percent of Qatar's terrestrial land was designated as protected area as of 2007 (State of Qatar, 2011). However, according to the World Bank, only 1.6 percent of terrestrial waters were classified as protected in 2014. This compares to around 8 percent in Bahrain and 21 percent in the UAE.

Importantly, the capacity of Qatar to manage marine ecosystems like coral reefs is limited by the lack of comprehensive data and baseline assessments (Burt et al., 2015), and also partly due to the inadequacy of onsite control mechanisms at sea and in local fish markets. Furthermore, with a very large expatriate population, Qatar lacks sufficient local personnel who are specialized in developing, assessing, and monitoring environmental laws (Othman and Clarke, 2014).

Responses in the Coastal Built Environment

As the majority of population centers and infrastructure in Qatar are located close to the coast, the vulnerability of the built environment to climate change risks is very high. Qatar is growing at a very fast rate in terms of population and economy. Construction on coastal areas for housing, desalination and power plants, ports, roads, and railways will be affected by increased sea and air temperatures, SLR, and a decrease in moisture available for local agriculture. Apart from direct impacts, energy demand for air conditioning used for cooling purposes in buildings, which is around 65 percent of total electricity consumption (Ayoub et al., 2014), will increase. Furthermore, seawater and reused water are being used for cooling purposes in power plants and industry. For this reason, it will be important to consider energy efficiency, water requirements, and heat generation in the design of power and desalination plants in order to accommodate expected changes in the climate. Apart from supply infrastructure, urban infrastructure will be affected by the increase of heat islands and extreme heatwaves, leading to more pressure on the electricity and water supply infrastructure. In terms of the effects of SLR on coastal and offshore zones, there have been no high-resolution scenarios made for the country. However, similar studies in Bahrain and initial studies in Qatar show high vulnerability even from small changes, potentially leading to losses to offshore infrastructure, coastlines, and vital industries like tourism and fisheries (State of Qatar, 2009: 111; 2011).

Qatar's responses to climate change needs to be seen in the context of a greater economic transition that aims to build a knowledge-based, and thus more resilient and diversified, economy while reducing the country's large per capita ecological footprint. For the built environment, essential reforms will be needed to public transportation and individual mobility, as will rethinking urban expansion and housing (e.g., high-rise houses instead of the large villas, which now accommodate around 40 percent of the population), building codes for energy efficiency, and eliminating generous water and electricity subsidies (see Richer, 2014, for an overview of these measures). Critical issues related to the promotion of environmental culture and awareness and phasing out subsidies that determine the shape of infrastructure are still not adequately represented in climate policies. In fact, the majority of climate responses in Qatar's national policies lie in hard measures covering the areas of water, waste, energy, and transport. These sectors are intertwined and, in light of water scarcity, growth pressures, and large footprints of resource use, will be become more interlinked. Farid et al. (2016) highlighted the scenarios of increased integration of water and energy supply systems in MENA countries in the sectors of water distribution, desalination, and water reuse. Although water reuse requires additional energy input, it is on the rise in the GCC. This is seen as a viable option to meet new demands, recharge depleted groundwater aquifers, and irrigate urban landscapes (Jasim et al., 2016; Ouda, 2016). The abundance of cheap energy will probably compensate for water and land scarcity in Qatar (see Darwish et al., 2016). However, additional energy increases need to be offset by more climate mitigation responses, such as renewables; resource conservation; and savings in household use of water, energy, and food.

The Outlook for Regional Cooperation

Regional cooperation with regard to climate change in coastal areas and the waters of the Gulf is important for two reasons. First, Qatar has a large share of the regional marine environment, and such ecosystems can only be protected regionally. Second, increased development and supply of infrastructure in coastal areas face common challenges and they use the same resources. There are thus significant opportunities to be gained from strengthening cooperation. While the first reason for cooperation is largely acknowledged, the second one is not fully appreciated. In terms of the marine environment, Qatar is a member of the Regional Organization for the Protection of the Marine Environment, established in 1979 in Kuwait as part of a regional marine pollution convention. The Mideast Coral Reef Society Initiative is another forum for cooperation among academics and other stakeholders interested in studying mainly the Gulf, the Red Sea, and reefs south to Oman and Yemen. The GCC as a regional organization also deals with environmental issues. However, its role is quite limited because it has mainly a coordination role. It issues mild guidelines and mainly organizes policy and scientific events, as is the case with other regional Arab organizations linked to environmental or climate change issues (Reiche, 2010b). These organizations have little financial resources in comparison to external organizations like the Organization of the Petroleum Exporting Countries, which has expressed willingness to mobilize funds for concrete mitigation projects.

With regard to supply infrastructure, the integration of water, energy, and food issues will be an important challenge, but it is also an opportunity for future growth strategies in the GCC region. Supply infrastructure shares similar risks, and there is little regional coordination in infrastructure design in terms of scale, sites, distribution networks, and the like. The region can benefit greatly from linking criti-cal infrastructure in order to accommodate external risks, increase storage and pre-paredness, and exchange expertise.

The intense political dispute between Qatar, on the one hand, and three other GCC countries (Saudi Arabia, Bahrain, and Kuwait, with support of Egypt), on the other, in 2017 revealed the fragility of regional cooperation frameworks. There is a sense of rivalry between GCC countries (e.g., Qatar and the UAE) in terms of becoming a global showcase and a regional leader in the adaptation of modern inno-vations in sectors like desalination, renewables, and education. With regard to cli-mate change responses, all countries lack national programs for climate change adaptation that can include comprehensive measures with regional linkages and joint action. Efforts like monitoring marine ecosystems, regional climate research, and vulnerability studies could be done jointly. Further, important technical areas of climate change responses could be targeted through joint policies and regulation. For example, regulations on carbon capture and storage, with relevant classifica-tions, regulation of transboundary transport, liabilities, incentives, and so forth, are largely missing in the GCC on both national and regional levels (Tsai, 2013). Other policy fields for potential cooperation include taxation, subsidies, and tariffs. Taxation is generally low to nonexistent in GCC countries, whereas subsidies are very high, despite plans to gradually eliminate them and to introduce environmental taxes. Here, the political pressure of rising costs and the negative economic as well as social impacts of reducing subsidies in individual GCC countries could be reduced through a regional approach. Importantly, coastal vulnerability, as well as mitiga-tion and adaptation responses specific to coastal infrastructure and maritime-based activities, need to be addressed in the context of national and regional policies. Most cultural and environmental heritage, as well as the people of Qatar and other coun-tries in the Gulf, depend on coastal development and ecosystems. However, these are not adequately understood and protected against external risks.

Conclusion

Risks from climate change in the Gulf can be associated with specific development challenges occurring along coastlines of GCC countries. Construction projects, pol-lution from fossil fuel industries, and high energy use have contributed to more fre-quent temperature extremes and heat islands. The climate challenge adds to environmental degradation problems and aggravates the natural scarcities of arable land and water. Anthropogenic drivers related to high rates of socioeconomic growth patterns, high coastal development ratios, and dependency on large-scale and coupled water and energy production plants are responsible for the heightened coastal vulnerability in the region. Climate-related policies are not advanced enough to face future risks arising from temperature increases, coastal erosion, desertifica-tion, bleaching of coral reefs, or flooding. Especially since the slip in oil prices in

2014, there has been a cautious approach toward climate change responses in order not to harm growth in the short run. Instead, climate change is being mainstreamed into ambitious national visions focusing on diversification, renewables, energy efficiency, and creating local entrepreneurs. Investments in these areas are set to mitigate some effects of climate change. With growing populations and economies, high consumption patterns per capita and low prices, as well as increased energy use for water production and reuse, the GCC countries might not be able to cut their emissions significantly. At the same time, GCC countries are responsible for only a small portion of global emissions and thus share a low responsibility for averting the effects of climate change. However, governments of the GCC recognize the expected negative effects of climate change and the need to address the issue. This is especially true for countries like Qatar, the UAE, and Bahrain, which are less climate-skeptic and exhibit significant coastal vulnerability. Key sectors that depend on marine ecosystems (e.g., tourism and fisheries) and critical infrastructure (e.g., water, energy, telecommunications) for coastal cities are clearly vulnerable. Such coastal vulnerability does not receive appropriate attention in the rather sketchy climate policies of GCC countries.

With the Paris Agreement, GCC countries might have an impetus to outline national climate programs that link their national efforts to regional and global commitments. Qatar has experience participating in and leading international climate negotiations. It prioritizes environmental sustainability, including climate change, in its national vision, and it seeks technology and expertise to implement concrete climate projects. The lack of technical capacity to establish comprehensive regulations and plans is a common problem with other GCC countries. This is a reason for advocating synergies through regional cooperation. Such cooperation could lead to common investments, policies, and regulations in key sectors like reclamation, carbon markets, energy subsidies and pricing, infrastructure planning, marine protection, climate research, and monitoring.

Governance of the Gulf waters in the context of climate change is in its infancy. The key issue is to jointly understand and quantify the future risks related to oceans and seas while bearing in mind the effects on coastal development. In the Gulf, the water and energy sectors have the potential to drive the direction of change due to their large carbon footprints and their impacts on environmental sustainability. In very dry environments, emissions will still be high if renewable energy is not used. Renewables can also decrease the centralized planning of supply infrastructure and the high density of plants on the coasts. The new design of infrastructure with renewables and integrated systems in predominantly coastal countries needs to be studied and debated among policy makers and academics. Development of coastal countries should be seen in light of environmental risks in political and cultural contexts. Considerations like meeting short-term supply needs, food security, promoting wealth, and even generosity should be weighed against future risks and needed resources. In the Gulf region, climate policies reflect societal and political values, as well as the rapid pace of change. This is something that needs to be debated more in the context of climate change and human development more broadly.

References

Ayoub, A., Musharavati, F., Pokharel, S., and Gabbar, H. A. (2014). Energy consumption and conservation practices in Qatar – A case study of a hotel building. *Energy and Buildings*, **84**, 55–69. http://doi.org/10.1016/j.enbuild.2014.07.050

Al-Ansi, M. A., Abdel-Moati, M. A. R., and Al-Ansari, I. S. (2002). Causes of fish mortality along the Qatari Waters (Arabian Gulf). *International Journal of Environmental Studies*, **59**(1), 39–71.

Al-Mamoon, A. A. and Rahman, A. (2017). Rainfall in Qatar: Is it changing? *Natural Hazards*, **85**(1), 453–70. doi:10.1007/s11069-016-2576-6

Almazroui, M., Hasanean, H. M., Al-Khalaf, A. K., and Basset, H. A. (2013). Detecting climate change signals in Saudi Arabia using mean annual surface air temperatures. *Theoretical and Applied Climatology*, **113**(3–4), 585–98.

Al-Saidi, M., Birnbaum, D., Buriti, R., et al. (2016). Water resources vulnerability assessment of MENA countries considering energy and virtual water interactions. *Procedia Engineering*, **145**, 900–7.

AlSarmi, S. and Washington, R. (2011). Recent observed climate change over the Arabian Peninsula. *Journal of Geophysical Research: Atmospheres (1984–2012)*, **116**(D11). doi: 10.1029/2010JD015459

BP (British Petroleum) (2016). *BP Statistical Review of World Energy*. Available from www.bp.com/content/dam/bp/pdf/energy-economics/statistical-review-2016/bp-statistical-review-of-world-energy-2016-full-report.pdf

Burke, L., Reytar, K., Spalding, M., and Perry, A. (2011). *Reefs at Risk*, Washington, DC: World Resources Institute.

Burt, J. A., Smith, E. G., Warren, C., and Dupont, J. (2015). An assessment of Qatar's coral communities in a regional context. *Marine Pollution Bulletin*, **105**, 473–79.

Burt, J. A., Bartholomew, A., and Feary, D. A. (2012). Man-made structures as artificial reefs in the Gulf. In B. M. Riegl and S. J. Purkis, eds., *Coral Reefs of the Gulf – Adaptation to Climatic Extremes*, Heidelberg: Springer.

Callen, T., Cherif, R., Hasanov, F., Hegazy, A., and Khandelwal, P. (2014). *Economic Diversification in the GCC: Past, Present, and Future*. IMF Staff Discussion Note, December 2014, International Monetary Fund. Available at www.imf.org/external/pubs/ft/sdn/2014/sdn1412.pdf

Cheng, W. L., Saleem, A., and Sadr, R. (2015). Recent warming trend in the coastal region of Qatar. *Theoretical and Applied Climatology*, **128**(1–2), 193–205. doi: 10.1007/s00704-015-1693-6

Darwish, M. A., Abdulrahim, H., Mohammed, S., and Mohtar, R. (2016). The role of energy to solve water scarcity in Qatar. *Desalination and Water Treatment*, **57**(40), 18639–67.

Depledge, J. (2008). Striving for no: Saudi Arabia in the climate change regime. *Global Environmental Politics*, **8**(4), 9–35.

Elagib, N. A. and Alvi, S. H. (2013). Moderate solar dimming in an accelerating warming climate of Bahrain. *International Journal of Global Warming*, **5**(1), 96–107.

Elagib, N.A. and Abdu, A.S.A. (2010). Development of temperatures in the Kingdom of Bahrain from 1947 to 2005. *Theoretical and Applied Climatology*, **101**(3–4), 269–79.

Elasha, B. O. (2010). *Mapping of Climate Change Threats and Human Development Impacts in the Arab Region*. United Nations Development Programme, Regional Bureau of Arab States, Arab Human Development Report. Research

Paper Series. Available at www.arabclimateinitiative.org/knowledge/background/
Balgis_mapping%20CC%20threats%20and%20human%20dev%20impacts%20in
%20Arab%20region.pdf

Fakhro, E. (2013). Land reclamation in the Arabian Gulf: Security, environment
and legal issues. *Journal of Arabian Studies*, **3**(1), 36–52.

Farahat, A. (2016). Air pollution in the Arabian Peninsula (Saudi Arabia, the
United Arab Emirates, Kuwait, Qatar, Bahrain, and Oman): Causes, effects,
and aerosol categorization. *Arabian Journal of Geosciences*, **9**(3), 196. doi: 10.1007/
s12517-015-2203-y

Farid Amro M., Lubega, W. N., and Hickman, W. (2016). Opportunities for
energy-water nexus management in the Middle East and North Africa.
Elementa: Science of the Anthropocene, 4. doi:10.12952/journal.elementa.000134

Feary, D., Burt, J., Bauman, A., et al. (2013). Critical research needs for identifying
future changes in Gulf coral reef ecosystems. *Marine Pollution Bulletin*, **72**,
406–16.

Hvidt, M. (2013). Diversification in GCC countries: Past record and future trends.
*The Kuwait Programme on Development, Governance and Globalisation in the
Gulf States*. London: The London School of Economics and Political Sciences.

IPCC (International Panel on Climate Change) (2011). Managing the risks of
extreme events and disasters to advance climate change adaptation. *Special
Report of the Intergovernmental Panel on Climate Change*, IPCC, United
Nations Framework Convention on Climate Change (UNFCCC).

IPCC (International Panel on Climate Change) (2007). *IPCC Report. The Fourth
Assessment Report (AR4)*, United Nations Framework Convention on Climate
Change (UNFCCC). Available at www.ipcc.ch/report/ar4/

Jasim, S. Y., Saththasivam, J, Loganathan, K., Ogunbiyi, O. O., and Sarp, S. (2016).
Reuse of treated sewage effluent (TSE) in Qatar. *Journal of Water Process
Engineering*, **11**, 174–82. Available at http://doi.org/10.1016/j.jwpe.2016.05.003

Lattemann, S. and Höpner, T. (2008). Environmental impact and impact assessment
of seawater desalination. *Desalination*, **220**, 1–15. doi:10.1016/j.desal.2007.03.009

Meltzer, J., Hultman, N., and Langley, C. (2014). *Low-Carbon Energy Transitions in
Qatar and the Gulf Cooperation Council Region*. Global Economy and Development
at Brookings Institute. Available from www.brookings.edu/wp-content/uploads/
2016/07/low-carbon-energy-transitions-qatar-meltzer-hultman-full.pdf

Mezher, T., Fath, H., Abbas, Z., and Khaled, A. (2011). Techno-economic assessment and environmental impacts of desalination technologies. *Desalination*,
266, 263–273. doi:10.1016/j.desal.2010.08.035

Milly, P. C. D., Dunne, K. A., and Vecchia, A. V. (2005). Global pattern of trends
in streamflow and water availability in a changing climate. *Nature*, **438**, 347–50.
doi: 10.1038/nature04312

Mondal, M. A. H., Hawila, D., Kennedy, S., and Mezher, T. (2016). The GCC
countries RE-readiness: Strengths and gaps for development of renewable
energy technologies. *Renewable and Sustainable Energy Review*, **54**, 1114–28.
Available at http://doi.org/10.1016/j.rser.2015.10.098

Murthy, A. (2016). *Qatar and the Climate Debate*. Earth Journalism Network.
Available at http://earthjournalism.net/stories/qatar-and-the-climate-debate

Othman, W. A. and Clarke, S. F. (2014). Charting the emergence of environmental
legislation in Qatar. A step in the right direction or too little too late? In
P. Sillitoe, ed., *Sustainable Development – An Appraisal form the Gulf Region*.
Bergbahn, New York: Oxford.

Ouda, O. K. M. (2016). Treated wastewater use in Saudi Arabia: Challenges and initiatives. *International Journal of Water Resources Development*, **32**(5). Available at http://dx.doi.org/10.1080/07900627.2015.1116435

Radhi, H. and Sharples, S. (2013). Global warming implications of façade parameters: A life cycle assessment of residential buildings in Bahrain. *Environmental Impact Assessment Review*, **38**, 99–108.

Reiche, D. (2010a). Renewable energy policies in the Gulf countries: A case study of the carbon-neutral "Masdar City" in Abu Dhabi. *Energy Policy*, **38**(1), 378–82.

Reiche, D. (2010b). Energy policies of Gulf Cooperation Council (GCC) countries: Possibilities and limitations of ecological modernization in rentier states. *Energy Policy*, **38**(5), 2395–403.

Riegl, B. and Purkis, S. (2012). *Coral Reefs of the Gulf: Adaptation to Climatic Extremes*. Dordrecht Heidelberg: Springer.

Riegl, B. M, Purkis, S. J., Al-Cibahy, A. S., et al. (2012). Coral bleaching and mortality thresholds in the SE Gulf: Highest in the world. In B. M. Riegl and S. J. Purkis, eds., *Coral Reefs of the Gulf – Adaptation to Climatic Extremes*. Heidelberg: Springer.

Richer, R. A. (2014). Sustainable development in Qatar: Challenges and opportunities. *QScience Connect*, **22**. doi: 10.5339/connect.2014.22

Sale, P., Feary, D., Burt, J., et al. (2010). The growing need for sustainable ecological management of marine communities of the Persian Gulf. *Ambio*, **40**, 4–17.

Siddiqi, A. and Anadon, L. D. (2011). The water-energy nexus in Middle East and North Africa. *Energy Policy*, **39**, 4529–40. doi:10.1016/j.enpol.2011.04.023

Smith, R., Purnama, A., and Al-Barwani, H. H. (2007). Sensitivity of hypersaline Arabian Gulf to seawater desalination plants. *Applied Mathematical Modelling*, **31**(10), 2347–54. doi:10.1016/j.apm.2006.09.010

Sowers, J., Vengosh, A., and Weinthal, E. (2011). Climate change, water resources, and the politics of adaptation in the Middle East and NorthAfrica. *Climate Change*, **104**, 599–627. doi: 10.1007/s10584-010-9835-4

State of Qatar (2008). *Qatar National Vision 2013*.

State of Qatar (2009) Advancing sustainable development. Qatar National Vision 2030. Second National Human Development Report. General Secretariat for Development and Planning. United Nations Development Programme. Available at http://hdr.undp.org/sites/default/files/qhdr_en_2009.pdf

State of Qatar (2011). *Initial Communication to the United Nations Framework Convention on Climate Change*, Ministry of Environment. State of Qatar. Available at http://unfccc.int/resource/docs/natc/qatnc1.pdf

State of Qatar (2012). *National Food Security Plan*. Unpublished document.

State of Qatar (2015). *Intended Nationally Determined Contributions (INDCs) Report*, Ministry of Environment. State of Qatar. Available at www4.unfccc .int/Submissions/INDC/Published%20Documents/Qatar/1/Qatar%20INDCs% 20Report%20-English.pdf

Tsai, I.-T. (2013). *Carbon Capture, Utilization and Storage Regulation in the Gulf Cooperation Council Countries: A Review of Current Status*. UN-ESCWA/ MASDAR Institute. Available at http://css.escwa.org.lb/SDPD/3303/Session5.pdf

UN Habitat. (2011). *Cities and Climate Change: Policy Directions. Global Report on Human Settlements*. United Nations Human Settlements Programme. Available at http://mirror.unhabitat.org/pmss/getElectronicVersion.aspx?nr=3086&alt=1

Wodon, Q., Burger, N., Grant, A., and Liverani, A. (2014). *Climate Change, Migration, and Adaptation in the MENA Region*, The World Bank. Available at https://mpra.ub.uni-muenchen.de/56927/1/MPRA_paper_56927.pdf

5

Adapting to Sea-Level Rise in the Indian Ocean
The Cases of India and Bangladesh

ARNE HARMS

Introduction

South Asia is considered to be particularly vulnerable to the impacts of anthropogenic climate change. The reasons for that fall into two broad categories. On one hand, the region is extremely diverse. Its hugely varied geomorphologic formations, ecosystems, and economies render the region susceptible to a range of adverse climate change impacts (Field et al., 2014; Stocker et al., 2014). At the same time, the region houses socially extremely vulnerable populations. Notwithstanding substantial growth rates, the region houses the world's second largest group of extremely poor persons on the globe (World Bank, 2016: 37–39). It is a matter of profound concern that South Asia's poor are predominantly dependent on natural resources for their day-to-day survival and that they occupy the most hazardous spaces, such as flood plains, arid areas, and coastlines. In concomitance with a range of further structural conditions, such as class, caste, and gender, these conditions render large swathes of South Asia's poor unusually and especially vulnerable to weather-related hazards of all kinds.

This chapter sheds light on the governance of climate change impacts on South Asia's coasts. It analyzes how hazards that emanate from warming and rising seas are managed. Drawing on the traditions of qualitative social sciences, the chapter accounts for relevant policy instruments from above and from below. That is, it offers an analysis based in the critical examination of policy documents *and* ethnographic evidence gathered during long-term original research on the subcontinent. The analysis is operating in the mode of "document view" and "field view," to adapt M. N. Srinivas's classic terminology (Srinivas, 1966). In order to follow through with this dual approach, the chapter concentrates on one region – the Bengal delta, comprising almost all of Bangladesh and the Indian state of West Bengal. This was chosen because this region has become iconic for climate change vulnerabilities and impacts, both within South Asia and across the globe. But the region is relevant for the purpose of this volume not only as an icon or an arbitrarily selected case study. The way climate change is unfolding and being encountered in the Bengal delta has something to say for coastal South Asia at large. Social vulnerabilities demonstrated by coastal populations, the hazards they face, and the particular alignment of policy instruments mirror the Bengal case to a considerable degree.

This chapter shows that climate change adaptation along South Asia's shores is *in practice* tilted toward disaster management. In spite of policy documents listing adaptation measures for addressing normalized effects of climate change, such as the reform of agricultural crop regimens, climate change is addressed on the ground largely in terms of protecting the coastal population from disasters, such as cyclones and storm surges. The chapter demonstrates that this orientation translates into the prioritization of one particular class of disasters – the "event" or "rapid-onset" type – in favor of another one – normalized or "slow-onset" disasters. This tilt toward the disastrous shapes South Asian climate change adaptation policy regimes. The chapter shows that state and nonstate actors responsible for the design, implementation, and maintenance of measures to absorb climate change impacts are largely concerned with disastrous events and, conversely to a much lesser degree, with normalized, everyday deteriorations.

This chapter argues that this orientation toward disaster management is predicated upon three sets of factors. First, policy follows from the relative invisibility of normalized degradations. Second, civil-society organizations, dedicated state agencies, and international nongovernmental organizations (NGOs) are well equipped to ease impacts of disasters but not of normalized changes. Third, the chapter suggests that in order to meaningfully address normalized transformations, substantial funds and structural changes will be necessary. The political will to release funds and rework policies seems to be lacking.

Extreme Weather Events and Normalized Landscape Transformations: Aggravating Coastal Hazards

For millennia, Bengal has been notorious for its natural hazards and recurring disasters. Most prominent among these are floods and storms plus ensuing storm surges. Societies removed from the sea are well adapted to cyclical large-scale floods occurring during monsoons, literally nourishing the land and invigorating intensive agriculture. In estuarine tracts, however, the nature of floods changes considerably due to tidal incursions. The tides push marine waters deep into river channels of the vast delta, rendering river waters brackish and thus harmful to crops and lands. Flood events here are a threat to subsistence and well-being, and local people struggle to avoid levee breaches. Efforts are ramped up during the spring tides of the monsoon seasons, when rivers brimming with waters from incessant rain upstream meet the tides pushing in with maximum force.

The other prominent extreme weather events – storms together with ensuing surges – have more dramatic impacts in the coastal belt of the delta. In contrast to floods, storms and surges are straightforward threats to survival, crops, and infrastructure. But only the coastal belt has to withstand the full force of the winds. Upon making landfall, the velocity of storms decreases rapidly, which means that the coastal belt has to weather the storms' greatest destructive impacts. In addition to that, surges whipped up and pushed into the land by the gale only affect coastal zones. Physical hindrances such as defense structures, settlements or forests, and the steadily reducing velocity of the storm after landfall weaken the surge. But embankments and defensive infrastructure keep surge waters enclosed long after the storm has passed. Water entrapped in devastated landscapes makes rescue operations more difficult, increases the likelihood of disease outbreaks, and thwarts the productivity of agricultural lands for long periods.

In addition to these extreme weather events, the Bengal delta has been notorious for its shifting state. Being a very dynamic delta formation, the landmass is reworked continuously and very swiftly by rivers and the sea. Sediments deposited by three of Asia's mightiest rivers transform the landscape continuously. River beds are rapidly elevated and banks transformed, frequently blocking streams and forcing rivers to shift course. As a consequence, river banks and islands have been washed away, razing villages and setting their inhabitants footloose almost overnight. Routinely, river branches are chocked, morphing into lakes or, alternatively, drying up. Again, Bengali societies have been fairly well adapted to such recurring shifts and transformations. Distinct societies have been populating ephemeral islands situated within streams, making the most of these fragile formations and shifting with the lands as they disappear and reappear again (Lahiri-Dutt and Samanta, 2013). Mainland farmers, on the other hand, have been mobile for centuries, colonizing jungles when their lands were being swallowed up by one of the mighty rivers.

In the coastal belt, the shifts of the rivers are complemented by transformations wrought by the sea. Waves, currents, and tidal dynamics also incessantly rework and remold the seafront. Through coastal erosion, islands disappear and shorelines retreat. Akin to erosion inland, at the coasts accretion occurs too. Yet in the western part of the delta, erosion has long been outweighing accretion (Bandyopadhyay, 1997; Hazra et al., 2010), translating into shrinking landscapes. In those parts of the delta where accretion occurs on a substantial scale, emerging lands are salty and thus unfit for cultivation.

Acknowledged across Bengal for centuries (Mukerjee, 1938), such normalized landscape transformations continue to spell out immense suffering. As the world is warming, future outlooks are bleak. Sea-level rise will only aggravate the situation, potentially resulting in the disappearance of entire landscapes and the displacement of substantial populations. Estimates have been floated for decades putting the number of potential refugees in the millions (see, i.e., Myers, 1993: 754). However, these estimates are far from reliable. Oftentimes they are based on flawed assumptions and data sets. Most importantly, they are underpinned by the assumption that sea-level rise will affect Bengal evenly and that human societies are passive. Instead, sea-level rise is affecting Bengal in highly context-specific ways, threatening one stretch of the coast more than another according to geomorphologic and hydrologic circumstance (Brammer, 2014), and human societies will likely come up with reinvigorated adaptation efforts slowing down regional effects of sea-level rise (Black, 2001: 7f). That being said, there is no reason to doubt the gravity of the situation. To date, the victims of coastal erosion have seen remarkably little assistance. One reason for this is that they do not fit into the framework of what is normally conceived of as disaster. While not fully invisible, they still remain under the radar of humanitarian assistance. In order to substantiate this claim, this chapter now briefly discuss patterns, impacts, and experiences of environmental hazards.

People of the Bengal delta have had to cope with and adapt to two different types of hazards and disasters. On one end of the spectrum we find extreme weather events, such as floods, storms, and surges. On the other end are normalized transformations, such as erosions and permanent inundations of landmass. Between these poles interlinkages occur, as, for instance, in the intensified erosion during storm surges or the frequent rerouting of river channels during annual floods. That being said, the two types of hazards are experienced differently, have divergent outcomes, and require distinct management and adaptation policies. The difference is that the first type of hazards conveniently falls under the category "disaster," while the other seems to resist an easy classification as such. This is not a purely academic matter but one that has profound consequences for dealing with these hazards in policy circles and by the humanitarian apparatus alike.

Floods, storms, and surges are extreme weather events affecting vast spaces and potentially large populations within a comparatively short time frame. Swollen rivers flood entire landscapes, while cyclonic formations span open seas, coastlines, and hinterlands. Within the rhythmic structure of onset, crest, and attenuation virtually all persons present in an affected area feel the impact. Some might sense only ripples, but substantial populations find themselves in a position where they have to weather nature's force *together* and *simultaneously* (Oliver-Smith, 1999a, 1999b). Normal routine comes to a standstill, with all efforts devoted to rescue operations. One of the key characteristics of disasters is that they engulf substantial spaces and populations in destructive and potentially deadly events, unfolding in fairly short time frames. The currency of disasters are fatalities, or at least massive evacuation efforts; destruction of infrastructure, things, and goods; and the disruption of routine. As such, extreme weather events lend themselves to be widely broadcast, inviting regional and transnational forms of solidarity. On the ground, we have seen a sophisticated humanitarian response being set into action consisting of diverse groups of actors, including civil society organizations, state disaster management institutions, and international NGOs.

Normalized transformations, such as coastal erosion and permanent inundation, all too easily recede from view. Coastal erosion is most often a drawn-out, chronic process. It consists of waves and currents tirelessly gnawing at a landmass, a process that intensifies with spring tides or rainy seasons but nevertheless works throughout the year. It might culminate in small flood events following the collapse of badly eroded embankments, after which coastal dwellers literally give up on the land and shift the outer ring embankment inland. Even while whole villages and islands fell prey to coastal erosion – pulverizing the land, washing away all immobile assets, and displacing inhabitants – at any given moment only a small number of people are directly affected. In other words, erosion consists of a series of events that are small scale. It always involves only the outer skin of a given coastline, and only individual stretches of that outer skin. Victims of coastal erosion are distributed along banks and shores, condensing into knots and hazard zones according to geomorphological and social specifics. Thus, the tight temporal and spatial framing of extreme weather events is inverted, giving way to processes that are as widely spread as they are chronic.

The social cost inflicted by erosions is limited, at least from the perspective of a macro-economist. And at any given moment, erosions affect "only" comparatively small numbers of people. At any given moment, erosions destroy assets that seem to be negligible, such as small plots of agricultural land of an inferior quality, thatched huts, and trees. But if amalgamated spatially and temporally, large populations and enormous assets have been affected. Finally, coastal erosions carry extremely little risks for survival. Against the background of intimate environmental knowledge, coastal dwellers know very well when and where to expect erosion and generally manage to move themselves and mobile belongings in time. Yet for affected people, coastal erosion amounts to a truly disastrous experience. Effortlessly and unforgiving, or so it seems, the waters literally dissolve the very basis of human existence. Land is gobbled up and with it the ground to walk upon, build upon, and live off. Along the impoverished and densely populated coasts of Bengal, coastal erosion turns peasants into paupers by stripping them from their agricultural lands, homesteads, and villages. Climate scientists agree that these transformations will only increase in velocity and spread, potentially affecting even larger populations in the immediate future.

Adaptations to coastal erosions across Bengal have proven to be more challenging than inland riverine erosions. On one hand, coastal dwellers are not in a position to shift with the islands as they disappear and re-emerge with the same ease as noted for their inland counterparts. Land emerging on the seafront is salty. Cultivation on those lands depends on a thorough diking of the land in order to keep marine waters out and let rain waters drain the salt from the soil. More important are political reasons. With but a few exceptions across Bengal, the seafront has been settled last. Colonists were aware that the soil quality was inferior and the availability of freshwater more insecure the closer one got to the sea. For centuries, a forested seafront also was seen to be a bulwark cushioning the impact of marine hazards – storms, surges, erosion – on the hinterland. Layers of mangroves were valued for tempering storms and blocking surges by virtue of being a massive wall and for decreasing erosion by sheltering and binding the soil. Therefore, the shorelines and banks of the coastal belt have been settled thoroughly only in the second half of the twentieth century when land became extremely scarce all over Bengal. But as the settlement of the coast was a matter of last resort, its inhabitants found it even more difficult to shift elsewhere once their homesteads began eroding because virtually all other land was either already settled, demarcated as nature reserves, or earmarked for industrial projects.

Framing Adaptation to Accelerating Coastal Hazards in India and Bangladesh

This section examines policies formulated and implemented as attempts to adapt to climate change in coastal Bengal. Climate scientists agree that both types of hazards will intensify in the coming years and decades. Although the quantity of storms and surges ravaging the Bengal delta might not increase, added velocity is very likely. Sea-level rise, on the other hand, will increase the pressure of tidal currents, increasing tidal incursions and coastal erosion along Bengal's coastal belt. Analyzing documents shows that climate change adaptation is bound up in a paradox. Although environmental transformations are acknowledged as potential factors, coastal Bengal policies are predominantly concerned with the adaptation to extreme weather events, both on paper and on the ground. This chapter suggests that the currency enjoyed by extreme weather events rests on the sophistication of dedicated disaster policies and a relative blind spot when it comes to chronic transformations.

In 2008, the government of India published the "National Action Plan on Climate Change" (NAPCC). Neighboring Bangladesh followed suit the next year, publishing its "Bangladesh Climate Change Strategy and Action Plan" (BCCSAP). To date these are the most comprehensive documents on climate change policies, both in terms of mitigation and adaptation, set forth by the respective countries. Concerning adaptation efforts, both documents put emphasis on the need to adapt to and cope with normalized changes wrought by climate change. Thus, in India's NAPCC crop improvement measures lead the list of existing adaptation programs. Other packages include drought-proofing and improvement of water management, clearly targeting the stabilization of the "production of crops and livestock" (Government of India, 2008: 17) under deteriorating conditions.

Similarly, Bangladesh refers to the successful implementation of climate-proofing, which is demonstrated by rising agricultural productivity and poverty reduction in the country (MOEF, 2009: 19). But Bangladesh's strategy paper puts a much stronger emphasis on the need to adapt to riverine and marine hazards. Measures to counter or cushion the detrimental effects of floods on rural and urban households dominate the list of adaptation programs in place (MOEF, 2009: 18f). This emphasis certainly rests in Bangladesh's overall extraordinary exposure to cyclones and floods and the dominant role they play within the country's portfolio of hazards. India, in contrast, is susceptible to a greater number of types of hazards, but cyclones and surges arguably do not have such a great hold on the country's overall population and economy.

To make a more meaningful comparison, it is useful to read Bangladesh's national plan alongside the plan recently formulated by India's state of West Bengal. In the sections devoted to climate change adaptation in West Bengal's coastal regions, the Sundarbans, the plan similarly and unequivocally prioritizes measures against floods and surges. It calls for massive investments in infrastructure, such as cyclone shelters and massive embankments, and early warning systems (Government of West Bengal, 2012: 224). Normalized impacts of climate change, such as increasing salinization or reduction of agricultural productivity, are mentioned but clearly considered to be of lesser importance.

Crucially, all reports acknowledge the likelihood of coastal erosion, permanent inundation, and salinization as sea levels rise. Yet concrete adaptation strategies remain vague. Thus, India's NAPPC mentions recent efforts to build coastal protection infrastructure and cyclone shelters and to plant coastal forests (Government of India, 2008: 18). However, as noted later, the coastal-protection infrastructure that is provided and coastal forests planted are in many places insufficient even within the current scenario, let alone a near future characterized by aggravated hazards. Bangladesh's action plan, on the other hand, refers to a number of measures that might be useful for successful adaptation to coastal erosion and permanent inundations, such as embankments and related defensive infrastructure. However, the plan introduces these as means to counter devastating floods, not normalized erosion (MOEF, 2009: 18f).

Summing up, the policy documents articulate the need to adapt to coastal impacts of climate change. Besides an emphasis on disastrous events, such as cyclones and surges, these documents also address normalized changes potentially making detrimental impacts across the subcontinent. But while sea-level rise and the threat of normalized inundations of large coastal tracts is a widely invoked scenario, and much feared, concrete adaptation measures brought forward seem to be insufficient. Strongest are adaptation measures gathered from the field of disaster management, and climate change adaptation operates in the mode of disaster management. Normalized landscape transformations and the social suffering they entail remain overshadowed. To further corroborate this claim, this chapter now turns to adaptation efforts in practice.

Implementing Adaptation

In coastal Bengal, climate change adaptation first and foremost means reducing the impact of storms, surges, and excessive seasonal floods. Climate change adaption thus builds on modern disaster management policies, which are in themselves partly in continuation of traditional management practices. Concrete measures fall into one or several of three fundamental categories: dissemination of early warning, provision of shelter, and maintenance of defense infrastructure. Beyond that a number of voices also call for planned relocations of vulnerable populations. This section will outline these one by one.

In the wake of the 1970 cyclone that left 350,000 people dead, the government of Bangladesh pushed for an effective early warning system. The program gained importance following the landfall of further cyclones (O'Donnell and Wodon, 2015: 154f). Since then it has matured into an effective system connecting meteorological institutes and government offices with villages via radio and, most importantly, volunteers carrying individual warnings from house to house. India followed suit but met lesser success. Indian officials failed to seamlessly integrate local counterparts, thereby jeopardizing the delivery of actual warnings to the populace. At the same time, warnings issued by the Indian Meteorological Department ended up being considered rather unreliable, and their warnings were not always heeded (Gupta and Sharma, 2009). At present, the Bangladeshi forecasts are eagerly attended to in the Indian part of the delta, and Indian forecasts are evaluated against those issued by their counterparts to the east.

But the problem always has been where to find a safe haven when a storm, cyclone, or super cyclone moved toward Bengal's coast. Storm fields are huge formations, easily covering several hundred square kilometers. Along these crowded and poor coasts, near-total evacuation out of the impact zone is a highly unrealistic scenario. For centuries, people have fled to higher grounds in order to increase chances for survival when storms struck. People routinely moved onto mounds, dikes, or roofs and, as a matter of last resort, onto trees to avoid being washed away by dreaded storm surges. In recent years, governments on both sides of the border have intensified attempts to provide safe havens to their citizens and erected a considerable number of cyclone shelters. Made from steel-enforced concrete, these structures provide a level of safety from storms that is far superior to that provided by thatched huts, embankments, or trees. Built either as two-floored or as stilted structures, they also offer refuge from surges and floods. In order to provide clean drinking water during a disaster and thus to reduce the risk of disease outbreaks in its wake, many cyclone shelters feature elevated hand pumps. In concomitance with early warning systems, cyclone shelters have significantly reduced the death tolls of recent storms.

Governments on both sides of the border have committed themselves to building more cyclone shelters as a means to adapt to climate change (Government of India, 2008; MOEF, 2009). There can be no doubt that these structures will help save lives and reduce suffering during extreme weather events. The proliferation of cyclone shelters and their propagation as crucial adaptation measures underscores the amount to which adaptation programs are informed by the logic of disaster management. At the same time, this proliferation also demonstrates the salience of infrastructure development for climate change adaption. Within Bengal's largely neglected and extremely impoverished coastal tracts, cyclone shelters are in fact major infrastructure development projects. As such, they embody substantial financial interests and are important vehicles of electoral politics. Local leaders use cyclone shelters to showcase their political clout and to strengthen their electoral base by bringing lifesaving and job-providing development projects to their respective constituencies.

Again, there hardly can be any doubt that such measures will be of tremendous help. But they embody an approach to climate change adaptation that is true to the logic of disaster management and is concerned with bringing down fatalities and with reducing damage to property and infrastructures. Normalized landscape transformations are still of another order. The fate of a flood shelter that this author visited occasionally during fieldwork in coastal West Bengal will help to clarify this point (the author conducted ethnographic fieldwork in West Bengal for more than 14 months between 2009 and 2017). The flood shelter was installed by a renowned international NGO as part of its efforts in the region to enhance resilience to natural disasters and climate change. The NGO had had valid reasons to include this particular stretch of coastline in its activities, since cyclones had in the past inflicted considerable damage on local villages. The NGO reached out to the population in meetings, educating villagers about what to do during impeding storms, and it facilitated training in civil rescue operations. In meetings, through music performances and via wall paintings, villagers were advised to stock drinking water, food, and medicine during high-risk months and to be generally prepared for emergencies. Life vests were provided to the community and their usage was explained. In addition to these efforts, the flood shelter was established on the second floor of a robust concrete building. Villagers were encouraged to take shelter therein during any kind of emergency.

Villagers appreciated these efforts. Yet during conversations they repeatedly stressed that this was all fair and good, but did not really help to deal with what they saw as the most pressing issue locally – the erosion of village lands by the sea. Indeed, the shoreline was rapidly eroding. Every year the ring embankment collapsed during the monsoon, letting seawater enter deep into the village. It had become routine to rebuild the embankment shortly before the monsoons, only for it to be eroded and washed away during one of the spring tides during the monsoons. For most of the year, tidal waters entered the village, attacking inner embankments, eroding homesteads, and rendering fields unfit for cultivation. Those who could move away did so. But many did not have the means and were forced to stay, struggling to make ends meet (Harms, 2017). As the sea was eroding the coast, the flood shelter became itself a victim of normalized landscape transformation. After the foundations had been eroded, it eventually collapsed and the remains were washed away.

It would be wrong to read this as an example of inefficient or faulty engineering – as if the shelter would have needed a safer locality (removed from the shore) or should have been of a more robust construction type (better reinforced). Both measures certainly would have extended the duration of the building's existence. Yet along these shores erosion was so rampant that the sea already washed away two consecutive villages that once stood between the site of the structure and the shoreline. And there was no end in sight. The dismantling of the flood shelter was only a matter of time and could not have been countered by the means favored by disaster management.

This example puts the spotlight on the third category of adaptation measures: defensive infrastructure. Embankments and sluice gates are a defining feature of the Bengal delta. Its settled parts are thoroughly shaped by outer ring embankments completely encircling islands, with mainland and inner embankments serving as secondary defensive structures. For the management of environmental hazards, they play a double role. On one hand, they are means to keep fields and homesteads safe from estuarine waters during periods of excessive rain or when a storm surge is hitting. On the other hand, embankments are erected to check normalized landscape transformation following coastal erosion. As this chapter has suggested, climate change policies in Bangladesh and West Bengal prioritize extreme weather events at the cost of normalized landscape transformations. As a consequence, there is little investment in defensive infrastructure that has the potential to slow down or end coastal erosions. In order to demonstrate the critical role of sophisticated defensive infrastructure in landscapes being transformed by coastal erosion, the chapter will now briefly illustrate the crucial hydrological, geomorphological, and social dynamics involved.

In coastal Bengal, coastal erosion progresses either by way of slope erosion or following embankment collapse. In the former case, chunks of land are undermined by estuarine waters and eventually break away. In the latter, sea or estuarine waters push against existing embankments until they collapse. Once the ring embankment is breached, tidal dynamics push sea or estuarine waters onto the landmass, producing small flood situations stretching between the breach and the next embankment line in the interior. As the distance between the outer and the next inner embankment typically does not exceed 400 meters, hardly more than one village is affected by these events. To stop erosion within the flooded area, the ring embankment is rebuilt after collapse as soon as funds are released for this purpose. Along particularly badly affected zones of the delta, as for instance, the village formerly housing the cyclone shelter, embankments collapse annually only to be rebuilt in the next months. If this happens for a few consecutive years and along a shore known for its erosion, the local administration eventually ceases to rebuild the breached embankment and declares the next inner embankment as the new outer ring embankment. This abandonment – a decision by bureaucrats rather than the result of a natural hazard – marks a threshold for the progression of coastal erosion. From now on the outer land is abandoned, subject to accelerating erosion and washed away within a few months to a few years. After that, the cycle typically restarts as the waters push once again forcefully against the outer embankment, eroding its fundament (see Figure 5.1).

At present, the large majority of people affected by coastal erosion in Bengal relies on embankments as their only defensive infrastructure. Most of these embankments are built entirely from mud, which is piled up manually and compressed by workers' feet. Only the embankments erected in zones considered to be extremely vulnerable to erosion are enforced by wooden frames and occasionally lined by sand sacks or bricks. It is justified to call the embankments dominating coastal Bengal's defensive infrastructure "primitive" (Kanjilal, 2000). Despite its obvious limitations in countering erosion, this type of embankment keeps being erected and re-erected. In due course, advances in the field of coastal engineering, now commonly propagating a sophisticated mix of soft and hard defensive structures, are being largely ignored. Coastal dwellers, local bureaucrats, politicians, and engineers are all well aware of the ineffectiveness of earthen embankments and the availability of other approaches to the menace of coastal erosion. But only rarely are other approaches implemented or even experimented with (see, e.g., Ghosh, Bhandari, and Hazra, 2003). Reasons for this are to be found in the sphere of political economy.

Fig. 5.1. Erosion of land, houses, and infrastructure following embankment collapse. Photo by the author.

Most observers agree that substantial funds are needed to ramp up efforts to protect coasts and populations from the threat of erosion. Given the delta's highly fissured nature, Bengal's coastline is very long. The length of West Bengal's embankments alone is estimated at 3,500 kilometers (Government of West Bengal, 2009: 310). Yet on both sides of the border, coastal tracts and agricultural subsistence are not state funding priorities. During fieldwork, villagers routinely demanded improved dike structures, praising the virtue of concrete structures, and comprehensive solutions. Local politicians and higher bureaucrats readily agreed to these demands, only to state that this would require funds far exceeding their budgets. As a consequence, inexpensive mud embankments are rebuilt, land abandoned, and people displaced.

While most voices on the ground seemed to favor hard defensive structures, such as sea walls, enforced embankments, and massive drainage systems, these measures have also come under intense criticism. Opponents argue that such structures actually have detrimental effects on water flows and geomorphologic dynamics, and that they worsen floods and intensify riverine and coastal erosions instead of stopping them. Bangladesh's Flood Action Plan – a large-scale project that implemented cutting-edge knowledge of Dutch engineers, among others – is a good example of the failure of technocratic approaches in dealing with the complexities of the Bengal delta (Cook and Wisner, 2010; Sultana, 2010). In order to deal with accelerating erosion events, and the immense suffering they entail, a good deal of experimentation, underpinned by substantial funds and political will, will be necessary.

A different approach to defensive structures rests in the implementation of a range of soft structures, following an approach to coastal management that works with nature, instead of working against it (Gesing, 2016). In coastal Bengal, this has meant the reinvigorated embrace of forests as shields. Thus, mangroves have been planted and revived as natural shielding for their hinterlands, not only from storms and surges but also from erosion. As promising as these and other measures, such as dunes or flood meadows, may be, they require substantial amounts of space – an extremely scarce resource in the overcrowded Bengal delta. With almost all available lands settled and people depending on agriculture, mangroves have been replanted in vulnerable zones either in thin lines along embankments or as patches on mudflats situated off the coast. However, mudflat mangroves in particular have become the object of contention. During my fieldwork, politicians and villagers repeatedly voiced hopes that those mangroves would flourish and attract actual accretion of soil. Growing mudflats into a proper landmass would eventually enlarge coastal islands and, most importantly, provide space for the growing population to colonize. In other words, they pinned their hopes to mangroves not as a soft structure, but as a transient land-producing machine. Soon, these interlocutors claimed, the mangroves would be felled and the land drained, embanked, and converted into agricultural plots. The mangroves' role as enduring soft structures was jeopardized from the beginning.

In the face of ongoing displacement by erosion and its amplification through climate change, a growing number of scientists, activists, and politicians call for planned relocation of persons vulnerable to erosion (see, for instance, Koslov, 2016; Barua et al., 2017). While some see relocation as the only truly viable option, most frame it as one measure within a broader portfolio of options also including the ones outlined earlier. All agree fundamentally that migration can be a useful adaptation strategy (see Black et al., 2011). Benefits of concerted and comprehensive relocation would be twofold. It would alleviate human suffering and realize citizenship rights by respective states. On the other hand, relocation would also decrease population density, even reversing settlement, thereby reducing detrimental impacts of human settlement on hydrological and geomorphological dynamics.

While planned relocation is a novel and intensely debated subject, coastal West Bengal has already seen a small number of successful resettlements of environmentally displaced coastal dwellers (Harms, 2015). However, these efforts always were decidedly small-scale and post hoc. What is more, resettlement efforts have been terminated since. Political priorities have shifted away from land reform and rural subsistence toward industrial development. At present, state officials do not seem willing to distribute scarce land to displaced persons. Instead, they try to attract multinational industrial endeavors by offering available land to investors (Rudra,

2007). In such a political climate, planned relocation is nowhere on the official agenda for climate change adaptation. To put it on the agenda will require profound structural changes and substantial funds, which depend on political will. It will need the reform of land governance in both states, and the reorientation of humanitarian assistance so as to include environmental displacement among its concern, and to follow through with pre-emptive measures, such as planned relocation.

Conclusion

This chapter has described current attempts to adapt to climate change in the coastal tracts of the Bengal delta. It has argued that adaptation in the delta is marred by a paradox. Climate scientists predict the intensification of extreme weather events and normalized landscape transformations, such as coastal erosion and permanent inundation. Yet current adaptation regimes are largely bypassing the latter type of hazard. Innovative and promising measures against storms and surges are implemented, while coastal erosion and permanent inundation are for the most part addressed by ineffective means, if they are even addressed at all. This predicament rests in a preoccupation with disastrous events, which strike large populations in an instant and involve large numbers of fatalities and enormous destruction. Normalized landscape transformations, on the other hand, remain under the radar of humanitarian assistance and public broadcasting, as they inflict suffering and damage at any given moment only to small groups of people. Adaptation policies addressing normalized landscape transformation face considerable hurdles. Concrete measures against transformations often involve huge funds and political will, but they do not lend themselves to be ideal vehicles of politics. That is, they are not instruments that would yield results to local politicians in terms of strengthening their hold on a given electorate.

The case study presented here is relevant for other coastal areas threatened by climate change, especially in South Asia at large, for two main reasons. First, the hazards and adaptation measures discussed here can be found along many of South Asia's coastlines. The Indian Ocean is but one of the global hotspots of cyclonic activity. Beyond Bengal, Chittagong in Bangladesh, Orissa, Tamil Nadu, and Gujarat in India, entire coastal Sri Lanka and the Sindh in Pakistan all have frequently been hit by tropical storms. In all of these states and regions, coastal areas largely house impoverished populations that are highly vulnerable to storms – storms that will be made more harmful by climate change. Similarly, normalized landscape transformations discussed here are not contained to the Bengal delta. Coastal erosion and permanent inundation affect low-lying coasts across South Asia. Second, humanitarian response and adaptation policies echo the Bengal scenario. With the exception of urban centers, such as Mumbai, Chennai, or Karachi, concerted and comprehensive efforts to adapt to accelerating normalized landscape transformations seem to be sidelined. It is very likely that this will result in immense suffering among populations across South Asia. The near invisibility of their plight, due to its normalized and temporally and spatially widely distributed nature, only adds to the drama.

References

Bandyopadhyay, S. (1997). Natural environmental hazards and their management: A case study of Sagar Island, India. *Singapore Journal of Tropical Geography*, **18**(1), 20–45.

Barua, P., Shahjahan, M., Rahman, M. A., Rahman, S. H., and Molla, M. H. (2017). Ensuring the rights of climate-displaced people in Bangladesh. *Forced Migration Review*, **54**, 88.

Black, R. (2001). *Environmental Refugees: Myth or Reality?* Geneva: United Nations High Commissioner for Refugees.

Black, R., Bennett, S. R. G., Thomas, S. M., and Beddington, J. R. (2011). Climate change: Migration as adaptation. *Nature*, **478** (7370), 447–9.

Brammer, H. (2014). Bangladesh's dynamic coastal regions and sea-level rise. *Climate Risk Management* 1 (Supplement C), 51–62. doi:10.1016/j.crm.2013.10.001.

Cook, B. R. and Wisner, B. (2010). Water, risk and vulnerability in Bangladesh: Twenty years since the FAP. *Environmental Hazards*, **9**(1), 3–7.

Field, C. B., Barros, V. R., Mach, K., and Mastrandrea, M. eds. (2014). *Climate Change 2014: Impacts, Adaptation, and Vulnerability*, Vol. I, Cambridge: Cambridge University Press.

Gesing, F. (2016). *Working with Nature in Aotearoa New Zealand: An Ethnography of Coastal Protection*. Bielefeld: transcript.

Ghosh, T., Bhandari, G., and Hazra, S. (2003). Application of 'bio-engineering' technique to protect Ghoramara Island (Bay of Bengal) from severe erosion. *Journal of Coastal Research*, **9**(2), 171–8.

Government of India. (2008). *National Action Plan on Climate Change*.

Government of West Bengal. (2009). *District Human Development Report: South 24 Parganas*. Kolkata: Development & Planning Department.

Government of West Bengal. (2012). *West Bengal State Action Plan on Climate Change*. Kolkata: Government of West Bengal, Government of India. Available at http://moef.nic.in/downloads/public-information/West-Bengal-SAPCC.pdf

Gupta, C. and Sharma, M. (2009). *Contested Coastlines: Fisherfolk, Nations and Borders in South Asia*. New Delhi: Routledge.

Harms, A. (2015). Leaving Lohachara: On circuits of displacement and emplacement in the Indian Ganges Delta. *Global Environment*, **8**(1), 62–85.

Harms, A. (2017). Citizenship at sea: Environmental displacement and state relations in the Indian Sundarbans. *Economic & Political Weekly*, **52**(33), 69–76.

Hazra, S., Samanta, K., Mukhopadhyay, A., and Akhand, A. (2010). *Temporal Change Detection (2001–2008) Study of Sundarban*. Kolkata: School of Oceanographic Studies, Jadavpur University.

Kanjilal, T. (2000). *Who Killed the Sundarbans?* Calcutta: Tagore Society for Rural Development.

Koslov, L. (2016). The case for retreat. *Public Culture*, **28**(279), 359–87. doi:10.1215/08992363-3427487.

Lahiri-Dutt, K. and Samanta G. (2013). *Dancing with the River: People and Life on the Chars of South Asia*. New Haven, CT: Yale University Press.

MOEF. (2009). *Bangladesh Climate Change Strategy and Action Plan (BCCSAP)*. Dhaka: Ministry of Environment and Forests (MOEF), Government of People's Republic of Bangladesh.

Mukerjee, R. (1938). *The Changing Face of Bengal: A Study in Riverine Economy*. Calcutta: The University of Calcutta.

Myers, N. (1993). Environmental refugees in a globally warmed world. *Bioscience*, **43**(11), 752–61.

O'Donnell, A. and Wodon, Q. (2015). Early warning systems. In A. O'Donnell and Q. Wodon, eds., *Climate Change Adaptation and Social Resilience in the Sundarbans*. London: Routledge.

Oliver-Smith, A. (1999a). The brotherhood of pain: Theoretical and applied perspectives on post-disaster solidarity. In A. Oliver-Smith and S. M. Hoffman, eds., *The Angry Earth: Disaster in Anthropological Perspective*. New York: Routledge.

Oliver-Smith, A. (1999b). 'What is a disaster?': Anthropological perspectives on a persistent question. In A. Oliver-Smith and S. M. Hoffman, eds., *The Angry Earth: Disaster in Anthropological Perspective*. New York: Routledge.

Rudra, K. (2007). The proposed chemical hub in Nayachar: Some argumentative issues of concern. *Counterviews Webzine*, **1**(1), 3–7.

Srinivas, M. N. (1966). *Social Change in Modern India*. Berkeley, CA: University of California Press.

Stocker, T., Qin, D., Plattner, G., et al., eds. (2014). *Climate Change 2013: The Physical Science Basis: Working Group I Contribution to the Fifth Assessment Report of the Intergovernmental Panel on Climate Change*. Cambridge: Cambridge University Press.

Sultana, F. (2010). Living in hazardous waterscapes: Gendered vulnerabilities and experiences of floods and disasters. *Environmental Hazards*, **9**(1), 43–53. doi:10.3763/ehaz.2010.SI02.

World Bank. (2016). *Poverty and Shared Prosperity 2016: Taking on Inequality*. Washington, DC: World Bank Publications.

6

Coastal Risks from Typhoons in the Pacific
The Case of the Philippines

NORALENE UY AND JOE-MAR PEREZ

Introduction

The Philippines, composed of over 7,000 islands, is surrounded by large bodies of water. Its location in the western side of the Pacific Ocean makes the archipelago naturally exposed to tropical cyclones, or typhoons (as severe tropical cyclones are called in the northwest Pacific). An average of 20 typhoons enter the Philippine Area of Responsibility (PAR) from July to November every year, with about seven to nine making landfall, often with accompanying strong winds and heavy rains. Storms and typhoons result in storm surges and waves, coastal flooding, erosion, saltwater intrusion, rising water tables, impeded drainage, and wetland loss and change (Wong et al., 2014; see also other chapters in Part II of this volume).

Over the years, damage and losses from typhoons have compromised the Philippines' hard-won development gains. Between 1980 and 2017, the Philippines experienced 261 typhoons that caused 36,217 deaths, affected more than 151 million people, and caused approximately $20.5 billion in direct physical damages (EM-DAT, n.d.). Probabilistic risk analysis estimates that the Philippines experiences an average annual loss (AAL) of $8.45 billion – equivalent to 2.69 percent of 2017 gross domestic product – as a consequence of natural hazards, including $4.07 billion from cyclonic wind, $2.54 billion from storm surges, and $545.43 million from flooding. The Philippines is ranked first for the highest tropical cyclone AAL in relation to capital investment among several countries assessed. The 100-year probable maximum loss is estimated at $21.89 billion due to cyclonic wind and $3.83 billion due to storm surges (UN, 2015).

Typhoon Trends and Attributions to Climate Change

Typhoons are increasing in number and strength to the east side of the Philippines according to data from 1945 to 2003 (Anglo, 2005). Such a trend indicates that strong typhoons may already be the "new normal." Frequent variations in atmospheric conditions, such as warmer sea surface temperatures, have the potential to produce significant changes in the risk posed by typhoons. However, there are complexities to the attribution of extreme climate events because it is difficult to isolate the influence of anthropogenic factors. Some studies show that atmospheric circulation changes are not the dominant factor where climate change is concerned. Hence, it is suggested to look at the influences of the changed large-scale thermodynamic environment on the extremes of temperatures and moisture associated with the event rather than on the synoptic event itself (Trenberth, et al., 2015).

Despite the difficulty of directly attributing strong typhoons to climate change, the link between sea surface temperature rise and increasing tropical cyclone intensity has been established (Elsner et al., 2008). Global warming will result in rising sea surface temperature, leading to potential increased typhoon intensity (Holland and Bruyère, 2014). While the Intergovernmental Panel on Climate Change's Fifth Assessment Report found low confidence in trends in typhoon frequency and intensity due to limitations in observations and regional variability, the occurrence of most intense typhoons is projected to increase (Wong et al., 2014). Strong typhoons attributed to climate change pose serious risks to the Philippines, being a coastal country that is prone to floods and other hydrometrological hazards. A study by Deltares and the Institute for Environmental Studies revealed that the risk associated with severe storms and typhoons is also increasing due to sea-level rise, thereby exposing around 7.5 million people in the Philippines between today and 2080 (Deltares, 2017). Further, storm surges from intense typhoons on higher sea levels are projected to affect 42 percent of the coastal population (Brecht et al., 2012).

Disaster Risk Reduction and Management: The Policy Process in the Philippines

Given the country's high risk of typhoons, floods, and other disasters due to its natural and geographic attributes, the government of the Philippines has undertaken proactive measures in strengthening its governance in disaster risk reduction and management (DRRM) and climate change adaptation (CCA) through legislation and institutional arrangements. This has been attained through enactment of the Republic Act (RA) 10121 or the Philippine DRRM Act of 2010. Enacted in 2010, RA 10121 is the legal basis for all DRRM-related activities in the country (Congress of the Philippines, 2010). It serves as "a boost to the development of policies and plans, implementation of actions and measures pertaining to all aspects of disaster risk reduction and management, including good governance, risk assessment and early warning, knowledge building and awareness raising, reducing underlying risk factors, and preparedness for effective response and early recovery" (NDRRMC, 2012: 11). The law also provides a definition for DRRM: "the systematic process of using administrative directives, organizations, and operational skills and capacities to implement strategies, policies and improved coping capacities in order to lessen the adverse impacts of hazards and the possibility of disaster."

With the enactment of RA 10121, DRRM has been the main approach of the Philippine government when it comes to managing disasters and climate change impacts. The act mandated the creation of the National Disaster Risk Reduction and Management Council (NDRRMC) to implement policies, programs, and activities on DRRM. Specifically, the NDRRMC is tasked with (1) policy making, coordination, integration, supervision, and monitoring and evaluation; (2) enforcement of various laws, guidelines, and technical standards; (3) managing and mobilizing resources for DRRM, including the National DRRM Fund; and (4) monitoring and providing guidelines and procedures on the Local DRRM Fund. Also in response to RA 10121, the Implementing Rules and Regulations (IRR) and the National DRRM Framework were developed by the government. The National DRRM Framework focuses on the integration of the four DRRM thematic areas: (1) Disaster Prevention and Mitigation; (2) Disaster Preparedness; (3) Disaster Response; and (4) Disaster Rehabilitation and Recovery. It considers factors such as hazard, exposure, vulnerability, and capacity, and it prescribes mainstreaming of disaster risk reduction (DRR) and CCA in planning and implementation.

The National Disaster Risk Reduction and Management Plan (NDRRMP), covering the period 2011–2028, was also formulated as a result of the enactment of the law. The NDRRMP outlines 14 objectives, 24 outcomes, 56 outputs, and 93 activities with implementing partners and corresponding short-, medium-, and long-term timelines. At the local level (i.e., regional, provincial, city, and municipal governments), local DRRM plans are developed following the NDRRMP. With regard to institutional setup, the NDRRMC is composed of members from 44 national government agencies, government financial institutions, the private sector, and civil society groups. The Secretary of the Department of National Defense serves as the chairperson of the NDRRMC with four vice chairpersons in the four DRRM thematic areas: (1) Disaster Prevention and Mitigation led by the Secretary of the Department of Science and Technology (DOST); (2) Disaster Preparedness led by the Secretary of the Department of Interior and Local Government (DILG); (3) Disaster Response led by the Secretary of the Department of Social Welfare and Development (DSWD); and (4) Disaster Recovery and Rehabilitation led by the National Economic and Development Authority (NEDA).

The NDRRMC is replicated at local levels to form a DRRM network of 17 regional DRRM councils, 81 provincial DRRM councils, 145 city DRRM councils, 1,489 municipal DRRM councils, and 42,029 *barangay* (village-level administrative unit) DRRM committees. The provincial, city, and municipal DRRM councils are chaired by local chief executives with membership from 18 agencies, while the *barangay* DRRM committee, chaired by the *barangay* chairperson, forms part of the *Barangay* Development Council. This network of DRRMCs allows widespread implementation of DRRM across all levels of governance. The Office of Civil Defense (OCD) functions as the secretariat and implementing arm of the NDRRMC and is tasked with administration of a comprehensive DRRM program in the country. It conducts the annual search for Gawad KALASAG (this acronym being developed from "*KAlamidad at Sakuna LAbanan, SAriling Galing ang Kaligtasan*"), which is NDRRMC's recognition program for excellence on DRRM and humanitarian assistance. The program aims to recognize outstanding performance of local DRRMCs (LDRRMCs); private and volunteer organizations; local, national, and international nongovernmental organizations; donor agencies; and communities in implementing significant DRRM projects and activities, and in providing humanitarian response and assistance to affected communities.

At the regional and global levels, the government of the Philippines actively participates in several DRRM mechanisms. As a member-state of the Association of Southeast Asian Nations (ASEAN), the Philippines is bound by the ASEAN Agreement on Disaster Management and Emergency Response to promote regional cooperation and collaboration in reducing disaster losses and enhancing joint emergency response in the region. The Philippines co-chairs the Risk Assessment, Early Warning, and Monitoring Working Group of the ASEAN Committee on Disaster Management. Further, DRRM was a priority pursued by the government when it hosted the Asia-Pacific Economic Cooperation (APEC) meetings in 2015. To this end, the 2015 Leaders' Declaration adopted the APEC DRR Framework for enhanced cooperation in building adaptive and disaster-resilient economies supporting inclusive and sustainable development. The Philippines also supports the operations of the Regional Integrated Multi-Hazard Early Warning System for Africa and Asia, which provides regional early warning services and builds capacity of its member states in the end-to-end early warning of tsunami and hydrometeorological hazards. Finally, the Philippines is a member of the Regional Consultative Committee on Disaster Management, which aims to identify disaster-related needs and priorities of Asia and the Pacific countries, promote regional and subregional cooperative programs, and develop regional action strategies for disaster reduction.

All climate change–related activities in the Philippines are governed by RA 9729, which is considered a "twin" law of RA 10121. In particular, RA 9729 seeks to integrate DRR into climate change programs and initiatives, including the assessment and management of risk to climate change, variability, and extremes. It provides for the establishment of the Climate Change Commission (CCC), the policy-making body on climate change chaired by the president of the Philippines and composed of three commissioners. The CCC has an advisory board consisting of 23 members from national government agencies, line agencies, leagues of government, academia, business, and nongovernmental organizations. One of the sectorial representatives comes from the DRR community. In addition, the CCC has a panel of technical experts consisting of practitioners from the climate change and DRR disciplines that provides technical advice on climate science and technology, best practices in risk assessment and management, and enhancement of adaptive capacity to potential impacts of climate change (Congress of the Philippines, 2009).

In 2011, RA 9729 was amended through RA 10174, establishing the People's Survival Fund to provide financing for adaptation activities based on the National Framework Strategy on Climate Change (NFSCC) (Congress of the Philippines, 2011). The CCC developed the NFSCC 2010–2022, which envisions a climate risk–resilient Philippines with healthy, safe, prosperous, and self-reliant communities and thriving and productive ecosystems. This framework serves as a guide in national and subnational development planning processes.

The National Climate Change Action Plan 2011–2028 was formulated following the NFSCC. It outlines the outcomes, outputs, and activities for adaptation and mitigation in strategic priorities, such as in food security, water sufficiency, environmental and ecological stability, human security, climate-friendly industries and services, sustainable energy, and knowledge and capacity development, as well as with respect to gender and development, technology transfer, education and communication, and capacity building. At the local level, local government units (LGUs) are tasked with the formulation, planning, and implementation of Local Climate Change Action Plans.

Experience of DRRM and CCA Governance in Coastal Areas: The Case of Typhoon Haiyan

Typhoon Haiyan, locally known as "Yolanda," struck the Philippines in 8 November 2013. Originating from the waters of the Pacific Ocean, the typhoon arrived with wind strength of 215 kilometers per hour and gusts of 250 kilometers per hour. It was one of worst typhoons ever recorded to hit the Philippines. Typhoon Haiyan resulted in about 6,300 causalities, with an estimated 16 million people affected. Its coverage extended to 12,139 *barangays* in 44 provinces, 591 municipalities, and 57 cities located in nine regions. Haiyan even ranked as first among the top ten worst typhoons in terms of damage to properties, amounting to around 93 billion pesos (PhP) (roughly $2 billion) (NDRRMC, 2013). The strong winds of Haiyan ravaged many coastal areas in Eastern Visayas, one of the hardest-hit islands. It also ravaged through Samar and Leyte islands, killing thousands of people with its five- to seven-meter-high storm surge (NDRRMC, 2014b: 27).

Disaster Preparedness

The devastation of Typhoon Haiyan led to realizations about the concept of storm surge. A storm surge involves "the sudden increase in sea water level associated with the passage of a tropical storm or typhoon." This is due to the push of strong winds on the water surface (wind setup), the piling up of the big waves (wave setup), storm central pressure (pressure setup), and astronomical tide moving toward the shore. The surviving locals described the storm surge as big waves that made disturbing sounds, like that of boiling water, which rushed toward communities (NDRRMC, 2014b: 25). One interesting revelation, based on anecdotes, is that the people in the coastal areas of Eastern Visayas generally did not know anything about storm surges. Some claimed that it was their first time experiencing such an event. However, historical data suggest that the Eastern Visayas had previously experienced storm surges. For example, in November 1984, Typhoon Undang, with a wind strength of 230 kilometers per hour, brought a two-meter storm surge, claimed 895 lives, and cost PhP 1.9 billion in damage. At that time, the term "storm surge" was not yet being used (NDRRMC, 2014b: 23).

Given the existing mechanisms for DRRM and CCA in the country, it is important to look at coastal governance during the onslaught of Typhoon Haiyan. With the enactment of RA 10121, the thematic area of disaster prevention and mitigation had been institutionalized prior to the typhoon's arrival. Headed by the DOST, this thematic area entails activities that focus on avoiding hazards and mitigating their potential impacts by reducing vulnerabilities and exposure and enhancing capacities of communities. Notably, according to the National DRRM Framework, the NDRRMC should bolster more efforts for disaster prevention and mitigation to achieve optimum resiliency (NDRRMC, 2012: 12).

To operationalize the thematic area of disaster prevention and mitigation, the NDRRMC undertook national-level projects to prevent disaster impacts. One of these was the READY Project, a joint effort of the NDRRMC, with the support of the United Nations Development Programme (UNDP) and Australian Aid (AusAID). The READY Project produced and disseminated multihazard maps covering 28 highly vulnerable provinces, including those areas that were affected by Typhoon Haiyan. Aside from the READY Project, the Mines and Geosciences Bureau (MGB), under the Department of Environment and Natural Resources (DENR), had produced multihazard maps with a 1:50,000 scale covering all municipalities and cities in the country. There was use of LiDAR technology, whereby high-accuracy and high-resolution digital elevation and surface maps were generated for more accurate risk assessment for floods, landslides, and earthquake hazards (NDRRMC, 2014: 5–6).

In addition to hazard and risk maps, another initiative under the thematic area of disaster prevention and mitigation is the mainstreaming of DRRM and CCA into local DRRM plans. In particular, all LGUs are required to produce comprehensive land-use plans. These plans aim to address disaster risks while pursuing developmental programs in their respective communities. Existing institutional mechanisms have ensured that DRRM is integrated in all government programs from the national to local level to protect the vulnerable areas in the country, including the coastal zones. Given the plethora of initiatives under the area of disaster prevention and mitigation headed by the DOST, the risk of the coastal areas to disasters is supposedly addressed and minimized.

With the advancement of technology, the advent of coastal-related disasters can now be detected. In fact, one week before Typhoon Haiyan entered the PAR, the NDRRMC, through its weather bureau, was already able to monitor and track its path. Several warning advisories were widely disseminated through various channels. Radio and television broadcasts were widespread, warning of potential damage to be brought by the typhoon, and specifically of the threat of a storm surge (NDRRMC, 2014: 32).

The regional government in Eastern Visayas admitted that, despite the availability of risk maps and the identification of no-build zones, public and commercial buildings were constructed in coastal areas that were found unsafe (NDRRMC, 2014: 47). Further, although the people had been warned about the arrival of the strong typhoon, many were not alarmed because they had experienced destructive weather events before. Some did not understand Haiyan's deadly potential. Even the emergency warning officers avoided using the term "storm surge" because they themselves could not explain it clearly (NDRRMC, 2014: 32). Clearly, people living in the coastal areas of Eastern Visayas were unaware of the risks of Typhoon Haiyan. Perhaps some were familiar with the risks but they chose not to take action.

The lack of awareness of the risks brought about by Typhoon Haiyan was also manifested during the implementation of emergency response activities. Specifically, the Emergency Operations Center (EOC) of the OCD in Eastern Visayas did not withstand the strong winds and floodwaters, rendering it heavily damaged and destroying all computers and communication equipment. Alarmingly, there were neither backup telecommunications nor radio and satellite phone systems in place, thereby making the coordination of response ineffective. The regional government was unable to provide updates to assisting forces outside Eastern Visayas due to a lack of means for communication (NDRRMC, 2014: 53).

DRRM and CCA Innovations Following Typhoon Haiyan

The experience with Typhoon Haiyan laid out several innovations in DRRM and CCA that aim to contribute to better risk reduction and resilience building as the dangers of typhoons increase with warming ocean temperatures. One of these innovations is the enforcement of land-use planning. Land-use planning is not only a question of geohazard mapping of safe and unsafe land as part of DRRM efforts. It also provides a policy framework for stakeholders to properly locate safe settlements, sound infrastructure, buffers and protection areas, and it identifies where agriculture production areas should be situated to address potential risks and conflicts arising from overlapping use of lands.

Following the devastation brought about by Typhoon Haiyan, a Joint Memorandum Circular was adopted in 2014 to prescribe hazard zone classifications in areas vulnerable to hydrometeorological hazards. The memorandum provides guidelines in determining appropriate activities and development in hazard-prone areas, particularly within the proximity of oceans and seas, among others. Such planning also informs affected persons where safe settlements should be placed and allows community participation in site selection and planning. This can support the increased resilience of disaster-affected communities, enabling stronger protection of livelihoods, greater community leadership in disaster management, and more robust recovery following a disaster.

In addition to land-use planning, the government initiated restoration of mangroves as natural protectors of coastal systems. Mangroves incurred extensive damage in many of the Typhoon Haiyan–affected localities. In response, the DENR planned to restore mangrove and beach forests along about 380 kilometers of coastline in Eastern Visayas and other affected areas, with an allocation of around $22 million for mangrove planting. This involves identifying the best areas for mangrove rehabilitation and for planning beach forests. These forests will be planted within the 20-meter zone along the shoreline reserved for public use under the Philippine Forestry Code.

At the strategic level, the lessons from Typhoon Haiyan also prompted the need for stronger DRRM implementation through institutional mechanisms. Specifically, the NDRRMC, in consultation with key stakeholders, has conducted a review of RA 10121 and its implementing rules and regulations (IRR), as well as the operational framework of the four thematic areas. The review produced an amendatory bill submitted to the House of Representatives that provides for the creation of an independent DRRM authority, establishment of a Public Assistance and Complaints Office, establishment of a DRRM Research and Training Center, institutionalization of a DRRM insurance and incentive scheme, and clarified procurement regulations, among other things. The IRR of RA 10121 was also revisited to further define the frameworks, roles, responsibilities, and activities of DRRM actors. Consultations and reviews have been conducted to finalize the amended provisions of RA 10121.

The DILG, being the vice chairperson of Disaster Preparedness of the NDRRMC, conducted a disaster preparedness audit of LGUs in 2014. Subsequently, an advocacy program – *Oplan Listo* – was initiated to strengthen the disaster preparedness of LGUs and communities using a whole-of-government approach. The program produced and disseminated Oplan Listo Disaster Preparedness Manuals for hydrometeorological hazards, particularly typhoons. These manuals describe the disaster preparedness minimum standards that should be implemented before, during, and after a disaster. The program also encouraged communities to start disaster preparedness in their homes through the *Gabay at Mapa para sa Listong Pamilyang Pilipino*, a family guide to action before, during, and after a disaster. This guide calls on families to make a household plan, determining safe places in their home, evacuation routes, and family meeting points, among other actions.

The NDRRMC further developed the National Disaster Response Plan (NDRP) for hydrometeorological disasters in 2014. This plan prescribes the activities for disaster response, whether as augmentation or for assumption of response functions in disaster-affected localities, and it identifies the roles and responsibilities of institutions and organizations during a disaster or emergency. In addition to the NDRP, the NDRRMC formulated the National Disaster Preparedness Plan (NDPP) in 2015. The NDPP identifies seven work areas critical to disaster preparedness: (1) information, education, and campaigns; (2) capacity building; (3) DRRM localization; (4) risk assessments and plans; (5) preparedness for emergency and disaster response; (6) continuity of essential services through the preparation of operations and continuity plans; and (7) partnerships (NDRRMC, 2015).

Managing Risks in Coastal Areas: Typhoon Haiyan and the Sendai Framework for Disaster Risk Reduction

Lessons from Typhoon Haiyan highlight the importance of strengthening coastal governance as climate change increases the dangers of typhoons. DRRM and CCA programs and interventions for coastal zones must be derived from scientific information and historical evidence. Coastal areas are generally prone to floods, typhoons, tsunamis, and rising sea levels. Interventions based on these risks are imperative. Otherwise, disaster managers could make poor decisions that are unresponsive to the needs of coastal communities. Following the Typhoon Haiyan experience in the Philippines and many other disasters around the world, in 2015 the global community collaborated to adopt the Sendai Framework for Disaster Risk Reduction (SFDRR). The SFDRR underscores the need to prioritize risk reduction to achieve optimum resiliency. Four priorities are identified in the framework. It is possible to draw lessons from the experience from Typhoon Haiyan using the lens of these four priorities. These lessons can be useful to other coastal countries as coastal risks increase due to climate change.

Priority 1: Understanding Disaster Risk

Managing disasters and climate change begins by thoroughly understanding the associated risks. The United Nations International Strategy for Disaster Reduction (2015) defines disaster risk as "the potential loss of life, injury, or destroyed or damaged assets which could occur to a system, society or a community in a specific period of time, determined probabilistically as a function of hazard, exposure, vulnerability and capacity." To effectively address the risks in coastal areas, disaster managers should be able to imagine the scope of damage that disasters may bring and the level of intervention required. With the geographic location of coastal areas near large bodies of water, one must expect the possibility of experiencing serious devastation due to common hazards such as floods, typhoons, and tsunamis. In the case of Tacloban City, most people failed to understand the risk that can be brought about by the storm surge despite the warnings from the national government. Hence, full understanding and imagining of such risk should prompt decision makers to take the appropriate interventions.

Priority 2: Strengthening Disaster Risk Governance to Manage Disaster Risk

Understanding disaster risk should be complemented by strong disaster risk governance. Institutional mechanisms from the national to local level must be integrated for more cohesive implementation of DRRM and CCA. During Typhoon Haiyan, the actions of the national government and the priorities of localities were notably disconnected. Although the top national leaders recognized in advance the devastation of the typhoon, such warning was not properly translated at the local level.

Indeed, disaster risk governance was ineffective. Knowing the gaps in local governance, the national government should take the lead in further intensifying its efforts to cascade DRRM and CCA down the line, especially to areas that are known to be at high risk, including coastal areas.

Priority 3: Investing in Disaster Risk Reduction for Resilience

Addressing risks and attaining resilience in coastal areas require investments. Disaster managers should realize that the costs of rebuilding and reconstruction in the aftermath of catastrophes are far more expensive than investing in DRRM and CCA programs. The glaring figures of damage and losses incurred during Typhoon Haiyan could have been minimized had there been enough investment in structural and nonstructural DRRM and CCA measures. An example is the investment for resettlement sites away from the coastal zones where communities can be relocated. Further, the construction of residential houses, facilities, and infrastructure should be compliant with water codes to avoid damage from floods, tsunamis, storm surges, or sudden rise in sea level. Further, localized early warning systems should be established. There should be investment in backup communication systems to ensure continuity of coordination among responders during emergencies.

Priority 4: Enhancing Disaster Preparedness for Effective Response

Aside from the long-term DRRM and CCA investments, it is also important to invest in preparedness. The onslaught of Typhoon Haiyan revealed that people at the time were generally unprepared. The emergency management facilities were destroyed. The disaster responders themselves became casualties. This highlights the need for localities to increase their capacities for effective disaster response. Learning from Typhoon Haiyan, even the establishment of emergency operations centers (EOCs) in locations that are safe from floods, tsunamis, storm surges, and other hazards is a must. Disaster responders should also be capable in terms of water search and rescue. Frequent simulation exercises, such as evacuation drills, should be implemented as well.

Conclusion

Given the archipelagic nature of the Philippines, much of its territory is prone to disasters that come from the sea. Through the years, the occurrence of extreme weather events, exacerbated by climate change, have become more frequent and devastating. With the enactment of RA 10121 and other parallel laws, institutional mechanisms are now in place to mainstream DRRM and CCA across all levels of governance and to protect people living in the most vulnerable locations, notably coastal areas. However, the devastation of Typhoon Haiyan in 2013 revealed significant gaps in coastal governance. This can be attributed to a lack of understanding of the risk associated with living in coastal zones, particularly given climate change.

Following the priorities of the SFDRR, lessons can be generated from the experience with Typhoon Haiyan. Foremost, it is important to have an understanding and imagination of the risks to disasters, especially for people in coastal areas. Through understanding of risk, coastal governance can be strengthened. To address the associated risks of living in coastal areas, investment in structural and nonstructural measures should be undertaken. Further, bolstering preparations for response to coastal-related disasters should be conducted.

The experience of the Philippines in preparing for the new normal of typhoons amidst climate change highlights the need to further strengthen coastal governance. Other coastal countries that are similarly vulnerable to typhoons can gain insights in revisiting policies, practices, and mechanisms to prepare and respond to coastal risks. The humanitarian community can likewise draw lessons in coordination and delivering more effective support to developing coastal countries. Overall, the key to effective coastal governance relies on focusing all efforts toward reducing risks to disasters made more likely, and potentially deadlier, by climate change.

References

Anglo, E. G. (2005). *Decadal Change in Tropical Cyclone Activity in the Western Pacific*. Paper presented at the Okinawa Typhoon Center Forum, Okinawa, Japan, 1 October 2005.

Brecht, H., Dasgupta, S., Laplante, B., Murray, S., and Wheeler, D. (2012). Sea-level rise and storm surges: High stakes for a small number of developing countries. *Journal of Environment and Development*, **21**(1), 120–38.

Congress of the Philippines. (2009). *Republic Act No. 9729 "An Act Mainstreaming Climate Change into Government Policy Formulations, Establishing the Framework Strategy and Program on Climate Change, Creating for This Purpose the Climate Change Commission, and for Other Purposes."*

Congress of the Philippines. (2010). *Republic Act No. 10121 "An Act Strengthening the Philippine Disaster Risk Reduction and Management System, Providing for the National Disaster Risk Reduction and Management Framework and Institutionalizing the National Disaster Risk Reduction and Management Plan, Appropriating Funds Therefore and for Other Purposes."*

Congress of the Philippines. (2011). *Republic Act No. 10174 "An Act Establishing the People's Survival Fund to Provide Long-term Finance Streams to Enable the Government to Effectively Address the Problem of Climate Change, Amending for the Purpose Republic Act No. 9729, Otherwise Known as the 'Climate Change Act of 2009', and for Other Purposes."*

Deltares. (2017). *Dutch Scientists Chart Coastal Flooding in the Future: 50% More People at Risk in 2080*. Available at www.deltares.nl/en/news/nederlandse-wetenschappers-brengen-toekomstige-kustoverstromingen-in-kaart-in-2080-50-meer-mensen-bedreigd/

Elsner, J. B., Kossin, J. P., and Jagger, T. H. (2008). The increasing intensity of the strongest tropical cyclones. *Nature*, **455**, 92–5. doi:10.1038/nature07234

EM-DAT: The Emergency Events Database – Université Catholique de Louvain (UCL) – CRED, D. Guha-Sapir – www.emdat.be, Brussels, Belgium.

Holland, G., and Bruyère, C. L. (2014). Recent intense hurricane response to global climate change. *Climate Dynamics*, **42**(3–4), 617–27. Available at https://doi.org/10.1007/s00382-013-1713-0

Joint Department of Environment and Natural Resources, Department of Interior and Local Government, Department of National Defense, Department of Public Works and Highways, and Department of Science and Technology. (2014). *Memorandum Circular No. 2014–1: Adoption of Hazard Zone Classification in Areas Affected by Typhoon Yolanda (Haiyan) and Providing Guidelines for Activities Therein.* Available at http://pcij.org/wp-content/uploads/2015/01/Joint-DENR-DILG-DND-DPWH-DOST-Adoption-of-Hazard-Zone-Classification.pdf

National Disaster Risk Reduction and Management Council. (2012). *National Disaster Risk Reduction and Management Plan, 2011–2028.*

National Disaster Risk Reduction and Management Council. (2013). *NDRRMC Update Final Report re EFFECTS of Typhoon "YOLANDA" (HAIYAN).*

National Disaster Risk Reduction and Management Council. (2014a). *National Disaster Response Plan for Hydrometeorological Disasters.*

National Disaster Risk Reduction and Management Council. (2014b). *Y It Happened: Learning from Typhoon Yolanda.*

National Disaster Risk Reduction and Management Council. (2015). *National Disaster Preparedness Plan.*

Prevention Web. (2014). *Basic Country Statistics and Indicators.* Available at www.preventionweb.net/countries/phl/data/

Trenberth, K. E., Fasullo, J. T., and Shepherd, T. G. (2015). Attribution of climate extreme events. *Nature Climate Change*, **5**, 725–30. doi:10.1038/nclimate2657

UN. (2015). *Global Assessment Report on Disaster Risk Reduction 2015: Making Development Sustainable: The Future of Disaster Risk Management.* Geneva.

United Nations International Strategy for Disaster Reduction. (2015). *Sendai Framework for Disaster Risk Reduction: 2015–2030.*

Wong, P. P., Losada, I. J., Gattuso, J.-P., et al. (2014). Coastal systems and low-lying areas. In C. B. Field, V.R. Barros, D.J. Dokken, et al., eds., *Climate Change 2014: Impacts, Adaptation, and Vulnerability. Part A: Global and Sectoral Aspects. Contribution of Working Group II to the Fifth Assessment Report of the Intergovernmental Panel on Climate Change*, Cambridge, UK and New York: Cambridge University Press, pp. 361–409.

7

Ocean Policy Perspectives: The Case of Indonesia

ACHMAD POERNOMO AND ANASTASIA KUSWARDANI

Introduction

Indonesia has 17,504 islands, 10,000 of which are small. Its coastline, the second longest in the world, spans 95,181 km and is home to more than 140 million people (out of a total national population of 250 million people). Indonesia's economy relies heavily on marine resources and coastal activities; the nation's "ocean economy" is estimated at US$800 million per year (Dahuri, 2011), which is about 20 percent of gross domestic product (GDP) (Huffard et al., 2012). Indonesia's oceans support 40 million jobs (Dahuri, 2011). Indonesia ranks second in global production of marine fish and cultured fish (FAO, 2016). Its coastal zone is home to 2,500 species of mollusk, 2,000 species of crustacean, 6 species of sea turtle, 30 species of marine mammals, and over 200 species of fish (ADB, 2009). The country's oceans provide an important source of protein that traditionally has been a major food source. Fish consistently contribute more than 10 percent of total food protein consumption in the country and more than 50 percent of animal food protein intake.

Located along the equator and important sea routes, Indonesia plays a significant role in the global supply chain. Providing three ocean passageways, known as the Indonesian Archipelagic Sea Lanes I, II, and III, has made Indonesian waters an important player in maritime traffic. It was estimated that 44 percent of global maritime traffic, and 95 percent of vessels in the Asia Pacific region, enter Indonesian waters at some point. The Strait of Malacca is the main shipping channel between the Indian Ocean and the Pacific Ocean, and almost half of the world's total annual seaborne trade tonnage, and 70 percent of Asia's oil imports, pass through the Malacca and Singapore straits (Mindef, 2016). According to Bateman et al. (2009), 8,678 ships make 75,510 passages through the Malacca and Singapore straits annually, carrying 3 billion tons of cargo worth a total of $390 billion.

Indonesia's marine and coastal vegetation are significant contributors to the global oxygen supply and to carbon dioxide absorption. Along Indonesia's coastline are the world's largest mangrove forests, with an area of 3.2 million hectares (ha). According to a United Nations Environment Programme report, the country's coastline is also home to the world's largest seagrass meadows, spanning 30,000 km². Being in the center of the Coral Triangle, which is also frequently called the Amazon of the Sea, Indonesia is blessed with nearly 50,000 km², or 18 percent, of the world's coral reefs (Gray, 1997; PEACE, 2007) (see Chapter 17).

Indonesia's position in currents of the so-called global ocean conveyor belt between the Indian and Pacific Oceans has made its seas of central global importance in terms of international waters, global climate, and biodiversity (Vantier et al., 2005). Indonesia is also a country vulnerable to climate change. In coastal areas where populations are high, impacts of climate change can be identified, with different levels of severity across the country. Income resources are under immediate threat from climate change, as well as in the longer term, while uncertainty in fishing activities, due to unpredictable seasonal changes and altered migration routes, have put added pressure on food security. This is worsened by human activities, which have deteriorated the coastal ecosystem. In the long term, not only will the economy of coastal communities be affected by climate change but also the national economy as a whole.

Starting with a review of Indonesia's ocean governance, this chapter describes the impacts of climate change in coastal areas of the country and discusses the approach of the Indonesian government to mitigating and adapting to those impacts.

Archipelagic Challenges: Ocean Governance and the Institutional Framework

A recent work by Butcher and Elson (2017) has extensively elaborated on the process of how Indonesia became an archipelagic state. It is interesting to note what Mochtar Kusumaatmadja, the head of the Indonesian Delegation to the Third United Nations Conference on the Law of the Sea in 1982, told reporters when he returned from the conference: "[W]hile Indonesia had every reason to be pleased with the international recognition of the archipelagic principle and the concept of an Economic Exclusive Zone (EEZ), the next problem was to take full advantage of what Indonesia had gained" (Butcher and Elson, 2017). Indeed after 35 years, it has been shown that Indonesia has not yet taken full advantage of its EEZ, and this remains a challenge. The following section attempts to unravel why this is so.

Indonesia unilaterally declared itself an archipelagic state on December 13, 1957, with what was then called the Djuanda Declaration, named after the Indonesian prime minister at that time. After a long and tiring deliberation, the concept of the archipelagic state came into existence when the Third United Nations Conference passed the United Nations Convention on the Law of the Sea (UNCLOS) in 1982 in Montana Bay (see Chapters 2 and 19 to 21 in this volume). The convention then came into force in 1994. The concept of the archipelagic state is contained in Part IV of the convention. The Indonesian Parliament, through Law Number 17 on the Ratification of UNCLOS 1982, ratified the convention in 1985. However, it was only after the second amendment of the country's 1945 constitution in August 2000 that Indonesia was legalized as an archipelagic country. This came 43 years after the Djuanda Declaration. UNCLOS also granted Indonesia an EEZ as vast as 2.9 million km², in addition to 3.1 million km² of archipelagic waters and 0.3 million km² of territorial sea.

In 1999, President Abdurrahman Wahid established a new ministry to deal with sea exploration. This ministry is now known as the Ministry of Marine Affairs and Fisheries (MMAF). More than 30 years earlier, President Soekarno, in his short-lived Dwikora Cabinet II, established a Coordinating Ministry of Maritime, which had three ministries under its coordination, namely, the Ministry of Sea Transportation, the Ministry of Marine Fisheries and Processing, and the Ministry of Maritime Industries. The cabinet only lasted for one month before dissolving with the fall of Soekarno in March 1966. From 1966 to 1998, under President Soeharto's administration, the orientation of Indonesian development was land based. Consequently, the ocean was almost neglected and the portfolio on ocean affairs was scattered across numerous ministries, thereby lacking in coordination. When MMAF was established, implementing the mandate of the ministry was the main challenge because ocean affairs had been under other ministries and there was a lack of strong leadership to relocate them under the new ministry. This situation forced the new ministry to explore only the tasks that did not already belong to other ministries. Fisheries and aquaculture were the only development sectors that were moved to the MMAF from the Ministry of Agriculture. Other tasks that were then self-created by the new ministry were coastal and small-island development and marine and fisheries surveillance, in addition to the usual tasks of a ministry, such as research and human resource development.

In 1960, President Soekarno formed the Indonesian Maritime Council, which was given a task of advising the president on maritime policy. After a long dormancy, the council was revitalized by President Soeharto in 1996 and was renamed the Indonesian Ocean Council. The council was chaired by the president, while the coordinating minister for political and security affairs was the vice chair and day-to-day chair. In addition to providing advice on ocean policy, the council was given the task of supervising maritime borders. In 1999 President Wahid dissolved the Indonesian Ocean Council and established the Indonesian Maritime Council, which was tasked with supervising policy on marine affairs. It was not clear why the council was named the Maritime Council when the main responsibility was to do with marine affairs. This shows that the two terms are considered interchangeable and differences between the two are ignored. This was again reflected when President Yudhoyono returned the name to the Indonesian Ocean Council in 2007. A decree was issued, and the names of some ministries were changed. The Ministry of Sea Exploration and Fisheries was changed to the Ministry of Marine Affairs and Fisheries (MMAF) during the President Wahid administration. Through all of this the task of the council remained the same: to provide advice to the president on marine affairs while being chaired by the president, giving the minister of Marine Affairs and Fisheries the task of being the day-to-day chair.

When Joko Widodo took the presidential office in 2014, he set a maritime vision for the country, established a Coordinating Ministry of Maritime Affairs (CMMA) and later dissolved the Indonesian Ocean Council in 2016. During the life of the councils, no significant actions were taken. The councils were trapped in routine activities, such as meetings; seminars; coordinating annual sailings; celebrating Nusantara Day; and publishing reports, documents, and manuscripts. One of the documents produced by the council was the Ocean Policy. However, it was never endorsed by the president before the council was dissolved.

Since the establishment of the MMAF, several laws on marine affairs and fisheries have been passed by the Parliament. These include the Law on Fisheries, the Law on Coastal Zone and Small Island Management, and the Law on the Ocean. The latter took almost ten years before it was passed in 2014, just a few hours before the members of Parliament of 2009–2014 finished their term. The long process was due to the fact that many oceans-related sectors had their own laws. To minimize overlaps, when passed by the Parliament, the Law on the Ocean was considered an umbrella for the existing laws pertaining to the oceans. Despite this, there are some new perspectives in the Law on the Ocean that open doors to strengthening the role of the ocean in Indonesia in the future.

It is clearly stipulated in the Law on the Ocean (no. 32/2014) that the law is to serve as a foundation for ocean governance, including the utilization of ocean resources. Ocean resources recognized by the law are fisheries, energy and minerals, coastal and small islands, and nonconventional resources, all of which should be utilized and managed using a blue economic approach. Blue economy in the law was adapted from a concept developed by Pauli (2010) and defined as the utilization and management of ocean resources using principles of community participation, resource efficiency, minimizing waste, and multiple revenues. However, the concept has not been implemented fully. The law also requires the government to prepare an Ocean Economy Policy, which is aimed at the use of the ocean as a basis for the economic development of the country. Thus, a clear direction for the role of the ocean in the economy of the country was set. This complements Law Number 17/2007 on the Long Term National Development Plan 2005–2025, which was issued by President Yudhoyono administration. As part of the national development mission, that plan stipulates the strengthening of the archipelagic state of Indonesia to help make the country self-reliant, advanced, strong, and able to protect national interests.

Having the Law on the Ocean in the Indonesian legal system, President Widodo, in his first speech after taking the oath of office as president in October 2014, stated the new vision of Indonesia to become a maritime nation and the Global Maritime Axis. He also stated this at the Ninth East Asia Summit on November 13, 2014, in Myanmar. President Widodo's statement created excitement in the country and led to what Ekawati (2016) called a maritime euphoria. However, there was no agreed perception on what a global maritime axis actually should be. No one could explain whether it was a vision, a strategy, or a policy. During his campaign for the presidency, Joko Widodo mentioned many times the term "toll sea," which again created confusion, as no clear definition was made available at the time. The public had misleadingly associated it with toll roads. Only after carefully reading the Medium Term National Development Plan 2015–2019, which was issued in early 2015, could one learn that "toll sea" was actually a program to improve maritime connectivity.

The Global Maritime Axis is said to have five key areas or pillars of development: maritime culture, maritime economy and resources, maritime infrastructure and connectivity, maritime diplomacy, and maritime security. Though good in name, it is actually lacking in implementation and coordination among ministries and other government institutions. This is due to the absence of a clear direction on how to implement these pillars. It could have been better if the CMMA was able to take the lead in the first place, but the ministry is still shaping itself and fine-tuning its portfolio. As a matter of fact, within 19 months of its establishment in October 2014, President Widodo had replaced the CMMA ministers three times. The new coordinating minister took office in July 2016. A draft of the Ocean Policy was submitted to the president, and in February 2017, Presidential Decree No. 16 on Ocean Policy was issued. A three-year action plan (2016–2019) was attached to the decree, listing activities for different ministries and other government institutions. As stipulated in the decree, the Ocean Policy is meant to accelerate the implementation of the Global Maritime Axis, and the CMMA was given the task to coordinate, monitor, and evaluate this implementation. However, given that there were barely two years left of President Widodo's term, significant achievements were unlikely. Nevertheless, a direction had been clarified.

Looking back at the last 60 years after the Djuanda Declaration in 1957, it was only after political reform in 1998, when President Soeharto had to step down, that gradually Indonesia became an archipelagic state in a political sense. This has affected ocean governance, which until today has not taken on an ideal form. The main causes are said to be overlapping portfolios, conflict among ministries and government institutions, and a lack of interest and concern by political leaders. Law Number 23 on Local Government, which was issued in 2014, has made the situation worse, as the authority to manage the sea is withdrawn from the district/city government and granted to provincial and central governments (0 to 12 miles and more than 12 miles from shore, respectively). In the old law, district/city governments were authorized to manage the sea between zero and four miles, and many local regulations had been issued based on this authority. At the same time, there are living laws such as the Law on Coastal Zone and Small Island Management that stipulate delegating authority to district/city governments. The new law has created mixed reactions from all levels of government, particularly district/city and provincial governments, because they are required to harmonize regulations and hand over official assets from district/city to provincial governments. This carries budget implications, and not all provincial governments have sufficient budgets to do this. It seems that it will take longer for Indonesia to develop effective institutions for ocean governance. This has implications for climate change mitigation and adaptation.

Coastal Areas and Climate Change: Impacts, Mitigation, and Adaptation

Impacts

As an archipelagic state, Indonesia faces serious threats from the adverse impacts of climate change. This is due to geographical realities: it is surrounded by the sea, with vulnerable landscapes being threatened with sea-level rise, acidification of Indonesian waters, and rising temperatures both on land and sea. Indonesia has also been experiencing extreme hydrometeorologically driven disasters that have exacted huge damage and losses. Yusuf and Francisco (2009) studied climate change vulnerability in Southeast Asia and concluded that sea-level rise is one of the dominant hazards in Indonesia, particularly in west and east Java. They showed that many coastal zones in Indonesia are subject to sea-level rise (within the five-meter inundation zone), including the east coast of Sumatera, the northern coast of Java, parts of the east coast of Kalimantan, and the southern coast of Papua. The Research Centre for Marine and Coastal Resources of the MMAF indicated in 2014 that sea-level rise in Indonesia is in the range of 0.7 cm per year, and that within 25 years sea level will rise by 19 cm, not taking into account land subsidence (cited in *Kompas Daily*, February 6, 2014: 13). Analyzing satellite altimetry data from 2009 to 2012, Zikra et al. (2015) indicated a sea-level rise of 1.18 to 14.1 mm per year in the coastal cities of Ambon (Maluku), Pemangkat (West Kalimantan), Medan (North Sumatera), and Manokwari (Papua). Based on five years (2004–2008) of sea-level observation at five main fishing ports (Ambon, Maluku; Bitung, North Sulawesi; Bungus, West Sumatera; Nizam Zachman, Jakarta; and Palabuhan Ratu, West Java), it was predicted that the sea would rise by 65 to 117 cm by 2026 (Poernomo et al., 2014). The Ministry of Environment (MoE) also reported that mean sea level increased by one to eight millimeters per year, with the highest increase registered in the area of Belawan (ADB, 2009). For Jakarta, the capital city of Indonesia, a report by PEACE (2007) indicates that the mean sea level in Jakarta Bay increased as much as 0.57 centimeters per year, and that coupled with land subsidence, makes Jakarta very vulnerable to flooding. According to Sudibyakto (2010), sea-level rise due to climate change in the north coast of Java is predicted to be six to ten millimeters per year. This will lead cities like Pekalongan and Semarang of Central Java to suffer 2.1 to 3.2 kilometers of inundation within 100 years.

Sea-level rise, together with extreme events, such as La Niña and tropical cyclones, contributes to a phenomenon locally referred to as "rob." Rob is the inundation of coastal areas during the spring tide. It has been observed in a number of coastal areas of Indonesia. Semarang, one of the biggest cities in the north of Java and the capital of Central Java Province, suffers from coastal flooding due to land subsidence, high-water tides, and inadequate infrastructure. In Demak, Central Java, the first rob occurred in 1995 and then followed in other districts, such as Banten, Jakarta, and other regions. A rob event in Demak affected more than 650 ha of coastal area in six villages and damaged infrastructure such as roads and railways. This has caused considerable problems to the country's transportation system, as well as in the economy, resulting in considerable losses to the tourism and aquaculture industries (Boer et al., 2007). PEACE (2007) reported an estimated loss of 7,000 tons of fish and 4,000 tons of prawns due to sea-level rise in the Karawang and Subang districts of the northern coast of West Java. Impacts are more severe in the lower Citarum-Basin, West Java, where 26,000 ha of ponds and 10,000 ha of crops are at risk. This could result in the loss of 15,000 tons of fish, shrimp, and prawns, as well as 940,000 tons of rice. Furthermore, the loss of small Indonesian islands due to sea-level rise has also been reported or predicted (Prabowo and Salahudin, 2016). (For the full description of the impacts of climate change on ocean and coastal areas of Indonesia, see the report prepared by the MoE [MoE, 2010].)

Although many authors and scientists have used different approaches in assessing the impacts of climate change on fisheries and aquaculture (Brugere and De Young, 2015), all demonstrate the certainty of their occurrence (Bene et al., 2016). The economic impact could be even worse than previously thought (Gaworecki, 2016). As described by Cohrane et al. (2009), "climate change is a compounding threat to the sustainability of capture fisheries and aquaculture development." They described two threats: impacts on ecosystems and impacts on livelihoods. The latter includes the impact on fish-food security, which is likely to occur in Africa and Asia, specifically in Indonesia, where fish is extraordinarily important to the diet of the people. Current fish consumption in Indonesia is 43.9 kg/capita per year, while total national production is 10.4 million tons (FAO, 2016). Climate change affects fish-food security through changes in fish availability, the stability of the fish supply, access to aquatic food, and aquatic product utilization. According to Huelsenbeck (2012), Indonesia ranks ninth in the list of the ten most vulnerable nations to fish-food security due to climate change. Allison et al. (2009) included Indonesia in a group of countries where the impacts of climate change on fisheries is moderate. Using 45 commonly utilized fish species as samples for analysis, Cheung et al. (2010) indicated that there will be a reduction of Indonesian fish catch potential of about 6 percent by 2100 if greenhouse gas concentrations remain at 2000 levels, or 22 percent if they are doubled. Gaol and Nababan (2012) indicated that pelagic fish production in Indonesia could decrease by 15 to 30 percent due to sea warming. Some types of adaptation exercised by local fishermen in Pelabuhan Ratu, Moro Island, and Panjang Island in response to climate change include fishing on new ground, replacing fishing gear, and reducing the number of fishing trips (Wahyudi, 2010; Rindayati et al., 2013; Wibowo and Satria, 2015).

In general, the average ocean temperature rise in Indonesia is estimated to be 0.5 to 3.92°C in 2100 compared to the level in 1981 (Boer and Faqih, 2005; Rataq, 2007). An increase of sea surface temperature above 1°C will shift the rainfall pattern of wet and dry months. This has been reported in some parts of Indonesia, including in northern Sumatra and Kalimantan, where higher intensity and shorter duration of rainfall has occurred. On the other hand, reverse conditions occurred in southern Java and Bali. This has brought about negative impacts to solar salt production, which is very much dependent on the availability of sun during the dry season. For example, in 2013 national salt production dropped to 577,917 tons, below the target of 700,000 tons, a target that had previously been revised downward from 1.845 million tons (Noviyanti 2016). The situation worsened in 2016 when production was only 144,000 tons or 4 percent of the target of 3 million tons. Again, weather anomalies, notably a wet summer, were said to be the cause (Ambari, 2017;

Trobos, 2017). A prolonged period of rain in Randutatah, Sampang, and Pamekasan, all in Madura Island, during 2010 and 2012 caused catastrophic losses for salt farmers (Kurniawan and Azizi, 2012; Wahyono et al., 2012). Salt farmers and fishermen have reported that in recent years the weather has been very difficult to forecast and local wisdom developed by their older generations has been practically useless in responding to the changing climate. The situation is worsened by the fact that not all fishermen and salt farmers have access to modern weather forecasts.

Coral bleaching is also a commonly reported impact of climate change. Various types of environmental stress, including temperature extremes and pollution, are the cause. Global warming by 1°C has been reported to have resulted in 0.3 g/cm² of coral calcification in the 40 colonies measured in Thailand, Australia, Hawaii, and Indonesia (PEACE, 2007). Coral bleaching has been observed in many parts of Indonesia, such as in the eastern part of Sumatra and in Java, Bali, and Lombok. In Thousand Islands (north of the Jakarta coast), about 90 to 95 percent of corals 25 meters below the surface were bleached (ADB, 2009).

Mitigation and Adaptation

Fisheries and aquaculture activities make a minor contribution to greenhouse gas emissions when compared to other sectors, and as such they have not been included in the international mechanisms for addressing climate change. However, in 2015 when Indonesia deregistered more than 1,300 fishing vessels in the effort to combat illegal, unreported, and unregulated fishing (IUU-F), 37 percent of fuel consumed in capture fisheries activity was saved. This equates to 1.3 million kiloliters, which amounts to a large reduction in CO_2 emissions. Related to this, since 2014 the government has provided more than 70,000 gas converter kits to small-scale fishermen in an attempt to reduce fuel consumption.

In the long term, Marine Protected Areas (MPAs) will help to address the impacts of climate change by providing a focus for area management and the application of scientific knowledge to reduce stressors and to monitor conditions and trends (see Chapter 11). At present, Indonesia's MPA covers 17.3 million ha at 154 locations. Types of MPAs include marine national parks, marine nature reserves, marine recreational parks, district-based MPAs, and special fishery management areas. Public engagement, especially from traditional "adat" communities, with their centuries-long wisdom in managing coastal resources, is of the utmost importance. The involvement of adat communities has added to the success of community-based management and climate change mitigation of ocean and coastal ecosystems in Sangihe and Talaud Islands, north Sulawesi (Crawford et al., 2004; Doaly, 2016; Wijayanto, 2016). Integration of indigenous-based management of marine resources with modern conservation has been conducted in Raja Ampat as a basic adaptation approach (Boli et al., 2014).

Mangrove rehabilitation is emerging as a solution, serving both as a carbon sink and offering coastal protection and food security. Since 2002, 14.9 million mangrove stalks have been planted by the government, covering 119.3 ha of coastal areas. Meanwhile, private companies, as part of their corporate social responsibility, have participated in a 1-million-hectare mangrove rehabilitation program (Maskur, 2013). Yagasu (an acronym for Yayasan Gajah Sumatera, which literally means Sumatera Elephant Foundation, a local nongovernmental organization [NGO] in North Sumatera), with the support of Livelihood Funds (a European NGO), replanted as many as 18 million mangrove trees in 5,000 ha at the end of 2014, which was reported to be able to sequester the equivalent of more than 2 million tons of CO_2 over 20 years (Livelihood Funds, 2016a). The replantation has been claimed to be able to rebuild fish and shrimp stocks and provide various saleable mangrove products to the local community, adding to its income (Livelihood Funds, 2016b). The government has established Indonesia Coastal Community Schools (CCCs) and the Climate Field School for Fishers to raise awareness in coastal communities, particularly among students and fishermen, and thereby foster enhanced capacities needed to mitigate and adapt to climate change. CCCs have been developed in 39 locations. Resilient coastal villages (47 of them) have also been developed to improve the preparedness of coastal communities to climate change–related disasters.

Faced with difficult choices in shaping the future of people in Indonesia and sustaining their livelihoods, adaptation is viewed as a matter of urgency. Mitigation is a strategic longer-term issue, in which case Indonesia will have to be integrated into the country's adaptation approaches and programs.

Oceans and Climate Change: The Policy Approach

Climate change is actually not a new issue for Indonesian lawmakers. Indonesia ratified the United Nations Framework Convention of Climate Change (UNFCCC) in 1994, and the Kyoto Protocol in 2004. Law 32/2009 on Environmental Protection and Management includes impacts of climate change in evaluating damage to the environment. Similarly, Law Number 32/2014 on Ocean has categorized impacts of climate change (global warming) as the cause of ocean disaster. Thus, taking these laws into account, climate change has actually been given substantial attention. However, climate change has attracted wider attention in Indonesia, especially since the 13th Conference of the Parties to the UNFCCC (CoP-UNFCCC), which was held in Denpasar, Indonesia, in 2007. Following the conference, the government started to initiate some moves to mitigate and adapt to climate change. The following describes some of these initiatives.

One year after the 13th CoP-UNFCCC, President Yudhoyono established a National Council for Climate Change (NCCC), tasking it with coordinating climate change policy making and strengthening Indonesia's position on climate change in international fora. The council is composed of 17 ministers and chaired by the president. There were six working groups under the council, namely those on

Adaptation, Mitigation, Transfer of Technology, Funding, Post-Kyoto 2012, and Forestry and Land Use Conversion. The Adaptation program focuses on agriculture, disaster risk reduction, data dissemination, and establishing an integrated development plan to improve climate resilience. With regard to the ocean, NCCC convinced President Yudhoyono to include in his statement to the UN Climate Summit in New York 2014 the importance of exploring the potential of blue carbon ecosystems as a carbon sink and to maintain the global temperature increase to less than 2°C. However, President Widodo dissolved the NCCC in 2015. The NCCC activities were then contained in the portfolio of the Directorate General of Climate Change (DGCC) in the Ministry of Environment and Forestry (MoEF). The NCCC reported to the president, while the DGCC reports to a sectoral minister in the ministry. This could bring bureaucratic problems, as the DGCC would have to coordinate other ministries. This is not easy in Indonesia due to a conflict of interests among ministries. At the same time, a very sectoral view is extremely likely, while mitigation and adaptation are cross-sectoral and multistakeholder tasks. This could be seen when the DGCC dissolved IPCC Indonesia in 2016 and established a new forum named the Climate Change and Forestry Expert Forum. Established in 2012, IPCC Indonesia was supposed to provide scientific advice to the Ministry of Environment on energy, food security, forestry, and marine and fisheries related to climate change. The merging of the Ministry of Environment and the Ministry of Forestry into one ministry by President Widodo in 2014 is problematic with regard to promoting concerted efforts in addressing climate change issues in Indonesia, let alone its mitigation and adaptation.

In February 2014, the Ministry of National Development Planning issued the National Action Plan on Climate Change Adaptation (RAN-API) of which the main objectives is to implement a sustainable development system that has a high resilience to climate change impacts. The ministry stipulated that RAN-API is not an autonomous document with formal legal power of its own, but instead it is the main input to, and an integral part of, national development planning documents and line-ministries planning. RAN-API is also a reference for local governments in developing local strategy and action plans for climate change adaptation.

There are five areas of RAN-API, namely economic resilience, livelihood resilience, ecosystem resilience, special areas resilience, and supporting systems. In each area there are subsectors, and coastal and small-island areas fall under special areas resilience. The Action Plan for Coastal and Small Islands Area is depicted in Table 7.1. RAN-API is a comprehensive document, designed to be included in short-, medium-, and long-term development plans, and thus it is meant to be implemented by the government at all levels. Kawanishi, Preston, and Ridwan (2016) have reviewed and analyzed the RAN-API. While acknowledging it as an important step taken by the government of Indonesia, they identified some discrepancies due to information or knowledge gaps among stakeholders, and they suggested the importance of communication and outreach in the implementation of the action plan.

Table 7.1. *National Action Plan on Climate Change Adaptation: Target, Strategy, and Action Plan for Coastal and Small-Island Areas (adapted from Ministry of National Development Plan, 2013)*

Target	Strategy	Clusters of Action Plan
• Improved capacity of coastal and small-island communities on climate change issues • Environment and ecosystem are managed and utilized for climate change adaptation • Structural and nonstructural adaptation measures are applied to anticipate the threat of climate change • Climate change adaptation is integrated into the management plan of coastal and small-island areas • Improved climate change adaptation supporting system in coastal and small-island areas	• Achievement of livelihood stability for coastal and small-island communities to the threat of climate change • Improvement of environmental quality in coastal and small-island areas • Implementation of adaptation structure development in coastal and small-island areas • Adjustment of coastal and small-island spatial plans to the threat of climate change • Development and optimization of research and information systems on climate in coastal and small-island areas	• Capacity building for coastal and small-island communities on climate change issues • Management and utilization of environment and ecosystem for climate change adaptation • Application of structural and nonstructural adaptation measures to anticipate the threat of climate change • Integration of adaptation efforts to the management plan of coastal and small-island areas • Improvement on climate change adaptation supporting systems in coastal and small-island areas

Indonesia's Nationally Determined Contribution: A Promise to Keep, a Long Way to Go

The Nationally Determined Contribution (NDC) of Indonesia, in accordance with the Paris Agreement on climate change, was submitted to the UNFCCC in September 2015. According to the NDC, Indonesia is committed to reduce greenhouse gas emissions by 29 percent below business as usual, and an additional 12 percent conditional reduction with sufficient international support, thus making a 41 percent total reduction by 2030. The intended contribution covers five sectors: energy (including transportation); industrial processes and product use; agriculture; land use, land-use change, and forestry; and the waste sector. How the 29 percent and 41 percent targets will be achieved, and what would be the reduction of each sector, are still unclear. Tumiwa and Imelda (2015) have critically reviewed Indonesia's NDC and concluded that it lacks clarity, while the basis for determining the 29 percent reduction below business as usual is questioned. Mitigation in the NDC is focused on these five sectors. Adaptation in the NDC is aimed at reducing risks on all development sectors, including maritime and fisheries, by 2030 through local capacity building, improved knowledge management, convergent policy on climate change adaptation and disaster risks reduction, and application of adaptive technology. Tumiwa and Imelda (2015) argued that the NDC, although rich with background thinking and processes, lacks measures to meet certain conditions as stipulated in the rationale of the document.

Several meetings hosted by the DGCC on the implementation of the NDC have focused only on mitigation in the five sectors mentioned earlier. This suggests that the DGCC is most concerned with how to meet the commitment to reduce the greenhouse gas emissions, while adaptation is left to other sectors to implement. This ignores the fact that coastal and marine ecosystem management is a multisectoral issue requiring multistakeholder collaboration. Integrated marine and coastal zone management, which is mandated to MMAF, has not yet been formulated, or at best has been scattered and not well planned, with activities being done by various institution lacking coordination. Even in MMAF, climate change has not yet been made a high priority and receives little attention.

In relation to the blue carbon ecosystem, Herr and Landis (2016) have reviewed 163 NDCs that have been submitted to UNFCCC and concluded that there were 151 NDCs containing at least one blue carbon ecosystem (seagrass, saltmarshes, or mangroves) and 71 contained all three. They further showed that 59 countries included their coastal ecosystems and coastal zones in adaptation strategies and 28 countries included a reference to coastal wetlands in terms of mitigation. It is interesting to note that Indonesia's NDC was not mentioned in the review. It is possible that Herr and Landis concluded that Indonesia's NDC did not reflect any clear message on ocean and coastal ecosystems, a view that was also shared by Tumiwa and Imelda (2015). Although references were made to the nature of the archipelagic state in Indonesia's NDC, the critical role of coastal ecosystems and the ocean in climate change was largely overlooked. Again, it seems that mainstreaming the marine and coastal sector in climate change mitigation and adaptation in Indonesia will take some time before more effective actions can be taken. Revision of Indonesia's NDC could be the first step in doing so.

Conclusion

Being an archipelagic state, Indonesia faces various challenges from climate change, especially due to its geographic position on the equatorial line between two continents and two oceans. Climate change is having serious impacts on Indonesia's coastal areas, with evidence of this identified in many parts of the country. Ecosystem management, food security, and maintaining livelihoods in coastal areas are the most serious issues, potentially bringing significant social and economic loss. Climate change has been an issue in Indonesia since the 1990s, and many policies have been made available since then. However, those policies are largely lacking in implementation. Different from those in the forestry sector, policies on ocean and coastal areas are very limited or not given sufficient attention. This is primarily due to the fact that ocean governance in Indonesia is still in its infancy. The situation worsens when government organizations are changed, as they often are. Furthermore, not all government levels have the same vision and approaches regarding climate change, especially with regard to oceans. Inconsistencies in laws and regulations have led to additional problems.

Improvement on Indonesia's climate change–related policies, notably those affecting its coasts and seas, will require strong political commitment and leadership within the government. Collaboration among government institutions is a must, while the role of the CMMA should be strengthened. It is also important that the DGCC be more open to ideas and suggestions from other sectors and stakeholders, especially because the mandate for climate change is given to them. The DGCC is the UNFCCC focal point in Indonesia and should function as a bridge to international fora. Furthermore, partnerships on oceans and climate change with island states – in particular, small-island and developing states – is unquestionably vital to helping Indonesia develop more effective policies for both mitigating the impacts on its own and the world's oceans and in developing effective forms of adaptation in the future.

References

Allison, E. H., Perry, A. L., Badjeck, M. C., et al. (2009). Vulnerability of national economies to the impacts of climate change on fisheries. *Fish and Fisheries*, **10**, 173–96.

Ambari, M. (2017). *Garam Nasional Gagal Produksi Sepanjang 2016, Kenapa Bisa Terjadi* (*Why 2016 Salt Production Dropped*). Available at www.mongabay.co.id/2017/01/16/garam-nasional-gagal-produksi-sepanjang-2016-kenapa-bisa-terjadi/

Asian Development Bank. (2009). *The Economics of Climate Change in Southeast Asia: A Regional Review*. Manila, The Philippines, Asian Development Bank, April. Available at: www.adb.org/Documents/Books/Economics-Climate-Change SEA/PDF/Economics-Climate-Change.pdf

Bateman, S., Ho, J., and Chan, J. (2009). *Good Order at Sea in Southeast Asia: RSIS Policy Paper*. Singapore, Nanyang Technological University.

Bene, C., Arthur, R., Norbury, H., et al. (2016). Contribution of fisheries and aquaculture to food security and poverty reduction: Assessing the current evidence. *World Development*, **79**, 177–96.

Boer, R., and Faqih, A. (2005). *Current and Future Rainfall Variability in Indonesia*. Technical Report of the AIACC Project Integrated Assessment of Climate Change Impacts, Adaptation and Vulnerability in Watershed Areas and Communities in Southeast Asia (AIACC AS21): Indonesia Component. Assessments of Impacts and Adaptations to Climate Change, Washington, D.C.

Boer, R., Sutikno, A. Faqih, B. D., Dasanto, and Rakhman, A. (2007). *Climate Change Detection and Socio-Economic Impact*. Project Report Indonesia Geophysics and Meteorology Agency and Bogor Agricultural University, Bogor.

Boli, P., Yulianda, F., Soedharma, D., Damar, D., and Kinseng, R. (2014). Integration of marine resources management based customary into modern conservation management in Raja Ampat, Indonesia. *International Journal of Sciences: Basic and Applied Research*, **3**(2), 280–91.

Brugere, C., and De Young, C. (2015). *Assessing Climate Change Vulnerability in Fisheries and Aquaculture: Available Methodologies and Their Relevance for the Sector*. FAO Fisheries and Aquaculture Technical Paper No. 597. Rome, Italy.

Butcher, J. G., and Elson, R. E. (2017). *Sovereignty and the Sea. How Indonesia Became an Archipelagic State*. Singapore, National University Press.

Cheung, W. L., Lam V. W. Y., Sarmiento, J. L., et al. (2010). Large scale redistribution of maximum fisheries catch potential in the global ocean under climate change. *Global Change Biology*, **16**(1), 25–34.

Cohrane, K., De Young, C., Soto, D., and Bahri, T. (2009). *Climate Change Implication for Fisheries and Aquaculture: Overview of Current Scientific Knowledge*. FAO Fisheries Technical Paper No 530. Rome, FAO.

Crawford, B. R., Siahainenia, A., Rotinsulu, C., and Sukmara, A. (2004), Compliance and enforcement of community-based coastal resource management regulations in North Sulawesi, Indonesia. *Journal Coastal Management*, **32**, 39–50.

Dahuri, R. (2011) *Laut dan Daya Saing* (*Ocean and Competitiveness*). Available at dahuri.wordpress.com/2011/10/22/laut-dan-daya-saing

Doaly, T. (2016). *Adaptasi Perubahan Iklim Berbasis Kearifan Lokal. Seperti Apa?* (*Climate Change Adaptation Based on Local Wisdom. How Is It?*). Available at www.mongabay.co.id/2016/10/17/adaptasi-perubahan-iklim-berbasis-kearifan-lokal-seperti-apa/

Ekawati, J. D. (2016). *Indonesia's Global Maritime Axis Askew*. Available at www.internationalaffairs.org.au/australianoutlook/indonesias-global-maritime-axis-askew/

FAO. (2016). *The State of World Fisheries and Aquaculture 2016. Contributing to Food Security and Nutrition for All*. Rome, FAO.

Gaol, J. L., and Nababan, B. (2012). Climate change impact on Indonesian fisheries. In J. Griffiths, C. Rowlands, and M. Witthaus, eds., *Climate Exchange*. Leicester, UK: Tudor Rose, pp. 72–74.

Gaworecki, M. (2016). *Economic Impacts of Climate Change on Global Fisheries Could Be Worse Than We Thought*. Available at https://news.mongabay.com/2016/10/economic-impacts-of-climate-change-on-global-fisheries-could-be-worse-than-we-thought/

Gray, J. S. (1997). Marine biodiversity: Patterns, threats and conservation needs. *Biodiversity and Conservation*, **6**, 153–75.

Herr, D. and Landis, E. (2016). Coastal blue carbon ecosystems. Opportunities for Nationally Determined Contributions. Policy Brief. Gland. Switzerland. IUCN and Washington DC, USA, TNC.

Huffard, C. L., Erdmann, M. V., and Gunawan, T. R. P. eds. (2012). *Geographic Priorities for Marine Biodiversity Conservation in Indonesia*. Ministry of Marine Affairs and Fisheries and Marine Protected Areas Governance Program. Jakarta, Indonesia.

Huelsenbeck, M. (2012). *Ocean-Based Food Security Threatened in a High CO_2 World: A Ranking of Nations' Vulnerability to Climate Change and Ocean Acidification*. Washington, DC, Oceana.

Kawanishi, M., Preston, B. L., and Ridwan, N. A. (2016). Evaluation of national adaptation planning: A case study in Indonesia. In S. Kaneko and M. Kawanishi eds., *Climate Change Policies and Challenges in Indonesia*. Japan, Springer, 85–107. doi 10.1087/978-4-431-55994-8

Kurniawan, T., and Azizi, A. (2012), Dampak perubahan iklim terhadap petani tambak garam di Sampang dan Sumenep (Climate change impact on salt pond farmers in Sampang and Sumenep districts). *Jurnal Masyarakat & Budaya*, **14**(3), 499–518.

Livelihood Funds. (2016a). *Indonesia: Mangroves Revitalizing Coastal Villages with Fishery & New Businesses.* Available at www.livelihoods.eu/projects/yagasu-indonesia/

Livelihood Funds. (2016b). *Indonesia: Mangroves Spurring New Businesses.* Available at www.livelihoods.eu/indonesia-mangroves-spurring-new-businesses/

Maskur, F. (2013). Ekosistem Mangrove: Rehabilitasi Kawasan 1 Juta Hektar, Pemerintah Gandeng Swasta (*Mangrove Ecosystem: Government Engaged Private Company for 1 Million Hectare Mangrove Rehabilitation*). Available at http://industri.bisnis.com/read/20130606/99/143121/ekosistem-mangrove-rehabilitasi-kawasan-1-juta-hektar-pemerintah-gandeng-swasta

Ministry of Defence, Singapore. (2016). *Fact Sheet: The Malacca Straits Patrol.* Available at www.mindef.gov.sg/imindef/press.../21apr15_fs.html

Ministry of Environment, Indonesia. (2010). *Indonesia Second National Communication Under the United Nations Framework Convention on Climate Change (UNFCCC)*, Indonesia

Ministry of National Development Plan, Indonesia. (2013). *National Action Plan for Climate Change Adaptation – Synthesis Report*, Indonesia.

Noviyanti, S. (2016). *Ketika Garam Tak Cuma Bikin Asin Makanan* (When salt is not only to make food salty). Available at https://ekonomi.kompas.com/read/2016/09/29/183336826/ketika.garam.tak.cuma.bikin.asin.makanan

Pauli, G. (2010). *The Blue Economy: 10 years, 100 Innovations, 100 million jobs.* Taos, NM, Paradigm Publication.

PEACE. (2007). *Indonesia and Climate Charge: Current Status and Policies.* Jakarta, The World Bank.

Prabowo, H. H., and Salahudin, M. (2016). Potensi tenggelamnya pulau-pulau kecil terluar wilayah NKRI (Threats drowning of NKRI outermost small islands). *Jurnal Geologi Kelautan*, **14**(2), 112–5.

Poernomo, A., Sulistiyo, B., and Pranowo, W. S. (2014). *Kenaikan Muka Air Laut terhadap Infrastruktur Perikanan Tangkap dan Perikanan Budidaya (Threats of Sea Level Rise on Fisheries and Aquaculture Infrastructures).* Jakarta, Agency for Marine and Fisheries Research and Development.

Rataq, M. A. (2007). *Climate Variability and Climate Change Scenario in Indonesia.* Paper presented at the Workshop on Issues on Climate Change and Its Connection with National Development and Planning, October 1, Jakarta.

Rindayati, H., Susilowati, I., and Hendrarto, B. (2013). *Adaptasi nelayan perikanan tangkap pulau Moro Karimun kepulauan Riau terhadap perubahan iklim* (*Climate Change Adaptation of Fisherman in Moro Island, Karimun, Riau Islands*). Proceedings of the National Seminar on the Management of Natural Resources and Environment, Semarang, Central Java, August 27, 2013.

Sudibyakto (2010). *Pesisir sebagai Daerah Terparah Perubahan Iklim (Most Severe Impact of Climate Change is in Coastal Area).* Available at www.ugm.ac.id/id/newsPdf/2121-dr.sudibyakto:.pesisir.sebagai.daerah.terparah.perubahan.iklim

Trobos (2017). *Anomali Cuaca Ganggu Produksi Garam (Weather Anomaly Disturbs Salt Production).* Available at www.trobos.com/detail-berita/2017/03/15/14/8598/anomali-cuaca-ganggu-produksi-garam

Tumiwa, F., and Imelda, H. (2015). *A Brief Analysis of Indonesia's Intended Nationally Determined Contribution (INDC)*. Jakarta, Institutes for Essential Services Reform (IESR).

Vantier, L., Wilkinson, C., Lawrence, D., and Souter, D., eds. (2005). *Indonesian Seas, GIWA Regional Assessment 57*. University of Kalmar, Kalmar, Sweden.

Wahyono, A., Imron, M., Nadzir, I., and Haryani, N. S. (2012). Kerentanan penambak garam akibat perubahan musim hujan di desa Randutatah Kabupaten Probolinggo (The vulnerability of the salt farmer due to change of rainy season in Randutatah Village Probolinggo District), *Jurnal Masyarakat & Budaya*, **14**(3), 519–40.

Wahyudi, D. P. (2010). *Pola Adaptasi Nelayan Terhadap Perubahan Iklim Dan Cuaca Pada Perikanan Payang Di Palabuhanratu, Sukabumi, Jawa Barat (Adaptation Pattern of Climate Change on Payang Fisheries in Plabuhan Ratu, West Java)*. Undergraduate thesis, Penerbit ITB, Bandung.

Wibowo, A., and Satria, A. (2015). Strategi Adaptasi Nelayan di Pulau-Pulau Kecil terhadap Dampak Perubahan Iklim (Adaptation strategies in small islands to the impacts of climate change – A case study in Pulau Panjang Village, Sub District, Natuna Regency, Riau Island), *Sodality: Jurnal Sosiologi Pedesaan*, August, 107–24.

Wijayanto, A. (2016). *Mengenalkan Pengelolaan Pesisir dan Laut Berbasis Masyarakat di Kepulauan Sangihe (Introducing Community Based Management of Ocean and Coastal Ecosystem in Sangihe Island)*. Available at www.mongabay .co.id/2016/06/04/mengenalkan-pengelolaan-pesisir-dan-laut-berbasis-masyarakat-di-kepulauan-sangihe/

Yusuf, A. A., and Fransisco, H. (2009). *Climate Change Vulnerability Mapping for Southeast Asia*. Singapore, EEPSEA.

Zikra, M., Suntoyo and Lukijanto (2015). Climate change impacts on Indonesian coastal areas. *Procedia Earth and Planetary Science*, **14**(2015), 57–63.

8

Politics of Rising Tides: Governments and Nongovernmental Organizations in Small-Island Developing States

IAN MCGREGOR AND HILARY YERBURY

Introduction

Much of the literature concerned with the politics of rising seas is about the involvement of states and others in the processes of the United Nations Framework Convention on Climate Change (UNFCCC) and its annual Conference of the Parties (COP). Although nonstate actors, including international nongovernmental organizations (NGOs), are acknowledged as being part of these processes, it has often been claimed that the voices of local NGOs are absent from debates on vulnerability and resilience to climate change, apparently as though there were no debates taking place elsewhere. This chapter addresses this absence by explicitly acknowledging that rising sea levels and the consequent changes to local lives are the focus of discussions beyond the UNFCCC. It shows how the politics of vulnerability are clear in messages around mitigation of climate change, where the audiences are foreign governments, institutions at a transnational level, and activists and interested citizens from around the world. It explores the politics of resilience in messages about adaptation to climate change disseminated by and through local NGOs, sometimes in collaboration with international NGOs, where the key audience is the population of island states.

The focus here is on small-island developing states (SIDS) in the Pacific (see also Chapters 3, 6, and 18), the engagement of their local NGOs with issues of sea-level rise and their involvement with other organizations and alliances, including international NGOs, specifically the Pacific Calling Partnership (PCP) and the Climate Action Network (CAN). CAN is a worldwide organization that aims to promote government and individual actions on climate change through information flow and the coordination of strategies on climate issues from local to international levels. It operates at the regional level through hubs such as the Pacific Island Climate Action Network (PICAN) and by coordinating local NGOs through organizations such the Kiribati Climate Action Network (KiriCAN) and the Tuvalu Climate Action network (TuCAN). Pacific Calling Partnership is a program of the Edmund Rice Centre for Justice and Community Education, an Australian development organization using community education to change the world, beginning with awareness raising and leading to advocacy and social action. PCP's focus is Pacific peoples affected by climate change, particularly in Kiribati and Tuvalu.

To understand how the politics of rising tides plays out, this chapter uses the framing of narratives to show and reveal messages communicated by the multiplicity of voices in the debates and actions. An analysis of the narratives of Pacific Island states shows that different groups express different understandings of what rising tides mean, that these narratives have specific audiences, and that they tend to exist in defined contexts, such as COP meetings or local settings (Fairclough, 2003). Unlike the perception of the outcomes of UNFCCC COP, where consensus is the expected outcome, in the broader context of local engagement with rising tides at a local level, it is apparent that there can be no consensus narrative. The messages embedded in these narratives of vulnerability and resilience are diverse, with each competing for the attention of its audience and evolving as new voices enter the debate and new technologies are used.

Vulnerability and Resilience

Vulnerability and resilience are key terms in debates on sea-level rise in SIDS. The work of the Intergovernmental Panel on Climate Change (IPCC) has become an authoritative source of definitions within the consensus decision-making approach of the UNFCCC (Adger, 2006: 273). In 2001, vulnerability was seen as "the degree to which a system is susceptible to, or unable to cope with, adverse effects of climate change, including climate variability and extremes. Vulnerability is a function of the character, magnitude, and rate of climate change and variation to which a system is exposed, its sensitivity and its adaptive capacity" (IPCC, 2001). This authoritative definition has been modified, with each IPCC reporting cycle showing slight variations, although always with a relationship between vulnerability and resilience. This is mediated through conceptions of adaptation and has more recently incorporated vulnerability resulting from loss and damage, which, it is recognized, cannot be mediated by adaptation but instead requires an approach based on insurance.

Otto et al. recently reviewed the extensive literature using the terminology of "social vulnerability" in order to highlight that vulnerability is more than the consequence of susceptibility of a system, being "heavily shaped by social, demographic, and institutional factors such as gender, age, culture, education and ethnicity" (Otto et al., 2017: 1658). The narrative of the Suva Declaration (Pacific Island Development Forum [PIDF], 2015) is one that uses the concept of *social vulnerability*, an approach that shifts the focus of concern to a local level and to the human scale and that introduces a sense of resilience. Here, although the populations of members of the PIDF may be victims of climate change, including sea-level rise, they are not powerless (Denton, 2017: 68). The declaration goes beyond the accepted narrative of *disproportionate impact*, an accepted assumption in understandings of vulnerability, to introduce *human rights violations* and *inequality* and *discrimination* into the debate.

Sea-level rise is linked to vulnerability in several ways. Although there are still major uncertainties over the extent of future sea-level rises under various levels of global temperature increase, the IPCC Fifth Assessment Report (2014b:11) shows that under the higher levels of temperature increase, sea-level rise of almost one meter may occur by 2100. Many of the SIDS are low lying. For example, in Tuvalu the average height of the land above sea level is less than two meters, and the country is therefore extremely vulnerable to the impact of rising sea levels. Here vulnerability is directly related to inundation and loss of land. As Kiribati states in their Intended Nationally Determined Contribution (INDC) to address climate change, prepared for COP21, "In the long term, the most serious concern is that sea-level rise will threaten the very existence of Kiribati as a nation" (Republic of Kiribati, 2015: 15). Sea-level rise also brings with it the threat of erosion and the consequent loss of land, both through the regular actions of the tides and through cyclones and storm surges. Inundation of the land is also a feature of these weather events, giving a foretaste of the predicted longer-term effects of sea-level rise. A third aspect of sea-level rise is salination of the soil and water, affecting food production and even the habitability of some islands. All three aspects of sea-level rise – inundation, erosion, and salination – can disrupt daily lives, affecting the societies and economies of these islands.

Resilience can also be conceptualized in several ways. As noted earlier, it is often linked to the notion of adaptive capacity, assuming that the system will be able to manage the impacts and keep working in more or less the same way (Gallopín, 2006). This has been the approach emerging from the IPCC, where "human systems" have "adaptive capacity," "the ability of a system to adjust to climate change (including climate variability and extremes) to moderate potential damages, to take advantage of opportunities, or to cope with the consequences" (IPCC, 2001). Cannon and Müller-Mahn express concern that the increasing linkage of resilience to vulnerability will make resilience part of the scientific debate, whereas in their view it is a concept from the social sciences and its concern for politics and economics should not be subsumed into technical concerns (2010: 633). Janssen et al. (2006), however, argue on the basis of an extensive bibliometric study, influenced by the IPCC reports, that whereas vulnerability is linked with adaptive capacity in the literature, resilience is separate.

Placing the concept of resilience firmly in the social sciences, Folke states that "[r]esilience is an approach, a way of thinking" (Folke, 2006: 260). He argues that in any consideration of resilience, social capital, including trust and social networks, is important, as are "social memory," including the experience of dealing with change, and knowledge systems, including a variety of types of knowledge (Folke, 2006: 259–62).

Addressing an International Audience

The successful use of a particular narrative of vulnerability by SIDS is illustrated by the concept's use over many decades within the global climate change policy process. The vulnerability of SIDS to rising sea levels was scientifically recognized early in the international climate process, with the United Nations Environment Programme (UNEP) producing a report in 1989 on "Assessing the Vulnerability to Sea Level

Rise" (UNEP, 1989). Vulnerability to sea-level rise was reflected in the first IPCC report (IPCC, 1990), and it has been acknowledged in a series of international agreements to address global warming, including the 1992 UNFCCC, the 1997 Kyoto Protocol, the 2010 Cancun Agreements (formalizing the 2009 Copenhagen Accord), and most recently, the 2015 Paris Agreement. However, it must be acknowledged that this recognition of vulnerability has not yet been matched by an effective worldwide agreement to cut greenhouse gas emissions to a level necessary to curtail temperature rise to 1.5°C, as called for in the Paris Agreement.

The successes in gaining an audience for their narratives of vulnerability have not been achieved by Pacific Islands in isolation. They are involved in several alliances and coalitions intended to give them greater influence in the UNFCCC processes and elsewhere. These include established coalitions, such as the Alliance of Small Island States (AOSIS), which has given the small states a "voice in the political arena" (Jaschik, 2014: 287) so that they "box way above their weight" (Betzold, 2010: 142) in ensuring that their concerns are on UN Climate Conference agendas and the Climate Vulnerable Forum, whose very name proclaims the way its members should be regarded. The formation of the Climate Vulnerable Forum, led by Maldives, in 2009, and its Male' Declaration made it possible for new voices to enter the international debates and be heard at COP15 in Copenhagen, where delegates delivered a message charging members of the developed world with having caused the problem of global warming and calling on them to take responsibility and take active steps to fix the problem. Before COP15, Tuvalu was virtually unknown on the world stage, even within global climate change politics, but the term *Tuvalu* entered the narrative of vulnerability, being mentioned in 542 stories related to climate change in December 2009, with 173 of these containing either "vulnerable" or "vulnerability" (according to a search on the global news database Factiva). In the same time period, posts in CAN International's email list-serv, CAN-talk, mentioned Tuvalu 532 times. At the same COP, Kiribati also rose to prominence, through what Webber (2013: 2728) refers to as an event that was "'scripted' and 'rehearsed' – 'performances' of vulnerability," with dances and traditional costumes.

Narratives of vulnerability must be continually performed for these small-island states to maintain the attention of their audiences. In the context of the UNFCCC, these states use their INDCs, which are statements from the government of a state, prepared in consultation with civil society, including NGOs. These statements have two audiences: the international audience within the UNFCCC and the local audience through national policy development and implementation. INDCs prepared for COP21 in Paris perpetuate a narrative that describes these islands as *low-lying, isolated* and *vulnerable, at risk from sea-level rise,* which is a *catastrophe* not of their making and for which they are *in no way responsible,* as their contribution to global warming is *insignificant,* their "emissions per capita being amongst the lowest in the world" (Republic of Kiribati, 2015). They further reinforce the narrative of vulnerability through references to the small populations of these states and their reliance on external funding to support strategies of adaptation.

The messages inherent in the INDCs and in the statements about the SIDS' plans for the future are, however, more complex than this. They go beyond the technical aspects of erosion and inundation, and in different ways they introduce the consequences of sea-level rise. Kiribati refers to "its highly vulnerable socio-economic and geographical situation" (Republic of Kiribati, 2015:15), noting that the state "has a right to develop its economy and improve the well-being of its population" and indicating its need for financial aid. The Republic of the Marshall Islands sees its people as being among the most vulnerable in the world to the impacts of climate change, which "inflict damage and impose substantial costs" and which threaten the livelihoods of communities, undermine food and water security, and put health at risk. Similarly, it "recognizes that it has a role to play in the global effort to combat climate change," even though its greenhouse gas emissions are "negligible on a global scale" (Republic of the Marshall Islands 2015 INDC). Tuvalu states that sea-level rise exacerbates existing cultural and socioeconomic vulnerabilities, which could "threaten the security of the nation" (Government of Tuvalu 2015 INDC).

The complexity of these narratives can be seen in the recent proliferation of coalitions, each with its own focus. Members of the Coalition of Low Lying Atoll Nations on Climate Change (CANCC) formed in 2014, which includes the nations of Kiribati, Maldives, Marshall Islands, Tokelau, and Tuvalu, state that they bring "a whole new meaning to human rights and the right to a secure future" (CANCC, 2014). Their priority is the resourcing of measures to protect the physical environment and to enhance the capacity of their populations through programs of education and awareness raising (CANCC, 2014). The High Ambition Coalition, created at COP21 in Paris in 2015, has been spearheaded by the Marshall Islands, a very small nation but one that is seen to have "moral authority and thought leadership" (Woodroofe, 2016). The rhetoric of ambitious action here is embedded in the group's title, and they have agreed to tackle some of the more intractable problems, such as taking the issue of shipping emissions to the International Maritime Organization, a significant step for countries like the Marshall Islands, which benefit from a flag of convenience shipping registry (see Chapter 2).

The V20 group, comprising the finance ministers of 20 members of the Climate Vulnerable Forum (including Kiribati, Tuvalu, and the Republic of Marshall Islands), was established in Lima, Peru, in 2015 at a meeting in conjunction with the 2015 annual meetings of the World Bank Group and International Monetary Fund. It similarly frames its narratives in terms of ambitious action in the face of dangers, including the health and economic impacts of climate change, seen to encompass human rights issues, as it seeks financial support for economic development and to cover loss and damage from severe weather events. This group took its message to the G20 meeting in April 2017, echoing the narrative of ambitious action expected from G20 members and reinforcing the difference in size and strength through the phrase "David meets Goliath" (Hansen, 2017).

In spite of similarities in the narratives presented to international audiences, it would be a mistake to assume that these Pacific Islands have identical narratives. The leaders of Kiribati, Tuvalu, and the Marshall Islands have presented conflicting narratives about their long-term futures, with Anote Tong, former president of Kiribati, "plan[ning] for the worst and hop[ing] for the best" (Weiss in Barnett, 2017: 8), whereas the former foreign minister of the Marshall Islands, Tony de Brum, said that "[w]e will operate on the basis that we can in fact help to prevent this from happening" (Mathieson in Barnett, 2017: 9). The prime minister of Tuvalu, Enele Sopoaga, stated that "[w]e do not want to move ... Our lives and culture are based on our continued existence on the islands of Tuvalu. We will survive" (Sopoaga in Barnett, 2017: 9).

Networking and NGOs

Networks such as Climate Action Network and NGOs such as Pacific Calling Partnership have complex relationships with the governments of islands in the Pacific and with NGOs in these countries. They are involved in the narratives with international audiences and narratives with local audiences. KiriCAN and TuCAN are members of CAN and its regional node, PICAN. Both CAN and PCP are enthusiastic users of online resources, especially Facebook, supporting the claim that NGOs concerned with climate change and with the environment are "champions of online climate communication" (Schäfer, 2012: 530–31). They use technologies to create their networks, thereby disseminating information, including to the media, increasing support for climate change action, and even mobilizing local citizens to take action. It is in this context that the narratives they use are considered here.

The Facebook page for PICAN, which provides information and links to resources, demonstrates that the narratives evident through online posts made in 2016–2017 are institutional, representing the UNFCCC agenda and promoting awareness raising and training to participate in debates at this level, with an emphasis on preparation for involvement in COP23. This is understandable because Fiji, a key country for PICAN, would hold the presidency for COP23. Unsurprisingly, given the strategy statement of PICAN, which includes an emphasis on fundraising, the Green Climate Fund and its application process is also given prominence, and Tuvalu is applauded for its success in receiving money through this fund. The narrative also demonstrates the significant engagement of women in the discussions and actions by them, with the emphasis on capacity building through women.

The PCP has a much higher profile online than PICAN, with more frequent postings and evidence of face-to-face work with local activists. The Kiribati-Tuvalu-Australia Exchange Program (KATEP) is a key focus of the work of PCP. It is a training program held in Australia and the Pacific that develops the climate change advocacy skills of emerging young leaders from Kiribati and Tuvalu. This is done through workshops and practical experience in Australia. This is an example of how PCP works in solidarity with the people of Kiribati and Tuvalu to build capacity to advocate for action on climate change and to raise awareness in Australia of how climate change is affecting the Pacific Islands. Its web page emphasizes its collaborative role in supporting communities in the Pacific to make their voices heard in climate change discussions. Its Facebook page shows this support in action, with photos and reports from Kiribati and Tuvalu by activists closely linked to PCP and its work, as well as from others undertaking capacity-building actions in the islands. The PCP's Facebook page often contains links to opinion pieces by prominent people, especially islanders. The narrative does not ignore institutional links but includes those relevant to the people of Kiribati and Tuvalu, such as the United Nations website's pages for International Day for the Conservation of the Mangrove Ecosystem and select scientific and other reports.

Involving the Local Population

There is no single narrative at a local level. The politics of rising tides at the local level means that government processes, NGOs, churches, local communities, traditional leaders, women, youth, and so on each have their own narrative that can exist independently of other local narratives. Cultures around belief systems and authority affect involvement in decision-making processes and in implementation of local decisions and national polices, reinforcing the existence of these separate narratives. Local knowledge is important, specifically in understanding engagement with issues of sea-level rise. For example, Christian churches in the Pacific Islands are highly influential, with more than 95 percent of the population being members, according to recent censuses in Tuvalu and Kiribati. Some Pacific Islanders do not believe in the narrative of rising tides because of their understanding of God's promise to Noah, documented in the Book of Genesis, that God would never again flood the Earth (Paton and Fairbairn-Dunlop, 2010; Donner, 2011). Countering this narrative, the Reverend Tafue Lusama, general secretary of the Congregational Christian Church of Tuvalu (Ekalesia Kelisiano Tuvalu – EKT) and chair of TuCAN, said in the context of Cyclone Pam: "Climate change represents a spiritual as well as a physical crisis for our people. We desperately need to educate communities about the fact that God has not abandoned us; climate change is caused by humans and requires a human response" (Uniting Church of Australia, 2015).

Another important influence on attitudes to rising sea levels and consequent actions is the politics of local decision making, which may mean that women and younger men (under the age of 50) are excluded from active participation in the decision-making process (Paton and Fairbairn-Dunlop, 2010). Island councils may be subordinated to higher-level authorities (Nunn et al., 2014), regardless of the legal status of this council of elders (Richardson, 2009). This assertion cannot be generalized across island states, or even within a given state, as the customs surrounding traditional leaders and their power differ from state to state and also may vary from island to island within a country (Nunn et al., 2014).

The traditional knowledge that has sustained life on the islands since time immemorial has recently been acknowledged as an integral part of the process of adapting to climate change (IPCC, 2014a: 26). In Tuvalu, for example, traditional forecasting techniques for anticipating extreme weather events are still relied on, and traditional knowledge has been adapted to match contemporary conditions – for example, when women from the Nanumea community on Funafuti bury germinating nuts and taro in plastic drums to keep them safe from rising saltwater (Nakashima, 2012: 93–95). Yet Maria Tiimon, Pacific Outreach Officer at the Edmund Rice Centre in Sydney and a native of Kiribati, noted that an important aspect of specialist traditional knowledge in Kiribati is that it is not part of a narrative for the community but remains secret in the custody of the family who have, over generations, developed this specialist expertise (Teaero, 2003). Such knowledge is lost if there is no one to pass it on to (personal communication, May 25, 2017).

The voices of scientists are rarely heard in local narratives. This may be because the islands lack the educational systems and sound knowledge-sharing infrastructure necessary to engage in climate change discussions in a nuanced way (Abeysinghe and Huq, 2016: 198). An exception may be seen in a book by i-Kiribati man, Riibeta Abeta, hailed as the "first i-Kiribati international publication on climate change by a single author" (Office of the President, Republic of Kiribati, 2014). Given the author's involvement at the time in the Ministry of Environment, Lands and Agricultural Development, it is clear that the work contributed to local debates.

NGOs may engage in so-called citizen science, a useful tool for collecting data and for awareness raising (Johnson et al., 2014). In Tuvalu, the NGO Alofa Tuvalu has used islanders to collect scientific data, both for local projects and for international projects. Further, Kelman (2010: 607–9) points out that peer-reviewed scientific journals do draw on SIDS perspectives and that SIDS scientists have played a crucial role in establishing and contributing to organizations such as the Secretariat of the Pacific Regional Environment Program (SPREP). People in government roles are involved in regional projects funded through organizations such as the Asian Development Bank and contribute their knowledge and expertise to a wider audience through the reports of these projects. However, there are obstacles to sharing scientific knowledge. The Asia Pacific Adaptation Network (APAN) notes the impact of "differences in language, perceptions and interests," with constraints on capacity and knowledge that are compounded by differences in the scale of member countries, ranging from Bangladesh and Indonesia to tiny Tuvalu (APAN, 2017). Occasionally, competing narratives of the government itself and its bureaucracy may arise as a barrier. An anonymous informant noted to us that "the work that I have published has not been utilized much. As the projects are mostly planned and executed by top down centrally controlled institutions, the expertise outside the government institutions [is] seldom utilized."

NGOs are important in the development of local narratives. Both Kiribati and Tuvalu have an active NGO network engaged in climate change actions. KiriCAN and TuCAN have the challenging tasks of both being the voices of NGOs in external fora and coordinating actions within these island states. Although the narratives of the NGOs may be different from those of the government, it would be naïve to claim that they are not significantly influenced by the wider context of the scientific community and UNFCCC debates (cf Rudiak-Gould, 2011). Although projects may be part of a state's plan for adaptation, and they may be funded as such by an external donor, the narrative is likely to be distinct from that of the government and the INDC. The narratives of local NGOs can be seen as narratives of resilience, tending to focus on improving conditions for communities.

Food security has been a significant narrative across the islands of the Pacific. In 2014, the Tuvalu Council of Women ran workshops on home gardening, and in 2015 it ran a competition on food security based on the knowledge and skills acquired in that workshop. They have also been involved in other projects to help villagers develop skills and expertise in growing staple foods such as *pulaka* (swamp yams) in pots, which help to protect them from salination and the effects of inundation from storm surge. In Tokelau, a non–self-governing territory of New Zealand, as in other parts of the Pacific, the development of keyhole gardens is being funded as a strategy to improve food security. This project, supported by the Fatupaepae (traditional local council of women) in the three villages of Tokelau, also engages youth.

Freshwater and sanitation constitute another narrative found in the work of NGOs. In Kiribati, the Kiribati Women's Centre (Aia Maea Ainen Kiribati – AMAK), the peak body for women's organizations, has been involved in a waste management project, the longer-term aim of which was to improve health outcomes and to return cultivable land to production. AMAK and other NGOs have been engaged in the planting of mangroves as part of a soft-barrier plan to prevent erosion by the sea. The Tobwaraoi Community Nanikaai (2017) established a committee to clean up the beach, which was being used as a public toilet, as well as a dump for rubbish, and to install a water tank. In Abaiang, KiLGA, the Kiribati Local Government Association (2016), an affiliate of KiriCAN, was involved in the installation of rainwater harvesting tanks to help solve problems of access to clean drinking water caused by the rising sea level, as well as by the pressure of population.

Globalizing and Localizing Ethical Practices

The politics of rising tides emerges from narratives described earlier. These narratives, which have changed over time, show messages aimed at a variety of audiences. The proliferation of coalitions involving small-island states, especially in the lead-up to COP21 in Paris, shows the complexity of the issue of rising tides and the importance of having access to a range of key audiences who have power in different aspects of global governance. At the international level, the small-island states have to acknowledge that their success in gaining acceptance of the statement at COP21 in Paris to hold global temperature rise to well below 2°C above preindustrial levels has not led to policies sufficient to prevent some currently inhabited atolls of Kiribati or Tuvalu from disappearing in the next 50 years. This is leading to the emergence of other narratives. The emphasis on isolation and distance from centers of power, which for so long set these islands aside and helped to create their vulnerability, is being replaced by the idea that they are the forerunners *on the frontline* of a catastrophe that will affect everyone in due course (CANCC, 2014). This is the *global scale* that the Republic of the Marshall Islands (2015) refers to in its INDC. Implicit in this narrative is the reminder that the threats faced by the populations of these islands will sooner or later affect the citizens of some of the heavy emitters.

This narrative is closely linked to the use of the *global citizen* narrative – that everyone has a responsibility toward the other and to the agreed processes that make a common or shared life possible. This narrative brings with it ideas of *global effort* in minimizing greenhouse gas emissions. It is this narrative that is in play for the High Ambition Coalition, whose message is clearly that many emitting countries are not ambitious enough in their targets and that they are not putting in enough effort. What that effort might be and how ambition might be measured is a moot point. Some members of the High Ambition Coalition have been charged with hypocrisy because of their continued high level of reliance on fossil fuels, for example, as they transition to other forms of energy. Yet the point remains that small-island nations such as the Marshall Islands, through the High Ambition Coalition for Shipping, are taking steps that others have so far found too difficult (Climate Policy Observer, 2017).

This sense of *high ambition* is matched with the notion of *moral leadership*, which was introduced by the CVF in its Male' Declaration at the same time as other terms in the narrative of vulnerability. *Moral leadership* does not seem to have been adopted as widely as other terms in the debates on rising tides, although Kiribati refers to its own *moral imperative* to contribute to limiting global temperature rise in its INDC 2015.

This proliferation of messages at the level of international institutions now approaches the multiplicity of narratives at the local level, but the strength of these narratives intended for international audiences could prevent local narratives of resilience and resourcefulness from being heard (Farbotko, 2005: 289). The question that might be asked about local narratives not being heard is whether they are narratives aimed at outsiders; if local narratives aimed at outsiders are not being heard, that would indeed be a cause for concern.

The concern here, then, is with the distinctions in content, purpose, and audience for narratives emerging from local populations and in particular from NGOs and intended for local audiences. Civil society arising through formal organizations is potentially at odds with traditional structures of decision making and implementation in Pacific Island cultures. Claims that women's voices are missing from local debates are not entirely borne out by the evidence that women's NGOs in the states considered here are influential in leading a range of adaptation strategies. However, it may be the case that women have been successful in using the channels of action available to them, not the channels of traditional leadership, but those opened through links to the international community, the channels of NGOs. In the urbanized areas of these island nations, many overseas NGOs have staff working or even offices which provide a support structure for women to be involved in debates and to take action. This gives women access to power and status of a kind not available to them through traditional structures.

Involvement with NGOs based overseas, such as PCP, influences local narratives in two ways. First, they bring the consensus narratives of climate science to the islands and train people in how to use them in local contexts; second, and importantly, they train young leaders to take messages based on their experiences to others in their community and into wider discussions. As Jill Finnane, the PCP coordinator, recently said: "We should never lose sight of the fact that this is about more than just science. It's about people, their cultures and their right to a just and secure future" (SciDev.Net, 2017).

This emphasis on daily life and the efforts involved in maintaining it are key elements in the local narrative. However, these efforts are not ones concerned, at one level, with preventing or minimizing rising tides; they are mainly concerned with ensuring a supply of safe drinking water or a staple food crop. In Tuvalu, in particular, they are concerned with recognition of the importance of traditional local knowledge and the use of *social memory*, a strong indication of a narrative of resilience (Folke, 2006). Local knowledge is not only knowledge about the environment and the skills in living between the land and the sea, it is also knowledge of the social and political processes through which decisions can be made and implemented (Lebel, 2013: 1071).

The young leaders in the PCP KATEP program bring their own experiences to the narrative of rising tides, ensuring that although key aspects of innovation and adaptability in daily life and an emphasis on the importance of traditional culture are part of this narrative, it is not a narrative that becomes ritualized. It is constantly renewed by these new voices, ones that have had the opportunity for experiences outside of the very local context. The KATEP program gives the opportunity for local experiences and practices to be shared so that innovations in one island can be considered for adoption in another island. It is a program dependent on the creation of personal and social relationships – social capital (cf. Folke, 2006) – which can be added to the network of relationships already existing within families, islands, and states.

This emphasis on sharing experiences is seen to build on local approaches to learning and building new knowledge. It is also a way to overcome the narratives of isolation and distance. Distance and isolation, caused in part by "intermittent and irregular boat trips" (Paton and Fairbairn-Dunlop, 2010: 689), have often meant that the inhabitants on individual islands must be able to meet their own basic needs. This reliance on self may have meant that people are unaware of the policies and plans of the government, located in some far-off atoll (Nunn et al., 2014), reinforcing the particularity of the very local narrative. The online presence of NGOs such as PCP and CAN helps to overcome this sense of separation, even though the Internet is rarely accessible to a majority of the population – 14.6 percent of the population in Kiribati, 37.6 percent in the Marshall Islands, and 50.1 percent in Tuvalu (Internet World Stats, 2017).

Social media and the Internet have the capacity to overcome perceptions of distance and isolation and to bring much greater immediacy to interactions, and through this facilitating the access of the very local and very personal into institutionalized discussions, as Mattlan Zackhras, a former government minister from the Marshall Islands, has argued. At a preparatory meeting for COP22 in Marrakech, he used Twitter to let his fellow attendees know about the situation at home: "This is personal. Hard being at Pre-COP as king tides hit my island home. Powerful reminder of why we all need to do more. Fighting for all those affected" and accompanied the tweet with photos showing the flooding (Koekoek, 2016). This use of the local in the context of global institutional meetings may signal the beginning of a new narrative for international audiences. This new narrative is developing in parallel with the conventional narratives, reinforcing the human rights narratives of food security and access to clean water, and demonstrating the ways that local populations use traditional knowledge, contemporary technical knowledge, and social customs to manage the problems caused by the rising tides of climate change.

Conclusion

There are two particularly important narratives that are widely used both by the governments of small-island developing states and NGOs in relation to the politics of rising tides. At a superficial level, it can be asserted that the first narrative relates to the SIDS' vulnerability. The focus for this is primarily international audiences. The second narrative relates to resilience, where the focus is primarily on local NGOs and citizens. That said, this overlooks the complex ways in which the narratives for international audiences have evolved since 2009, from ones that brought a spotlight to the notion of victim, through a series of moves that give these *victims* the opportunity to point out that they are just part of *an early warning system* for the rest of the world and to claim the moral high ground as they call for global effort and greater ambition. It also overlooks the forces at play in the narratives for local audiences, which focus on everyday life issues and where conflict may arise between beliefs based on religion and the secular, traditional leadership and modern legislated processes of government, between traditional community and civil society enshrined in NGOs, the roles of men and women, and traditional knowledge and scientific knowledge.

The claims of *high ambition* and *moral leadership* may signal a significant shift in the politics of rising sea levels. Understanding local issues and problems as human rights issues will inevitably broaden both audiences and the development of narratives, especially through the immediacy of social media. These changes highlight the importance of maintaining a social justice perspective on sea-level rise.

References

Adger, W. N. (2006). Vulnerability. *Global Environmental Change*, **16**(3), 268–81.
Abeysinghe, A., and Huq, S. (2016). Climate justice for LDCs through global decisions. In C. Heyward and D. Roser, eds., *Climate Justice in a Non-Ideal World*. Oxford, UK: Oxford University Press, pp. 189–207.

APAN. (2017). *About APAN.* Available at www.asiapacificadapt.net/about-us

Barnett, J. (2017) The dilemmas of normalising losses from climate change: Towards hope for Pacific atoll countries. *Asia Pacific Viewpoint*, **58**(1), 3–13.

Betzold, C. (2010) "Borrowing" power to influence international negotiations: AOSIS in the climate change regime, 1990–1997. *Politics*, **30**(3), 131–48.

Cannon, T., and Müller-Mahn, D. (2010). Vulnerability, resilience and development discourses in context of climate change. *Natural Hazards*, **55**(3), 621–35.

Climate Policy Observer. (2017). *IMO Takes a Step Forward to Tackle GHG Emissions from the Shipping Sector.* Available at http://climateobserver.org/ imo-takes-step-forward-tackle-ghg-emissions-shipping-sector/

CANCC. (2014). *Coalition of Low Lying Atoll Nations on Climate Change (CANCC) Known as the Global Early Warning System.* Available at www .sprep.org/climate-change/coalition-of-low-lying-atoll-nations-on-climate-change-cancc-known-as-the-global-early-warning-system

Denton, A. (2017). Voices for environmental action? Analyzing narrative in environmental governance networks in the Pacific Islands. *Global Environmental Change*, **43**(March 2017), 62–71.

Donner, S. (2011). Making the climate a part of the human world. *Bulletin of the American Meteorological Society*, **92**(10), 1297–302. Available at http://journals .ametsoc.org/doi/pdf/10.1175/2011BAMS3219.1

Fairclough, N. (2003). *Analysing Discourse: A Social and Critical Approach.* London: Routledge.

Farbotko, C. (2005). Tuvalu and climate change: Constructions of environmental displacement in *The Sydney Morning Herald. Geografiska Annaler*, **88**(4), 279–93.

Folke, C. (2006). Resilience: The emergence of a perspective for social-ecological systems analyses. *Global Environmental Change*, **16**(3), 253–67.

Gallopín, G. (2006). Linkages between vulnerability, resilience and adaptive capacity. *Global Environmental Change*, **16**(3), 293–303.

Government of Tuvalu. (2015). *Intended Nationally Determined Contribution.* Available at http://www4.unfccc.int/ndcregistry/PublishedDocuments/Tuvalu% 20First/TUVALU%20INDC.pdf

Hansen, G. (2017). *David Meets Goliath: First Ever V20 – G20 Meeting Highlights the Mutual Benefits of Climate Action.* Blog post, April 2017. Available at www.v-20.org/david-meets-goliath-first-ever-v20-g20-meeting-highlights-mutual-benefits-climate-action/

Internet World Stats. (2017). *Oceania and Pacific.* Available at www.internetworldstats .com/pacific.htm

Intergovernmental Panel on Climate Change (IPCC). (1990). *IPCC First Assessment Report.* Available at www.ipcc.ch/pub/reports.htm

Intergovernmental Panel on Climate Change (IPCC). (2001). *Climate Change 2001: Impacts, Adaptation, and Vulnerability, Contribution of Working Group II to the Third Assessment Report of the Intergovernmental Panel on Climate Change.* Cambridge, UK: Cambridge University Press. Available at www.ipcc .ch/ipccreports/tar/wg2/index.php?idp=0

Intergovernmental Panel on Climate Change (IPCC). (2014a). *Climate Change 2014: Impacts, Adaptation, and Vulnerability, Working Group II to the Fifth Assessment Report of the Intergovernmental Panel on Climate Change.* Cambridge, UK: Cambridge University Press. Available at www.ipcc.ch/report/ar5/wg2/

Intergovernmental Panel on Climate Change (IPCC). (2014b). *Climate Change 2014: Synthesis Report Summary for Policymakers.* Available at www.ipcc.ch/ pdf/assessment-report/ar5/syr/AR5_SYR_FINAL_SPM.pdf

Janssen, M., Schoon, M., Ke, W., and Börner, K. (2006). Scholarly networks on resilience, vulnerability and adaptation within the human dimensions of global environmental change. *Global Environmental Change*, **16**(3), 240–52.

Jaschik, K. (2014). Small states and international politics: Climate change, the Maldives and Tuvalu. *International Politics*, **51**(2), 272–93.

Johnson, M. F., Hannah, C., Acton, L., et al. (2014). Network environmentalism: Citizen scientists as agents for environmental advocacy. *Global Environmental Change*, **29**(Nov 30), 235–45.

Kelman, I. (2010). Hearing local voices from Small Island Developing States for climate change. *Local Environment*, **15**(7), 605–19.

Kiribati Local Government Association. (2016). *KiLGA Supports UNICEF's WASH Programme.* Available at www.kilga.org.ki/?p=451

Koekoek, P. (2016). *Heartfelt Tweets from the Marshall Islands' Fearless Climate Warrior.* Available at https://dailyplanet.climate-kic.org/heartfelt-tweets-from-the-marshall-islands-fearless-climate-warrior/

Lebel, L. (2013). Local knowledge and adaptation to climate change in natural resource-based societies of the Asia Pacific. *Mitigation and Adaptation Strategies for Global Change*, **18**(7), 1057–76.

Nakashima, D. J., Galloway McLean, K., Thulstrp, H. D., Ramos Castillo, A., and Rubis, J. T. (2012). *Weathering Uncertainty: Traditional Knowledge for Climate Change Assessment and Adaptation.* Paris: UNESCO, and Darwin: UNU.

Nunn, P., Aalbersberg, W., Lata, S., and Gwilliam, M. (2014). Beyond the core: Community governance for climate-change adaptation in peripheral parts of Pacific Island Countries. *Regional Environmental Change*, **14**(1), 221–35.

Office of the President, Republic of Kiribati. (2014). *I-Kiribati First International Publication on Climate Change.* Available at www.climate.gov.ki/tag/riibeta-abeta/

Otto, I., Reckien, D., Reyer, C., et al. (2017). Social vulnerability to climate change: A review of concepts and evidence. *Regional Environmental Change*, **17**(6), 1651–62.

Pacific Island Development Forum (PIDF). (2015). *Suva Declaration on Climate Change.* Available at http://pacificidf.org/wp-content/uploads/2013/06/PACIFIC-ISLAND-DEVELOPMENT-FORUM-SUVA-DECLARATION-ON-CLIMATE-CHANGE.v2.pdf

Paton, K., and Fairbairn-Dunlop, P. (2010). Listening to local voices: Tuvaluans respond to climate change. *Local Environment*, **15**(7), 687–98.

Republic of Kiribati. (2015). *Intended Nationally Determined Contribution.* Available at http://www4.unfccc.int/ndcregistry/PublishedDocuments/Kiribati%20First/INDC_KIRIBATI.pdf

Republic of the Marshall Islands. (2015). *Intended Nationally Determined Contribution.* Available at http://www4.unfccc.int/ndcregistry/PublishedDocuments/Marshall%20Islands%20First/150721%20RMI%20INDC%20JULY%202015%20FINAL%20SUBMITTED.pdf

Richardson, P. (2009). Governing the outer islands: Decentralisation in Kiribati and Tuvalu. *Commonwealth Journal of Local Governance*, **2**(January), 120–27.

Rudiak-Gould, P. (2011). Climate change and anthropology: The importance of reception studies. *Anthropology Today*, **27**(2), 9–12.

Schäfer, M. (2012). Online communication on climate change and climate politics: A literature review. *Wiley Interdisciplinary Reviews: Climate Change*, **3**(6), 527–43.

SciDev.Net. (2017). *Sea-Level Rise Accelerates as Adaptation Turns Urgent.* Available at www.scidev.net/asia-pacific/climate-change/news/sea-level-rise-accelerates-climate-change-adaptation-urgent.html

Teaero, T. (2003). Indigenous education in Kiribati. In K. H. Thaman, ed., *Educational Ideas from Oceania: Selected Readings*. Suva: Institute of Education, University of the South Pacific, pp. 106–15.

Tobwaraoi Community Nanikaai. n.d. *Tobwaraoi Community Nanikaai Website*. Available at http://tobwaraoicommunity.webs.com/

UNEP. (1989). *Criteria for Assessing Vulnerability to Sea Level Rise: A Global Inventory to High Risk Areas*. Delft: UNEP and the Government of the Netherlands.

Uniting Church of Australia. (2015). *Can God Fix Climate Change?* Available at http://uca.org.au/can-god-fix-climate-change/

Webber, S. (2013). Performative vulnerability: Climate change adaptation policies and financing in Kiribati. *Environment and Planning A*, **45**(11), 2717–33.

Woodroofe, T. (2016). High ambition coalition plays key role in securing the Paris Agreement. *Climate and Development Outlook*, October 2016. Available at https://cdkn.org/wp-content/uploads/2016/10/Negotiations_Support_Outlook.pdf

Part III

Marine Fisheries and Pelagic Seas

9

Climate Change and Fisheries Politics
Case Studies from the United States, New Zealand, and Norway

ANDREW TIRRELL

Introduction

Many scholars have noted the significant impacts that anthropogenic changes to the global climate will likely cause in fish stocks, including those that make up important commercial fisheries across the world (Kell et al., 2005; Brander, 2007; Cochrane et al., 2009; Cheung et al., 2010; Fulton, 2011; Richardson et al., 2012; Poloczanska et al., 2013; Lajus et al., 2017). Moreover, changes not only to stock size but also to the ranges of species have already begun to affect social, political, and economic dynamics between and within countries (Drinkwater, 2005; Sumaila et al., 2011; Vaidyanathan, 2017). Although there is a growing literature on how international law (Jeffers, 2010; Glass-O'Shea, 2011) and national systems of fisheries management should address these issues (Kutil, 2011; Melnychuk et al., 2014; Weatherdon et al., 2016), additional focus on the concerns of community-level fisheries stakeholders is warranted. Few have considered the impacts of climate change on inshore, largely owner-operator, fisheries from the perspective of the stakeholders themselves.

The nexus of climate change and fisheries politics brings us to the intersection of two "wicked" problems, to use Rittel and Webber's (1973) term for complex challenges whose solutions require input from multiple perspectives and fields of expertise. The same intertwined social, political, ecological, and economic systems that complicate responses to climate change on various levels of governance pose similarly daunting challenges in fisheries management (Hilborn, 2007: 293). The sociopolitical complications found in fishing communities raise particularly thorny issues, because "fishermen respond to regulation in ways that often surprise managers, and managers must understand the motivation and incentives for fishermen to understand how they respond" (Hilborn, 2007: 286, citing Larkin, 1988). This same attention to motivation and incentives can explain how fishing communities perceive climate change and how these perceptions shape community responses to their chronic and escalating climate challenges.

Building off previous research into the role that sociocultural institutions, such as norms and values, play in community responses to fisheries management, this chapter will explore the various positions taken by stakeholders in small- and medium-sized inshore fisheries communities when asked about climate change (see Tirrell, 2017 for an in-depth analysis of the Norwegian case). In particular, interview data collected between 2013 and 2017 in dozens of fishing towns and villages within three countries – the United States (specifically the New England region), New Zealand (communities spread across both the North and South Islands), and Norway (specifically the three northernmost counties of Nordland, Troms, and Finnmark) – will be used to explain the social and political conditions that allow stakeholders, who are often antagonistic to environmental advocacy, to acknowledge (and, at times, even advocate for) the need for climate change adaptation and mitigation. This chapter will consider each of these national cases separately before engaging in a comparative analysis in the concluding section.

The United States

Fisheries Context

The fishing industry in the United States traces back to the earliest colonial times and, despite steadily decreasing economic importance, fishing remains culturally and economically important, especially in the New England region (Georgianna and Amaral, 2000; National Marine Fisheries Service, 2014). Catch share fisheries management in the United States is distinct from the other catch share management systems described in this chapter because a uniform system has yet to materialize on a national level in the United States. This difference is due to each region's unique interplay of fisheries markets, technology, policy and decision making, government regulation, community institutions, and marine ecosystems. In the United States, most fisheries management is conducted at the state level, and federal jurisdiction is limited to territories, the high seas, international treaty management and administration, and protection of certain species in navigable waters (Atkinson, 1988: 111). Overlaid on these state systems are regional councils (including the one in New England) that have authority over federal waters through a voting body composed of federal agencies, state management agencies, and public members appointed by the federal government (Heinz, 2000: 8). The interplay between these conflicting authorities is challenging, particularly as policy motivations change with the shifting political, social, economic, and environmental considerations of the times.

By the 1990s, fish stocks in New England were in dramatic decline, and the industry was in economic peril. Due to industry resistance and political action, previous management strategies were not ecologically sound enough to counter the drastic overfishing that had occurred since the 1970s. The 1996 Magnuson-Stevens Act (MSA) aimed to strike a balance between promoting fishing for economic gain and recognizing the immediate and dire need for conservation. As environmental issues came to the forefront, fisheries councils faced the difficulty of abating tensions between industry stakeholders, conservation groups, and fisheries managers. As exports of seafood fell and domestic demand rose, the New England council recognized both social and cultural impacts of domestic fisheries. This increasing awareness of impacts beyond economic and conservation concerns inspired a call for a more "formal treatment" of the sociocultural ramifications of fisheries regulations (Heinz, 2000: 30). The new social dynamic of coastal populations in New England, unstable oceanic and climate conditions, and a rise in conservation pressure from nongovernmental organizations (NGOs), deeply modified public views of fisheries accountability. The MSA addressed these changing policy considerations by introducing the category of the "community" as an unequivocal issue to be considered in fisheries management council plans, mandating "impact assessments of management plans on 'fishing communities'" (St. Martin, 2006: 176). Regional councils followed suit by allowing greater participation in the system from environmental organizations through their membership on the council and their engagement in advisory panels. The difficulty in engaging the varied considerations of each level and sector of fisheries interest, from industry stakeholders, to conservation NGOs and the general public, has only grown since establishment of the MSA.

Climate change has added further stress to the already beleaguered fish stocks that form the commercial fisheries in New England. Significantly warmer ocean temperatures along the coast have been recorded since 1950 (Melnychuk et al., 2014: 565–69), affecting both anadromous species, such as salmon and alewife (Lynch et al., 2016: 351–52), and groundfish stocks, such as cod and haddock (Fogarty et al., 2008; Lavelle, 2015; Meng et al., 2016). Models suggest that these impacts will greatly increase over the course of the next century (Fogarty et al., 2007), possibly driving Atlantic cod permanently north of the Georges Bank, the rich fishing ground outside the Gulf of Maine, and causing a significant decline in the Gulf of Maine itself (Drinkwater, 2005: 1333).

Stakeholder Responses

In general, the fishermen (though gendered, this is the nomenclature preferred by the majority of both men and women in the industry in New Zealand and New England) interviewed for this study have low confidence in the science, as produced by the National Oceanic and Atmospheric Administration (NOAA), behind management of New England fish stocks. Surprisingly, however, many fishermen who questioned fisheries science and groused over government overreach into fisheries management admitted the dangers posed by climate change. This willingness to embrace climate science, if not fisheries science, seemed to occur under particular conditions. For example, there were a few communities where openness to longer-term thinking had taken hold. This relatively unique sensitivity to the future resulted in progressive ideas about fisheries ecology, climate change, and collaboration with conservation NGOs, seemingly without significant changes in attitude toward the government or its fisheries management systems.

Those communities where longer-term, ecologically minded thinking had gained a foothold seemed to have one particular attribute in common: the establishment of a local stakeholder organization that served as both a forum for these new ways of thinking and a bridge to outside conservation groups. Many scholars of environmental policy have identified the "bridge organization" phenomenon (Lewin and

Stephens, 1993; Brown, 2001; Prasad and Elmes, 2005; Gupta, Pistorius, and Vijge, 2015). Such organizations negotiate the divide between stakeholder groups that may otherwise be skeptical of conservation and organizations that provide expertise, technological innovation, and other strategic assistance to resource-dependent communities. In this case, several New England communities formed collective fishing groups that gradually came to work with outside organizations on fisheries' sustainability issues. These same communities exhibited higher degrees of climate change acknowledgment and awareness.

"I don't think the relationships we've built could have happened without our partner organizations," one NGO fisheries specialist told me, referring to community-based organizations representing fishermen. That specialist added: "These local organizations are usually staffed by people who work with the fishermen daily and sometimes are from the communities themselves, and have a huge amount of trust and social capital in those communities. At the same time, they have mostly trained as environmental scientists or biologists, at least at the undergraduate level, so an ecosystem approach is natural to them, and working with [our organization] is thought of as a good thing." In this sense, the bridge between fishermen and conservationists is not just organizational but very much relies on individual human connections. As one staff member of a local fishermen's community organization put it: "We know how to speak both languages. In organizing parlance, we can change registers depending on who we are communicating with. In some ways, we're like interpreters, but with fisheries expertise, as well." For these "interpreters," decoding climate change is part of the job. As one respondent to my interviews noted:

It's actually not that hard to talk to fishermen about climate change, as long as it's coming from the right source. They see the changes every time they go out on the water, and they hear how their fathers, uncles, and grandfathers describe the way things used to be. They know things are changing, and those personal experiences match what they see in the news about climate change. For them, the only problem with talking about climate change is the concern that admitting that climate science might be right might also mean admitting that fisheries science is right. It's a dilemma that has led many to avoid the subject of climate change altogether. We can help to start those conversations in a positive way.

As another community organizer warns, however, it's not as easy as it sounds:

Am I trusted? Yeah, sure. The guys like me. They think I'm trying to help. But, I'm not one of them, and I never will be, I suppose. They rib me all the time about my fancy degree, how I'm better suited behind a laptop at Starbucks than on a fishing boat. But at the end of the day, if I suggest something, they might listen. They won't adopt a new technology wholesale, but one guy might try it, and others will follow if it helps them to fish better. They're willing to shift some of their thinking to longer-term issues. But, it still has to make financial sense. This isn't really about making them into environmentalists, a label they would absolutely hate. It's convincing them that conservation means more money over time.

When it comes to climate change, this endpoint involves linking climate mitigation or adaptation directly to community outcomes. One fisherman who was open to suggestions from outside organizations, including electronic jigging machines and onboard video monitoring, echoes this observation:

I'll do anything that helps the bottom line. I suspect there are more fish out there than scientists say, but if I can avoid catching codfish both because I'll stay out of trouble and because it may lead to better stocks in the future, why not. If there's some technology that can help me do that, and somebody's going to help me get it for free or pretty cheap, sign me up.

Still, many communities as a whole remain deeply skeptical of collaborating with conservation organizations, let alone admitting that NOAA might be on the right side of any issue. One long-time fisherman from Gloucester, Massachusetts, was quick to disparage fishermen in Cape Cod for their relationships with both conservation organizations and regulators. As he described it:

Those hookers (a joke based on the name of the Cape Cod Hook Fishermen's Alliance) in Chatham sold us out. First, they were in bed with the government to get a catch history benefitting them. Now they've teamed up with [outside organizations] and are getting all kinds of enviro money. They aren't really fishermen anymore.

The animosity, and underlying differences in perspective, run much deeper than this single issue, however. It is clear that the fishermen's organizations in New England fishing communities that reject collaboration with conservation groups are not only ill equipped to serve as bridge organizations because they draw staff only from the industry side, usually people with little to no training in environmental conservation. They are also groups that often serve to intensify industry hostility toward ecological concerns, including climate change. In other words, they serve as walls, rather than bridges, a fitting distinction given the election rhetoric circulating around these fishing communities when the bulk of these interviews were held during the summer before the 2016 US presidential election. Contrary to the sentiment in most of New England, the majority of fishing communities were loud and proud of their support for then-candidate Donald Trump. Many interviewees brought up Trump's campaign spontaneously, believing that he would work to restore the economic viability of the industry by disempowering NOAA, the National Marine Fisheries Service, and other administrative bodies. For many stakeholders who admitted that they had been apathetic about politics for most of their lives, Trump's support for industries that had been maligned as bad environmental actors (e.g., coal, oil, and mining, in addition to fishing) tapped into a well of frustration in fishing communities that had been building up for decades.

An owner-operator whose truck was adorned with Trump campaign stickers confirmed this assessment: "We're hardly surviving … How can you talk about fish stocks when I might lose my house next month? How can I worry about climate change when sometimes I can't afford the fuel to go fishing?" At the mention of climate change, I become curious about his thoughts on the issue. I ask him if he meant to say that climate change is indeed happening, but he said that he does not have the capacity to worry about it. According to him, most everyone he knows is aware, but they are torn between an unwillingness to agree with environmentalists on any issue and the desire to be relieved of some of the blame for declining fish stocks. As the owner-operator put it:

Any long-time fisherman paying attention should have noticed climate change 25 years ago. It's been happening all around us. We're out on the water so much of our lives, of course we see the changes. In fact, that's part of what gets me so mad. We always get blamed for destroying fish stocks, while practically every other industry is behind climate change, which is *really* destroying the fish stocks.

A fisheries sector manager from another community confirmed the paradoxical climate politics of New England fishermen:

New England fishermen are the largest Republican voting bloc that believes in climate change, partially because it redirects the blame over fish stock declines. Lobsters are already disappearing from southern Maine. In fact, Maine fishermen recently held a big climate change conference, which was very well attended. The tenor of that meeting was "how will we adapt," with very little discussion of whether climate change is actually happening.

Although some fishermen interviewees did question climate science, these outliers often seemed more attached to the notion of environmental science being generally shaky, and they were not specifically concerned with climate change denial.

As it was with other conservation issues, age was a relevant factor in how New England fishermen viewed climate change. Very few younger fishermen question climate science, just as they are less likely to dispute declining groundfish stocks. As one older fisherman put it: "There are huge differences in approach[es] to sustainability between younger and older fishermen. That's one reason we need young blood in the industry. Preferably locals. They are the future of our communities." That same fisherman was making plans to turn his boats over to two young crewmembers. He would have preferred that the business stay in the family. However, at the age of almost 60, he had not found time to marry and start a family. As he elaborated, "[T]hese young guys are like family to me now anyhow. They came here looking for jobs because they love being on the water. It's a calling for them, like it was for me. They have their whole future ahead of them." When the conversation turned to climate change and its potential impacts on fish stocks, a somber look flashed across his face, and he shook his head and said, "Hopefully there will still be fish here in 10 or 20 years for them to catch."

New Zealand

Fisheries Context

Fishing as a livelihood dates to the first days of the Maori settlement of New Zealand, beginning sometime during the eleventh to thirteenth centuries. Although the 1839 *Treaty of Waitangi* guaranteed Maori trusteeship over their "treasures" (*taonga*), understood to include natural resources, English settlers quickly overtook the Maori in the exploitation of New Zealand's fisheries. The threat of overexploitation was recognized within a couple decades of the signing of the treaty, prompting the first efforts in the mid-nineteenth century to centrally govern fisheries in New Zealand under the control of the Marine Department (Johnson, 2004: 55).

By the middle of the twentieth century, however, fishermen had come to see the department as a "bureaucratic machine that stifled the industry," and, conversely, the department was "anxious about the industry and felt that without regulation fishermen would plunder the sea" (Haworth, 2008: 15). As a response to these shifting attitudes, fishermen began to take overtly political actions. Until this point, they were "little influenced by government attempts to regulate the industry" (Johnson, 2004: 163). In 1952, however, six fishermen formed a federation of owner-operators of fishing vessels to discuss common-interest issues and to influence marine policy in New Zealand. By 1958, the federation had grown into the New Zealand Federation of Commercial Fishermen, with fishermen, wholesaler, and retailer representatives making up a national fisheries advisory committee (Haworth, 2008: 25).

According to some stakeholders in the New Zealand fisheries, pride in "rugged individualism" exposes the root of current challenges facing many inshore fishing boat owner-operators. It is fishermen's hesitance to assert themselves, they claim, that has allowed the industry to pass into the hands of fewer individuals and more corporate entities over the past several decades. They would argue that, although New Zealand fishing communities used to organize and demand their rights, their recent history has instead been one of decline and dissolution.

Others, however, believe that the New Zealand fisheries management system should be celebrated as one that successfully balances sustainability, individual rights, and economic vitality. It is only through limiting the number of stakeholders, they argue, that New Zealand has been able to provide enough fish for the fishermen who continue to cast their lines and nets. They maintain that buy-out programs initiated in the mid-1980s, negotiated through the advocacy of fishermen's organizations, provided a safe exit for those who might have otherwise faced economic struggles throughout their lives, and a path toward sustainability for those who remained in the industry in the 30 years since.

As a high-latitude nation, New Zealand's economy is particularly dependent on fisheries export income (Allison et al., 2009: 185). That revenue stream will face threats from anthropogenic climate change in the coming decades, as sea-surface temperatures are expected to rise sharply (Hurst et al., 2012). Because most marine animals are adapted to particular temperature ranges (Pörtner and Knust, 2007), a generally southerly shift in the current ranges of New Zealand commercial fish stocks is expected, and this will be exacerbated by a strengthening of southward-flowing currents along the eastern coast of the North Island (Willis et al., 2007: 35). Species ranges may also shift in response to cascading changes in the populations of habitat-forming species (Schiel et al., 2004).

Stakeholder Responses

For inshore fishermen in New Zealand, the issue of climate change comes with a lot of baggage. Whether true or not, many fishermen in New Zealand perceive themselves to be societal outcasts in a country where environmentalism holds sway. According to one respondent, fishermen are "villains" in the eyes of the public, criticized with an intensity otherwise reserved for "Japanese trawlers and offshore oil rigs." For him, in addition to the fisheries management reforms of 30 years ago, the rise of environmentalism in the 1970s and 1980s in New Zealand is largely to blame for the alienation between fishermen and their communities. As he described it:

You wouldn't believe the ugly things some of those people have said about me, sometimes to my face. Of course, they want to go to the market and buy fish. They just don't want to know where it comes from. And many of them fish themselves! But they act as though that's an entirely separate thing. Like we aren't fishing from the very same waters.

Recreational marinas abound in New Zealand, especially along the Northland Peninsula, where the same fisherman lives. Marinas in ports like Kerikeri, Paihia, Opua, and Whangarei are dotted with small day-boats ready to take Kiwis and foreign visitors out for the morning or afternoon. According to the Ministry for Primary Industries, 31 percent of New Zealanders engage in some kind of recreational fishing, hauling in more than 25,000 tons per year (Ministry for Primary Industries, 2010). In contrast, one former commercial fisherman now struggles to make a living repairing fishing gear on the northwest coast of the South Island, home to the once-bustling commercial ports of Greymouth and Westport. Fishing has nearly died out in these ports, giving way to economic depression. This respondent claims that the double standard, along with the quota buy-out scheme enacted by the government in the 1980s, drove him out of the industry:

It's not us inshore guys. Most of us haven't got enough quota to make a dent, anyhow. I had to get out. I had so few permits, and the quota for each was so small, that it wasn't worth keeping it up. I sold it all for a pittance, and I make more money doing this [repairing gear] than I ever could with my quota. Most of the guys who were bought out are still working, like me, but at a worse job. It wasn't enough to retire on for most. And by selling their quota back, they allowed the government to go forward with their plan to put the entire industry in the hands of a few companies .... Now all that's left is SeaLords, Leigh, and a few other companies, half of whom are partnered with foreigners. Then there's all of the inshore guys scraping out a living doing contract work for them. It's really sad to see where things have gone in my lifetime.

That alienation, in turn, has caused many to reject talk of climate change impacts on fisheries as just another attack on what remains of the small-scale fishing industry. In shrugging off climate concerns, many reference government encouragement of investment in the late 1970s and early 1980s. There was significant backlash from owner-operators, when, after increasing inshore permits from 4,000 to 14,000 between 1975 and 1979, fisheries managers determined that the inshore fleet was overcapitalized by as much as 20 percent, and fish stocks began to show signs of serious decline (Ministry for Primary Industries, 2010). Defending his distrust of climate-related government warnings, one respondent recalled:

Some of us were just starting out almost 40 years ago [and] the government told us to "go big." Guys bought new gear, and even new boats in some cases, but then they decided that it's the companies who would go big, and we should just go away. That was a devastating time for all of us. Unfortunately, the worst stories are from people you'll probably never find. They're long out of the industry now, and some long dead.

Given these experiences, many inshore fishermen in New Zealand find themselves at odds with both the government and civil society. This breakdown of trust has tainted climate change in the eyes of many respondents as an issue pushed by hostile government or environmental interests to further degrade small-scale fishermen's livelihoods. Even those who begrudgingly admit that the evidence for climate change is convincing find themselves unmoved by the long-term threats to fish stocks. As one fisherman quipped, "I don't expect small-scale owner-operators to stay in business long enough to have to worry about climate change impacts."

Norway

Fisheries Context

The counties of Nordland, Troms, and Finnmark harbor the most active inshore fishing fleets in Norway. Many species of fish, including mackerel, herring, cod, saithe, halibut, and redfish, migrate along the nearby continental shelf (Statistics Norway, 2017). Commercial fisheries industrialized rapidly over the course of the twentieth century, causing many species to be overfished by the 1970s (Holm, 1995; Hersoug, 2005). This modernization created greater efficiency, delivering increased catches to fewer fishermen, at lower costs. Consequently, fishing now makes up a smaller percentage of the Norwegian economy than in the past, decreasing from 5.7 percent of gross national product (GNP) in 1930 (OECD, 2004) to just 0.45 percent of GNP in 2009 (Sandberg et al., 2009).

Although Norway's fishing fleet has industrialized significantly in recent decades, many small coastal communities remain socioeconomically dependent on fishing. The majority of boats leaving these harbors are owner-operator ventures, relying on a captain and small crew of one or two hands. These towns and villages endure despite the challenges posed by increasingly strict fishing management policies and economically neoliberal restructuring of fisheries rights. Consequently, these communities are increasingly vulnerable, with their cultural, social, and economic ways of life threatened by anthropogenic changes to the global climate.

Studies indicate that many commercially important species of fish in Norway are sensitive to changes in climate (Stenevik and Sundby, 2007). For example, Arcto-Norwegian cod (*Gadus morhua*) shift spawning sites northward during warm climatic events, delaying the decline of fish stocks (or even creating a temporary boon) in the northern Norwegian counties of Nordland, Troms, and Finnmark (Sundby and Nakken, 2008). Because of these effects, the possibility of sea-surface temperature increases of more than 8.5°C by the year 2100 threatens significant decline in the stock of Norwegian cod, among other critical species (Drinkwater, 2005: 1333).

Stakeholder Responses

In contrast with stakeholders interviewed in the other two cases in this study, Norwegian fisheries stakeholders were quick to acknowledge climate change as an important issue. Warming oceans, they maintained, are a very serious threat to their livelihoods, in the long term, if not the short term. This distinction between present and future risk makes sense in Norway (especially in the northern Norwegian communities that were the focus of this research) because of the potential delay in climate consequences for Arctic fisheries near Norway. Nevertheless, many fisheries respondents were willing to accept recommended changes to current fisheries policies based on climate projections, setting them well apart from their counterparts in New England and New Zealand. This willingness to embrace science-based policy making informed by climate change data echoes the sentiments these communities have about fishing policies in Norway generally. Even in cases where changes to fishing quotas, gear allowances, or fishing grounds access would result in economic losses to the community, most respondents trusted government managers to make the "right" decision. In many cases, interviewees articulated an understanding that such policy decisions might be justified because of the importance to a larger cause – the industry, nation, or even the environment – even when the results would be devastating for the community. Indeed, some fishing communities have been crippled to the point of abandonment or precipitous decline in recent decades.

Norwegian fishing communities often afford a huge measure of trust to their government, including with respect to the issue of climate change. As one respondent put it, one's outlook "depends what you trust [the government] to do." The most optimistic interviewees voiced trust that the government will work to restore the small-scale fishing industry in villages and towns that were dependent on fisheries. One respondent from the fishing village of Bleik doubted that would ever happen, and he did not believe that many other fishermen would fail to see that even if they do not say so. As he put it, "Some of them are just playing optimist; it's a nice story for old men to tell each other because we miss the old days, but we're all retired or nearly retired now anyhow. Still, I do trust the government. They will do what's best for the people, and for the fish. It's just not likely to be that good for Bleik," which he said with a weary chuckle.

For many interviewees, following the government's lead amounts to good citizenship. This understanding of what it means to be a good citizen – and especially the willingness to sacrifice individual or community interests for the good of a larger cause – seems to be why climate change is approached differently by fisheries stakeholders in Norway. Climate change might be the ultimate example of a challenge whose solution requires communitarian values to prevail. Norwegian norms around sociopolitical behavior expect such a communitarian outlook by citizens, beginning with turning the well-being of the country over to government expertise. It should not be surprising, then, that respondents were able to distinguish individual and global interests regarding climate change action, usually prioritizing the latter. One *fiskerlag* (fishermen's union) leader expressed this sentiment when considering the short-term benefits of local increases in sea-surface temperatures: "Sure, the fishing has actually been a bit better here in Vesterålen [a district in Nordland county] in recent years. But, how can I be happy about that when it means others are having worse catches in Lofoten [another district south of Vesterålen] or even further south? And, I certainly can't be happy about climate change in general, which will cause a lot of suffering all over the world."

Although it would be convenient for someone in the *fiskerlag* leader's position to simply deny that climate change is happening and attribute his increased catches to good fortune, such an individually motivated response seemed alien to Norwegian respondents. Moreover, ingrained trust in the government and its scientific authorities relegates climate change denial to the fringes of thought for most Norwegians. One interviewee-related acceptance of the reality of climate change to the bevy of other challenges faced in northern Norway. He referenced the local expression "Vi står han av," translating the phrase to something like "we'll grin and bear it." He noted that it often refers to the harsh and unpredictable weather above the Arctic Circle in Norway, but it could easily apply to fishing, climate change, or any other volatile set of challenges. As he said, "Better to cope with bad times the best you can than to grumble. That's the Norwegian way."

Even when norms, values, and other informal sociopolitical institutions are favorable for the government, as they are in Norway, trust can only be stretched so far. Although these developments occurred too recently to be adequately studied within this project, Norway has been experiencing what the national fishing trade newspaper *Fiskeribladet* has been calling a "coastal roar" of dissent against fisheries policies that favor liberalization of quotas, threatening the viability of small-scale fishing communities. The major grievances include the relaxation of regulations requiring trawlers to land catches in small fishing ports and the creation of a liberalization task force that raised so many objections that it was abruptly scuttled by the Directorate of Fisheries.

This erosion of trust has blurred the distinctions between government and governance that once benefitted fisheries managers in Norway. That is, where once stakeholders could disagree with a specific policy but remain supportive of a government they believed to be acting in the best interests of the people, now the critiques personally target government actors who promote those policies. For example, a new candidate for Parliament from the Coastal Party not only spoke out against the new policies for trawlers, but he put his critiques in terms of government "fraud and crime," accusing Minister of Fisheries Per Sandberg of "selling Norway piece by piece."

The Norwegian government has done much to build trust with its people, even those in struggling fishing communities. This trust, coupled with a favorable institutional context, has afforded policy makers much latitude in setting fisheries management policies and reacting to related challenges posed by climate change. Nevertheless, although institutions are durable, they are "products of action" as much as they are "frameworks for action" (Holm, 1995: 399). Failure to recognize this may result in fisheries managers taking institutions for granted and allowing the reservoir of trust between community and state to be drained.

Conclusion

Many of the insights provided in this chapter revolve around the importance of norms; values; and other informal social, cultural, and political institutions in interactions with fisheries stakeholders. The willingness of fishing communities to engage with climate change as a critical issue that may affect their livelihoods is often tied up in the relationships that those communities have with whomever delivers the climate change message. These relationships can supersede factual understandings of potential climate change impacts, and they seem to influence the risk perceptions of stakeholders, even in cases of climate vulnerability. In New England, for example, the vexed relationship between fishing communities and the government comes as no surprise to those familiar with American norms of individuality and distrust of the government. Only a quarter of Americans believe that they can trust Washington "to do what's right," and people increasingly believe that government is "run by crooks or for a few big interests, and that it wastes a lot of money, without paying much attention to what people think" (Popkin and Dimock, 2010: 135). Lacking the social capital and norms of reciprocity necessary for cooperative ventures (Putnam, 2000: 348), such as securing sustainable fisheries or mitigating climate change, fishermen in New England would rather be left to their own devices than to collaborate with the government. In other words, Alexis de Tocqueville's nearly 200-year-old observation on the United States still rings true: "Hence arises the maxim, that everyone is the best and sole judge of his own private interest, and that society has no right to control a man's actions ... This doctrine is universally admitted in the United States" (de Tocqueville, 1969 [1835]: 22).

However, even in New England, these limitations may be overcome when fishermen are connected to *nongovernmental* sources of information that frame climate action as a matter of personal interest. In the case of New Zealand, on the other hand, the strained relationship between small-scale fishermen and both government *and* civil society seems to preclude engagement with issues related to climate change impacts. Although potentially influenced by norms of individualism common to other settler colonies (such as the United States), this alienation seems to have occurred in recent decades as a result of both neoliberal fisheries management policies that brought financial troubles and the perceived vilification of fishermen by civil society. Such alienation is not unheard of in New Zealand, as small political parties sometimes organize around "the theme of distrusting the government," creating loyal partisans but little broad support (Banducci, Donovan, and Karp, 1999: 551). Moreover, the hopeless financial outlook for many inshore owner-operators has rendered the long-term fisheries implications of climate change moot for many stakeholders.

In contrast, the case of Norway demonstrates both the power of social capital and the importance of citizenship norms for successful government engagement of fisheries stakeholders. That goodwill between the government and communities, however, should not be taken for granted and must be safeguarded. Until recently, Norwegian fisheries managers seemed to be in an enviable position that assured stakeholder cooperation with fisheries management and acceptance of the need to mitigate climate change risks. This was largely due to "social integration and social resources [that] seem to foster general trust and confidence in governmental ability," reinforcing a cooperative stance toward government initiatives, including fisheries management and climate mitigation and adaptation (Christensen, Fimreite, and Laegreid, 2011: 579).

The cases of the United States, New Zealand, and Norway offer lessons for fisheries policies amidst climate change. Whereas good citizenship in Norway rests on institutions such as communitarianism, social integration, and respect for leadership and expertise, citizenship norms in the United States focus on individualistic acts such as voting, speaking out, and party identification (Torney-Purta and Richardson, 2004). In recent years, however, the decline of small-scale fishing communities as a result of neoliberal reforms benefitting larger industrial players has eroded the relationship of trust between community stakeholders and the government, clouding future responses of fishing communities to climate-driven fisheries policies. Effective governance of fisheries in a future of climate change will require a kindling or nurturing of trust between fisherman, on the one hand, and experts, officials, and agencies, on the other. Trust may be as crucial as science.

References

Allison, E. H., Perry A. L., Badjeck, M., et al. (2009). Vulnerability of national economies to the impacts of climate change on fisheries. *Fish and Fisheries*, **10**(2), 173–96.

Atkinson, C. E. (1988). Fisheries management: A historical overview. *Marine Fisheries Review*, **50**(3), 111–23.

Banducci, S. A., Donovan, T., and Karp, J. A. (1999). Proportional representation and attitudes about politics: Results from New Zealand. *Electoral Studies*, **18**(4), 533–55.

Brander, K. M. (2007). Global fish production and climate change. *Proceedings of the National Academy of Sciences*, **104**(50), 19709–14.

Brown, L. D. (2001). *Practice-Research Engagement and Civil Society in a Globalizing World*. Cambridge, MA: Hauser Center for Nonprofit Organizations.

Cheung, W. W. L., Lam, V. W. Y., Sarmiento, J. L., et al. (2010). Large-scale redistribution of maximum fisheries catch potential in the global ocean under climate change. *Global Change Biology*, **16**(1), 24–35.

Christensen, T., Fimreite, A. L., and Lægreid, P. (2011). Crisis management: The perceptions of citizens and civil servants in Norway. *Administration & Society*, **43**(5), 561–94.

Cochrane, K., De Young, C., Soto, D., and Bahri, T. (2009). Climate change implications for fisheries and aquaculture. *FAO Fisheries and Aquaculture Technical Paper*, No. 530.

Drinkwater, K. F. (2005). The response of Atlantic cod (Gadus morhua) to future climate change. *ICES Journal of Marine Science*, **62**(7), 1327–37.

Fogarty, M., Incze, L., Wahle, R., et al. (2007). Potential climate change impacts on marine resources of the northeastern United States. *Report to Union of Concerned Scientists*.

Fogarty, M. Incze, L., Hayhoe, K., et al. (2008). Potential climate change impacts on Atlantic cod (Gadus morhua) off the northeastern USA. *Mitigation and Adaptation Strategies for Global Change*, **13**(5–6), 453–66.

Fulton, E. A. (2011). Interesting times: Winners, losers, and system shifts under climate change around Australia. *ICES Journal of Marine Science*, **68**(6), 1329–42.

Georgianna, D., and Amaral, P. (2000). *The Massachusetts Marine Economy*. Hadley: University of Massachusetts Donahue Institute.

Glass-O'Shea, B. (2011). Watery grave: Why international and domestic lawmakers need to do more to protect oceanic species from extinction. *Hastings West-Northwest Journal of Environmental Law and Policy*, **17**, 191.

Gupta, A., Pistorius, T., and Vijge, M. J. (2015). Managing fragmentation in global environmental governance: The REDD+ partnership as bridge organization. *International Environmental Agreements: Politics, Law and Economics*, **16**(3), 1–20.

H. John Heinz II Center for Science, Economics and the Environment. (2000). *Fishing Grounds: Defining A New Era for American Fisheries Management.* Washington, DC: Island Press.

Haworth, J. (2008). *Tides of Change: The Story of the New Zealand Federation of Commercial Fishermen, 1958–2008.* Christchurch, NZ: Wily Publications.

Hersoug, B. (2005). *Closing the Commons: Norwegian Fisheries from Open Access to Private Property*. Delft, Netherlands: Eburon Uitgeverij BV.

Hilborn, R. (2007). Managing fisheries is managing people: What has been learned? *Fish and Fisheries*, **8**(4), 285–96.

Holm, P. (1995). The dynamics of institutionalization: Transformation processes in Norwegian fisheries. *Administrative Science Quarterly*, **40**(3), 398–422.

Hurst, R. J., Renwick, J. A., Sutton, P. J., et al. (2012). Climate and oceanographic trends of potential relevance to fisheries in the New Zealand fisheries region. *New Zealand Aquatic Environment and Biodiversity Report No. 90.*

Jeffers, J. (2010). Climate change and the Arctic: Adapting to changes in fisheries stocks and governance regimes. *Ecology Law Quarterly*, *37*, 917–77.

Johnson, D. (2004). *Hooked: The Story of the New Zealand Fishing Industry.* Christchurch, NZ: Hazard Press.

Kell, L. T., Pilling, G. M., and O'Brien, C. M. (2005). Implications of climate change for the management of North Sea cod (Gadus morhua). *ICES Journal of Marine Science*, **62**(7), 1483–91.

Kutil, S. M. (2011). Scientific certainty thresholds in fisheries management: A response to a changing climate. *Environmental Law*, **41**(1), 233–75.

Lajus, D., Stogova, D., and Lajus, J. (2017). Importance of consideration of climate change at managing fish stocks: A case of northern Russian fisheries. In *The Interconnected Arctic – UArctic Congress 2016*, Cham, Switzerland: Springer International Publishing, pp. 127–34.

Larkin, P.A. (1988). The future of fisheries management: managing the fisherman. *Fisheries*, **13**(1), 3–9.

Lavelle, M. (2015). Collapse of New England's iconic cod tied to climate change. *Science*, **358**(6362).

Lewin, A. Y., and Stephens, C. U. (1993). Designing postindustrial organizations: Combining theory and practice. In G. P. Huber and W. H. Glick, eds., *Organizational Change and Redesign*, Oxford: Oxford University Press, pp. 393–409.

Lynch, A. J., Myers, B., Chu, C., et al. (2016). Climate change effects on North American inland fish populations and assemblages. *Fisheries*, **41**(7), 346–61.

Melnychuk, M. C., Banobi, J. A., and Hilborn, R. (2014). The adaptive capacity of fishery management systems for confronting climate change impacts on marine populations. *Reviews in Fish Biology and Fisheries*, **24**(2), 561–75.

Meng, K. C., Oremus, K. L., and Gaines, S. D. (2016). New England cod collapse and the climate. *PLOS ONE*, **11**(7), e0158487.

Ministry for Primary Industries. (2010). *New Zealand Fisheries at a Glance.* Wellington, NZ: Ministry for Primary Industries.

National Marine Fisheries Services. (2014). *Fisheries Economics of the United States, 2012.* NOAA Technical Memorandum NMFS-F/SPO-137. Washington, DC: NMFS.

Organization for Economic Cooperation and Development. (2004). *Country Note on National Fisheries Management Systems – Norway*. Paris: OECD.

Putnam, Robert D. 2000. *Bowling Alone: The Collapse and Revival of American Community*. New York: Simon and Schuster.

Poloczanska, E. S., Brown, C. J., and Sydeman, W. J. (2013). Global imprint of climate change on marine life. *Nature Climate Change*, **3**(10), 919–25.

Popkin, S. L., and Dimock, M. A. (2010). Political knowledge and citizen competence. In S. L. Elkin and K. E. Soltan, eds., *Citizen Competence and Democratic Institutions*, State College, PA: Penn State Press, pp. 117–46.

Pörtner, H. O., and Knust, R. (2007). Climate change affects marine fishes through the oxygen limitation of thermal tolerance. *Science*, **315**(5808), 95–7.

Prasad, P., and Elmes, M. (2005). In the name of the practical: Unearthing the hegemony of pragmatics in the discourse of environmental management. *Journal of Management Studies*, **42**(4), 845–67.

Richardson, A. J., Brown, C. J., Brander, K., et al. (2012). Climate change and marine life. *Biology Letters*. doi:10.1098/rsbl.2012.0530.

Rittel, H. W. J., and Webber, M. M. (1973). Dilemmas in a general theory of planning. *Policy Sciences*, **4**(2), 155–69.

Sandberg, M. G., Volden, G. H., and Aarhus, I. J. (2009). Betydningen av fiskeri og havbruksnæringen for Norge i 2007: en ringvirkningsanalyse. *Trondheim, SINTEF*.

Schiel, D. R., Steinbeck, J. R., and Foster, M. S. (2004). Ten years of induced ocean warming causes comprehensive changes in marine benthic communities. *Ecology*, **85**(7), 1833–39.

St. Martin, K. (2006). The impact of 'community' on fisheries management in the U.S. northeast. *Geoforum*, **37**(2), 169–84.

Statistics Norway. (2017). *Fisheries, 2016, Preliminary Figures*.

Stenevik, E. K., and Sundby, S. (2007). Impacts of climate change on commercial fish stocks in Norwegian waters. *Marine Policy*, **31**(1), 19–31.

Sumaila, U. R., Cheung, W. W. L., Lam, V. W. Y., et al. (2011). Climate change impacts on the biophysics and economics of world fisheries. *Nature Climate Change*, **1**(9), 449–56.

Sundby, S., and Nakken, O. (2008). Spatial shifts in spawning habitats of Arcto-Norwegian cod related to multidecadal climate oscillations and climate change. *ICES Journal of Marine Science*, **65**(6), 953–62.

Tirrell, A. (2017). Sociocultural institutions in Norwegian fisheries management. *Marine Policy*, **77**(1), 37–43.

Tocqueville, Alexis de. 1969 (1835). "Democracy in America," edited by JP Mayer, translated by George Lawrence. *Garden City, NY: Anchor* Press.

Torney-Purta, J., and Richardson, W. K. (2004). Anticipated political engagement among adolescents in Australia, England, Norway and the United States. In J. Demaine, ed., *Citizenship and Political Education Today*, London: Palgrave Macmillan, pp. 41–58.

Vaidyanathan, G. (2017). Inner workings: Climate change complicates fisheries modeling and management. *Proceedings of the National Academy of Sciences*, **114**(32), 8435–7.

Weatherdon, L. V., Ota, Y., Jones, M. C., et al. (2016). Projected scenarios for coastal first nations' fisheries catch potential under climate change: management challenges and opportunities. *PLOS ONE*, **11**(1), e0145285.

Willis, T. J., Handley, S. J., Chang, F. H., et al. (2007). Climate change and the New Zealand marine environment. *NIWA Client Report NeL2007–025 for the Department of Conservation, Wellington*.

10

Policy Options for Marine Fisheries
Potential Approaches in a Changing Climate

WENDY E. MORRISON AND VALERIE TERMINI

Introduction

As the climate changes, marine ecosystems will continue to undergo significant physical and chemical changes. Some anticipated shifts include increases in temperature, decreases in pH (acidification), changes in currents, rises in sea level, and changes to freshwater inputs (Stock et al., 2011; Rhein et al., 2013). These changes will, in turn, affect species abundances and/or distributions, species phenology, marine habitats, marine communities, and the resulting ecosystem productivity and function (Badjeck et al., 2010; Stock et al., 2011). The effects of the changing climate on marine ecosystems are already being observed. Shifting distributions and abundances of fish species are being documented worldwide (Perry et al., 2005; Cheung et al., 2010; Hoegh-Guldberg and Bruno, 2010), including several important commercial and recreational fish species in the United States (Nye et al., 2009; Mills et al., 2013; Pinksy et al., 2013). In addition, shifts in productivity have already been identified for some fish stocks (i.e., populations) (Klaer et al., 2015). Fishery managers and scientists need to be prepared to anticipate these changes even though they are typically making decisions on a much shorter time frame.

Environmental change is not new to fisheries. For example, anchovy and sardine populations have well-known oscillations that began prior to human exploitation (Baumgartner et al., 1992). However, climate change differs from past environmental change because, even though there will be variations in magnitude across years, the overall trends are drifting away from recent average conditions. The future may bring environmental conditions not yet experienced in the historical record, and the speed of these changes is predicted to be faster than at any time in history, challenging the adaptation capabilities of marine species. In addition, the uncertainty associated with when and how climate change will affect stocks results in decreased confidence in stock projections through time. Biological, ecological, and social responses to these changes are hard to predict but have the potential to occur at any time.

It is unclear whether the common approaches currently being used in fisheries management will be sufficient in the future (Field and Francis, 2002; Plaganyi et al., 2011; Tingley et al., 2014). The goal of this chapter is to review scholarly literature in order to outline suggested or implemented ideas for adapting fisheries to a changing climate.

Climate Change Adaptation Can Be Proactive or Reactive

Climate change adaption can be divided into two main alternatives: proactive or reactive management. However, these labels are somewhat arbitrary; what we describe as reactive options can also be used proactively, and vice versa. Reactive management focuses on responses to changes that have already occurred, whereas proactive management prepares for changes before their effects occur. Proactive management is often aimed at increasing the resilience of species, ecosystems, or fishermen. Because adaptation actions can include costs (resources to implement adaptation actions, etc.), there are trade-offs between adapting now versus in the future (Grafton, 2010), and there is a balance between actions that occur too early or too late (Fussel, 2007). No matter how proactive managers are in planning for climate change, there will still be a need for reactive management because scientists will be unable to predict all future changes (Schindler and Hilborn, 2015).

Integrating fisheries monitoring into existing management frameworks is important to making more informed decisions and could be an important component of many strategies related to managing fisheries in a changing climate (Madin et al., 2012; Plaganyi et al., 2013). Current monitoring programs could benefit from a reevaluation of their design and scope in light of climate change (e.g., Bates et al., 2014) and integration of newer technologies, such as cellular phone "apps" (software applications) that allow real-time catch or sightings information and satellites that measure primary productivity.

We divide the management options into five main categories (see Table 10.1). Managers are encouraged to investigate the trade-offs associated with each approach. They may determine that a mixed approach is the best option. For the most part, the alternatives discussed here are within the control of fisheries managers. For a recent review that covers fisheries adaptation options for specific climate changes, see Pinsky and Mantua (2014).

Reactive Management: Accounting for Observed Changes

Following a reactive (also called adaptive) management approach – where management is rapidly updated (e.g., emergency fishery closure) in response to identified changes in the environment or resource – may be more successful than undertaking a proactive management approach that depends on predictive modeling of future environmental–fisheries relationships that may or may not be accurate (Plaganyi et al., 2011; Schindler and Hilborn, 2015). However, even in a reactive management framework, it can be useful to consider predictions about possible future states and to develop management measures that are robust to future changes.

Creating Flexible, Nimble Management Systems

The key to successful fisheries management in a changing climate is a flexible, nimble, and adaptive management system at all levels of management (Johnson and

Table 10.1. *Summary of management alternatives*

	Possible options					
Reactive management	Creating flexible, nimble management systems	Adjusting reference points	Adjusting fisheries allocations	Adjusting fishing practices or gear	Enhancing or translocating stocks	Managing to promote adaptive capacity
Proactive management to increase resilience of stocks or species	Incorporating environmental parameters into stock assessments and management	Managing for uncertainty: scenario planning	Protecting age structure of species and/or old females	Decreasing existing stressors		
Proactive management to increase resilience of ecosystems	Protecting key habitats and species	Designing appropriate marine reserves	Applying ecosystem models to create robust management			
Proactive management to increase resilience of fishermen	Providing insurance for fishermen	Expanding flexibility in fisheries permitting	Improving flexibility in the supply chain			

Welch, 2010; Stein et al., 2013): within fisheries, between fisheries, and across jurisdictions. Creating a system that identifies when management changes are needed and is able to implement these changes in a timely manner may be an effective strategy. Management decisions need to be made with a clear understanding of the goals and priorities associated with the fishery and its ecosystem. To understand trade-offs and prioritize adaptation options, management goals should be current, clear, measurable, and forward-looking (Grafton, 2010; Stein et al., 2013).

Dynamic ocean management (DOM) is an example of a flexible management system. DOM is "management that uses near real-time data to guide spatial distribution of commercial activity," and examples of DOM exist around the world (Lewison et al., 2015). In the United States, DOM is primarily used to reduce unwanted bycatch, for example, to protect turtles in the tropical Pacific and yellowtail flounder in the northeastern Atlantic (Lewison et al., in press). For sea turtles and yellowtail flounder, bycatch advisories are provided to participating fishermen, who can then voluntarily avoid areas with higher probability of bycatch (Lewison et al., in press). Although there can be a disconnect between the time frames associated with climate change and DOM, the approach may provide a mechanism for quickly adjusting fisheries management.

Adjusting Reference Points After Changes in Species Productivity or Stock Structure

When environmental changes affect the long-term productivity of a stock, biological reference points (such as catch limits and rebuilding biomass targets) that are based on historical conditions may no longer be relevant, achievable, or appropriate (MAFMC, 2014). Changes in biological stocks, such as splitting or merging of fish populations, may affect stock identification, productivity, evaluation, and management success (Link et al., 2011). In theory, adjusting management to account for changes in productivity or stock structure makes sense and seems simple. However, it can be difficult to discern when changes in stock dynamics are simply due to normal variation around a historical average or the start of a more permanent regime shift that will cause the stock to drift away from previously observed states (Punt et al., 2014). If a regime shift is underway, waiting too long to implement management changes could result in negative effects on the resource. Brown et al. (2012) used simulation models to demonstrate how delaying management responses more than five years after a decrease in stock productivity occurs results in a greater probability of stock collapse. Alternatively, overreacting to normal environmental variability could result in management practices that are unstable and "chasing noise" (MAFMC, 2014). Both yellowtail flounder in the northeastern United States and jackass morwong in Australia have had their biological reference points adjusted to account for changes in stock productivity. An analysis of the evidence used to determine whether these changes in productivity occurred found lower weight of evidence to support the management shift in yellowtail flounder than in jackass morwong (Klaer et al., 2015).

Recent research suggests that certain biological parameters could provide predictions about which stocks are experiencing a change in productivity due to changes in climate, habitat, or community interactions. For example, declines in mean weight of fish at a given age often precede stock collapse by several years, suggesting that monitoring age and length of an indicator species could predict future changes in stock abundance (Brander, 2010). Research has found that spatial variability in catches increased prior to stock collapse (Litzow et al., 2013). However, the researchers caution that more studies are necessary to determine whether this indicator would be useful to fisheries management.

Adjusting Fisheries Allocations After Changes in Species Abundance or Distribution

Changes in species distributions can create management challenges, particularly when they cross jurisdictional boundaries. As the abundance or distribution of fish species changes in the future, following the common practice of basing catch allocations on historical catch rates (Bailey et al., 2013) may no longer be appropriate. Fish may be in a new location because their distribution has shifted, or they might be more abundant and have expanded into new habitat (Bell et al., 2014). Bates et al. (2014) have outlined a methodology for identifying species that have significantly shifted their geographical range. They noted their methodology only applies to abundant species and that improved monitoring would be needed to document range shifts in rare species.

Fishermen who have become reliant on particular fish species are unlikely to willingly decrease their allocation as stock distributions shift out of their areas where they operate. Similarly, fishermen in areas that are being colonized by shifting stocks may want to increase exploitation of the species that are now readily available in their regions. Both sides may have valid claims to the fish, creating a political situation that could lead to overexploitation and thus compromise the sustainability of fish stocks. For example, as water around the European Union (EU) increased in temperature, Atlantic mackerel stock shifted their migration pattern to the north and into the waters around Iceland and the Faroe Islands (Jensen et al., 2015). In response, Iceland and the Faroe Islands increased their harvest levels, while the EU (especially Scotland) continued their historic catch levels. The combined quota demands exceeded the sustainable harvest levels, creating a politically charged allocation dispute that included trade restrictions, landing bans, and ultimately a rejection of Iceland's application for full EU membership (Jensen et al., 2015). Multiple failed negotiations were attempted before a 2014 agreement was reached between Norway, the EU, and the Faroe Islands (Iceland chose not to participate) (Jensen et al., 2015).

The research literature mentions multiple options for adjusting allocations as the distribution of fish species moves into and out of adjacent areas or across jurisdictions of fisheries management areas or states. First, managers could create an allocation mechanism where annual allocations are based on the distribution of the stock that year (Bailey et al., 2013). For example, an agreement between tuna fishermen in the Western Pacific Ocean determines annual allocations based on a combination of historical effort in a zone (50 percent of the total) and estimated biomass in each zone each year (50 percent of the total) (Bell et al., 2013). Similarly, prearranged management responses (Brander, 2010) can be negotiated that "clearly articulate a set of rules for adjusting quotas and allocations as a function of mutually agreed upon indicators of changes in the shared stock" (Miller and Munro, 2004: 388). Finally, countries with abundant capital might purchase rights to access resources within another country's exclusive economic zone (Sumaila et al., 2011). Payments for those access rights can either be direct payments for fishing rights or side payments in an allocation agreement (Miller and Munro, 2004; Sumaila et al., 2011; Bailey et al., 2013; Pinsky and Mantua, 2014). Side payments – whereby parties that will benefit the most from an allocation agreement compensate the parties that would benefit from breaking the agreement – can help improve compliance (Munro et al,. 2004; Bailey et al., 2013). Munro et al. (2004) clarify that side payments do not have to be monetary. For example, they can include trade concessions on products other than fish.

Adjusting Fishing Practices

As species shift abundances and distributions in response to a changing climate, the mix of species caught in specific fishing gear will also change, potentially creating management issues when the overlap of targeted stocks with either nontarget stocks or protected species increases. Fishermen can adjust their fishing practices or gear to minimize these interactions. Adjusting fishing practices or gear could be considered reactive or proactive, depending on the situation. It could be reactive if the changes in practice are in response to a change in fish abundance that has already occurred, or proactive if it is in response to either a predicted change in catch composition or a need to improve resilience of the ecosystem by reducing habitat or bycatch impacts (see later). Even though there is limited discussion in the literature related to adjusting fishing practices as an adaptation to climate change, there is considerable literature available on how changes in fishing behavior (Abbott et al., 2015; Wallace et al., 2015) or changes in gear (Glass, 2000; Senko et al., 2014) can decrease habitat impacts, bycatch of sensitive stocks, or interactions with marine megafauna. Trade-offs associated with switching behaviors or gear need to be considered (Jenkins and Garrison, 2013), and compliance for using new fishing gear can be low (Orphanides and Palka, 2013). Therefore, involving fishermen early in discussions and analyses of socioeconomic and biological trade-offs can improve the success of these initiatives (Jenkins and Garrison, 2013) (see Chapter 9).

Proactive Management: Increasing Resilience of Individual Stocks or Species

Resilience is defined as the "capacity of an ecosystem [species, or industry] to absorb recurrent disturbances or shocks and adapt to change while retaining essentially the same function and structure" (McClanahan et al., 2012). Most discussions of resilience are at the scale of the ecosystem (McClanahan et al., 2012). However, the idea can be expanded to include resilience of a species, community, or industry to resist change as it occurs, and to recover after a change has occurred.

Managing for Uncertainty: Scenario Planning

Scenario planning is a methodology for identifying uncertainties, articulating possible future scenarios, and determining options that will meet management goals across multiple plausible futures (Moore et al., 2013). Scenario planning includes low-tech and high-tech options for analyzing the success of management options in meeting specific management goals. At a relatively simplistic level, written descriptions of plausible scenarios can be used to describe how climate change could affect fisheries and to determine management responses that could be successful across multiple plausible futures. In four fisheries in Australia, stakeholders used short descriptions of possible futures to brainstorm possible adaptation options, including moving fisheries to more productive areas and improving the quality of their fishery products (Lim Camacho et al., 2014).

Management strategy evaluation (MSE) is a more technical approach to answering the same questions. Management strategy evaluations normally use simulations and models to quantify how often management alternatives meet quantifiable management goals under given uncertainties. For example, Ianelli et al. (2011) simulated and tested multiple management alternatives for the Alaska pollock fishery, given expected changes in sea surface temperature and a predicted functional relationship between sea surface temperature and fish recruitment. Results showed that the status quo management alternative resulted in lower catches and a higher risk of fishery closures than alternatives that allowed for adjustments in carrying capacity (Ianelli et al., 2011). Punt et al. (2016) provide a good summary of MSEs completed to date and suggest best practices for completing future MSE analyses.

Managing to Promote Adaptive Capacity

Management decisions can have a direct impact on a fish population's ability to adapt as the climate changes (Munguia-Vega et al., 2015). Little is known about the adaptive capacity of marine fish and invertebrate populations, suggesting a priority for future research. There are three components to adaptive capacity: (1) ability of species to adjust to new conditions (i.e., plasticity in phenotypic response), (2) ability of species to relocate if or when conditions change (i.e., dispersal), and (3) ability of species to evolve strategies to survive in new conditions (which requires sufficient genetic diversity) (Beever et al., 2015). Management approaches are largely restricted to efforts that improve the genetic diversity of managed species, but management aimed at improving or assisting movement of adult fish may also be useful.

Given the high rate of expected environmental change in the future, genetic adaptation to climate change may be necessary. An aim of management should therefore be to increase or preserve current genetic diversity. For example, genetic diversity at the edges of a population may differ from the rest of the population. The rear edge often has high genetic diversity as it preserves historical alleles (Pauls et al., 2013). Conversely, even though the leading edge of a population can contain lower diversity, it can also contain the alleles best adapted to the new conditions and could thus be the "source for most of the surviving lineages and persisting alleles" (Pauls et al., 2013: 927). Therefore, scientists have suggested that as fish distributions advance or contract, it may be prudent to limit fishing of populations located at both edges of the species' distribution to help individuals with advantageous adaptations establish in new areas or to remain viable in their historical areas (Brander, 2010; Pinksy and Fogarty, 2012).

Protecting Age Structure

Protecting or recovering the full age structure of a fish stock or population can increase that population's resilience to a changing environment. The importance of a full age structure, including the presence of large females, to a population is not new (Palumbi, 2004; Planque et al., 2010). Large females tend to have larger, healthier eggs and more of them, all of which contribute to the subsequent recruitment success of a population (Planque et al., 2010). Recent research suggests that a full age structure is even more important for a population's persistence if it experiences environmental changes (Field and Francis, 2002; Planque et al., 2010; Rouyer et al., 2011). Management options that may improve a population's age structure include increased use of marine protected areas (see Chapter 11), maximum size or slot limits, grates that stop large fish from entering nets, using gear that improves postrelease survival, or closures of fisheries during times when, and over areas where, large individuals congregate.

Incorporating Environmental Parameters into Stock Assessments and Management Measures

Because the productivity of many fish stocks is directly influenced by environmental variables, it may be beneficial to integrate appropriate environmental parameters into stock assessments and management decisions. For example, the management of US Pacific sardine accounts for temperature-dependent productivity by using an environmentally dependent maximum sustainable yield parameter and adjusting harvest levels to match expected population size (PFMC, 2014). In another example, scientists used water temperature to improve predictions of butterfish habitat along the US east coast, which reduced uncertainty in the butterfish stock assessment and increased catch limits for this species (NEFSC, 2014). (For an overview of analytical methods for incorporating environmental variables into fisheries models, see Keyl and Wolff, 2008.)

However, multiple studies highlight challenges and indicate that there may be limited advantages for incorporating environmental parameters into stock assessments and management. Punt et al. (2014) reviewed simulation studies as well as current fisheries that incorporate the environment into assessments and found limited success: "modifying management strategies to include environmental covariates did not improve the ability to achieve management goals over time scales relevant to short and medium-term fisheries management decision making much, if at all. They did so only if information on environmental factors driving the system was well known" (Punt et al., 2014: 2217). Therefore, Punt et al. (2014), suggest that a better option is to assess which management strategies are successful across multiple possible future conditions, a process known as scenario planning. However, as knowledge of relationships between managed fish stocks and environmental dynamics continues to improve, there will be more justification for incorporating environmental factors into stock assessment and management (Keyl and Wolff, 2008). Also, as the ability to forecast environmental conditions on the seasonal to decadal time scales that are most germane to fisheries management continue to improve, fisheries managers will be better equipped to facilitate climate-ready fisheries management within existing management frameworks.

Decreasing Existing Stressors

One strategy for increasing resilience of fish stocks (and their ecosystems) to climate change is to decrease stressors already affecting the stocks (Sumaila et al., 2011; Stein et al., 2013; Pinsky and Mantua, 2014). Scientists theorize that species experiencing other stressors (that is, subject to cumulative impacts) are more likely to have faster and more acute reactions to climate change (Sumaila et al., 2011; Stein et al., 2013). Stress on living marine resources can come from many sources, including high mortality from fishing, habitat degradation, invasive species, disease, pollution, and hypoxia. Fishing mortality is one source of stress on stocks that is within the control of fisheries managers. Given that scientists are uncertain about how climate change will affect fish stocks, it has been suggested that managers be more precautionary in their decisions regarding catch limits. Increased precaution can include increasing the buffer around allowable catch limits to account for increased uncertainty (Johnson and Welch, 2010; Pinksy and Mantua, 2014). However, not everyone supports this idea (Schindler and Hilborn, 2015). Trade-offs between economic, ecological, and social impacts need to be considered (Johnson and Welch, 2010).

Location-specific planning can be used to assess what stressors could be reduced and to develop regional management responses. For example, although ocean acidification is increasing globally, there are also local contributors to the chemistry of coastal waters where acidification effects may disproportionately affect certain coastal communities (Kelly et al., 2011; Ekstrom et al., 2015). Developing regional planning bodies or using existing regional frameworks might be helpful in developing a management response to the many effects on the marine environment from land-based sources. For example, a 2012 Washington State Blue Ribbon Panel on Ocean Acidification identified local actions to combat ocean acidification.

Enhancing or Translocating Stocks

As climate change affects economically or ecologically important fish species, fisheries managers may want to consider more resource-intensive options for sustaining these important species, including active enhancements and translocations of stocks (Madin et al., 2012; Lorenzen et al., 2013; Tingley et al., 2014). To date, there have been limited analyses of these options for marine systems. Translocation, also called assisted migration, is "the intentional translocation or movement of species outside of their historical ranges in order to mitigate actual or anticipated biodiversity losses caused by anthropogenic climate change" (Hewitt et al., 2011). For example, in Tasmania, active translocation of rock lobster has been considered as a methodology for reducing the abundance of sea urchins that have recently expanded their range and are denuding important kelp habitats (Madin et al., 2012). Translocation is controversial because it creates complex scientific, policy, and ethical questions (Hewitt et al., 2011). There are concerns about unintended consequences, such as the spreading of diseases, parasites, and invasive species (Hoegh-Guldberg et al., 2008; Tingley et al., 2014).

Stock enhancement is defined as "a set of management approaches that involve the release of cultured organisms to enhance or restore fisheries" and can include sea ranching (release for direct recapture), stock enhancement (continued release into wild stocks), or restocking (temporary releases aimed at building up wild stock) (Lorenzen et al., 2013). The use of stock enhancements as a response to climate change has not received much discussion in research literature. Lorenzen et al. (2013) have outlined a responsible approach to stock enhancement, and they created modeling approaches that allow for appraisal of the costs and benefits of possible enhancement projects before implementation. Given the magnitude of expected climate change, stock enhancements of important species may increase in the future.

Proactive Management: Managing to Increase Ecosystem Resilience

Climate change is expected to increase the need to account for ecosystem interactions as productivity, distributions, species interactions, and habitats adjust to changing environmental conditions. Scientists recommend moving toward ecosystem-based fisheries management (EBFM) to increase the health and resilience of ecosystems (a full discussion of EBFM is beyond the scope of this chapter). Management that maintains ecosystem heterogeneity may better retain ecosystem functionality and improve the stability of the resources (Schindler and Hilborn, 2015). As the climate changes, it may be unrealistic to preserve ecosystems in their historical conditions (Stein et al., 2014). Instead, it may be more realistic to manage fisheries to maintain key ecological functions (Moore et al., 2013).

Protecting Key Habits and Species

Because healthy marine ecosystems are more resilient to environmental changes, managers should consider regulations that protect key habitats and species. Focusing management on species that fulfill key ecological functions may be necessary for maintaining both ecosystem function and resiliency as the environment changes (MacNeil et al., 2010; Tingley et al., 2014). Improving our understanding about which species have important functional roles in an ecosystem and how these roles may change as the environment adjusts should be prioritized.

Gear modifications that reduce effects on habitats will result in a less stressed, more resilient ecosystem (Sumaila et al., 2011). Where habitats have become degraded, active restoration or creation of new habitat may be a viable management option. For example, creating reserve zones behind current stands of mangrove trees (Gilman et al., 2008) could allow larval habitats, such as wetlands, to migrate inland as sea level rises. In a more extreme example, van Oppen et al. (2015) suggest assisted evolution in habitat-forming coral species may be necessary to help these vulnerable species adapt to climate change.

Designing Appropriate Marine Reserves

Marine reserves are a tool that can help maintain ecosystem resilience in a changing climate (Bellwood et al., 2004; Micheli and Halpern, 2005). In theory, protecting a section of habitat and subset of individuals from the effects of fishing or other human uses can increase the resiliency to climate effects of both the species being protected and the associated ecosystem (Bellwood et al., 2004). For example, reserves protect multiple trophic levels and help retain the functional diversity of an area (Micheli and Halpern, 2005), thereby improving its ability to maintain basic ecosystem functions as the environment changes (Bellwood et al., 2004). Marine reserves also provide locations to observe and study how ecosystems react to climate change without the added stress of fishing.

There are three options for creating marine reserves that are effective even in the face of a changing climate: (1) Managers can locate reserves to include the habitat or species we currently want to protect, in addition to the areas where we expect habitats or species to move (Hobday, 2011). (2) Managers can periodically reexamine and modify closures to ensure that they remain centered on core areas of stock distribution (Campbell, 2013) and that they are maintaining management goals. (3) Managers can create reserves to be dynamic, whereby boundaries are tied to current environmental conditions (Hobday et al., 2010; Campbell, 2013; Pinsky and Mantua, 2014).

Proactive Management: Managing to Increase Resilience of Fishing Businesses

The idea of resilience can be expanded to include not only fish species but also the resilience of fisheries and fishing businesses (Pinsky and Mantua, 2014). As fish stocks adjust their distributions and abundance due to climate change, fishing effort may also have to adjust by changing the species targeted and the locations and times fished, as well as changing the landing or processing locations (Haynie and Pfiefer, 2012). Communities that are reliant on more than one fishery, or that have alternative livelihood options, will be better able to adjust as climate change affects marine resources (Mathis et al., 2015). Similarly, overcapitalized fisheries operating with a thin profit margin will be limited in their ability to experiment with new practices or to remain viable as the environment changes.

Expanding Flexibility in Fisheries Permitting

To adapt to a changing climate, fishermen will need the flexibility to adjust where, when, and what they catch, as well as the associated permits, depending on conditions (Miller and Munro, 2004; MacNeil et al., 2010; Campbell, 2013; Mills et al., 2013; Schindler and Hilborn, 2015). Systems locking fishermen into specific species, locations, or gears may reduce fishing flexibility (Grafton, 2010; Schindler and Hilborn 2015), and options for amending or switching permits should be considered (Mills et al., 2013). For example, policies may need to be created that facilitate access to loans for operators who want to diversify by purchasing quota in other fisheries (Kasperski and Holland, 2013) or by expanding their business to include aquaculture.

Some experts suggest that changing to rights-based management (such as individual fishing quotas) could increase the adaptability of fishermen (Grafton, 2010; Sumaila et al., 2011), whereas others argue the opposite (Campbell, 2013). Differences in interpretation may depend on the design elements of programs. Examples that limit flexibility include requirements to land fish in specific areas or to use specific processors, high entrance costs into a fishery, and single-species permits. Alternatively, rights-based management can provide fishermen incentives to experiment with alternative fishing methods that could increase efficiency or decrease environmental impacts. Even though rights-based management can decrease the diversification of fishermen, it provides the opportunity for fishermen to build a portfolio of harvest privileges that can reduce their income risk (Kasperski and Holland, 2013). In addition, scientists have suggested that policies that anchor quota within communities rather than individual businesses could create a robust system better able to adapt to environmental change (Schindler and Hilborn, 2015).

Providing Insurance for Fishermen

Extending insurance to fishermen to cover years with low catch could increase their financial stability. Similar to crop insurance for farmers, an insurance program for fishermen run by the US federal government has been discussed (Mumford et al., 2009). In brief, fishing insurance provides a program whereby fishermen pay an annual insurance premium that guarantees them some proportional payout if catch rates are low. Positive benefits to this stability include decreasing the incentives for fishermen to keep fishing when biomass is at low levels, as well as improved fishing practices if eligibility for insurance were to be tied to sustainable fishing practices (Mumford et al., 2009). Cons of fishing insurance include the possibility of fishermen reducing effort and "gaming" the system, and a decreased incentive to test new fishing practices that decrease catch on limiting species (Mumford et al., 2009).

Improving Flexibility in the Supply Chain

Changes to the composition, magnitude, and timing of landings could be complicated if the shore-side processing and supply chain are not adaptable. For example, a 2012 heat wave in the Gulf of Maine led to an increase in American lobster landings, and the processing facilities and markets were unable to respond appropriately (Mills et al., 2013). Due to the lack of flexibility in the supply chain, fishermen experienced an unexpected decrease in the price they received for their catch, which resulted in severe economic hardship and the need to immediately reduce fishing effort and landings (Mills et al., 2013). Identifying key elements along a supply chain could help identify adaptation strategies that would improve the resilience of the markets to changes in supply or demand (Plaganyi et al., 2014).

Conclusion

Climate change will continue to affect the abundance and distribution of fisheries resources. Because fisheries management occurs in an uncertain and changing environment, the foundation already exists for management to adjust as the climate changes. The list of management alternatives provided here is not comprehensive, and there is no one "right" answer to how marine fisheries can be adapted to climate change. Suitable approaches will differ depending on local conditions, such as life history traits of the species being managed, the type of management being implemented, the fishing community, and the resources available for monitoring and modeling.

In general, management actions that seek to increase management flexibility and provide incentives to the fishing industry to try new approaches, while preserving genetic diversity of the fished populations, should prove to be beneficial. Managers should consider the advantages and disadvantages of different alternatives when determining the appropriate mix of options for their situation. This should include estimating the risks associated with each action (or inaction), where risk includes the probability of different consequences. Other questions that should be considered include what the associated costs and benefits will be, which options would be more acceptable to fishermen, which options require an update to government mandates, and which of them are feasible under current legal regimes. Managers should involve constituents and discuss what mix of existing and new management options would be most appropriate for their fisheries. The analysis in this chapter should assist managers to consider various management approaches and options. New research, ideas, and options should emerge as fisheries management across the globe grapples with climate change and determines what works and what does not work in each region. As with other ocean governance issues, new research, ideas, and options should emerge as management across the globe grapples with climate change and determines what works and what does not work in each region.

Acknowledgments

We appreciate the constructive feedback from K. Abrams, B. Arnold, Y. deReynier, M. Johnson, J. Link, P. Lynch, S. McAfee, F. Pflieger, and M. Nelson on earlier drafts of this chapter. The views expressed herein are the authors and do not necessarily reflect the views of the National Oceanic and Atmospheric Administration or the state of California or any of its subagencies. Ideas in this chapter have not been evaluated for feasibility and applicability under current US legal authorities. Inclusion of policy alternatives in this chapter does not represent endorsement by the National Marine Fisheries Service or the state of California.

References

Abbott, J., Haynie, A., and Reimer, M. (2015). Hidden flexibility: Institutions, incentives, and the margins of selectivity in fishing. *Land Economics*, **91**(1), 169–95.

Badjeck, M., Allison, E., Halls, A., and Dulvey, N. (2010). Impacts of climate variability and change on fishery-based livelihoods. *Marine Policy*, **34**(3), 375–83.

Bailey, M., Ishimura, G., Paisley, R., and Sumaila, U. (2013). Moving beyond catch in allocation approaches for internationally shared fish stocks. *Marine Policy*, **40**, 124–36.

Bates, A., Pecl, G., Frusher, S., et al. (2014). Defining and observing stages of climate-mediated range shifts in marine systems. *Global Environmental Change*, **26**, 27–38.

Baumgartner, T., Soutar, A., and Ferreira-Bartrina, V. (1992). Reconstruction of the history of Pacific sardine and northern anchovy populations over the past two millennia from sediments of the Santa Barbara Basin, California. *CalCOFI Reports*, **33**, 24–40.

Beever, E., O'Learly, J., Mengelt, C., et al. (2015). Improving conservation outcomes with a new paradigm for understanding species' fundamental and realized adaptive capacity. *Conservation Letters*, **9**(2), 131–37.

Bell, J., Ganachaud, A., Gehrke, P., et al. (2013). Mixed response of tropical Pacific fisheries and aquaculture to climate change. *Nature Climate Change*, **3**(6), 591–99.

Bell, R., Hare, J., Manderson, J., and Richardson, D. (2014). Externally driven changes in the abundance of summer and winter flounder. *ICES Journal of Marine Science*, **71**(9), 2416–28.

Bellwood, D., Hughes, T., Folke, C., and Nystrom, M. (2004). Confronting the coral reef crisis. *Nature*, **429**(6994), 827–33.

Brander, K. (2010). Impacts of climate change on fisheries. *Journal of Marine Systems*, **79**(3), 389–402.

Brown, C., Fulton, E., Possingham, H., and Richardson, A. (2012). How long can fisheries management delay action in response to ecosystem and climate change? *Ecological Applications*, **22**(1), 298–310.

Campbell, C. (2013). Implications of changing ecosystems for fisheries managers. In *Managing Our Nations Fisheries 3: Advancing Sustainability. Proceedings of a Conference on Fisheries Management in the United States.* Portland, Oregon. pp 160–67. Available at https://static1.squarespace.com/static/534429e9e4b0cc91 fc021a15/t/534572f6e4b012983d5d9606/1397060342418/MONF+Proceedings+-+ Session+2.pdf

Cheung, W., Lam, V., Sarmiento, J., et al. (2010). Large-scale redistribution of maximum fisheries catch potential in the global ocean under climate change. *Global Change Biology*, **16**(1), 24–35.

Ekstrom, J., Suatoni, L., Cooley, S., et al. (2015). Vulnerability and adaptation of US shellfisheries to ocean acidification. *Nature Climate Change*, **5**(3), 207–14.

Field, J. and Francis, R. (2002). Cooperating with the environment: Case studies of climate and fisheries in the Northern California current. In N. McGinn, ed., *Fisheries in a Changing Climate*. American Fisheries Society Symposium 32, Bethesda, Maryland, pp. 245–60.

Fussel, H. (2007). Adaptation planning for climate change: Concepts, assessment approaches and key lessons. *Sustainability Science*, **2**(2), 265–75.

Gilman, E., Ellison, J., Duke, N., and Field, C. (2008). Threats to mangroves from climate change and adaptation options. A review. *Aquatic Botany*, **89**(2), 237–50.

Glass, C. (2000). Conservation of fish stocks through bycatch reduction: A review. *Northeastern Naturalist*, **7**(4), 395–410.

Grafton, R. (2010). Adaptation to climate change in marine capture fisheries. *Marine Policy*, **34**(3), 606–15.

Haynie, A. and Pfiefer, L. (2012). Why economics matters for understanding the effects of climate change on fisheries. *ICES Journal of Marine Science*, **69**(7), 1160–7.

Hewitt, N., Klenk, N., Smith, A., et al. (2011). Taking stock of the assisted migration debate. *Biological Conservation*, **144**(11), 2560–2.

Hobday, A. (2011). Sliding baselines and shuffling species: Implications of climate change for marine conservation. *Marine Ecology*, **32**(3), 392–403.

Hobday, A., Hartog, J., Timmiss, T., and Fielding, J. (2010). Dynamic spatial zoning to manage southern Bluefin tuna (*Thunnus maccoyii*) capture in a multi-species longline fishery. *Fisheries Oceanography*, **19**(3), 243–53.

Hoegh-Guldberg, O., Hughes, L., McIntyre, S., et al. (2008). Assisted colonization and rapid climate change. *Science*, **321**, 345–6.

Hoegh-Guldberg, O. and Bruno, J. (2010). The impact of climate change on the world's marine ecosystems. *Science*, **328**(5985), 1523–8.

Ianelli, J., Hollowed, A., Haynie, A., Mueter, F., and Bond, N. (2011). Evaluating management strategies for eastern Bering Sea walleye pollock (*Theragra chalcogramma*) in a changing environment. *ICES Journal of Marine Science*, **68**, 1297–304.

Jensen, F., Frost, H., Thogersen, T., Andersen, P., and Anderson, J. (2015). Game theory and fish wars: The case of the Northeast Atlantic Mackerel Fishery. *Fisheries Research*, **172**, 7–16.

Jenkins, L. and Garrison, K. (2013). Fishing gear substitutions to reduce bycatch and habitat impacts: An example of social-ecological research to inform policy. *Marine Policy*, **38**, 293–303.

Johnson, J. and Welch, D. (2010). Marine fisheries management in a changing climate: A review of vulnerability and future options. *Reviews in Fisheries Science*, **18**(1), 106–24.

Kasperski, S. and Holland, D. (2013). Income diversification and risk for fishermen. *Proceedings of the National Academy of Sciences*, **110**(6), 2076–81.

Kelly, R., Foley, M., Fischer, W., et al. (2011). Mitigating local causes of ocean acidification with existing laws. *Science*, **332**(6033), 1036–37.

Klaer, N., O'Boyle, R., Deroba, J., et al. (2015). How much evidence is required for acceptance of productivity regime shifts in fish stock assessments: Are we letting managers off the hook? *Fisheries Research*, **168**, 49–55.

Keyl, F. and Wolff, M. (2008). Environmental variability and fisheries: What can models do? *Reviews in Fish Biology and Fisheries*, **18**(3), 273–99.

Lewison, R., Hobday, A., Maxwell, S., et al. (2015). Dynamic ocean management: Identifying the critical ingredients of dynamic approaches to ocean resource management. *Bio Science*, **65**(5), 486–98.

Lim-Camacho, L., Hobday, A. J., Bustamante, R. H., et al. (2014). Facing the wave of change: stakeholder perspectives on climate adaptation for Australian seafood supply chains. *Regional Environmental Change*, **15**, 595–606.

Link, J., Nye, J., and Hare, J. (2011). Guidelines for incorporating fish distribution shifts into a fisheries management context. *Fish and Fisheries*, **12**(4), 461–9.

Litzow, M., Mueter, F., and Urban, J. (2013). Rising catch variability preceded historical fisheries collapses in Alaska. *Ecological Applications*, **23**(6), 1475–87.

Lorenzen, K., Agnalt, A., Blankenship, H., et al. (2013). Evolving context and maturing science: Aquaculture-based enhancement and restoration enter the marine fisheries management toolbox. *Reviews in Fisheries Science*, **21**(3–4), 213–21.

MacNeil, M., Graham, N., Cinner, J., et al. (2010). Transitional states in marine fisheries: Adapting to predicted global change. *Philosophical Transactions of the Royal Society B*, **365**(1558), 3753–63.

Madin, E., Ban, N., Doubleday, Z., et al. (2012). Socio-economic and management implications of range-shifting species in marine systems. *Global Environmental Change*, **22**(1), 137–46.

MAFMC (Mid Atlantic Fisheries Management Council). (2014). East Coast Climate Change and Fisheries Governance Workshop. March 19–21, Washington, D.C. Available at www.mafmc.org/council-events/(2014)/east-coast-climate-change-and-fisheries-governance-workshop

Mathis, J., Cooley, S., Lucey, N., et al. (2015). Ocean acidification risk assessment for Alaska's fishery sector. *Progress in Oceanography*, **136**, 71–91.

McClanahan, T., Donner, S., Maynard, J., et al. (2012). Prioritizing key resilience indicators to support coral reef management in a changing climate. *PlosOne*, **7**(8), e42884.

Micheli, F. and Halpern, B. (2005). Low functional redundancy in coastal marine assemblages. *Ecology Letters*, **8**(4), 391–400.

Miller, K. and Munro, G. (2004). Climate and cooperation: A new perspective on the management of shared fish stocks. *Marine Resource Economics*, **19**(3), 367–93.

Mills, K., Pershing, A., Brown, C., et al. (2013). Fisheries management in a changing climate: Lessons from the (2012) ocean heat wave in the Northwest Atlantic. *Oceanography*, **26**(2), 60–4.

Moore, S., Seavy, N., and Gerhart, M. (2013). Scenario planning for climate change adaptation: A guidance for resource managers. Point Blue Conservation Science and California Coastal Conservancy. Available at www.prbo.org/refs/files/12263_Moore(2013).pdf

Mumford, J., Leach, A., Levontin, P., and Kell, L. (2009). Insurance mechanisms to mediate economic risks in marine fisheries. *ICES Journal of Marine Science*, **66**(5), 950–9.

Munguia-Vega, A., Saenz-Arroyo, A., Greenley, A., et al. (2015). Marine reserves help preserve genetic diversity after impacts derived from climate variability: Lessons from the pink abalone in Baja California. *Global Ecology and Conservation*, **4**, 264–76.

Munro, G., VanHoutte, A., and Willman, R. (2004). The conservation and management of shared fish stocks: Legal and economic aspects. FAO Fisheries Technical Paper 465. Rome, FAO. Available at ftp://ftp.fao.org/docrep/fao/007/y5438e/y5438e00.pdf

NEFSC (Northeast Fisheries Science Center). (2014). *58th Northeast Regional Stock Assessment Workshop (58th SAW) Assessment Summary Report*. US Department of Commerce, Northeast Fisheries Science Center. Available at http://nefsc.noaa.gov/publications/crd/crd1403/crd1403.pdf

Nye, J., Link, J., Hare, J., and Overholtz, W. (2009). Changing spatial distribution of fish stocks in relation to climate and population size on the Northeast United States continental shelf. *Marine Ecology Progress Series*, **393**, 111–29.

Orphanides, C. and Palka, D. (2013). Analysis of harbor porpoise gillnet bycatch, compliance, and enforcement trends in the U.S. northwestern Atlantic, January 1999–May 2010. *Endangered Species Research*, **20**(3), 251–69.

Palumbi, S. (2004). Why mothers matter. *Nature*, **430**, 621–2.

Pauls, S., Nowak, C., Balint, M., and Pfenninger, M. (2013). The impact of global climate change on genetic diversity within populations and species. *Molecular Ecology*, **22**(4), 925–46.

Perry, A., Low, P., Ellis, J., and Reynolds, J. (2005). Climate change and distribution shifts in marine fishes. *Science*, **308**(5730), 1912–5.

PFMC (Pacific Fishery Management Council). (2014). Status of the Pacific Coast coastal pelagic species fishery and recommended acceptable biological catches. Stock Assessment and Fishery Evaluation for 2014. Appendix C: 2014 Pacific Sardine Stock Assessment. Available at http://www.pcouncil.org/wp-content/uploads/(2014)_CPS_SAFE_Text_FINAL.pdf

Pinksy, M. and Fogarty, M. (2012). Lagged social-ecological responses to climate and range shifts in fisheries. *Climatic Change*, **115**(3–4), 883–91.

Pinsky, M. and Mantua, N. (2014). Emerging adaptation approaches for climate-ready fisheries management. *Oceanography*, **27**(4), 17–29.

Pinsky, M., Worm, M., Fogarty, M., Sarmiento, J., and Levin, S. (2013). Marine taxa track local climate velocities. *Science*, **341**, 1239–42.

Plaganyi, E., Weeks, S., Skewes, T., et al. (2011). Assessing the adequacy of current fisheries management under changing climate: A southern synopsis. *ICES Journal of Marine Science*, **68**(6), 1305–7.

Plaganyi, E., Skewes, T., Dowling, N., and Haddon, M. (2013). Risk management tools for sustainable fisheries management under changing climate: A sea cucumber example. *Climate Change*, **119**(1), 181–97.

Plaganyi, E., vanPutten, I., Thebaud, O., et al. (2014). A quantitative metric to identify critical elements within seafood supply networks. *PlosOne*, **9**(3), e91833.

Planque, B., Fromentin, J., Cury, P., et al. (2010). How does fishing alter marine populations and ecosystem sensitivity to climate? *Journal of Marine Systems*, **79**(3), 403–17.

Punt, A., A'mar, T., Bond, N., et al. (2014). Fisheries management under climate and environmental uncertainty: Control rules and performance simulation. *ICES Journal of Marine Science*, **71**(8), 2208–20.

Punt, A., Butterworth, D., de Moor, C., De Oliveira, J., and Haddon, M. (2016). Management strategy evaluation: Best practices. *Fish and Fisheries*, **17**(2), 303–34.

Rhein, M., Rintoul, S., Aoki, S., et al. (2013). Observations: Ocean. In T. Stocker, D. Qin, G. Plattner et al., eds., *Climate Change (2013): The Physical Science Basis. Contribution of Working Group I to the Fifth Assessment Report of the Intergovernmental Panel on Climate Change.* Cambridge University Press, New York, pp. 255–315. Available at www.ipcc.ch/pdf/assessment-report/ar5/wg1/WG1AR5_Chapter03_FINAL.pdf

Rouyer, T., Ottersen, G., Durant, J., et al. (2011). Shifting dynamic forces in fish stock fluctuations triggered by age truncation? *Global Change Biology*, **17**(10), 3046–57.

Tingley, M., Darling, E., and Wilcove, D. (2014). Fine- and course-filter conservation strategies in a time of climate change. *Annals of the New York Academy of Sciences*, **1322**(1), 92–109.

Schindler, D. and Hilborn, R. (2015). Prediction, precaution and policy under global change. *Science*, **347**(6225), 953–4.

Senko, J., White, E., Heppell, S., and Gerber, L. (2014). Comparing bycatch mitigation strategies for vulnerable marine megafauna. *Animal Conservation*, **17**(1), 5–18.

Stein, B., Staudt, A., Cross, M., et al. (2013). Preparing for and managing change: Climate adaptation for biodiversity and ecosystems. *Frontiers in Ecology and Environment*, **11**(9), 502–10.

Stein, B., Glick, P., Edelson, N., and Staudt, A., eds. (2014). *Climate-Smart Conservation: Putting Adaptation Principles into Practice.* National Wildlife Federation, Washington, D.C. Available at www.nwf.org/pdf/Climate-Smart-Conservation/NWF-Climate-Smart-Conservation_5-08-14.pdf

Stock, C., Alexander, M., Bond, N., et al. (2011). On the use of IPCC-class models to assess the impact of climate on living marine resources. *Progress in Oceanography*, **88**(1), 1–27.

Sumaila, U., Cheung, W., Lam, V., Pauly, D., and Herrick, S. (2011). Climate change impacts on the biophysics and economics of world fisheries. *Nature Climate Change*, **1**(9), 449–56.

Van Oppen, M., Oliver, J., Putnam, H., and Gates, R. (2015). Building coral reef resilience through assisted evolution. *Proceedings of the National Academy of Sciences*, **112**(8), 2307–13.

Wallace, S., Turris, B., Driscoll, J., et al. (2015). Canada's pacific groundfish trawl habitat agreement: A global first in an ecosystem approach to bottom trawl impacts. *Marine Policy*, **60**, 240–8.

11

Large Marine Protected Areas and Ocean Resilience
Stakeholder Conflict in Pelagic Seas

JUSTIN ALGER

Introduction

The existential threat to marine ecosystems posed by climate change, overfishing, and coastal development has led many states to significantly expand the scope of their conservation efforts. States have protected more ocean space in the previous decade than in all of history. This unprecedented rate of ocean protection is the result of just a handful of large marine protected areas (MPAs), defined here as those exceeding 200,000 km^2. In 2017 there were over 11,000 designated MPAs, but just the largest 19 of these account for more than 80 percent of the total area protected globally. States have established all but one of these large MPAs – Australia's Great Barrier Reef Marine Park – since 2006. This expanded scope of conservation reflects a new norm of large MPAs in global politics (Alger and Dauvergne, 2017a, 2017b). One of the defining features of this norm involves greater protection of pelagic seas, which historically have been the least protected ecosystems on the planet (Game et al., 2009). As states target these previously neglected pelagic waters for marine protection, newfound tension is being created between commercial fishers increasingly worried about losing their fishing grounds and governments looking to strike a balance between industry concerns and the necessity for better environmental protections.

The fishing industry feels under siege as these large-scale protections threaten to encroach on areas traditionally controlled by usually industry-friendly fisheries management organizations. Industry proponents argue that many of these fisheries are already sustainably managed and that large MPAs are an unscientific fisheries management tool (Hilborn, 2015). They claim that there is a disconnect between the main threat to ocean ecosystems – climate change – and the target of government regulations – the fishing industry. But there is growing scientific support to suggest that well-managed MPAs enhance the resilience of marine ecosystems to the effects of climate change. In large MPA consultations and negotiations, industry stakeholders not only need to demonstrate that they are fishing sustainably but also that it is more important that a given area remain open to fishing than closed for conservation. The imperative of insulating marine ecosystems from climate impacts is leading to heightened tension between environmentalists, conservation-minded governments, and the commercial fishing industry.

This chapter makes two arguments about these stakeholder dynamics. First, governments have so far been highly responsive to fishing industry concerns in how they set boundaries for and regulate large MPAs. And second, despite this government responsiveness and the minimal impact of large MPAs on the commercial fishing industry, its opposition to them has been disproportionately high. Minor industry interests are proving influential in large MPA negotiations, despite the ongoing push for more ambitious marine protections that large MPAs represent. These interests are often overshadowing the growing imperative of a more precautionary approach to ocean governance. The arguments in this chapter are based on 74 field interviews with key individuals from government, civil society, and industry in five large MPA case studies: the Papahānaumokuākea and the Pacific Remote Islands Marine National Monuments in the United States, the Coral Sea Commonwealth Marine Reserve (CMR) in Australia, the Palau National Marine Sanctuary (PNMS), and the Pitcairn Islands Marine Reserve in the UK Overseas Territories.

The Environmental Case for Large MPAs

Despite protections that cover millions of kilometers of previously unprotected ocean space, governments have strategically targeted areas with low fisheries output. One of the primary criticisms of large MPAs has in fact been that they are too remote from commercial activity, and therefore do not meaningfully address the root causes of ocean decline (De Santo, 2013; Devillers et al., 2015; Jones and De Santo, 2016). As one marine scientist put it, governments have focused their efforts on "low-hanging fruit" – areas that are biodiversity rich but not heavily commercially exploited (interview with Program Director, Sea Around Us, University of British Columbia, Vancouver, BC, August 17, 2015). Many prominent environmental groups, such as Greenpeace, do not pursue large MPAs in national waters, preferring to focus their efforts on commercially exploited marine spaces. Environmental groups that do campaign for large MPAs prioritize identifying target areas that are "politically feasible" or "pristine." Large MPA advocates and critics alike recognize that large MPAs have so far minimized short-term commercial fisheries impact by design. Their remoteness is what has allowed states to create such massive reserves, the majority of which are either fully no-take or have large no-take zones within them.

That remoteness has also fueled skepticism about the desirability and effectiveness of large MPAs in combatting ocean decline. Advocates argue in response that even if they are remote, large MPAs prevent the future commercial exploitation of biodiversity-rich areas. The growing technological sophistication of the fishing industry has increased its ability to fish in increasingly remote areas, so this argument is not without merit (DeSombre and Barkin, 2011; Barkin and DeSombre, 2013). Protecting remote marine ecosystems is therefore what one prominent marine campaigner called an "insurance policy" (interview with Program Director, National Geographic Society, Washington, DC, September 17, 2015). They act as a bulwark against future encroachments, ensuring that pristine areas already threatened by climate change are not further undermined by extraction. Studies around the Line Islands and Pitcairn Islands suggest that even low levels of extractive activity can lead to significant reductions in biomass, undermining ecosystem health and resilience (DeMartini et al., 2008; Sandin et al., 2008; Friedlander et al., 2014). The scientific evidence for a precautionary approach to protecting pristine marine ecosystems is growing, strengthening the case for fully no-take MPAs.

The primary purpose of large MPAs is to limit or prohibit commercial activity within their boundaries, but they also play an increasingly important role in climate mitigation and adaptation (Roberts et al., 2017). Large, well-managed MPAs promote healthy, intact marine ecosystems with a greater abundance and diversity of species (Edgar et al., 2014). These healthy ecosystems are essential for the ocean's prominent role in carbon sequestration and storage, with one-third of human CO_2 emissions to date absorbed by oceans. Limiting direct human stressors on marine habitats also greatly increases their resilience to the effects of climate change. Ecosystems insulated from commercial fishing and other direct stressors are better positioned to recover from the shocks associated with warming oceans. This is particularly true for the recovery of fragile coral reef ecosystems, which many large MPAs jointly protect alongside pelagic waters (Graham et al., 2008; Hughes et al., 2003). Area protections are not a substitute for measures that directly combat out-of-control carbon emissions, but they are a relatively politically feasible and cost-effective way of increasing ocean resilience. Their attractiveness to governments as a partial and imperfect response to climate change is what has, in fact, led to growing resistance from the commercial fishing industry to large-scale area protections.

In short, the case for protecting large ocean spaces in part to address climate change is growing. The climate justification for large MPAs is twofold: healthy marine ecosystems are able to absorb greater levels of carbon, contributing to climate mitigation, and they are also more resilient to climate impacts. Governments and environmental groups routinely cite these benefits in their promotion of a particular large MPA, in addition to the more obvious and immediate benefits of limiting or prohibiting extractive activity. This climate justification does not convince the commercial fishing industry that they deserve to be more heavily regulated, however, particularly in sustainably managed fisheries.

Fishing Industry Opposition

Efforts by governments and environmental groups to target politically feasible large MPA sites that minimize impact on fisheries have done little to assuage fishing industry concerns. The response from the industry in many cases has been vehement. In all but one of the five case studies examined here – presented in Table 11.1 – the fishing industry strongly opposed large MPA regulations. The only exception was the Pitcairn Islands Marine Reserve, due simply to the fact that no legal commercial fishing industry exists around the highly remote islands. Even in this case, environmental

Table 11.1. *Case study overview*

Name	Country	Designated/ Expanded	Size (km^2)	Location	Management Type
Papahānaumokuākea Marine National Monument	US	2006/2016	1,508,870	Overseas Territories	No-Take
Pacific Remote Islands Marine National Monument	US	2009/2014	1,270,000	Overseas Territories	No-Take
Coral Sea Commonwealth Marine Reserve	Australia	2012	990,000	Eastern Expanse of EEZ	Mixed Use (24% No-Take)
Palau National Marine Sanctuary	Palau	2015	500,000	80% of EEZ	No-Take
Pitcairn Islands Marine Reserve	UK	2015	834,300	Overseas Territories	No-Take

groups alongside the local island council produced an economic analysis to convince the UK's Foreign and Commonwealth Office (FCO) that the area could not sustain a commercial fishery in an effort to garner its support for the reserve (Pew Charitable Trusts et al., 2013). In the other four cases, the task was decidedly more difficult.

United States

Industry stakeholders in the United States were vocal in their opposition to Papahānaumokuākea and the Pacific Remote Islands Marine National Monument. Presidents George W. Bush and Barrack Obama skirted around both the US Congress and the National Oceanic and Atmospheric Administration (NOAA) in creating and expanding these reserves, expediting the process and minimizing the legal requirements for consultations with industry and local stakeholders. The Western Pacific Regional Fishery Management Council (Wespac) and fishing lobby groups condemned the process as an overreach of executive authority that silenced the voices of local fishers (interview with Manager, Western Pacific Regional Fishery Management Council, phone, Honolulu, Hawaii, October 1, 2015). They labeled the process "undemocratic," and they lamented the loss of fishing grounds that resulted from White House efforts to conserve marine ecosystems. The 1976 Magnuson-Stevens Fishery Conservation and Management Act created eight regional fishery management councils, including Wespac, which were to be responsible for the sustainable management of US fisheries. These industry-friendly bodies have had mixed success in doing so, but nonetheless they aggressively defend their jurisdiction over fisheries management in their respective regions.

Wespac and industry supporters had some success in pushing back against large MPAs in the western Pacific. They helped convince President Bush to extend both MPAs only to 50 nautical miles from shore rather than the full 200-nautical-mile limit of the US exclusive economic zone (EEZ). During consultations on the expansion of the Pacific Remote Islands, Wespac produced a report that helped to convince President Obama to scale back the initially proposed expansion of boundaries (Western Pacific Regional Fishery Management Council, 2014). This scaling back left the two most productive fishing zones around the Pacific Remote Islands out of the reserve. Congressional Republicans have also taken up the cause through sustained hostility to US marine national monuments, through oversight hearings and public opposition. These sustained efforts may have achieved their goal, with President Donald Trump proving a more willing industry partner than his predecessors. Trump put the US reserves under review with an executive order. As of September 2017, Secretary of the Interior Ryan Zinke recommended scaling back the boundaries around the Pacific Remote Islands, but up to this writing he has reportedly remained silent on whether the iconic and potentially more divisive Papahānaumokuākea Marine National Monument will also face changes. It remains unclear whether the White House has the legal authority to diminish monuments established by previous presidents, so legal battles will inevitably follow any attempt to do so.

Australia

Industry opposition to MPAs in the United States pales in comparison to industry opposition to the Coral Sea CMR in Australia. The push for a large no-take marine reserve in the Coral Sea began in 2007, and the reserve was formally designated in 2012, and yet as of 2017 there were still no new protections in effect on the water. The original management plan for the reserve was cancelled in 2013 by the Abbott government shortly after taking office, necessitating a new consultation process. This new review process recommended decreasing the no-take area in the 990,000-km^2 reserve from the original 51 percent to 41 percent (Buxton and Cochrane, 2016). But the subsequent Turnbull government considered this to be too much, and it proposed decreasing the no-take area further to just 24 percent, covering predominantly the most remote areas of the Australian Coral Sea.

The Coral Sea CMR has seen what is most likely the most extensive stakeholder consultations of any large MPA to date, taking over a decade and consisting of multiple rounds of bargaining. The reason for such a long bargaining process was the vocal opposition of the commercial fishing industry. There were a handful of businesses in the region whose survival was tied directly to the zoning of the Coral Sea. Environmental groups also acknowledge that Australian fisheries are sustainably managed (interview with Campaigner, Greenpeace Australia, Sydney, NSW, March 29, 2016; interview with Program Director, Pew Charitable Trusts, Canberra, ACT, May 6, 2016). Businesses were feeling particularly hard-pressed as a result of the push for extensive no-take zoning in the Coral Sea. The government's initial intention was to buy out businesses that would not be able to remain operational, but the objective of the review process was to keep all companies in business. One local business owner expressed exasperation at the idea that the government would want to undermine a local fishery that was sustainably managed (interview with Owner, commercial fishing business, phone, Cairns, QLD, May 30, 2016). Commercial fishers were in a state of constant uncertainty, and toward the end of nearly ten years of consultations had "become numb to the process" (interview with owner, commercial fishing business, May 30, 2016). As with politicians in the United States, their cause was taken up by politicians in Canberra, who advocated vociferously on behalf of the commercial fishing industry.

The most vocal opposition to the Coral Sea CMR ironically came from the recreational fishing lobby. The irony is that the Coral Sea is too remote for an overwhelming majority of recreational fishers in Queensland, so recreational fishing in the Coral Sea is virtually nil. Recreational fishers themselves tend to be divided on the desirability of marine reserves in Australia, so the lobby's position did not necessarily reflect that of its constituents (Meder, 2016). The rationale for their opposition was that the proposed "lock-outs" were unscientific and set a precedent for future closures in areas that are frequented by recreational fishers. These lobbyists were still reeling from the 2004 rezoning of the Great Barrier Reef Marine Park, which did displace recreational fishing. The issue had become highly politicized, leading to unreserved opposition to any new no-take zones.

In both the United States and Australia, these industry stakeholders rarely take seriously the contribution of large, no-take marine reserves to building ecosystem resilience. This is understandable, as they are more concerned about the direct, immediate impact of closures on their profitability and livelihoods than they are on the diffuse, long-term, and uncertain benefits of improving ecosystem resilience. But in some cases, the connection is more salient. The Great Barrier Reef – which extends into the Coral Sea CMR – is in ecological decline due to ocean warming and acidification, and poor water quality caused by coastal runoff (Furnas, 2003; De'ath et al., 2012). Coral bleaching events that are growing in frequency and intensity are the most visible manifestation of climate impacts on marine ecosystems, and these impacts extend to pelagic waters as well. Many marine species are highly mobile and migratory, forming links between reefs and pelagic ecosystems that make protecting both of these areas crucial for building resilience (McCauley et al., 2012; Roberts et al., 2017). As noted earlier, the case for a more precautionary approach to protecting marine ecosystems is growing, and commercial fishing industries increasingly find themselves embattled by it.

Palau

In Palau, one of the primary drivers of a large MPA is to protect the tourism industry, which accounted for 54 percent of national gross domestic product (GDP) in 2015. The legislation for the new sanctuary bans foreign fishing fleets, who, along with illegal fishers, are the main extractive resource users of Palau's pelagic waters. Political resistance from these foreign fleets was manifested in the form of opposition from Palauan politicians, some of whom benefitted financially from foreign fishing. Eventually these politicians were shamed into supporting the sanctuary and a strictly domestic fishing zone, but only after delaying the process in the Palauan Congress for two years. The mainly Japanese and Taiwanese fleets that frequent Palau's EEZ had little political clout outside of a few sympathetic politicians. Their only ties to the national economy came in the form of fishing licenses, because they were taking both their catch and revenues outside of Palau. Moreover, these fleets had regularly underreported their catch, so they were not able to make a compelling economic case without revealing their true catch value (interview with Consultant, Pew Charitable Trusts, Koror, June 17, 2016). Domestic operators also stand to benefit from the ban on foreign fleets. Quincey Kuniyoshi, a Palauan senator and owner of one of Palau's two domestic commercial fishing companies, became a key supporter for this reason.

A common criticism of the sanctuary within Palau is that it does not actually encompass the major dive sites that drive the tourism industry and that the protections are therefore misplaced. But once again this criticism neglects the growing scientific evidence of the linkages between reef and pelagic marine ecosystems and the importance of wholesale ecosystem protection to building resilience. As with Australia, these connections are especially salient in Palau. Palau is a world-class dive destination, in large part because of its abundance of highly mobile sharks. The lifetime tourism value of a single shark in Palau is roughly $1.9 million – over 16,000 times greater than the value of its fin on the open market (Vianna et al., 2010). Protecting sharks in pelagic waters is equally important to the country's tourism industry as is protecting them at dive sites. Palau's inland tourist attractions – most notably the iconic Jellyfish Lake – have already been decimated by climate change. The country's economic stability depends on the resilience of its reefs, so pelagic protections in Palau are especially important for both environmental and human reasons.

To summarize, in each of the large MPA case studies examined here, governments took into account the fisheries impact of a new reserve. They made efforts to minimize this impact, often by designing boundaries and regulations to allow businesses to continue operating. The only exception was in Palau, where predominantly foreign fleets did little to benefit the local economy. These large MPAs were attractive to governments because of their minimal commercial impact, making them a path of least resistance for governments struggling with balancing environmental regulation with commercial interests. Government efforts were nonetheless not well received by the commercial fishing industry, which opposed any restrictions that would limit their activities. This opposition has been vocal and protracted, in some cases continuing long after a government formally enacted new MPA legislation. The next section will outline why this opposition is disproportionate, given the minimal impact of large MPAs on fisheries in these five cases.

Fisheries Impact of Protections

The fishing industry's responses to Papahānaumokuākea, the Pacific Remote Islands, and the Coral Sea – the three cases with developed domestic fishing industries – were disproportionately antagonistic, given the minimal impact of government regulations. In both the United States and Australia, industry expressed greater concern with short-term access to fishing grounds over the long-term benefits to fisheries of a large MPA. This claim that the industry response was disproportionate nonetheless needs to be qualified in two ways. First, businesses pursue profitability and autonomy above all, and typically with a short time horizon (Cutler et al., 1999). It is perhaps idealistic to expect businesses to weigh the long-term fisheries benefits of large MPAs more heavily than they did in these cases. And second, not all businesses were affected equally by proposed or enacted regulations. In some cases, regulations did or would have put an operator out of business (albeit usually with compensation). It is important to differentiate between the impact of a large MPA on the industry writ large and its impact on the livelihoods of individual business owners who may bear a heavier share of the burden.

US Marine Monuments

The currently embattled Papahānaumokuākea and Pacific Remote Islands Marine National Monuments are perhaps the clearest examples of a disproportionate anticonservation crusade from the fishing industry. Conservation is a bipartisan tradition in the United States, with US presidents from both parties using executive authority through the 1906 Antiquities Act to protect valuable cultural and environmental sites. This holds true for the United States' large MPAs. Both monuments were designated by President Bush, a Republican, and later expanded by President Obama, a Democrat. This bipartisanship from the executive branch has not muted industry opposition, nor opposition from congressional Republicans. What is particularly striking about the industry response to these monuments is how small of an impact they have on commercial fishing.

In 2006, President Bush limited the boundaries of Papahānaumokuākea to encompass 362,000 km^2 of ocean space, which at the time was still the largest MPA on the planet. The fishing industry impact of this decision was limited to just nine individual fishers that had permits for the region, one of which happened to be in jail at the time (interview with Former Executive, Pew Charitable Trusts, phone, Juneau, AK, October 7, 2015). All nine were offered compensation for the loss of their permits, which according to one environmental campaigner they were enthusiastic about, given the low profit margins for fishing in the area (interview with Former Executive, Pew Charitable Trusts, October 7, 2015). President Obama's dramatic expansion of this reserve to roughly 1.5 million km^2 in 2016 extended its boundaries farther into pelagic waters, where longer-range fleets can conceivably fish. But the waters around Papahānaumokuākea are not a productive fishery,

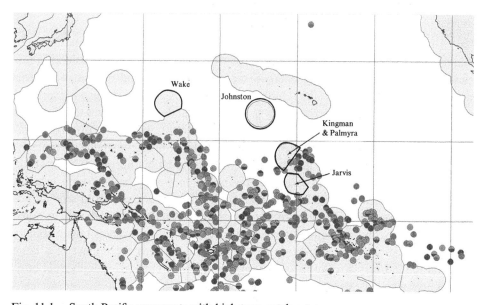

Fig. 11.1.   South Pacific seamounts with high tuna catch rates.
Note: The initially proposed Pacific Remote Islands reserve expansion is outlined in light shading. President Obama only expanded the three zones around Jarvis Island, the Johnston Atoll, and Wake Island. Circles depict seamounts with catches of yellowfin, bigeye, or albacore tuna.
Sources: Morato et al. (2010); Sala et al. (2014).

including for the all-important tuna that local fleets rely on (Morato et al., 2010). Moreover, the US western Pacific fleet is subject to quotas that stem from regional fishery cooperation agreements. In 2015 and 2016, the US fleet hit its annual quota for its target species – bigeye tuna – in August and July, respectively, forcing fishing vessels to remain docked for months. Papahānaumokuākea has done virtually nothing to impede the industry from reaching its quota, and the western Pacific fleet is seemingly reaching it more quickly every year.

The waters around the Pacific Remote Islands are more productive than Papahānaumokuākea's, but not by much (see Figure 11.1). The latest fisheries economics report from the US Department of Commerce credits the western Pacific fleet with revenue of $91.5 million in 2012, a modest 1.8 percent of the national total (US Department of Commerce, 2014). The Hawaii-based longline fleet and American Samoa–based purse seine fleet only set 4 percent and 5 percent of their hooks, respectively, within the boundaries of the initially proposed expansion (Sala et al., 2014). As noted earlier, President Obama ultimately chose to exclude the two most productive fisheries zones from this expansion, so these numbers overestimate historical catch in the now-protected zones. Industry notes that these waters have seen higher catch levels in certain years based on migratory patterns of tuna (Western Pacific Regional Fishery Management Council, 2014). Industry representatives also note that warming oceans will alter migratory patterns and that locking the industry out of these waters could take a greater toll in the future (interview with Manager, Western Pacific Regional Fishery Management Council, October 1, 2015). Neither of these arguments were enough to convince the Obama administration to not go ahead with the expansion, although they did contribute to reducing the initial proposal by roughly 600,000 km^2.

The US fishing industry is under pressure not only from marine monument designations. The 37-boat western Pacific fleet currently faces a host of pressures that are threatening its viability, including from the following: fluctuating tuna prices, with a substantial drop in prices in 2015 due to oversupply; a minimum wage increase for American Samoa; losing traditional fishing grounds around Kiribati to Chinese fishers; and the perpetual challenge of securing higher quotas through the Western and Central Pacific Fisheries Commission. Each of these pressures individually poses a greater threat to the industry than marine monuments, which have had virtually no impact. These pressures have already contributed to the shuttering of one of American Samoa's two canneries – the largest industry on the island and a major source of employment. This hostile business climate has galvanized industry opposition to marine monuments. Unlike global market conditions or regional cooperation arrangements, monuments are an exclusively domestic target. Much in the same way that large MPAs are a relatively easy policy victory for governments in combatting climate change, they are also an easy target for opposition by fishing industry stakeholders with little control over the other factors affecting their industry.

A difficult business climate coupled with what industry sees as an infringement on their jurisdiction has contributed to industry's hostility toward marine monuments in the United States. There is a conversation to be had about the importance of well-managed, large MPAs to long-term fisheries sustainability, notably in the context of climate change. However, this conversation is not happening in the US western Pacific due to the polarization that has occurred around marine monuments. The benefits of large MPAs to fisheries, including building climate resilience and increasing fish biomass, are simply rejected by vocal industry representatives who see no role for large-scale conservation in the region. Large MPAs are a conservation tool rather than a fisheries management tool, but the two objectives are, of course, interconnected. Discussion over how the two might coexist between government, environmental groups, and the commercial fishing industry has not been constructive, despite efforts from subsequent US presidents to minimize the commercial impact of new and expanded monuments.

The Coral Sea Commonwealth Marine Reserve

In Australia, stakeholder negotiations with the commercial fishing industry were decidedly more difficult than they were in the United States. Unlike in the United States, new zoning regulations in the region would have a direct and immediate impact on some businesses. Two fisheries were vulnerable: the Commonwealth Eastern Tuna and Billfish Fishery (ETBF), and the Commonwealth Coral Sea

Fishery, which targets aquarium species, among others. Whereas the western Pacific fleet in the United States could simply relocate their efforts with relatively little added cost, Coral Sea CMR regulations threatened to put some (primarily) tuna fishers out of business entirely. And although it was never actually on the table, the reserve also had the potential to shut down the entire Coral Sea Fishery. This fishery was not the target of government regulations as long as it maintained established practices to limit localized depletion (Buxton and Cochrane, 2016). Businesses in both fisheries were tied to local stocks and vertically integrated with local facilities and infrastructure, so their ability to relocate their efforts was limited (ABARES, 2012). The Australian commercial fishing industry's opposition to new regulations was more reasoned and grounded in justifiable concerns about the immediate viability of the operations of certain businesses.

The industry-wide impact of even the initially proposed 52 percent no-take zoning proposed by the Gillard government was small. The ETBF is an expansive, but unproductive, fishery. It encompasses the entirety of the eastern portion of Australia's EEZ, with operators based all along the coast. Only one operator – Great Barrier Reef Tuna based in Cairns – was at risk due to Coral Sea regulations. Great Barrier Reef Tuna holds 9 percent of the quota for the entire ETBF, and its operations are exclusively within the Coral Sea CMR boundaries. The Gillard plan earmarked A$100 million to compensate businesses affected by the plan, including Great Barrier Reef Tuna, and the recommendation of the Bioregional Advisory Panel rezoned the region to enable their continued operations. The owners of this family-run business were consulted extensively throughout the Coral Sea process, and they had become numb to the constant rezoning and renegotiations over the decade-long process (interview with Owner, commercial fishing business, phone, Cairns, QLD, May 30, 2016). Their most notable frustration was that their activities were already well regulated and sustainable, which environmental groups acknowledged (interview with Campaigner, Greenpeace Australia, Sydney, NSW, March 29, 2016; interview with Program Director, Pew Charitable Trusts, May 6, 2016). The Australian government faced an undeniably difficult balancing act in expanding protections while deciding how best to accommodate a long-standing locally owned business.

This justifiable concern, however, became the impetus for a pendulum swing in the opposite direction – a "paper park" with very little conservation benefit. As noted earlier, the Turnbull government scaled back protections quite dramatically (Buxton and Cochrane, 2016; Director of National Parks, 2017). The Bioregional Advisory Panel's recommendation after conducting its review was to designate 41 percent of the Coral Sea as no-take. This recommendation was the result of an extensive consultation process, with the expressed purpose of ensuring that not one single business would go under as a result of new regulations. The panel's recommendation was an industry-friendly outcome, and even triggered a backlash from environmental groups. The Turnbull government nonetheless rejected this recommendation, opting for just 24 percent no-take instead and dropping any pretense that the reserve would meaningfully limit extractive activity in the Coral Sea.

Any recognition of the importance of large, no-take zones for marine ecosystem resilience was gone by this stage. The Coral Sea CMR had become a high-profile and thoroughly politicized environmental issue. A hardline and vocal recreational fishing lobby had galvanized some local opposition in Queensland, and that lobby was a constant presence in Canberra. The 2016 election result saw Turnbull more reliant on the more radical elements of his coalition government, so these voices had grown more influential over time. The viability of a few businesses and the opposition from a recreational fishing lobby not affected by the reserve won out over the potential conservation and climate benefits of more comprehensive protection. Put simply, minor business interests ultimately overruled building resilience into a vulnerable marine ecosystem that is already being decimated by climate change, with 2016's record coral bleaching event being just the latest example of this.

Disproportionality in MPAs

The common thread across the three MPA cases with prominent domestic fishing industries is a disproportionate level of antagonism from industry. Given the anticipated impacts of climate change on the oceans, the conservation benefits of large-scale ecosystem protections have the potential to far outweigh the short-term gains of avoiding regulation, and not just for biodiversity protection but also for fisheries sustainability and profitability. Declining fish stocks due to overfishing is putting further pressure on profit margins in an industry with low margins to begin with. Large MPAs can provide what many marine campaigners and scientists refer to as a "fish bank." They are intended to protect pristine regions in which species can flourish absent the pressures of extraction. They can serve to mitigate the shifting baselines problem, in which the reference point for what is considered a healthy fish stock degrades over time (Pauly, 1995). The long-term benefits to fisheries is one of the main arguments put forward by marine campaigners and scientists in favor of large MPAs.

These arguments have not convinced most fishing industry stakeholders that large MPAs are not just unscientific regulations that undermine their profitability. This antagonism to conservation and governments' responsiveness to it has perpetuated a global regime for fisheries management that has been catastrophic to date. Despite assertions of sustainability by the UN Food and Agriculture Organization, enhancing short-term fisheries productivity tends to guide global fisheries management (Lobo and Jacques, 2017). Most fisheries management organizations, including Wespac and the Australian Fisheries Management Authority (AFMA), have mandates of achieving maximum sustainable yield (MSY). Their stated policy is effectively to push fish stocks to their limit, leaving little room for error. This global regime has contributed to overfishing and widespread fisheries decline. In an era in which the impact of climate-related and other stressors to marine ecosystems is growing, a more precautionary approach to fisheries management is needed.

Large MPAs can be a component of this more precautionary approach, but the stakeholder dynamics surrounding them suggest an ongoing denial of the ecological crisis facing the world's oceans. The United States and Australia case studies discussed in this chapter highlight the influence that industry opposition can have over government policy, even when the impact is minimal. The two zones that President Obama did not expand in the Pacific Remote Islands total roughly 600,000 km^2 – space with limited fisheries productivity that otherwise would have been protected. The spaces that he and President Bush did protect are also now threatened due to President Trump being excessively responsive to minor industry interests. The Australian government ultimately wasted a lot of time, money, and effort over ten-plus years on a marine reserve that allows for business as usual on the Coral Sea. In the Pitcairn Islands, the British government still wanted evidence that commercial fishing was not viable in the future to go ahead with their reserve, despite industry seeming to have never caught a single fish there.

Only in Palau, where ecotourism is paramount and commercial fishing gains were mostly exported, was a government able to take a strong stance. This contrast highlights that the success of efforts by environmental groups and governments to increase marine conservation through large MPAs depends on their ability to navigate, and mitigate, the disproportionate influence of the fishing industry.

Conclusion

At the origin of stakeholder tension around large MPAs is a normative divide over whether marine ecosystems should be *conserved* or *managed* for human uses. Scholars of marine conservation are devoting increasing attention to questions of how to bridge that divide (Gruby et al., 2016; Christie et al., 2017; Gray et al., 2017). Protected areas, by definition, need to prohibit, or at least limit, extractive activity within their boundaries. Industry stakeholders often refer to this goal as "nonscientific." They reject that there is not a nonzero level of extractive activity that is consistent with a healthy ecosystem. They instead advocate more sophisticated management practices, such as regulations on equipment, catch limits, and vessel limits. They do not reject that no-take areas can lead to healthier ecosystems, but rather make the case that most of the time those areas are extreme measures that undermine businesses and livelihoods, and that governments should manage marine resources to maximize their benefit to society. These claims are particularly salient when new protections might directly lead to business closures, as was the case in the Coral Sea. Large MPAs therefore face a Catch-22: they are either criticized for being too remote, and therefore not addressing the root causes of ocean decline, or that they are less remote and run the risk of undermining business and livelihoods.

The challenge is that managing an area for sustainable commercial activity and conserving it to combat a global ecological crisis exacerbated by climate change, particularly in the world's oceans, are very different goals. Governance of pelagic seas has long been based on a perception of the oceans as boundless, and current norms and practices reflect that. Closing hundreds of thousands of square kilometers of contiguous ocean space to extractive activity is a new practice, barely a decade old. It threatens long-standing ocean governance practices and the industries that have relied upon them. This challenge to the existing paradigm of ocean governance has, perhaps understandably, triggered a defensive industry response.

Nevertheless, there is growing scientific evidence of the benefits of large, no-take marine reserves. The case to be made for a precautionary approach to oceans governance has never been stronger because of the intensifying impacts of climate change. This leaves the fishing industry beset by climate change on two fronts. Warming oceans not only undermine the health of ecosystems that the industry relies upon, but the fishing industry also bears the brunt of the regulatory burden for mitigating and adapting to climate change at sea. Climate mitigation and adaption for the oceans nonetheless require limiting commercial fishing activity. Doing so through large MPAs has, for now at least, exacerbated divisions between certain marine resource users and environmentalists. This highlights the power of politics to influence ocean governance amidst climate change.

References

Australian Bureau of Agricultural and Resource Economics (ABARES). (2012). *Coral Sea Commonwealth Marine Reserve: Social and Economic Assessment of the Impacts on Commercial and Charter Fishing*, Canberra: Commonwealth of Australia.

Alger, J. and Dauvergne, P. (2017a). The politics of pacific ocean conservation: Lessons from the Pitcairn Islands Marine Reserve. *Pacific Affairs*, **90**(1), 29–50.

Alger, J. and Dauvergne, P. (2017b). The global norm of large marine protected areas: Explaining variable adoption and implementation. *Environmental Policy and Governance*, **27**(4), 298–310.

Barkin, J. S. and DeSombre, E. R. (2013). *Saving Global Fisheries: Reducing Fishing Capacity to Promote Sustainability*, Cambridge: MIT Press.

Buxton, C. and Cochrane, P. (2016). *Commonwealth Marine Reserves Review: Report of the Bioregional Advisory Panel*, Canberra: Department of the Environment.

Christie, P., Bennett, N. J., Gray, N. J., et al. (2017). Why people matter in ocean governance: Incorporating human dimensions into large-scale marine protected areas. *Marine Policy*, **84**, 273–84.

Cutler, A., Haufler, C. V., and Porter, T. (1999). *Private Authority and International Affairs*, Albany: SUNY Press.

De Santo, E. M. (2013). Missing marine protected area (MPA) targets: How the push for quantity over quality undermines sustainability and social justice. *Journal of Environmental Management*, **124**, 137–46.

De'ath, G., Fabricius, K. E., Sweatman, H., and Puotinen, M. (2012). The 27–year decline of coral cover on the Great Barrier Reef and its causes. *Proceedings of the National Academy of Sciences*, **109**(44), 17995–9.

DeMartini, E., Friedlander, A., Sandin S., and Sala E. (2008). Differences in fish-assemblage structure between fished and unfished atolls in the Northern Line Islands, Central Pacific. *Marine Ecology Progress Series*, **365**, 199–215.

DeSombre, E. R. and Barkin, J. S. (2011). *Fish*, Cambridge: Polity Press.

Devillers, R., Pressey, R. L., Grech, A., et al. (2015). Reinventing residual reserves in the sea: Are we favouring ease of establishment over need for protection? *Aquatic Conservation: Marine and Freshwater Ecosystems*, **25**(4), 480–504.

Director of National Parks. (2017). *Draft Coral Sea Commonwealth Marine Reserve Management Plan 2017*, Canberra: Director of National Parks.

Edgar, G. J., Stuart-Smith, R. D., Willis, T. J., et al. (2014). Global conservation outcomes depend on marine protected areas with five key features. *Nature*, **506**, 216–20.

Friedlander, A. M., Caselle, J. E., Ballesteros, E., et al. (2014). The real bounty: Marine biodiversity in the Pitcairn Islands. *PLoS One*, **9**(6).

Furnas, M. J. (2003). *Catchments and Corals: Terrestrial Runoff to the Great Barrier Reef*. Townsville City: Australian Institute of Marine Science & CRC Reef Research Centre.

Game, E. T., Grantham, H. S., Hobday, A. J., et al. (2009). Pelagic protected areas: The missing dimension in ocean conservation. *Trends in Ecology & Evolution*, **24**(7), 360–9.

Graham, N. A. J., McClanahan, T. R., MacNeil, M. A., et al. (2008). Climate warming, marine protected areas and the ocean-scale integrity of coral reef ecosystems. *PLoS One*, **3**(8), e3039.

Gray, N. J., Bennett, N. J., Day, J. C., et al. (2017). Human dimensions of large-scale marine protected areas: Advancing research and practice. *Coastal Management*, **45**(6), 407–15.

Gruby, R. L., Gray, N. J., Campbell, L. M., Action, L. (2016). Toward a social science research agenda for large marine protected areas. *Conservation Letters*, **9**(3), 153–63.

Hilborn, R. (2015). Marine protected areas miss the boat. *Science*, **350**(6266), 1326.

Hughes, T. P., Baird, A. H., Bellwood, D. R., et al. (2003). Climate change, human impacts, and the resilience of coral reefs. *Science*, **301**(5635), 929–33.

Jones, P. J. S. and De Santo, E. M. (2016). Viewpoint: Is the race for remote, very large marine protected areas (VLMPAs) taking us down the wrong track? *Marine Policy*, **73**, 231–4.

Lobo, R. and Jacques, P. J. (2017). SOFIA'S choices: Discourses, values, and norms of the world ocean regime. *Marine Policy*, **78**, 26–33.

McCauley, D. J., Young, H. S., Dunbar, R. B., et al. (2012). Assessing the effects of large mobile predators on ecosystem connectivity. *Ecological Applications*, **22**(6), 1711–7.

Meder, A. (2016). 'The issue was that big, I swear!': Evidence for the real impacts of marine protected areas on Australian recreational fishing. In J. Fitzsimons and G. Wescott, eds., *Big, Bold and Blue: Lessons from Australia's Marine Protected Areas*, Clayton: CSIRO Publishing, pp. 349–62.

Morato, T., Hoyle S. D., Allain, V., and Nicol, S. J. (2010). Tuna longline fishing around west and central Pacific Seamounts. *PLoS One*, **5**(12).

Pauly, D. (1995). Anecdotes and the shifting baseline syndrome of fisheries. *Trends in Ecology & Evolution*, **10**(10), 430.

Pew Charitable Trusts, National Geographic Society and Pitcairn Island Council. (2013). *Is Offshore Commercial Fishing a Prospect in the Pitcairn Islands?* Available at www.pewtrusts.org/~/media/legacy/uploadedfiles/peg/publications/report/ioffshorecommercialfishinginthepitcairnislandspdf.pdf

Roberts, C. M., O'Leary, B. C., McCauley, D. J., et al. (2017). Marine reserves can mitigate and promote adaptation to climate change. *Proceedings of the National Academy of Sciences*, **114**(24), 6167–75.

Sala, E., Morgan, L., Norse, E., and Friedlander, A. (2014). *Expansion of the U.S. Pacific Remote Islands Marine National Monument*, Washington, DC: Marine Conservation Institute.

Sandin, S. A., Smith, J. E., DeMartini, E. E., et al. (2008). Baselines and degradation of coral reefs in the Northern Line Islands. *PLoS One*, **3**(2).

US Department of Commerce. (2014). "Fisheries Economics of the United States 2012." *NOAA Technical Memorandum NMFS-F/SPO-137*, Silver Spring, MD: National Oceanic and Atmospheric Administration.

Vianna, G. M. S., Meekan, M. G., Pannell, D., Marsh, S., and Meeuwig, J. J. (2010). *Wanted Dead or Alive? The Relative Value of Reef Sharks as a Fishery and an Ecotourism Asset in Palau*, Perth: Australian Institute of Marine Science and University of Western Australia.

Western Pacific Regional Fishery Management Council. (2014). *An Ocean Legacy the US Pacific Island Way*. Honolulu: Western Pacific Regional Fishery Management Council.

12

Climate Change and Contested Marine Areas in the Arctic

The Case of Svalbard

RACHEL TILLER AND DOROTHY J. DANKEL

Introduction

Changing climate will affect the marine environment, including the contested marine area surrounding the archipelago of Svalbard, an area known as the Svalbard Fisheries Protection Zone. In this chapter, we hypothesize about the effects of climate change on this management regime, particularly with respect to fishing, and we consider how resilient the regime will be in the face of exogenous shocks and perturbations. We are particularly interested in exploring what defines a resilient management regime and whether the fisheries regime encircling Svalbard is resilient enough to cope with the effects of climate change. To make this assessment, we draw upon the experiences of two different international environmental management regimes that experienced shocks in the 1980s. Specifically, we look at how the ozone hole affected the regime for protecting Earth's stratospheric ozone layer and at the impact of Black Forest death on the Convention on Long-Range Transboundary Air Pollution (LRTAP). We compare lessons learned from these treaties, and especially their responses to exogenous factors, to better understand the impact of the creation of a 200-mile fisheries protection zone around Svalbard. This zone represents an area where changes in the spatial distribution and relative abundance of commercially valuable fish species are likely. The dynamics of fishing activities and exploitation patterns may change when commercial fishers from farther south follow fish northward as their ranges change due to warming seas.

Given that the existing management regime is weak in terms of international acceptance, climatic stressors could have dramatic impacts on its workings. This could change the dynamics of international relations and customary good communication in the Northeast Atlantic. To explore these issues, in this chapter we will proceed as follows. We start by looking at climate changes in the Arctic that will have the largest effect on the Svalbard Fisheries Protection Zone. We then focus on the application of regime theory and the questions of vulnerability, resilience, and adaptive capacity. This is followed by a look at different environmental management regimes that have experienced different kinds of stresses or perturbations and considering to what extent these regimes were able to adapt. We apply these findings to the fisheries zone around Svalbard, assessing mechanisms by which this regime may manage climatic stressors. We look at the patterns of change we may expect to see in regimes that are undergoing substantial shocks. These patterns reveal options that may be available when exogenous shocks brought forth by climatic stressors lead to major changes in existing institutional arrangements in the High North.

The Effects of Climatic Stressors on the Svalbard Fisheries Protection Zone

The term "climatic stress" can refer to how organisms or ecosystems relate to climatic changes such as variations in temperature, salinity, air pressure, or atmospheric emissions. Climatic stressors can have effects on the growth rate of an organism or food or habitat availability and/or the quality of a product made from that organism. Scientists have observed and reported many changes in the Arctic regions of the ocean over the past decade, but of all the climate-induced changes in the ocean, the steady increase in sea temperature is perhaps the most influential and most easily observed consequence for fish stocks. Sea temperature is a climatic factor to which all marine organisms are subject and have adapted to accordingly throughout their evolution. Since fish are cold-blooded ectotherms, they are very sensitive to temperature and must seek out colder waters when the sea around them warms beyond their comfort level. This is the basis for the scenarios of mass exodus of fish species toward the High North (Gattuso et al., 2015). Another critical factor influencing fish distribution and size is food availability. Marine phytoplankton and zooplankton represent the all-important biomass at the bottom of the marine food web. As marine organisms, these plankton are also adapted to certain temperatures and light availability. Fish must find their own food, so adaptations that plankton make in response to changing temperature will in turn also affect their predator species.

To what degree will these environmental factors have an effect the Svalbard fisheries regime? To answer this, we must first turn to the Svalbard Treaty itself, which is at the root of the disagreements over the fisheries regime beyond the archipelago. The Svalbard Treaty was signed in Versailles after World War I on February 9, 1920. Norway was named sovereign of the archipelago as part of war reparations following its suffering during the war. When Norway, under Article 1 of the treaty, was granted territorial sovereignty over the archipelago of Svalbard, it effectively meant that it was free to regulate all activities on the islands. There were far-reaching restrictions to this sovereignty, however. The most important of these was the requirement of nondiscrimination toward other signatories. This exclusion represented the wish of the signatories to preserve the previous *terra nullius* status of the area. This was specifically stated in Article 2(1) of the treaty: "Ships and nationals of all the High Contracting Parties shall enjoy equally the rights of fishing ... in the territories specified in Article 1 and in their territorial waters" (Arlov, 2003).

Norway has never contested the limitations to its sovereignty under the Svalbard Treaty. However, it asserts that the treaty does not extend beyond Svalbard's 12-mile territorial waters and is furthermore silent on Norway's right to create an exclusive economic zone (EEZ) beyond this. However, this is not an argument that has gained accolade from the other signatories. In fact, some signatories of the Svalbard Treaty consider the waters beyond the 12-mile limit, now included in the zone, to be international waters. They accordingly argue that fisheries in these waters need to be regulated by international regimes only, or be subject to free fishing without limitations (Hoel, 2005). Other signatories, supported by some scholars, emphasize that the treaty makers could not have foreseen future international laws and restrictions, most specifically the Law of the Sea and the development of EEZs (see Chapter 2). They see the treaty as a fluid and living document that has evolved over time in response to new needs and challenges, and therefore argue that any EEZ, or zone, would have to belong to Svalbard and be subject to the articles of the Svalbard Treaty in the same manner as territorial cases would be (Hoel, 2005).

Facing these protests, Norway therefore did not outright extend a Norwegian EEZ of 200 miles outside Svalbard, but instead created a fisheries protection zone. To this zone they extended the principles of nondiscrimination from the Svalbard Treaty in an effort to appease the signatories and prevent a dispute that could land Norway in an international court. This principle involves allocating quotas based on a prior history of traditional fishing in the Svalbard Area. Such a principle has been found to be sufficiently nondiscriminatory by the European Court. The reasons behind this logic are that those nations that have traditionally fished in the area are more dependent upon this fishery than are those that have only recently discovered the benefits of the area. Because their economies have not traditionally been dependent upon the area, they arguably would not suffer substantially by being withheld from it (Ulfstein, 1995). Given this, Norway has accepted a halfway solution to the problems with the zone to avoid outright conflict between nations.

Despite the protests from other nations since the fisheries zone was created, development of the zone was grounded in the principles set forth by a recognized regime. Since that time, the zone has gradually become institutionalized into a legitimate Norwegian management area by practice, with heavy coast guard presence, and other fisheries nations have mostly acquiesced.

Given rising temperatures as a climatic stressor, three major questions arise for the Svalbard fisheries zone: (1) What will be the main consequences of the current rate of change in the surrounding ocean on marine organisms? (2) Will marine organisms be able to adapt quickly enough to these changes to keep ecosystems and their services intact? (3) When new organisms move into the Svalbard zone due to climate change, will they destabilize the regime because they do not have historical fishing rights attached to them and thereby may be subject to equal treatment among signatories to the Svalbard Treaty?

Northeast Atlantic mackerel is a recent fascinating example of a fish species that has adapted to changing climatic stressors in the High North. Since 2007, scientists and fishermen have witnessed an extreme increase in the distribution of Northeast Atlantic mackerel from 1.3 km^2 summer feeding distribution to 2.9 km^2 in 2014 (ICES, 2016; Nøttestad et al., 2016b). Nøttestad et al. (2016a) also note observations of a new swimming strategy where mackerel, which have no swim bladder and must constantly be in motion to not sink in the water column, use the Northeast Atlantic current to reach far-off regions to feed. In fact, due to the massive expansion of the mackerel, Iceland has been able to fish the stock within their EEZ since around 2008. Furthermore, and more important for the purposes of this chapter, in 2013, the first mackerel from the Northeast Atlantic stock was observed off the coast of Svalbard, perhaps as a result of the marked decrease of summer ice cover in the region positively affecting the habitat suitability for both plankton and fish (Haug et al., 2017). Today, locals in both Reykjavík, Iceland, and in Longyearbyen, Svalbard, report that they can cast their fishing lines from shore and catch mackerel.

Another example of a northwards-moving marine resource is the snow crab (aka the queen crab, *Chionoecetes opilio*). The snow crab fishery first started in Canada in the 1960s and was registered in the Barents Sea for the first time in 1996 (Fisheries and Oceans Canada, 2014; Murray and Ings, 2015). In 2003, the first two snow crabs were caught in Norwegian waters near the northernmost region of Finnmark (Alvsvåg et al., 2009). In the Barents Sea, snow crab is today mostly spread west of Novaja Semlja, but based on the yearly stock assessment carried out by the Institute of Marine Research of Norway (IMR), the prediction is that the snow crab stock will continue growing and spreading in the Barents Sea, moving toward colder waters around Svalbard. Speculation is that the value of this fishery may exceed that of the Norwegian cod fisheries in the future, with a projected annual catch value of 1 to 5 billion NOK by 2020 (Hvingel and Sundet, 2014). As a new species, no country has any historical fishing rights to the snow crab, and it may as such become the first climate change–related shock to the fisheries protection zone regime, or possibly to the Svalbard Treaty itself.

Regime Vulnerability, Resilience, and Adaptive Capacity

International regimes can be characterized as sets of principles, norms, rules, and procedures for making decisions about specific issues of importance to actors in the international arena. A regime ultimately has as its aim to solve or ameliorate socioeconomic, environmental, and/or political issues that are or could be a problem for stakeholders in the area within the framework of a specific topic area, such as fisheries, pollution, or migration of birds (Krasner, 1982; Hasenclever et al., 2000; Stokke, 2007). International regimes are sometimes also equated semantically with international institutions by researchers in the field (Young, 1998; Bohman, 1999; Young, 2002; Peters, 2006; Ostrom, 2007). Ostrom (2007) defines regimes as "shared concepts used by humans in repetitive situations organized by rules, norms and strategies." Others have defined institutions as "persistent and connected sets of rules and practices that prescribe behavioral roles, constrain activity and shape expectations" (Keohane et al., 1993).

Snow crab and the mackerel are both examples of marine species that are changing their distribution patterns because of changing climatic stressors. As such, they may become catalysts for changes in existing regimes, depending on those regimes' *vulnerability, resilience, and adaptive capacity* under different levels of climatic stress.

In this section, we will adapt these interlinked concepts – vulnerability, resilience, and adaptive capacity (Gallopín, 2006) – and apply them to environmental management regimes. In assessing vulnerability, resilience, and adaptive capacity to exogenous shocks, notably from the effects of climate change on the ocean, we turn to accepted definitions of these terms that are applicable to management regimes. The term "vulnerability" has been extensively studied (Adger, 2006), but there is no coherent interdisciplinary definition of it. According to Cutter et al., vulnerability can be defined as "those characteristics of the population that influence the capacity of the community to prepare for, respond to, and recover from hazards and disasters" (Cutter et al., 2009). Another definition of vulnerability is "the characteristics of a person or group and their situation that influence their capacity to anticipate, cope with, resist and recover from the impact of a natural hazard" (Wisner, 2004). Vulnerability is always a specific disturbance to a state of equilibrium and about the potential for transformation when confronted with external or internal stressors.

Since our aim is to assess a marine management regime resilience under exogenous shocks brought forth by climate change, the differences between stressors and perturbations should also be discussed. Stress is considered to originate within a system itself and is usually continuous and even sometimes within the range of normal variability for that system. It does apply pressure to the system, but not dramatically so. Perturbations, in contrast, are not normally within the range of the normal variability of a system, and usually they originate outside the system. However, for institutions, the sources of perturbations may be internal as well, or a combination of both internal and external. They can potentially produce improvements to a given institutional setting (Young, 2010). Young (2010) gives the example of the United Nations Convention on the Law of the Sea (UNCLOS). He shows how additions and changes to the treaty have not significantly transformed it or changed the character of the arrangements originally agreed upon. This occurred even though there have been major additions to the treaty over the years, such as those dealing with pollution, straddling fish stocks, marine transport, and so forth. These additions would be stresses to the institutional arrangements, but not perturbations with large transformative effects on the regime itself. A fundamental difference between vulnerability and resilience is that the former refers to the ability to preserve regime structures, such as that of UNCLOS in the example earlier, in light of external stressors. The more it is stressed, however, the more vulnerable a regime becomes, and a stress situation may even become a perturbation as a result thereof.

Resilience, on the other hand, refers to a regime's ability to recover from structural changes to its original functionality resulting from vulnerability. It is sometimes related to a history of past exposures to vulnerability that increase the resilience of a given system (Gallopín, 2006). We will apply both vulnerability and resilience to comparative cases later in this chapter. In addition, we will consider the popular term *adaptive capacity*. This term can be framed in a number of ways (Smit and Wandel, 2006), but in its most raw form it is defined as "an ability to become adapted (i.e., to be able to live and to reproduce) to a certain range of environmental contingencies" (Gallopín, 2006). In terms of its adaptation to institutional dynamics, adaptive capacity includes being able to take advantage of opportunities that arise as a result of climate change, to adjust to damage from it, and to cope with the consequences that arise from it (Intergovernmental Panel on Climate Change, 2001).

Before an institution can be considered resilient, or to have high adaptive capacity, it first must experience a perturbation, or exogenous shock, that affects it to an extent that brings forth a change to the system, sometimes after a long period of institutional stability, where continuity and reproduction of the institutions have taken place. Only a radical change can destabilize the system and cause the institutional structure to be refashioned (March and Olsen, 2006), in line with the differences between stresses and perturbations noted earlier. Because not all changes to institutions are brought about by shocks to the system, two concepts that add dimensions to the theoretical body of institutional change are institutional *layering* and institutional *conversion* (Thelen, 2003; 2004). Institutional layering is the grafting of new elements onto an already stable institutional framework. This may alter the institutional trajectory of a given institution. Institutional conversion, on the other hand, is the adaptation of new rules or new goals into the institution in question. The function that the institution has historically served can thereby change, as can the role that the institution performs. In other words, the institution will convert from one function to another.

For any of these changes to happen, whether institutional layering or conversion, in order for the topic to rise to the top of the policy makers' agenda, there has to be a public pressure that arises from a sense of urgency. This urgency can be based on public perceptions of risks and hazards, such as that of a hole in the ozone layer or dying forests, as implied by the cases introduced later. One factor is that the threat or hazard must be considered globally catastrophic, involuntary and threatening to people, and be backed by scientific evidence as well as consensus. The threat also must be observable, and there has to be an immediacy to it, both temporal and geographical, rather than it being a delayed threat that is geographically distant. Finally, risks have to have an element of exposure, and increased exposure will be less palatable to actors, especially if the risk is considered personally threatening, involuntary, and immediate (Morrisette, 1989). For a global agreement to be reached, however, there often also has to be other elements to it, such as sufficient and accepted leadership moving the agreement forward, added value of international collaboration, and measures available that allow the network to address the challenge in an affordable manner (Sliggers et al., 2004). Given this, we will now turn to examples of two international management regimes that experienced perturbations, stressors, or exogenous shocks. We will compare these examples to potential future scenarios for the Svalbard fisheries protection zone.

The Effect of the Ozone Hole on the Ozone Regime

In the early 1970s, the first scientific publications on stratospheric ozone depletion as a result of chlorofluorocarbons (CFCs) and other ozone-depleting substances were published (Molina and Rowland, 1974). Subsequently, there was a public fear of millions of new cases of skin cancer and other human health effects, such as suppression of the immune system and eye disorders (Morrisette, 1989). Despite antiregulatory campaigns by the CFC-producing and -consuming industries in the United States, by the mid-to-late 1970s several countries banned the use of CFCs in aerosols, including the United States, Canada, Sweden, Norway, and Denmark (Morrisette, 1989). In 1985, the Vienna Convention for the Protection of the Ozone Layer, ratified by 193 parties, declared that the ozone layer needed protection, though there were no obligations yet concerning the use and production of CFCs. Subsequently, in 1987 governments agreed to a protocol to the Vienna Convention – the Montreal Protocol on Substances that Deplete the Ozone Layer – which placed restrictions on the production of CFCs.

After the Vienna Convention was agreed, scientists realized that there appeared to be a hole in the ozone layer over Antarctic (Farman et al., 1985). This discovery increased pressure, both nationally and internationally, to resume negotiations to strengthen the Montreal Protocol so as to more effectively protect the ozone layer. Although the cause of the Antarctic ozone hole had not been proven by the time the Montreal Protocol was agreed, it nevertheless had a tangible impact. The hole was a powerful symbol of the need to take action, and it helped to galvanized public opinion at the global level on the importance of banning CFCs (Morrisette, 1989). There was never an actual "hole" in the ozone layer in the Antarctic. The term was used as a metaphor to describe areas where ozone concentrations had dropped below historical thresholds. Measurements of this "hole" started in 1979, when the concentration was at 194 Dobson units, just below the historical threshold of 220 Dobson units. In 1983, this fell, and it continued to fall rapidly to 124 in 1985, and in 1991 the concentrations fell below 100 Dobson units for the first time. The lowest measured concentrations were in 1994 when they fell to 73 Dobson units. Since the mid-1990s, the area and depth of the Antarctic ozone "hole" have stabilized, and scientists predict that it will recover around 2040 (Lindsey, 2016) after 90 percent of global production of CFCs were eradicated by the late 1990s (Solomon, 2004), thanks to the Montreal Protocol and subsequent amendments to eliminate many ozone-destroying chemicals.

The Montreal Protocol on Substances that Deplete the Ozone Layer entered into force in 1989. The protocol was the first international agreement that appears to have solved a global atmospheric challenge (Velders et al., 2007). The metaphorical "hole" in the ozone layer was a *perturbation* that pushed the international community from the noncommittal Vienna Convention to the Montreal Protocol and beyond by galvanizing world opinion on the importance of protecting the ozone layer.

Black Forest Death and the Convention on Long-Range Transboundary Air Pollution

The 1979 Convention on Long-Range Transboundary Air Pollution (LRTAP) has a regional focus. It is based on cooperation among 49 parties that are members of the United Nations Economic Commission for Europe. Formation of the convention was a result of concerns about acid rain, focusing on emissions of sulfur oxide (SO_2) initially linked to the acidification of Scandinavian lakes and rivers. The issue was put on the international agenda by Norway and Sweden, and by an unlikely ally, the Soviet Union. In its initial form, the convention was a compendium of broad principles and promises of joint research activities. It lacked concrete measures that would actually curb acid rain (Levy, 1995). It was not until the 1985 Sulphur Protocol that concrete measures were taken to curb emissions. Prior to this, the problem was not deemed serious enough by the European nations (with the exception of Norway and Sweden) because only the Scandinavian countries were affected by the results and only Germany was seen as a major source of the pollution.

As such, acid rain was initially not a priority for most signatory countries (Sliggers et al., 2004). However, from 1982 onwards it was found that long-range air pollution was damaging not only Scandinavian lakes but German forests as well. Dying forests in Germany were coined "Waldsterben" (forest death) by German biologist Bernhard Ulrich. In a *New York Times* article from 1984, the author poetically described the phenomena in the following way: "As fir needles turn yellow and spruce branches sag limply, the industrial poisoning of West Germany's forests is assuming the dimensions of a spiritual, as well as an environmental, catastrophe" (Markham, 1984). A "Green Movement" ensued and consequently influenced the environmental narratives of most German political parties. Germany, one of the largest producers and polluters of SO_2, therefore announced at the 1982 Stockholm Conference on Acidification of the Environment that they were joining the Norwegian, Swedish, and Soviet efforts to seek SO_2 reductions (Levy, 1995). In addition to the outrage and the Green Movement, technological solutions were also in development at that time that would allow governments and industries to solve the problems of not only SO_2 emissions but also those of nitrogen oxides.

The Sulphur Protocol, signed by 19 parties in Helsinki in 1985, had clear goals and measures to verify that parties upheld their obligations for major reductions in SO_2 emissions. Once again, a *perturbation*, this time a realization that the problem was more widespread than first realized, resulted in a regime response with environmental impact.

Implications for the Svalbard Fisheries Protection Zone

In what way are the experiences of ozone depletion and acid rain applicable or transferable to the fisheries protection zone around Svalbard? Recalling factors that are critical for an issue to reach the agenda of policy makers and the public, the issue must be considered globally catastrophic, involuntary and personally threatening to people, and backed by scientific evidence and scientific consensus. In addition, the people should observe the threat and have a sense that there is an immediacy to the problem, both temporal and geographical, rather than it being a delayed threat

Table 12.1. *Differences and similarities among three regimes*

	Vienna Convention	LRTAP	Svalbard Fisheries Protection Zone
Original issue	**Protect the ozone layer from CFCs**	**Protect the environment from acid rain**	**Protect marine environment from overfishing**
Institutional elements	Little obligation	No concrete measures	Clear measures and strong obligation
Original parties	193	33	1 (Norway)
Exogenous shock/ perturbation	Ozone hole over the Antarctic	Dying forests in Germany	Shifting fish stocks in contested waters
Scientific evidence and consensus	Yes	Yes	Yes
Observable threat	Yes	Yes	Yes
Added value to international collaboration	Yes	Yes	Yes
Measures to address challenge	Yes	Yes	Yes
Global impact	Yes	Yes	No
Threat to people	Yes	Yes	No
Temporal immediacy	Yes	Yes	No
Geographical immediacy	Yes	Yes	No
Accepted leadership	Yes	Yes	No
Adaptation measure	Montreal Protocol	Sulphur Protocol	None to date

that is geographically distant. Finally, risks have to be perceived to have an element of exposure that would affect humans (Morrisette, 1989). There would also have to exist sufficient and accepted leadership to move agreement on an issue forward, a sense of added value of international collaborations, and, preferably, measures available that allow the states to address the challenge in an affordable manner (Sliggers et al., 2004). As noted in Table 12.1, the Vienna Convention and the LRTAP are affirmative in all variables of importance when it comes to agenda setting and urgency of issue. For the Svalbard Fisheries Protection Zone, we see that some variables that were of great importance in the ozone hole and Waldsterben cases are not affirmative.

Institutional layering is the grafting of new elements onto an already stable institutional framework. These new elements may or may not alter the institutional trajectory of the institution in question. Institutional *conversion*, on the other hand, is the adaptation of new rules or new goals into the institution in question. The function that the institution has historically served can thereby change, as can the role that the institution performs. In other words, the institution will convert from one function to another. The Antarctic hole in the ozone layer was a perturbation to the Vienna Convention system, in that it was beyond the normal variability of the system and originated outside the system. As such, it was a force that spurred international negotiations and pushed the public into demanding action from policy makers. There was a potentially direct effect to people's health, but people were not required to change their lifestyles because there were, or shortly would be, alternatives to CFCs available. Though the 1985 Vienna Convention did have binding treaty obligations, it was a loose framework convention, with no institutional framework and no independent authority to enforce the commitments in the treaty. As such, it was a soft form of international legalization with little "bite"; alone it could not solve the environmental threat in the form of a "hole" in the ozone layer. The Montreal Protocol was a much stronger agreement, arguably in part due to concerns among negotiators of a potential ozone hole. It has had a real impact on state behavior (Abbott et al., 2000).

The Montreal Protocol changed the functionality of the underlying regime. It is an example of institutional conversion as a result of the perturbation. This conversion was necessary because of people's perceptions of the threat of the ozone hole. When there are perceptions of global impact and the effects on people are involuntary, individuals demand action from policy makers. This has to happen in conjunction with an issue having a temporal and geographical immediacy to it, as well as accepted leadership, for the issue to move onto the agenda of environmental management. Under such circumstances, the regime is likely to adapt. These elements were natural for the ozone hole and the dying forests as well. They could be visualized, and indeed were in news reports. In both cases, there was a sense of temporal immediacy, the feeling that (to imagine the pleas), "This has to be fixed now, or the trees will all die/we will all get skin cancer." The solutions in both cases were near to hand or easy to envision, making them another driving force. In addition, there was leadership by influential states (e.g., the United States for ozone, West Germany for acid rain) that pushed the agenda toward transformation into adapted regimes that would have more "bite" in both cases.

As shown in Table 12.1, a number of these elements do not apply to the Svalbard Fisheries Protection Zone. Officially, the zone was developed to protect the marine resources in the 200-mile zone around Svalbard, and as such it is not perceived to have global significance. Furthermore, this is an issue that only has one party to it, namely Norway. It is an agreement that other nations tacitly accept by complying with Norwegian regulations, paying their citations, and signing coast guard reports. But they do not accept it officially. Norway unilaterally decided to establish a large marine protected area with limited resource extractions in the area. It has shared the resource with Russia because of the element of historical fishing rights being a determinant of quota allocations, with only a small share going to third parties. It was not a joint global or even regional decision to set aside this area for marine protection. No state formally accepts Norwegian sovereignty of this area. However, with shifting fish stocks, a perturbation or exogenous shock to the institution may push it to adapt, much as was the case with regimes associated with ozone hole and forests dying due to acid rain.

Conclusion

In this chapter, we have looked at the patterns of change in environmental regimes that have undergone substantial shocks and successfully adapted to these perturbations. The patterns revealed by studying these regime changes revealed options for ocean governance due to implications of climate change in the High North. The changes, with perturbations in the form of a warming ocean, would be an exogenous shock for the Svalbard Fisheries Protection Zone and could lead to major changes in existing institutional arrangements for managing marine fisheries in the area.

We have considered institutional layering – the grafting of new elements onto an already stable institutional framework. We understand that climate change may bring on effects that could alter the regime for Svalbard fisheries because of scenarios of new fish species moving into the seas of the High North. Within the framework of this scenario, it may be that the Fisheries Protection Zone could graft onto its practices the management of new resources, such as mackerel or snow crab. In doing so, however, it would disable the provisions of distributing quotas based on historical fishing rights. Norway may bypass the current self-imposed decision to use the Svalbard Treaty element of historical fishing rights as a criterion for quota allocations. The zone itself has over the years become institutionalized as a management zone under Norwegian authority. The years of Norwegian presence as sovereign in this area have created a situation of stability that no nation wishes to challenge or disturb. Informal understandings regarding the appropriate behavior of the actors in a given setting are critical, as are activities that have sprung up as a result of implementation attempts (Young, 2002). Therefore, instead of ignoring the problems that arise due to climate change, the institution can adapt to the change and layer new elements on top of existing structures.

Given the uncertainty about fish distribution and migratory patterns under different climate change scenarios, however, creating a separate regime for managing the up-and-coming species may still be the only solution that most nations will agree to. Future research on the topic must therefore concentrate on what type of regime these migratory trends must allow for and how it can be effective in protecting not only the species themselves but also all other ecosystem goods and services they depend on. It is a task that will demand a lot of research, but is necessary to ensure the future of the High North as a rich fishery area for future generations that is accessible to more than just the few climate change "winners." The Arctic nations, in collaboration with Japan, South Korea, Iceland, and the EU (all of which have large fishing fleets) have reached an historic agreement to abstain from any future unregulated fishing in the international waters of the Arctic Ocean, just north of Svalbard and its Fisheries Protection Zone. This suggests that it may be possible to respond to the uncertainty of migrating fish stocks, thereby achieving a degree of ocean governance in the High North despite climate change.

References

Abbott, K. W., Keohane, R. O., Moravcsik, A., Slaughter, A.-M., and Snidal, D. (2000). The concept of legalization. *International Organization*, **54**(3), 401–19.

Adger, W. N. (2006). Vulnerability. *Global Environmental Change*, **16**(3), 268–81.

Alvsvåg, J., Agnalt, A. L., and Jørstad, K. E. (2009). Evidence for a permanent establishment of the snow crab (Chionoecetes opilio) in the Barents Sea. *Biological Invasions*, **11**(3), 587–95.

Arlov, T. B. (2003). *Svalbards Historie (The History of Svalbard – in Norwegian)*. Trondheim: Tapir.

Bohman, J. (1999). International regimes and democratic governance: Political equality and influence in global institutions. *International Affairs*, **75**(3), 499–513.

Cutter, S. L., Emrich, C. T., Webb, J. J., and Morath, D. (2009). Social vulnerability to climate variability hazards: A review of the literature. *Final Report to Oxfam America*, pp. 1–44.

Farman, J. C., Gardiner, B. G., and Shanklin, J. D. (1985). Large losses of total ozone in Antarctica reveal seasonal ClOx/NOx interaction. *Nature*, **315**(6016), 207–10.

Fisheries and Oceans Canada. (2014). *Snow Crab*. Available at www.dfo-mpo.gc.ca/fm-gp/sustainable-durable/fisheries-peches/snow-crab-eng.htm

Gallopín, G. C. (2006). Linkages between vulnerability, resilience, and adaptive capacity. *Global Environmental Change*, **16**(3), 293–303.

Gattuso, J.-P., Magnan, A., Billé, R., et al. (2015). Contrasting futures for ocean and society from different anthropogenic CO_2 emissions scenarios. *Science*, **349**(6243), aac4722-1–aac4722-10.

Hasenclever, A., Mayer, P., and Rittberger, V. (2000). Integrating theories of international regimes. *Review of International Studies*, **26**, 3–33.

Haug, T., Bogstad, B., Chierici, M., et al. (2017). Future harvest of living resources in the Arctic Ocean north of the Nordic and Barents Seas: A review of possibilities and constraints. *Fisheries Research*, **188**, 38–57. Available at https://doi.org/10.1016/j.fishres.2016.12.002

Hoel, A. H. (2005). Er Fiskevernsonen Internasjonalt Farvann? (*Is the Fisheries Protection Zone International Waters? – in Norwegian*). Nordlys.no. Available at www.nordlys.no/debatt/kronikk/article1849275

Hvingel, C. and Sundet, J. (2014). *Snow Crab – A New Substantial Resource in the Barents Sea* (In Norwegian: Snøkrabbe – en ny stor ressurs i Barentshavet). Nordlys. Available at www.imr.no/publikasjoner/andre_publikasjoner/kronikker/2014_1/snokrabbe_en_ny_stor_ressurs_i_barentshavet/nb-no

ICES. (2016). *Report of the Working Group on Widely Distributed Stocks* (WGWIDE), August 31–September 6, 2016, ICES HQ, Copenhagen, Denmark. ICES CM 2016/ACOM:16.

Intergovernmental Panel on Climate Change. (2001). *Technical Summary: Climate Change 2001: Impacts, Adaptation and Vulnerability. A Report of Working Group II of the Intergovernmental Panel on Climate Change*. Available at www.grida.no/climate/ipcc_tar/wg2/pdf/wg2TARtechsum.pdf

Keohane, R. O., Haas, P. M., and Levy, M. A. (1993). The effectiveness of international environmental institutions. In R. O. Keohane, P. M. Haas, and M. A. Levy, eds., *Institutions for the Earth: Sources of Effective International Environmental Protection*. Cambridge, MA: MIT Press.

Krasner, S. D. (1982). Structural causes and regime consequences: Regimes as intervening variables. In S. D. Krasner, ed., *International Regimes. Ithaka and London*, Ithaca, NY: Cornell University Press.

Levy, M. A. (1995). International cooperation to combat acid rain. *Green Globe Yearbook*, **1995**, 59–68.

Lindsey, R. (2016). *Antarctic Ozone Hole*. Available at https://earthobservatory.nasa.gov/Features/WorldOfChange/ozone.php

March, J. G. and Olsen, J. P. (2006). Elaborating the "new institutionalism." In R. A. W. Rhodes, S. A. Binder, and B. A. Rockman, eds., *The Oxford Handbook of Political Institutions*. Oxford: Oxford University Press, 3–22.

Markham, J. M. (1984). In a 'dying' forest, the German soul withers too. *The New York Times*. Available at www.nytimes.com/1984/05/25/world/in-a-dying-forest-the-german-soul-withers-too.html?mcubz=3.

Molina, M. J. and Rowland, F. S. (1974). Stratospheric sink for chlorofluoromethanes: chlorine atom-catalysed destruction of ozone. *Nature*, **249**(5460), 810–2.

Morrisette, P. M. (1989). The evolution of policy responses to stratospheric ozone depletion. *Natural Resources Journal*, **29**(3), 793–820.

Murray, G. D. and Ings, D. (2015). Adaptation in a time of stress: A social-ecological perspective on changing fishing strategies in the Canadian snow crab fishery. *Marine Policy*, **60**, 280–6.

Nøttestad, L., Diaz, J., Penã, H., et al. (2016a). Feeding strategy of mackerel in the Norwegian Sea relative to currents, temperature, and prey. *ICES Journal of Marine Science*, **73**(4), 1127–37. Available at https://doi.org/10.1093/icesjms/fsv239

Nøttestad, L., Utne, K. R., Óskarsson, G. J., et al. (2016b). Quantifying changes in abundance, biomass and spatial distribution of Northeast Atlantic (NEA) mackerel (Scomber scombrus) in the Nordic Seas from 2007 to 2014. *ICES Journal of Marine Science*, **73**(2), 359–73. doi:10.1093/icesjms/fsv218.

Ostrom, E. (2007). Institutional rational choice: An assessment of the institutional analysis and development framework. In P. A. Sabatier and C. M. Weible, eds., *Theories of the Policy Process*. Boulder, CO: Westview Press, 35–72.

Peters, B. G. (2006). *Institutional Theory in Political Science: The 'New Institutionalism'*. London: Continuum.

Sliggers, J., Kakebeeke, W., and United Nations Economic Commission for Europe (2004). *Clearing the Air: 25 Years of the Convention on Long-range Transboundary Air Pollution*, Geneva: United Nations Economic Commission for Europe.

Smit, B. and Wandel, J. (2006). Adaptation, adaptive capacity and vulnerability. *Global Environmental Change*, **16**(3), 282–92.

Solomon, S. (2004). The hole truth. *Nature*, **427**(6972), 289–91.

Stokke, O. S., ed. (2007). *Examining the Consequences of Arctic Institutions. International Cooperation and Arctic Governance: Regime Effectiveness and Northern Region Building*. London: Routledge.

Thelen, K. (2003). How institutions evolve: Insights from comparative historical analysis. In J. Mahoney and D. Rueschemeyer, eds., *Comparative Historical Analysis in the Social Sciences*. Cambridge: Cambridge University Press, 208–40.

Thelen, K. (2004). *How Institutions Evolve: The Political Economy of Skills in Germany, Britain, the United States, and Japan*. Cambridge: Cambridge University Press.

Ulfstein, G. (1995). *The Svalbard Treaty: From Terra Nullius to Norwegian Sovereignty*. Oslo: Scandinavian University Press.

Velders, G. J. M., Andersen, S. O., Daniel, J. S., Fahey, D. W., and McFarland, M. (2007). The importance of the Montreal Protocol in protecting climate. *Proceedings of the National Academy of Sciences*, **104**(12), 4814–19.

Wisner, B. (2004). *At Risk: Natural Hazards, People's Vulnerability and Disasters*, London: Routledge.

Young, O. R. (1998). *Creating Regimes: Arctic Accords and International Governance*. Ithaca, NY and London: Cornell University Press.

Young, O. R. (2002). *The Institutional Dimensions of Environmental Change: Fit, Interplay and Scale*. London: MIT Press.

Young, O. R. (2010). Institutional dynamics: Resilience, vulnerability and adaptation in environmental and resource regimes. *Global Environmental Change*, **20**(3), 378–85.

Part IV
Changing Polar Seas

13

Climate Change and the Southern Ocean
The Regime Complex for Regional Governance

MARCUS HAWARD

Introduction

The Southern Ocean has an enormous influence on the global climate system. This unique environment, with its seasonal sea-ice zone, also has a major role in global ocean circulation, and it is a significant carbon sink. The region is experiencing the impacts of climate change, as evidenced by warming ocean temperatures and changes in ocean salinity. These impacts have created marine climate hotspots off the Antarctic Peninsula and in the Southern Indian Ocean (Hobday and Pecl, 2014). The Southern Ocean is undergoing significant changes, at times with spectacular outcomes. Between July 10 and 12, 2017, a 5,800-km^2 iceberg calved off the Larsen B ice shelf in West Antarctica. The new iceberg, named A68, was noted in different media reports as being the size of the Indonesian island of Bali, the country of Luxembourg, or the state of Delaware in the United States (Voosen, 2017). The size and scale of the event creating A68, together with its occurrence in a recognized global climate change hotspot, focused attention on the Southern Ocean, in particular, its ice shelves and ice sheets. It also served as a useful focus on the impacts of anthropogenic climate change on this ocean and raised questions about the region's governance.

The key institution for governing the Southern Ocean is the Antarctic Treaty System (ATS). Drawing on regime theory, in this chapter the ATS is characterized as an environmental and resource *regime,* that is, "an assemblage of rights, rules and decision-making procedures that influence the course of human interactions" (Young, 2010: 1). It is important to note that there are other instruments and associated regimes applicable to the Southern Ocean, including those related to the law of the sea (see Chapter 2 and chapters in Part IV of this volume), international shipping (see Chapter 23), marine pollution (see Chapter 21), safety of life at sea, protection of the marine environment and biodiversity, and international fisheries (see chapters in Part III). Climate change cuts across each of these issues and is likely to lead to increased interactions, creating what is termed a regime complex – "a loosely coupled set of specific regimes" (Raustiala and Victor, 2004) "that pertain to the same issue domain or spatially defined area ... and interact with one another in the sense that the operation of each affects the performance of the others" (Young, 2014: 394).

201

This chapter outlines the key institutions and actors involved in the governance of the Southern Ocean. It briefly considers the role of the Southern Ocean in global and regional climates. The influence of climate change on the Southern Ocean governance is explored by examining how institutions and actors have responded to, and have sought to address, climate change. The chapter also considers the extent to which climate change provides a driver for ongoing analysis of the resilience of Antarctic regimes and instruments, and for the shaping of the existing regime complex.

Climate Change and the Southern Ocean

The Southern Ocean, linking the Earth's great ocean basins through the Antarctic Circumpolar Current, has a key role in global ocean circulation and is a major influence in the global climate system. Ocean currents are a key vehicle for transferring the heat that is absorbed by the oceans at its surface. Although wind-driven currents such as the Antarctic Circumpolar Current are critical, movement of water, due to differences in temperature and salinity, is also a major influence on global ocean circulation. The interaction between atmosphere and ocean means that "winds from the Antarctic continent help create ice-free marine areas (polynyas), while at the same time the differential in salinity between surface and deep water in the sea-ice zone creates bottom water that helps drive circulation" (Haward and Jabour, 2014: 27). Cold, saline Antarctic waters help moderate water temperatures as dense bottom water moves away from the Antarctic. As colder water moves away from the Southern Ocean and rises from the bottom, the water mass loses its relative salinity and increases in temperature. This oceanographic feature, where colder bottom water rises and forces warmer surface water to move, is known as the overturning circulation.

The world's oceans have absorbed 30 percent of CO_2 released by human activities over the past 50 years (Gleckler et al., 2016). The Southern Ocean is an important component of this global carbon sink as CO_2 is absorbed more efficiently in colder waters (Haward and Jabour, 2014). Increasing levels of CO_2 in the Southern Ocean have key consequential impacts as a result of the chemical reaction caused by CO_2 absorption in seawater. Oceans are weakly alkaline. The addition of CO_2 changes the carbonate chemistry of the ocean and drives ocean pH (the extent to which a solution is acid or alkaline) lower, meaning oceans becomes less alkaline (or, in simple terms, more acidic). Impacts of ocean acidification are projected to vary, but the cold Southern Ocean waters – with high levels of the global oceans' store of CO_2 – are considered to be much closer to carbonate saturation than other oceans. Although the ecological impacts of ocean acidification in the Southern Ocean are not well known, it is likely that they will affect rates of primary production by organisms that have calcium carbonate shells as coccolithophorid species, a key microalgal primary producer in the ocean ecosystem (Antarctic Climate and Ecosystems Cooperative Research Centre, 2008). Changes in primary production are likely to have impacts through the food web. The impacts of ocean acidification on higher trophic orders in the Southern Ocean may lead to significant changes in ecological communities (see for example Vizzini et al., 2017).

The third key element in the Southern Ocean's role in the global climate system relates to the significance of the annual freeze–thaw cycle of the Antarctic sea-ice zone. This region, a major component of the global cryosphere, is important for both the physical environment and the ecosystems that exist within this environment. Warming temperatures are likely to modify the seasonal freeze–thaw pattern in the Southern Ocean and influence the extent, thickness, and duration of sea ice (Haward and Jabour, 2014). Considerable interannual variability has been observed in the extent of sea ice around Antarctica, and current research is focused on identifying climate change signals in this variability (Haward, 2014). Changes in sea-ice extent and thickness have a range of potential impacts. A loss of sea ice will reduce the albedo effect (the capacity of sea ice to reflect sunlight), contributing to further warming. Changes in sea ice are also likely to have ecosystem impacts, reducing the habitat for krill, the keystone species of the Southern Ocean food web, which utilizes algae growing on sea ice for winter food. These impacts on the ecosystem are likely to be heighted by changes in carbonate chemistry of the Southern Ocean through ocean acidification.

Governing the Southern Ocean: The Antarctic Treaty and Its System

The Antarctic Treaty of 1959 forms the basis for an extensive and evolving regime governing the Antarctic and the Southern Ocean, a region comprising approximately 10 percent of the Earth's surface area. There are 53 states that are party to the Antarctic Treaty: 29 Consultative Parties (ATCPs) with formal decision-making power within the Antarctic Treaty Consultative Meeting, and an additional 24 contracting parties that have acceded to the treaty. Consultative-party status is open to contracting parties, in addition to original signatories, that have committed to significant research programs (Haward, 2017a).

The Antarctic Treaty's preamble includes two key principles:

Recognizing that it is in the interest of all mankind that Antarctica shall continue forever to be used exclusively for peaceful purposes and shall not become the scene or object of international discord; and

Acknowledging the substantial contributions to scientific knowledge resulting from international cooperation in scientific investigation in Antarctica.

The commitment to scientific cooperation in the Antarctic is elaborated in the wording of Article II of the treaty: "Freedom of scientific investigation in Antarctica and cooperation toward that end ... shall continue."

The ATS comprises institutions, instruments, and arrangements governing the Antarctic and the Southern Ocean. The ATS is centered on the Antarctic Treaty, but it also includes the following subsidiary and related instruments: the Convention for the Conservation of Antarctic Seals (CCAS) of 1972; the Convention on the Conservation of Antarctic Marine Living Resources (CCAMLR) of 1980; and the Protocol on Environmental Protection to the Antarctic Treaty, also known as the Madrid Protocol, which was agreed in 1991. The ATS also includes the following institutions:

- Antarctic Treaty Consultative Meetings (ATCM)
- Scientific Committee on Antarctic Research (SCAR)
- Committee for Environment Protection (CEP)
- Standing Committee on Antarctic Logistics and Operations (SCALOP)
- Council of Managers of National Antarctic Programs (COMNAP)
- The CCAMLR Secretariat
- The Antarctic Treaty Secretariat

These institutions are supplemented by Special Antarctic Treaty Consultative Meetings (SATCMs) and the Antarctic Treaty Meeting of Experts (ATME). SATCMs are called in special circumstances, for example, when negotiating new instruments, such as the Protocol on Environmental Protection. Meetings of experts are a part of intersessional work of the ATCM and can help to focus attention on key issues. In the context of this chapter, the most salient example is the ATME on Climate Change, which was held in Norway in 2010 (ATCM, 2010).

Antarctic Treaty Consultative Meetings

The ATCPs meet annually at the Antarctic Treaty Consultative Meeting, a cornerstone of the Antarctic regime. The ATCM, mandated through Article IX of the Antarctic Treaty, provides opportunities for information exchange (Article III) through reports on activities (Article IX), reports on inspections (Article VII), and presentation of working and information papers prepared by parties. The ATCM advances collaborative action and governance of the region through a range of Measures, Decisions and Resolutions, which in 1995 replaced the earlier designation of ATCM Recommendations. Measures, Decisions, and Resolutions are outcomes of the ATCM. A Measure contains provisions that are designed to be legally binding on all ATCPs once the provision has been approved by all parties. A Decision is a matter related to internal organizational issues, and a Resolution is a hortatory text agreed at an ATCM (see ATCM, 1995).

Since 1983, non-Consultative Parties have been invited to take part in ATCM meetings, but they do not participate in decision making. In addition, three organizations participate as formal observers: CCAMLR, SCAR, and COMNAP. ATCMs also provide opportunities to receive reports from SCAR that have helped maintain a focus on current Antarctic climate change–related science. The SCAR Lecture, delivered by an eminent Antarctic scientist, has provided a focus for discussions on environmental and climate change–related issues to be raised within the ATCM. Other organizations, such as the Antarctic and Southern Ocean Coalition (ASOC) – the peak body for Antarctic and Southern Ocean environmental nongovernmental organizations – and the International Association of Antarctica Tourist

Operators (IAATO) – the peak body for tourism operators – are invited by the parties to attend as experts. In recent years, the ATCM has also invited experts from a range of international organizations to attend the meeting, reflecting the broadening of the ATCM's agenda but also the increasing intersection between the work of the ATCM and that of other international organizations.

The Protocol on Environmental Protection to the Antarctic: The Madrid Protocol

The negotiation and entry into force of the 1991 Protocol on Environmental Protection to the Antarctic Treaty, known colloquially as the Madrid Protocol, is a significant element in the governance of Antarctica and the Southern Ocean (Stokke and Vidas, 1996). The Madrid Protocol was developed in response to opposition (both external and internal) to the ATCPs' attempt to develop a regime to regulate potential mineral resource activity and mining in Antarctica. The Protocol reiterates the core values of the Antarctic Treaty, but it has also broadened the base of science from solely traditional disciplinary based activities to encompass multidisciplinary environmental science (Dudeney and Walton, 2012). In doing so, the Madrid Protocol provides support to climate impact and adaptation research that necessarily links biological and physical sciences. The Madrid Protocol applies to the Antarctic Treaty Area, and thus applies to the Southern Ocean south of latitude 60° South.

The Committee for Environmental Protection

The CEP provides advice and formulates recommendations to the ATCM under Article 12 of the Protocol. The CEP's work plan for 2014–18 includes a focus on climate change. The CEP's Climate Change Response Work Program developed from the ATME on Climate Change in 2010. It "provides specific actions by the CEP to support efforts within the Antarctic Treaty System to prepare for, and build resilience to, the environmental impacts of a changing climate and the associated implications for the governance and management of Antarctica" (ATCM, 2016: 174). The Climate Change Response Work Program was adopted by the ATCPs in Resolution 4 of 2015 (ATCM, 2015).

Commission for the Conservation of Antarctic Marine Living Resources

The CCAMLR is an international organization established in 1982 through the entry into force of the Convention on the Conservation of Antarctic Marine Living Resources. This Convention, and the work of its Commission, is directly linked to the Antarctic Treaty and is a central element of the Antarctic Treaty System. Attention to marine resource management within the ATS was driven by Article IX of the Antarctic Treaty, which recognized that the ATCM could consider preservation and conservation of living resources (see Orheim et al., 2011: 209–10). The development of CCAMLR was initiated by an ATCM Recommendation that centered on protection of the environment and wise use of resources.

The Convention on the Conservation of Antarctic Marine Living Resources applies to Antarctic marine living resources (with the notable exception of whales, which are the responsibility of the International Whaling Commission) in an area bounded to the north by a line that approximates the position of the Antarctic Polar Front (previously termed the Antarctic Convergence) – which is where cold waters of the Southern Ocean, driven by the Antarctic Circumpolar Current, meet and mix with warmer waters of lower latitudes – and to the south by the Antarctic continent. The CCAMLR Commission and its Scientific Committee are charged with conservation of Antarctic marine living resources. CCAMLR pioneered the use of an ecosystem approach to marine resources management that "does not concentrate solely on the species fished, but also seeks to avoid situations in which fisheries have a significant adverse effect on dependent and related species" (Kock, 2000: iii). There are currently 25 members of the Commission.

Climate Change and the Antarctic Treaty System

Although the Southern Ocean has long been a major focus for climate research, it is only in the last decade that the ATCM has formally given attention to climate risk management (Comba, 2013). In 2007, after "extensive consideration," ATCM XXX "agreed to revise Agenda Item 13 to read Scientific Issues, including Climate Related Research, Scientific Cooperation and Facilitation." The CEP had earlier agreed to include a similar item on its agenda (ATCM, 2007: 43). Climate change is an element of the CEP work plan and has been a focus of a number of contributions to the ATCM from SCAR. SCAR's contribution has included influential lectures at the ATCM in 2006 and 2007 (Masson-Delmotte, 2006; Rapley, 2007) and a number of SCAR information or background papers, for example, SCAR's paper on the Paris Climate Agreement presented at the 2017 ATCM (SCAR, 2017).

A major impetus for increased attention to climate change in the formal workings of the CEP and ATCM arose through the convening of an ATME on climate change in 2010. The proposal for a climate-change ATME initiative was first raised by Norway and the United Kingdom at ATCM XXXI in 2008, through working paper (WP) 35: "Antarctic Climate Change Issues" (ATCM, 2008: 56). In their joint paper, Norway and the United Kingdom recalled Resolution 3 (2003): "Support of the ATCM for the International Polar Year 2007/8," which, in its preamble, recognized "the important role of the Polar Regions both in driving and responding to Global Climate Change." The proposal from Norway and the United Kingdom for greater attention to climate change with the ATS was supported by information papers (IPs) submitted at the 2008 ATCM by the Antarctic and Southern Ocean Coalition (IP56) and SCAR (IP62) (ATCM, 2008: 56).

The formalization of the ATME on climate change occurred through the ATCPs' support for ATCM Decision 1 (2009), the "Meeting of Experts of Climate Change"

(ATCM, 2009), which, in addition to committing to the meeting, provided the scope and topics for consideration. Decision 1 (2009) also encouraged participation by consultative and nonconsultative parties to the Antarctic Treaty, representative ATS organizations and observers, and other key bodies. Invitations were sent to the International Union for the Conservation of Nature, the International Maritime Organization, the International Hydrographical Organization, the World Meteorological Organization, the Intergovernmental Panel for Climate Change, and the United Nations Environment Programme (ATCM, 2009: 213). The ATME was held in Norway in early April 2010, and its outcomes were reported to the ATCM XXXIII, held in Punta Del Este later in 2010 (ATCM, 2010). As noted earlier, the meeting's outcomes led directly into the CEP's Climate Change Response Work Program and served to maintain the ATCPs' focus on climate change.

At ATCM XXXV held in Hobart, Australia, in 2012, the host nation tabled WP32, titled "ATCM Interests in International Climate Change Discussions – Options for Enhanced Engagement" (ATCM, 2012: 68). This working paper was supported by a secretariat paper (SP8): "Actions Taken by the CEP and the ATCM on ATME Recommendations" (ATCM, 2012: 70). WP32 noted that a changing climate had implications for the parties' shared objectives under the Antarctic Treaty. It noted that recommendations of the 2010 ATME had focused on "opportunities for enhanced engagement with international organizations discussing climate change action" (ATCM XXXV/CEP XV, 2012: 3). Specific recommendations included registering the Antarctic Treaty Secretariat as an observer organization with the United Nations Framework Convention on Climate Change (UNFCC), issuing a statement to the UNFCCC conference of parties, engaging more directly with the UNFCCC's Nairobi work program on scientific and technical advice as a partner organization, and hosting an event on the Antarctic at a UNFCCC conference of parties.

WP32 noted that "pursuing closer engagement with the UNFCCC would be consistent with the provisions of the Antarctic Treaty and with the practice of establishing effective working relations with other international organizations where necessary to advance the protection and management of the Antarctic region" (ATCM XXXV/CEP XV, 2012: 3). The communique from ATCM XXXV committed the ATCM to continue to focus on understanding and addressing implications of climate change for Antarctica. The communique identified specific actions, such as identifying areas of conservation importance on account of their resilience to climate change. "Parties reaffirmed their commitment to undertake and promote scientific research in Antarctica, [and] to enhance understanding of global climate change and its implications for our planet" (ATCM, 2012: 167).

One outcome was reengagement with the World Meteorological Organization (WMO) on climate change–related science and increased direct linkages with the Intergovernmental Panel on Climate Change (IPCC). Although the WMO had been an invited expert to the ATCM, its engagement had not been consistent (WMO, 2015: 2). Reengagement began at ATCM XXXV and continued through subsequent meetings (WMO, 2015). As the ATCM has sought links with organizations such as the WMO in relation to climate change, so, too, has the WMO seen the meaningful and constructive benefits for engagement. Increased direct engagement between the ATCM and the IPCC has also occurred. At ATCM XXXIX in 2016, the IPCC representative spoke about an information paper on its Fifth Assessment Report, focusing on matters relevant to the Antarctic area (ATCM, 2016: 22).

The Antarctic Regime Complex

A number of legal instruments cover the Southern Ocean and address issues such as species protection and trade, whaling, marine pollution and wastes, natural heritage, and the marine environment. The regimes that emerge through the rights, rules, and decision-making procedures (Young, 2010: 1) of these instruments together contribute to an emergent regime complex in the Southern Ocean (Haward, 2017b). In addition to the ATS discussed earlier, key legal instruments include the Convention on Migratory Species of Wild Animals (CMS), the Convention on International Trade of Endangered Species Wild Flora and Flora (CITES), the International Convention for the Regulation of Whaling (ICRW), the Convention for the Prevention of Pollution from Ships (MARPOL), the Convention on the Prevention of Marine Pollution by Dumping of Wastes and Other Matter (London Convention/ Protocol), the Law of the Sea Convention (LOSC), the Convention on Biological Diversity (CBD), the World Heritage Convention, the Safety of Life at Sea Convention (SOLAS), and the UNFCCC.

The interplay between CCAMLR and the CMS regime is a useful example of the regime complex in action. CMS provides the legal base for the Agreement in the Conservation of Albatrosses and Petrels (ACAP). ACAP was developed from concern over the depletion of populations of sub-Antarctic seabirds through impacts of fishing. These impacts were first recorded in the CCAMLR area, which includes the sites of the major breeding colonies and foraging areas for the vast majority of the world's albatrosses and petrel species (Haward et al., 1997; Hall and Haward, 2001). Addressing and combating the incidental mortality of seabirds associated with fishing activities became the focus of concerted action and conservation measures under CCAMLR (Hall, 2007).

Another example of the interplay that has shaped the Antarctic regime complex are the linkages between the ATS and the Law of the Sea Convention. The Antarctic Treaty (1959) was negotiated after the first United Nations Conference on the Law of the Sea (UNCLOS I) in 1958 (see Chapter 2). The Antarctic Treaty specifically recognizes high seas rights within the treaty area, which is south of latitude 60° South. The nexus between these regimes has been important and is likely to remain a significant driver in shaping the regime complex in the future (see Haward, 2009). The sub-Antarctic islands – which are under national jurisdiction and are outside the treaty area and therefore not subject to Article IV of the Antarctic Treaty – can legitimately generate exclusive economic zones and continental shelves (Baird,

2004). Potential differences between treaty parties over declaration of maritime zones offshore from Antarctica were set aside by Australia's innovative solution to the interplay between the ATS and LOSC (Jabour, 2009) in which data related to delimitation of the continental shelf off the Australian Antarctic Territory was submitted to the Commission on the Limits of the Continental Shelf, although the Commission asked not to consider these data.

There are, however, examples that suggest that the provisions of some regimes have not been as clearly integrated (see Chapter 16). For example, interplay between CCAMLR and CITES highlights challenges arising over questions of jurisdiction and competence between regimes, especially when commercial activities are involved. Other examples of challenges include interplay with the Convention on Biological Diversity, the World Heritage Convention, and the Whaling Convention, all examples of competing regimes.

Although there have been calls to designate Antarctica and the Southern Ocean under the World Heritage Convention (Mosely, 2007), this initiative has gained little traction, although sub-Antarctic islands within the CCAMLR area but outside the Antarctic Treaty area have been designated as sites under this convention (Haward, 2008). There has been strong resistance from within the ATCM to engage more fully with the World Heritage Convention, and there are practical constraints to the use of the World Heritage Convention in the Antarctic Treaty area. Nomination of sites for consideration under the World Heritage Convention need to be made by the state on whose territory the site is located. Given the particular status of territorial claims under the Antarctic Treaty, World Heritage nominations in the Antarctic Treaty area remains problematic (Press, 2012). United Nations negotiations over conservation of areas beyond national jurisdiction have raised consideration and debate over the nexus between the World Heritage Convention and the high seas (UNESCO, 2016).

Whaling is deliberately not addressed by either the Antarctic Treaty or CCAMLR, with competence in this issue clearly directed to the International Whaling Commission through the ICRW (1946). The ICRW was negotiated, and the convention entered into force, prior to the Antarctic Treaty. There was a concern by some states that greater regulation of whaling was needed, as harvesting was likely to increase in the postwar period. The whaling issue has salience in the Southern Ocean, too, as the regime shifted from its original focus on harvesting to conservation of stocks, with a moratorium on commercial whaling (see Chapter 2).

While discussion on whaling is quarantined within the IWC, and thus not discussed in the ATCM, the issue continues to have impact within the broader Antarctic regime complex. In recent years, the Scientific Committee of CCAMLR has met with the IWC's Scientific Committee. More controversially, the intersection between whaling and the Antarctic regime has centered on Japan's self-described "scientific" whaling program in the Southern Ocean. This whaling, conducted under a "special permit" provision under the ICRW, has been controversial and led Australia to initiate proceedings against Japan's scientific program in the International Court of Justice (ICJ) on May 31, 2010. The case was heard at the ICJ in The Hague between June 26 and July 16, 2013, with the court's decision brought down on March 31, 2014. The ICJ found that Japan's whaling program contravened the schedule to the ICRW, that it ran against Japan's obligation to the ICRW, and that Japan had not complied with obligations to the IWC in "good faith." The ICJ decision was a significant (if short-term) "win" for Australia and New Zealand, but rather than banning scientific whaling, the court decision set clearer conditions for lethal research whaling. Japan cancelled its existing whaling program in 2014–15 and recast it as a new program, and it resumed scientific whaling in the Southern Ocean in December 2015.

The Southern Ocean regime complex's interaction with the ATS is a key influence in the future governance of the region. The ATS is a robust and successful regime in its own right. It has successfully addressed key issues that gave impetus to its establishment and has broadened its scope to issues such as marine resource management and environmental protection. Climate change is, however, likely to test the ATS's robustness and resilience (Chown et al., 2012; Haward et al., 2012).

Climate Change and the Resilience of Antarctic Regimes and Instruments

Science is a core value of the ATS, providing a focus for national programs and for engagement within the Antarctic and Southern Ocean regime (Herr and Hall, 1989). The Southern Ocean is a critical laboratory for climate change research, with the cryosphere and oceans remaining areas of significant uncertainty in recent assessments of climate change by the IPCC. The impact of warming Southern Ocean waters on the regional and global climate systems, and the impacts of anthropogenic climate change on this region, have led to increased research efforts. Key objectives of current Southern Ocean research include improving understanding of environmental conditions and ecosystem vulnerabilities.

The consequences of biophysical change in the Southern Ocean are potentially significant, and "the most obvious physical changes to date include increasing ocean temperatures, poleward shifts of ocean fronts and regionally contrasting changes in the extent and seasonality of sea ice" (Constable et al., 2014: 3005). The poleward movement of marine species (range shifting) away from areas of increasing warming temperatures is likely yet not fully understood (Constable et al., 2014) (see Chapters 9 and 10). These changes are likely to increase the interplay between different regimes. As Young (2010) notes, the ATS has in the past shown institutional adaptation and evolution in response to significant internal or external stressors. While the ATS traditionally has addressed emergent issues, such as marine resource management (CCAMLR), tourism (ATCM), and biological prospecting (CCAMLR), solely within its own decision-making forums (see, generally, Haward and Griffiths, 2011), climate change is likely to increase the salience of a broader regime complex, first in the work of the ATS and second in terms of future governance of the Southern Ocean.

Conclusion

This chapter has highlighted the role of climate change in Southern Ocean governance. The ATS has provided the framework for governance in the Southern Ocean for six decades. Notwithstanding the longevity and effectiveness of this regime, climate change tests its resilience in two complementary ways: first, how the ATS's constituent institutions and instruments respond, and second, how climate change shapes regime development and the interplay of an emergent regime complex. The ATS in response has increased its attention to climate change and also appears now more willing to interact with other regimes, as shown by the engagement with the UNFCCC and the IPCC discussed earlier.

Recent changes in the Southern Ocean, most notably the calving of iceberg A68, have drawn attention to the potential impacts of anthropogenic climate change on the ice shelves and ice sheets of the Southern Ocean. Warming waters and changes in ocean chemistry will affect the region's ecosystems. Climate change impacts will cut across multiple regulatory instruments and regimes that govern the Southern Ocean. The Southern Ocean, facing direct impacts of climate change, is an important laboratory for understanding this global climate-change laboratory. Climate change will also increasingly influence deliberations shaping governance in this fragile region.

References

Antarctic Climate and Ecosystems Cooperative Research Centre. (2008). *Position Analysis: CO2 and Climate Change: Ocean Impacts and Adaptation Issues.* Hobart: Antarctic Climate and Ecosystems Cooperative Research Centre.

Antarctic Treaty Consultative Meeting. (1995). *Decision 1 (1995). Measures, Decisions and Resolutions.*

Antarctic Treaty Consultative Meeting. (2007). *Final Report of the Thirtieth Antarctic Treaty Consultative Meeting.* New Delhi, India, April 30–May 11, 2007. Secretariat of the Antarctic Treaty: Buenos Aires. Available at http://ats.aq/devAS/ats_meetings.aspx?lang=e

Antarctic Treaty Consultative Meeting. (2008). *Final Report of the Thirty-First Antarctic Treaty Consultative Meeting.* Kyiv, Ukraine, June 2–13, 2008. Secretariat of the Antarctic Treaty: Buenos Aires. Available at http://ats.aq/devAS/ats_meetings.aspx?lang=e

Antarctic Treaty Consultative Meeting. (2009). *Final Report of the Thirty-Second Antarctic Treaty Consultative Meeting.* Baltimore, United States, April 6–17, 2009. Secretariat of the Antarctic Treaty: Buenos Aires. Available at http://ats.aq/devAS/ats_meetings.aspx?lang=e

Antarctic Treaty Consultative Meeting. (2009). *Decision 1 (2009). Meeting of Experts on Climate Change.*

Antarctic Treaty Consultative Meeting. (2010). *Final Report of the Thirty-Third Antarctic Treaty Consultative Meeting.* Punte del Este Uruguary. Secretariat of the Antarctic Treaty: Buenos Aires. Available at http://ats.aq/devAS/ats_meetings.aspx?lang=e

Antarctic Treaty Consultative Meeting XXXV/Committee for Environmental Protection XV. (2012). *ATCM Interests in International Climate Change*

Discussions – Options for Enhanced Engagement. WP032. Available at http://
ats.aq/devAS/ats_meetings_doc_database.aspx?lang=e&menu=2

Antarctic Treaty Consultative Meeting. (2012). *Final Report of the Thirty-Fifth
Antarctic Treaty Consultative Meeting.* Hobart, June 11–20, 2012. Secretariat of
the Antarctic Treaty: Buenos Aires. Available at http://ats.aq/devAS/ats_meetings
.aspx?lang=e

Antarctic Treaty Consultative Meeting. (2015). *Final Report of the Thirty-Eighth
Antarctic Treaty Consultative Meeting.* Sofia, Bulgaria, June 1–20, 2015.
Secretariat of the Antarctic Treaty: Buenos Aires. Available at http://ats.aq/
devAS/ats_meetings.aspx?lang=e

Antarctic Treaty Consultative Meeting. (2016). *Final Report of the Thirty-Ninth
Antarctic Treaty Consultative Meeting.* Santiago, Chile, May 23–June 1, 2016.
Secretariat of the Antarctic Treaty: Buenos Aires. Available at http://ats.aq/
devAS/ats_meetings.aspx?lang=e

Baird, R. (2004). Can Australia assert an Extended Continental Shelf off the
Australian Antarctic Territory consistent with the Law of the Sea and within
the Constraints of the Antarctic Treaty? *Maritime Studies*, 13, 1–19.

Chown, S. L., Lee, J. E., Hughes, K. A., et al. (2012). Challenges to the future conservation of the Antarctic. *Science*, 337(6091), 158–9.

Comba, D. (2013). An analysis of the impact of climate change for management
and governance of the Antarctic region. In W. Leal Filho, ed., *Climate Change
and Disaster Risk Management*. Berlin: Springer, pp. 311–19.

Constable, A. J., Melbourne-Thomas, J., Corney, S. P., et al. (2014). Climate change
and Southern Ocean ecosystems I: How changes in physical habitats directly
affect marine biota. *Global Change Biology*, 20(10), 3004–25.

Dudeney, J. R., and Walton, D.W.H. (2012). Leadership in politics and science
within the Antarctic Treaty. *Polar Research*, 31(1), 11075. doi: 10.3402/polar
.v31i0.11075.

Gleckler, P. J., Durack, P. J. Stouffer, R. J. Johnson, G. C., and Forest, C.E.
(2016). Industrial-era global ocean heat uptake doubles in recent decades.
Nature Climate Change, 6(4), 394–8. doi:10.1038/nclimate2915.

Hall, R. (2007). Saving seabirds. In L. Kriwoken, J. Jabour, and A. Hemmings, eds.,
Looking South: Australia's Antarctic Agenda. Sydney: The Federation Press,
pp. 117–32.

Hall, R., and Haward, M. (2001). Enhancing compliance with international legislation
and agreements mitigating seabird mortality on longlines. *Marine Ornithology*,
28(1), 183–90.

Haward, M. (2008). *Governance and Management of the Southern Ocean:
Approaching Assessments of Regime Effectiveness.* Paper presented at ISA
Conference, San Francisco, March 26–30, 2008.

Haward, M. (2009). The Law of the Sea Convention and the Antarctic Treaty
System: Constraints or complementarity. In S.-Y. Hong and J. van Dyke, eds.,
Maritime Boundary Disputes, Settlement Processes, and the Law of the Sea.
Leiden: Martinus Nijhoff Publishers, pp. 231–51.

Haward, M. (2014). The Southern Ocean, climate change and ocean governance. In
C. Schofield, S. Lee, and M.-S. Kwon, eds., *The Limits of Maritime
Jurisdiction*. Leiden: Martinus Nijhoff Publishers, pp. 507–23.

Haward, M. (2017a). The originals: The role and influence of the original signatories to
the Antarctic Treaty. In A. Hemmings, K. Dodds, and P. Roberts eds., *Handbook on
the Politics of Antarctica*. Cheltenham: Edward Elgar United Kingdom, pp. 232–40.

Haward, M. (2017b). Contemporary challenges to the Antarctic Treaty and Antarctic Treaty System: Australian interests, interplay and the evolution of a regime complex. *Australian Journal of Maritime & Ocean Affairs*, **9**(1), 21–4.

Haward, M., and Griffiths. T. (2011). *Australia and the Antarctic Treaty System: Fifty Years of Influence*. Sydney: UNSW Press.

Haward, M., and Jabour, J. (2014). Environmental change and governance challenges in the Southern Ocean. In T. Stephens and D. VanderZwaag, eds., *Polar Oceans Governance in an Era of Environmental Change*. Cheltenham: Edward Elgar, pp. 21–41.

Haward, M., Jabour, J., and Press, A. J. (2012). Antarctic Treaty System ready for a challenge. *Science*, **338**(6107), 603.

Herr, R. A., and Hall, H. R. (1989). Science as currency and the currency of science. In J. Handmer, ed., *Antarctica: Policies and Policy Development*. Canberra: Centre for Resources and Environmental Studies, ANU, pp. 13–23.

Hobday, A. J., and Pecl, G. T. (2014). Identification of global marine hotspots: Sentinels for change and vanguards for adaptation action. *Reviews in Fish Biology and Fisheries*, **24**(2), 415–25.

Jabour, J. (2009). The Australian Continental Shelf: Has Australia's high latitude diplomacy paid off? *Marine Policy*, **33**(2), 429–31.

Kock, K.-H., ed. (2000). *Understanding CCAMLR's Approach to Management*. Hobart: CCAMLR.

Masson-Delmotte, V. (2006). Climate change – An Antarctic Perspective. SCAR Lecture ATCM XXIX.

Mosely, J. G. (2007). *Antarctica: Securing Its Heritage for the Whole World*. Canterbury, NSW: Envirobook.

Orheim, O., Press, A., and Gilbert, N. (2011). Managing the Antarctic environment: The evolving role of the committee for environmental protection. In P. A. Berkman, M. A. Lang, D. W. H. Walton, and O. R. Young, eds., *Science Diplomacy: Antarctica, Science and the Governance of International Spaces*. Washington DC: Smithsonian Institution Scholarly Press, pp. 209–22.

Press, A. J. (2012). An own goal in political ice hockey? World Heritage listing for Antarctica could reopen minerals debate. *The Conversation,* June 16. Available at https://theconversation.com/an-own-goal-in-political-ice-hockey-world-heritage-listing-for-antarctica-could-reopen-minerals-debate-7683

Rapley, C. (2007). Climate change and the Antarctic: What's next? SCAR Lecture ATCM XXX.

Raustiala, K., and Victor, D. (2004). The regime complex for plant genetic resources. *International Organization*, **58**(2), 277–310.

Scientific Committee on Antarctic Research. (2017). What does the United Nations Paris Climate Agreement mean for Antarctica? ATCM XL BP20.

Stokke, O. S., and Vidas, D. (1996). *Governing the Antarctic: The Effectiveness and Legitimacy of the Antarctic Treaty System*. Cambridge: Cambridge University Press.

UNESCO. (2016). World Heritage in the High Seas: An Idea Whose Time Has Come. World Heritage Reports 44, Paris, UNESCO.

Vizzini, S., Martinez-Crego, B., Andolina, C., et al. (2017). Ocean acidification as a driver of community simplification via the collapse of higher order and rise of lower order consumers. *Nature Scientific Reports*, **7**, 4018. doi: 10.1038/s41598-017-03802-w.

Voosen, P. (2017). Delaware sized iceberg splits from Antarctica. *Science*, July 12. Available at www.sciencemag.org/news/2017/07/delaware-sized-iceberg-splits-antarctica

World Meteorological Organization. (2015). *The Antarctic Treaty Consultative Meeting (ATCM)*. World Meteorological Organization Executive Council Panel of Experts on Polar and High Mountain Observations, Research and Services. Sixth Session Reykjavik, September 8–11, 2015.

Young, O. R. (2010). *Institutional Dynamics: Emergent Patterns in International Environmental Governance*. Cambridge, MA: MIT Press.

Young, O. R. (2014). Building an international regime complex for the Arctic: Current status and next steps. *The Polar Journal*, **2**(2), 391–407.

14

Policy Responses to New Ocean Threats
Arctic Warming, Maritime Industries, and International Environmental Regulation

BENJAMIN HOFMANN

Introduction

A warming Arctic Ocean faces new environmental threats from maritime industries. Climate change enhances the accessibility of the northernmost ocean by diminishing its sea-ice cover. The wintertime maximum extent of Arctic sea ice had never been smaller than in 2017 (Garner, 2017). Better accessibility facilitates maritime shipping and offshore oil and gas production. Trans-Arctic liner shipping may not become commercially viable before mid-century (Ørts Hansen et al., 2016), but ever larger ships navigate Arctic waters: In 2013, *Nordic Orion* carried 73,500 tons of coking coal from Vancouver via Greenland to Finland. In 2016, *Crystal Serenity*, with a capacity of over 1,000 passengers, became the largest cruise ship to have ever transited the Northwest Passage. Offshore hydrocarbon exploitation is also expanding even though low oil prices have slowed down or halted many projects (Harsem et al., 2011). In 2013, Russia's first Arctic offshore oil platform came on stream in the Pechora Sea. Three years later, the world's northernmost offshore oil field, Goliat, was launched in the Norwegian Barents Sea. Both ships and drilling platforms can have numerous negative transboundary external effects on ecosystems and human health. They discharge oil, chemicals, and other harmful substances into the sea; emit greenhouse gases and air pollutants; introduce invasive species; and cause disturbance to marine life.

How have states responded to these new threats to the Arctic Ocean in terms of international regulation? Broadly defined, the realm of international regulation comprises formal and explicit rules, standards, and guidelines agreed by two or more states (Braithwaite and Drahos, 2000: 19–20). Regulations differ greatly in design, yet data on regulatory variance in Arctic shipping and offshore oil and gas remain scarce. Existing datasets focus on larger, relatively stable regime elements or on single international agreements rather than on regulations as more flexible, intermediate units of analysis. Moreover, they take into account only part of all maritime regulations that apply to the Arctic. The International Environmental Agreement Database (Mitchell, 2017) lists conventions, protocols, and their amendments, but it excludes many relevant resolutions and decisions of intergovernmental organizations. The International Regime Database (Breitmeier et al., 2006) covers selected regimes, such as to the regime for the prevention of oil pollution from ships, but it leaves out most other sets of regulations applicable to maritime industries in the Arctic. More comprehensive descriptions of legal instruments are found in qualitative work (e.g., Koivurova and Molenaar, 2009; Jensen, 2016). However, these contributions often face limitations in systematizing and comparing regulatory data. The data gap resulting from these observations becomes increasingly problematic as economic and regulatory activities in the Arctic grow.

This chapter contributes to closing this gap by introducing a new dataset of international regulations that address operational environmental impacts of shipping (see also Chapter 23) and offshore hydrocarbon production in the Arctic. The chapter proceeds in five steps: First, it demonstrates why international regulation is relevant in the Arctic context. Second, it outlines the novel concept of stringency for assessing regulations from an international relations perspective. Third, it describes the scope and structure of the regulation database. Fourth, it empirically assesses and compares regulatory stringency and identifies emerging patterns across four dimensions: time, industry, international organization, and external effect. Finally, it discusses implications for international maritime regulation amidst climate change in the Arctic and beyond, and for future research.

International Maritime Regulations in the Arctic

International regulation will play a vital role in protecting a warming Arctic. I focus on three intergovernmental organizations whose regulations cover the entirety or parts of the Arctic Ocean. The International Maritime Organization (IMO) is a specialized agency of the United Nations that globally regulates environmental impacts of ships (see Chapter 23) and, to some extent, offshore structures. Its main legal instrument is the International Convention for the Prevention of Pollution from Ships (MARPOL), which works beside other issue-specific instruments (Tan, 2005). The Polar Code contains additional regional regulations for an area including the central Arctic Ocean and coastal waters of the "Arctic Five" (Canada, Denmark-Greenland, Norway, Russia, and the United States). Among the other global instruments that apply to the Arctic is the London Convention and Protocol on the dumping of wastes, whose secretariat is hosted by the IMO.

The Arctic Council is an intergovernmental forum of eight Arctic states that has developed the Arctic Offshore Oil and Gas Guidelines. Compared to the IMO Polar Code, the intended area for application of the guidelines is larger. They additionally include Iceland; the Faroe Islands; and larger parts of the Barents, Bering, and Norwegian Seas. Other offshore initiatives that have been carried out in the Arctic Council context are the Arctic Offshore Regulators Forum and the Russia-USA-Norway-Arctic Offshore Oil and Gas Regime project. However, these initiatives have not produced any international regulations, as their focus has rather been on exchanging best practices and on developing Russian national legislation, respectively. The Arctic Council does not regulate shipping, even though it might facilitate regulatory advances in the IMO (Stokke, 2013).

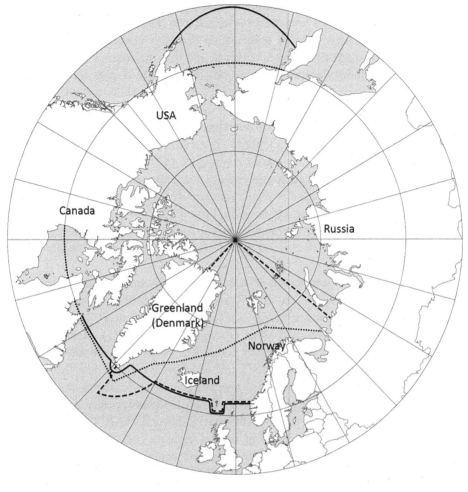

.... **IMO:** International Code for Ships Operating in Polar Waters (Polar Code)
————— **Arctic Council:** Arctic Offshore Oil and Gas Guidelines
– – – – – **OSPAR Commission:** North-East Atlantic Region I (Arctic waters)

Basemap by Jürnjakob Dugge. Adapted by the author.

Fig. 14.1. International regulatory frameworks in the maritime Arctic

The OSPAR Commission is a regional organization with 16 contracting parties that deals with marine pollution in the northeast Atlantic. The organization succeeded the formerly separate Paris and Oslo Commissions (hence the name OSPAR). Its geographic definition of the northeast Atlantic includes the region that covers the European part of the Arctic. Within this region, OSPAR decisions and recommendations apply to Denmark, Iceland, and Norway but not to the nonmember Russia. In this area, the three intergovernmental organizations also have the largest geographic overlap (see Figure 14.1).

Despite this presence of international regimes, the role of international regulation in a warming Arctic is not uncontested (see Chapter 12). Critics would emphasize that most maritime activity takes place within exclusive economic zones of states, that coastal state jurisdiction is more effective, and that the Arctic Council lacks formal regulatory authority. Although these observations are largely correct, they do not diminish the importance of international regulation. Many arguments can be made for such regulation.

First, the transboundary nature of maritime activities and their impacts call for international regulation. Arctic shipping today concentrates along the coastlines, but international regulation helps to ensure compliance of ships flying non-Arctic flags. Offshore hydrocarbon exploration and exploitation only take place within exclusive economic zones, but international regulation can create a level playing field in the Arctic. A potential role model is the regional framework for North Sea oil and gas (see Leeuwen, 2010). International regulation can also reduce transboundary environmental pressures from border-zone oil and gas fields, for instance, in the Barents Sea (Bambulyak et al., 2015).

Second, coastal state jurisdiction alone does not always have greater leverage than international regulation. Canada and Russia have adopted maritime regulations exceeding international standards based on the "Arctic exception" in Article 234 of the Convention of the Law of the Sea (Rayfuse, 2014: 238–41). However, Arctic coastal state jurisdiction is challenged by continued dispute over the international or internal status of the Northwest Passage and the Northern Sea Route (McDorman, 2014; Rayfuse, 2014: 243). Moreover, regulatory experience shows that at times national jurisdictions might have limited leverage in technologically globalized sectors. For instance, equipment requirements that go beyond MARPOL standards are unenforceable if foreign administrations do not approve such equipment (e.g., US EPA, 2011: 5, 78). International regulation can synchronize the flag, coastal, and port states needed for proper enforcement (Tan, 2005: 201–29) (see Chapter 2). National authority can be effective in banning certain activities, as illustrated by recent moratoria for Arctic offshore drilling in Canada and US federal waters. It is not clear, though, how enduring these bans will be given strong opposition from industry and subnational jurisdictions. Besides, hydrocarbon exploration and exploitation continue in other parts of the Arctic, in particular, Norway and Russia.

Third, while the Arctic Council largely remains a soft-law forum (Young, 2009: 79), its outputs fall within a broad definition of regulation, which includes nonbinding standards and guidelines. The contents of the Arctic Offshore Oil and Gas Guidelines are structurally similar to sets of nonbinding IMO guidelines and OSPAR recommendations, in that they aim at incorporation into national legislation (Koivurova, 2018: 25). Their nonbinding character should thus not prevent us from comparing these guidelines to other, possibly more stringent regulations. Research has shown that under certain conditions soft law can be effective too (Skjærseth et al., 2006; Soltvedt, 2017). In sum, this and other arguments justify a more thorough engagement with existing international environmental regulations of maritime industries in the Arctic.

The Concept of Regulatory Stringency

International environmental regulations can be assessed and compared in terms of stringency. Stringency is defined as the function of formal tightness *and* substantive ambition of a regulation. Stringency considers both formal and substantive design elements because both affect intermediaries and targets of international maritime regulation in the Arctic, namely, states and corporations, respectively. Tightness can elicit corporate compliance through shaping domestic implementation. Ambition specifies the substance of compliance in terms of required behavioral changes. Furthermore, formal and substantive design properties are interrelated, considering their potential mutual trade-off in negotiations (Guzman, 2010: 156). For instance, a regulation with demanding greenhouse gas emission standards would not be considered stringent if it was nonbinding and not subject to compliance monitoring. The following subsections briefly present formal and substantive dimensions of stringency, and from that derive a stringency index.

Formal Tightness

The tightness of regulation depends on its legality, precision, and compliance system. Legality matters because binding regulations "tend to ... have a higher capacity to facilitate implementation and compliance" (Andresen et al., 2013: 427). This chapter follows Raustiala (2005: 586–91) in his critique of continuous legality concepts and conceives of legal status as binding or nonbinding. Legality is high when regulations are binding, that is, when states have expressed their consent to be bound by them (Abbott et al., 2000: 410). This is usually the case for conventions, protocols, mandatory codes, and decisions of intergovernmental organizations. Legality is low when regulations are nonbinding, that is, when the aforementioned consent is absent or explicitly negated. Examples are recommendations, guidelines, or reservations on treaties. Legality is intermediate when an environmental impact is regulated by a set of binding and nonbinding instruments.

Precision can enhance regulatory effectiveness by producing clarity about substantive requirements (Böhmelt and Pilster, 2010: 257). Precise regulations leave little room for interpretation by providing clear definitions, stating specific limit values, prescribing technical equipment and procedures, and defining unambiguous conditions for any exemptions (Breitmeier et al., 1996: 82; Abbott et al., 2000: 413–15). Imprecise regulations lack these properties, thus leaving substantial discretion in interpretation. Precision is at an intermediate level when precise and imprecise traits coexist.

The strength of the compliance system affects compliance rates of regulatory targets. Compliance can be achieved through incentives or threats (Reiss, 1984: 91). However, tangible incentives are not yet found in international maritime regulation in the Arctic. Threat-based monitoring and enforcement systems are strong when coercing compliance *ex ante* by means of equipment, certification, and review requirements or specific permit schemes (Mitchell, 1994: 433–35, 454–56). The deterrence of noncompliance by *ex post* detection through often patchy surveillance, prosecution, and sanctioning of violations, as well as general permit schemes, are on an intermediate level (Mitchell, 1994: 454). Some regulations lack monitoring and enforcement provisions altogether.

In a nutshell, international regulation is tight when it is legally binding, highly precise, and endowed with strong compliance mechanisms. Such regulation is most likely to elicit effective domestic implementation that pushes business actors to comply with its substantive provisions.

Substantive Ambition

Ambition results from the scope of the regulation and its requirement levels in relation to the external effect and compared with other international regulations. Scope reflects how comprehensively a regulation covers the addressed "pollution source population" of contracting parties. The pollution source population is the universe of a source–effect combination, for example, nitrogen oxide emissions from ships. This definition is more instrumental to an analysis of many individual regulations compared to notions of scope that focus on the number of issues covered (Breitmeier et al., 1996: 84; Koremenos et al., 2001: 770–71). Regulations are large in scope when applying to the entire pollution source population without any relevant exemptions. Intermediate scope exempts a limited part of this population, for instance, pollution from smaller engine classes or existing vessels. Scope is small when only a minor part of the present and future pollution source population is covered.

Substantive requirements differ in their potential to reduce external effects. I benchmark these reductions for a single compliant polluter, for example, the decrease of oil content in water discharges from an offshore platform. This corporate benchmark differs from state-centered notions of depth (Downs et al., 1996: 383; Bernauer et al., 2013: 480–81). It also seeks to reduce ambiguities found in other operationalizations of depth (Breitmeier et al., 1996: 85). Requirement levels are high when virtually eliminating the externality, for instance, by means of zero-discharge standards or bans. Intermediate requirements are performance or equipment standards that demand major reductions in the range of 30 to 90 percent compared to a situation without regulation. Low requirement levels hardly demand behavioral changes from corporate actors compared to unregulated activity.

The international comparison of requirement levels is useful because multinational corporations care about regulatory differences that could affect their competitive position (Falkner, 2008: 33). Regionally differentiated regulations also raise race-to-the-bottom-or-top questions, similar to those in the area of national flagging standards (e.g., DeSombre, 2006) (see Chapter 2). Requirement levels are high in cross-regional perspective when they are among the most demanding ones globally at the time of entry into force (only considering other effective international regulations). For instance, emission limits are lower and technical standards less permissive than in almost all other sea regions. Conversely, low requirement levels are among the least demanding ones internationally. Intermediate requirements are more advanced than those in laggard regions, but they do not reach up to those in leading regions.

In short, ambitious international regulation possesses a large scope as well as high requirement levels in relation to the external effect and other international regulation. In case of full compliance, such regulation would virtually eliminate the external effect addressed.

Stringency Index

In this section, the qualitative description of the six design elements is translated into a compound stringency index to assess and compare regulations in absolute terms. The index is formed in three steps: First, the six design elements are coded (high = 1, intermediate = 0.5, low = 0). Second, the design element scores are aggregated separately for form and substance; the scores are added, assuming a family resemblance structure with partial compensation and equal weight of all design elements (Goertz, 2006: 60, 115). This produces two dimension scores ranging from loose (0) to tight (3) and from unambitious (0) to ambitious (3). Third, the two dimension scores are aggregated, assuming that formal tightness and substantive ambition are noncausal necessary conditions of stringency. Regulatory stringency is achieved when overcoming the trade-off between tightness and ambition, but high ambition can also make a difference if tightness is larger than zero, and vice versa. However, regulations are clearly lax if formally loose and/or unambitious. The aggregation rule most accurately reflecting this structure is multiplication of unweighted dimension scores (Goertz, 2006: 111). A final index score is obtained on a scale from zero to nine, which is divided here into five stringency classes to ease interpretation: regulation is lax (0.00 to 0.75), somewhat lax (1.00 to 2.00), intermediate (2.25 to 3.00), somewhat stringent (3.75 to 5.00), or stringent (6.00 to 9.00).

This index orders regulations hierarchically according to absolute levels of stringency and facilitates broader comparative assessments. It is intended to complement rather than to replace descriptive accounts that pay close attention to the particularities of individual regulations. The index scores can neither reflect the size of stringency differentials nor record regulatory changes below coding thresholds. Notwithstanding these limitations, the stringency index is a useful tool for assessing the state and variance of regulation.

International Maritime Regulation Database

The assessment of regulatory stringency in international maritime regulation builds on a new dataset compiled for the Arctic region. The dataset covers international regulations of IMO, the Arctic Council, and OSPAR that apply to the entire Arctic Ocean or to part of it. The analysis of relevant nonmaritime instruments, such as the

United Nations Economic Commission for Europe (UNECE) Convention on Long-Range Transboundary Air Pollution, is left to future research. Within the scope of the database are regulations addressing operational environmental impacts of seagoing ships and fixed offshore oil and gas platforms from 1950 to 2017. Beyond its scope are other maritime activities, such as dumping of onshore wastes, dredging, fisheries, offshore renewable energy, and seabed mining. Also excluded are regulations dealing with nonoperational environmental impacts of industry, for example, accidental harm, such as oil spills, and resource overappropriation. This chapter focuses on operational impacts because they can serve as a policy-relevant and empirically diverse test case for future data collection and analysis in other regulatory areas.

The database seeks to complement existing descriptive accounts by providing a broad and systematic picture of the regulatory development. The unit of analysis is the regulation of an environmental externality from a specified operational source. A single impact can arise from multiple, often differently regulated, sources as the diverse oil pollution sources in offshore hydrocarbon production illustrate. These operational impacts were compiled inductively from regulations and secondary sources (see Table 14.1). The large number of externalities implies that although the data analysis in this chapter can provide a general overview, it cannot discuss individual regulatory issues in depth. It is important to bear in mind that most of these externalities are likely to increase as the Arctic becomes more accessible due to climate change.

The data compilation and coding proceeded in three steps: Each industry was first coded as to whether an impact and source were regulated in a given year. If so, basic regulatory information was gathered next, drawing on existing datasets where available (Breitmeier et al., 1996; Mitchell, 2017). Basic information includes names of regulatory documents, full texts, international organization, year of adoption, entry into force or effect, geographic scope, and applicability to Arctic states. Applicability benchmarks are the Arctic boundaries as defined in the contexts of shipping (Arctic Five) and offshore oil and gas (Arctic Five plus Iceland). The stringency of formally effective regulations was then assessed according to the stringency index. These data allow for analyzing the number and stringency of international regulations across four dimensions: time, industry, international organization, and externality.

The Stringency of Arctic Maritime Regulations

This section analyzes empirical patterns in the number and stringency of international regulations addressing operational environmental impacts of Arctic shipping and offshore hydrocarbon production. It identifies four major patterns: First, the body of applicable regulations has successively grown, with dedicated Arctic regulation emerging in the 1990s and 2000s in response to accelerating sea-ice melt. Second, regulation is generally more stringent and less fragmented for shipping than for offshore oil and gas. Third, stringent regulations have been adopted by IMO and OSPAR but not by the Arctic Council. Fourth, although some issues, such as oil

Table 14.1. *Operational impacts of ships and offshore oil and gas platforms*

Type	Ships	Platforms	General issue	Specific externality
Air emissions	•	•	Pollutants	Black carbon, carbon monoxide, nitrogen oxides, particulate matter, metals in particulate matter, polycyclic aromatic hydrocarbons, sulfur oxides, ozone, volatile organic compounds, incineration of ship-generated wastes
		•	Pollutants | GHG	Flaring, venting, and dehydration
	•	•	GHG	Carbon dioxide, methane, hydrofluorocarbons, etc.
	•	•	Ozone depletion	Ozone-depleting substances
Discharges into the sea	•	•	Oil	Bilge water
	•		Oil	Ballast water, oil from cargo areas of tankers, sludge
		•	Oil	Oil-based drilling fluids and cuttings, displacement water, drainage water, well testing
	•		Oil | pollutants	Lubricants on underwater hull components
		•	Oil | pollutants	Produced water, sand and scale, cutting piles
	•	•	Pollutants	Antifouling systems, washwater and washwater residues from exhaust gas cleaning systems
	•		Pollutants	In packaged form, noxious liquid substances
		•	Pollutants	Offshore chemicals, synthetic-based drilling fluids and cuttings, naturally occurring radioactive materials, carbon dioxide sequestration
	•	•	Sewage	Black water, gray water
	•	•	Garbage	Cleaning agents and additives, food wastes, plastic, etc.
	•		Garbage	Animal carcasses, cargo residues
	•	•	Debris	Removal and disposal of vessels/units
		•	Other	Cooling water, desalination brine
Other	•		Invasive species	Ballast water, biofouling, sediments
	•	•	Disturbance	Light, noise
	•		Disturbance	Mammal strikes, icebreaking

GHG: Greenhouse gases

Note: Many externalities have more than one source that can be regulated. For instance, sulfur oxide emissions can be addressed by regulating fuel oil quality or engine exhaust.

Compiled by the author based on regulations and on information from AMAP (2010: chapter 5), Arctic Council (2009a: 136–51), IFC (2015), and OSPAR (2010: 63–70).

and garbage discharges, are relatively well covered, regulatory gaps persist in both industries. The subsections subsequently illustrate each of these patterns.

Historical Development

The international regulation of Arctic shipping and offshore hydrocarbon production has historically evolved in two spheres: (1) global and regional regulations *applicable to* the Arctic and (2) regulations specifically *designed for* the Arctic Ocean. Regulations in these spheres differ in their timing and stringency. Regulatory sphere (1) emerged in the 1950s and accelerated in the 1970s and 1980s. The regulation of vessel-source pollution initially focused on oil discharges. It was boosted by the MARPOL annexes on oil, noxious liquid substances, harmful substances in packaged form, sewage, garbage, and air pollution that successively entered into force between 1983 and 2005. Many of these regulations were made more stringent in the 1990s and 2000s. Since then, IMO has continued to strengthen existing regulations and to address emerging issues, such as the introduction of invasive species. The regulation of offshore oil and gas, which has centered on discharges of oil and chemicals, as well as on the removal and disposal of disused platforms, exhibits a similar pattern. The Paris Commission developed a growing body of regulations for the northeast Atlantic between the late 1970s and early 1990s. Regulatory stringency increased under its successor OSPAR in the 1990s and early 2000s (Figure 14.2). Relevant offshore regulations of IMO and the London Convention and Protocol on dumping follow the shipping pattern even more closely.

An Arctic regulatory sphere (2) did not emerge until the 1990s when climate change awareness and prospects of a more accessible circumpolar region were rising in most Arctic states (Borgerson, 2008). The first regulations in this sphere were the Arctic Offshore Oil and Gas Guidelines adopted by the Arctic Council in 1997. These guidelines addressed numerous operational impacts of hydrocarbon production but did not regulate them stringently (see Figure 14.2). Updates in 2002 and 2009 brought some changes in issue coverage and content but no major advances in stringency. Therefore, several more stringent, non-Arctic regulations remain relevant in this sector. Dedicated Arctic shipping regulation initially also took the form of guidelines. The IMO issued Guidelines for Ships Operating in Arctic Ice-Covered Waters in 2002 and updated them in 2009, including an extension to all polar waters. These guidelines primarily dealt with safety issues and only contained a general provision on minimizing operational pollution without mentioning specific pollution sources. The IMO Polar Code, which entered into force in 2017, has been a dual turning point because it addresses specific impacts *and* regulates most of them rather stringently. However, although it covers discharges of oil and other substances into the sea, emissions into the air remain subject to global IMO regulations.

In short, regulations applicable to the Arctic have successively increased in numbers and stringency since the regulatory take-off in the 1970s and 1980s. Regulations designed for the Arctic emerged in the 1990s and 2000s in conjunction with Arctic

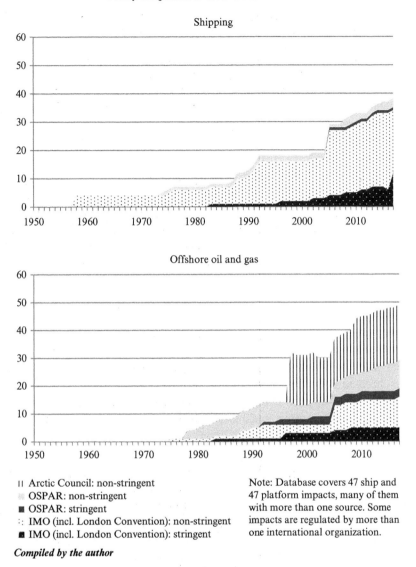

Shipping

Offshore oil and gas

| | Arctic Council: non-stringent
OSPAR: non-stringent
■ OSPAR: stringent
∷ IMO (incl. London Convention): non-stringent
■ IMO (incl. London Convention): stringent

Note: Database covers 47 ship and 47 platform impacts, many of them with more than one source. Some impacts are regulated by more than one international organization.

Compiled by the author

Fig. 14.2.   Number and stringency of regulations, years of entry into force, 1950–2017

warming. Since then, shipping and offshore hydrocarbon regulations have developed differently in terms of stringency (see Figure 14.2).

Industries

The historical analysis in this chapter identified different regulatory trajectories for Arctic shipping and offshore hydrocarbon production. Indeed, present international regulation of both industries varies in its fragmentation and stringency. Fragmentation denotes the extent to which an industry is covered by more than one regulatory framework. International regulations for offshore oil and gas are more fragmented than shipping regulations (see Figure 14.2; see also Lyons, 2015). Offshore hydrocarbon production is subject to regulations of IMO, the Arctic Council, OSPAR, and other international regimes. Their regulations partially overlap; in particular, the Arctic Offshore Oil and Gas Guidelines address various issues already regulated by other regimes. By contrast, almost all international shipping regulation is developed by IMO. Issues addressed by other regimes have been the dumping of vessels (London Convention and OSPAR), as well as antifouling paints and ballast water exchange (IMO and OSPAR). The IMO thus faces little competition from other international organizations in the regulation of Arctic shipping.

Regulatory stringency in the two industries varies, too. Shipping is subject to more stringent Arctic-wide regulation than offshore oil and gas. More than 30 IMO regulations cover Arctic Ocean shipping, including all or most of the Arctic Five. Over one-third of these regulations are stringent. A few additional OSPAR rules or guidelines apply in the European Arctic. Operational impacts of platforms are covered by more than 35 international regulations (IMO and Arctic Council) applying to all or most of the Arctic Five and Iceland. However, only every seventh of these regulations qualifies as stringent. Another 13 OSPAR regulations exist in the European Arctic, of which 3 are stringent. Unlike in shipping, stringent offshore oil and gas regulation thus concentrates in one Arctic subregion. Also, the number of regulated impacts differs across industries. A meaningful comparison is difficult, however, because both the operations of ships and platforms and the regulatory delineation of source–impact combinations differ. Therefore, the chapter turns to another analytical dimension that partly accounts for the cross-industry variance in regulatory stringency: the international regulators.

International Organizations

The international organizations that regulate maritime industries in the Arctic differ in their levels of regulatory activity and their ability to adopt stringent regulations. The IMO and Arctic Council address large numbers of impacts; IMO and OSPAR stand out in terms of regulatory stringency. The IMO and related global agreements regulate large numbers of operational impacts, and many of them rather stringently. They address 50 impacts in the two industries and regulate one-third of them stringently. Stringency largely arises from formally tight MARPOL and Polar Code regulations that are typically binding, precise, and endowed with compliance systems. Variance in stringency is usually due to different ambition levels. Ambition is high for issues covered by the Polar Code, such as discharges of oil, noxious liquid substances, and garbage. It is lower for issues under global MARPOL regulations; in particular, air pollutant standards are more demanding in special areas outside the Arctic. Additional IMO guidance tends to be less stringent because of its formal looseness.

The Arctic Council (2009b) addresses many impacts of offshore hydrocarbon exploitation but does not regulate them stringently. The nonbinding status of the

Arctic Offshore Oil and Gas Guidelines limits their formal tightness. Imprecision and the absence of strong compliance systems increase their looseness even further. In particular, the lack of proper monitoring programs weakens the effectiveness of nonbinding instruments (Soltvedt, 2017). Moreover, the guidelines only contain modest requirements, especially when compared to other sea regions, such as the northeast Atlantic.

The OSPAR Commission covers fewer maritime industry impacts than IMO but has adopted several stringent regulations. Out of 17 present regulations, 4 are stringent. The focus of OSPAR is on offshore oil and gas production for which it has adopted both binding decisions and nonbinding recommendations. Both tend to be very precise, except for recommendatory guidelines on recently addressed issues, such as platform lighting impacts on birds. The organization has also developed extensive monitoring and improvement programs for priority issues, such as oil discharges and offshore chemicals. Ambition levels are generally high, but some requirements are less restrictive than in other sea regions, notably the Baltic Sea where only minor offshore hydrocarbon activity takes place.

Overall, regulatory activity and stringency vary considerably across the three intergovernmental organizations. Arctic Council regulations are broad and lax, OSPAR regulations are more stringent, and IMO aligns high regulatory activity with stringency. The analysis also suggests that variance in stringency within the regulatory frameworks developed by each organization should be examined more closely.

External Effects

International regulatory coverage and stringency not only vary over time, across industry, and across intergovernmental organization. Shipping and offshore petroleum regulations are also diverse when compared across external effects. Five groups can be distinguished for analytical purposes. A first group of externalities is stringently regulated in the entire Arctic Ocean. Many ship impacts fall into this group, including most types of oil and garbage discharges, all discharges of noxious liquid substances, and antifouling paints (see Table 14.2). The Polar Code notably prohibits discharges of oil sludge, oily bilge water, oily cargo tank washings, and noxious liquid substances (IMO, 2015: part II-A, chapters 1–2). The implementation of these discharge bans depends on appropriate reception facilities (PAME, 2017). In offshore hydrocarbon production, only antifouling paints as well as discharges of oily bilge water and garbage are subject to stringent Arctic-wide regulation.

A second group of external effects is subject to Arctic-wide, yet nonstringent, regulation. Ship impacts in this group include oily ballast water, lubricants used on underwater hull components, some garbage discharges, biofouling, underwater noise, and the dumping of vessels into the sea. Applicable global standards for energy efficiency and air pollutant emissions fall into this group, too (see Chapter 23). Platform impacts are laxly regulated under the Arctic Offshore Oil and

Table 14.2. *Stringency and coverage of regulations for selected impacts from ships, as of January 2018*

Operational environmental impacts		Applicability to Arctic states		
		All (5)	Most (3–4)	Some (1–2)
Oil discharges	Sludge, bilge water, oil from cargo area of tankers	■■■■■		
	Ballast water	■■■■		
Discharges of harmful substances	Lubricants on underwater hull components	■■■■		
	Noxious liquid substances	■■■■■		
	Antifouling systems	■■■■■		
Air emissions	Nitrogen oxides, sulfur content of fuel, black carbon from heavy fuel oil use	■■■		
Invasive species	Ballast water and sediments		■■■■	▨▨▨
	Biofouling	■■		
Garbage discharges	Animal carcasses, cargo residues not contained in hold washing water, plastic	■■■■■		
	Food waste, cargo residues, and cleaning agents contained in hold washing water	■■■■		
Sewage discharges	Black water		■■■■	
	Gray water			
Disturbance	Underwater noise	■■		
Debris	Disposal of vessels at sea	■■■■		▨▨▨▨▨

Regulation

■■■■■	Stringent
■■■■	Somewhat stringent
■■■	Intermediate
■■	Somewhat lax
■	Lax
	None

International Organization

■	IMO (incl. London Convention)
▨	OSPAR

Note: Aggregated impacts share same level of regulatory stringency but may be subject to different regulatory provisions.

Compiled by the author

Gas Guidelines, including air pollutant emissions; noise; and discharges of oil, chemicals, cuttings, and sewage (see Table 14.3). The global London Convention rules and IMO guidelines for the removal and disposal of disused offshore structures are not stringent either.

Table 14.3. *Stringency and coverage of regulations for selected impacts from platforms, as of January 2018*

Operational environmental impacts		Applicability to Arctic states		
		All (6)	Most (4–5)	Some (1–3)
Oil discharges	Produced water	▦▦		■■■■■
	Displacement water			▦▦
	Bilge water	■■■■■ ▦▦▦		
	Oil-based drilling fluids and cuttings	▦		■■■■■
Discharges of harmful substances	Synthetic-based drilling fluids and cuttings	▦▦		■■■■■
	Chemicals, produced water	▦▦▦		■■■■
	Antifouling systems	■■■■■		
Air emissions	Air pollutants from flaring of gas	▦▦		
	Nitrogen oxides in engine exhaust	▦	■■	
Garbage discharges	Garbage: food wastes, cleaning agents, plastic, all other	■■■■■		
Sewage discharges	Black water, gray water	▦▦▦		
Disturbance	Light			▦▦
	Underwater noise	▦		
Debris	Disposal of disused platforms at sea	■■■■ ▦▦		■■■■■

Regulation		**International Organization**	
■■■■■	Stringent	■	IMO (incl. London Convention)
■■■■	Somewhat stringent	▦	OSPAR
■■■	Intermediate	▧	Arctic Council
■■	Somewhat lax		
■	Lax		Note: Aggregated impacts share same level of regulatory stringency but may be subject to different regulatory provisions.
	None		

Compiled by the author

A third group of impacts is stringently regulated but not in the entire Arctic. In shipping, this is true for the OSPAR ban on the dumping of vessels into the sea, which only applies to the European Arctic. The group of stringent international regulations with limited geographic application is somewhat larger in offshore oil and gas. It consists of OSPAR decisions and recommendations on the use and discharge of drilling fluids and cuttings, as well as on the disposal of disused platforms.

A fourth group of regulations is neither stringent nor Arctic-wide. In shipping, both the regulatory frameworks for black water discharges and ballast water management have a limited ambition. In addition, the United States is not a party to the relevant MARPOL annex IV and to the global Ballast Water Management Convention and enforces its own demanding requirements instead. More issues abound for offshore hydrocarbon production. They include nonbinding OSPAR regulations on impacts, such as oil and harmful substances in produced water, radioactive substances, the management of cuttings piles, and lighting. The stringency of these recommendations and guidelines is also weakened by the fact that, often for technological reasons, they do not aim to completely eliminate the externality for all installations. Nevertheless, OSPAR regulations have proven successful in reducing the amount of oil discharged from offshore platforms (OSPAR, 2017). In addition, MARPOL annex VI provisions on air pollutant emissions from platforms will soon apply to all Arctic coastal states (Iceland from February 2018) but are not stringent. However, the larger limitation here is that annex VI does not apply to operational emissions directly arising from hydrocarbon extraction, such as flaring and venting.

A fifth group of operational pollution sources is unregulated. Regulatory gaps in Arctic shipping concern black carbon exhaust emissions and gray water discharges. Black carbon is an air pollutant and climate-forcing agent. The IMO Polar Code only indirectly addresses black carbon, as it recommends not using heavy fuel oil (see Chapter 15). Gray water refers to all wastewater streams except for those from toilets. It is not covered by the present sewage provisions of the Polar Code. Both issues are thus candidates for consideration in a second phase of Polar Code negotiations. Another unaddressed issue is disturbance from icebreaking activity. Unregulated operational impacts of offshore oil and gas include the venting of gas and minor discharges into the sea (cooling water and desalination brine). However, knowledge gaps persist regarding different chemicals and compounds contained in water produced during hydrocarbon extraction (OSPAR, 2016: 7). Future scientific evidence for negative environmental effects may bring additional issues to the fore.

In sum, variance in the regulatory coverage of external effects and in the stringency of these regulations suggests that gaps persist in both industries. The five groups of externalities identified here can be matched with different priorities for future action: (1) enforcement, (2) increases in regulatory stringency, (3) geographic extension of existing regulations, (4) improvements in stringency and extension, and (5) additional regulation. Whether or not these routes shall and will be pursued for individual impacts depends on environmental, economic, and societal considerations that go beyond the scope of this chapter.

Implications for Arctic Regulatory Governance

Climate change poses threats to the Arctic Ocean beyond warming. A warming Arctic Ocean will be a more accessible ocean – for cruise ships, cargo ships, and offshore oil and gas production. Maritime industries can cause different types of operational pollution in the fragile Arctic environment. This chapter examined how states have responded to these new threats in terms of international regulation. International regulation plays a role in mitigating maritime industry impacts in the Arctic for three reasons: the transboundary nature of the industries and their impacts, the additional leverage gained by international regulation, and the regulatory function of both hard and soft law. The analysis used new data on the stringency of all relevant international maritime regulations that are applicable to the Arctic Ocean or part of it and that have entered into force between 1950 and 2017. Stringent regulations were defined to be tight in form and ambitious in substance as measured on an index with six design elements.

The empirical analysis revealed that regulatory activity and stringency vary over time and across industries, international organizations, and externalities. Regulations designed for the Arctic emerged in the 1990s and 2000s in response to accelerating sea-ice melt. Since then, unified and rather stringent regulations have evolved for shipping with the IMO Polar Code, while offshore oil and gas regulation has remained fragmented and less stringent. These differences are rooted in the regulatory approaches pursued by the three dominant intergovernmental organizations. The IMO has been an active and rather stringent shipping regulator and an occasional offshore regulator. The OSPAR Commission has been a partly stringent, yet less broad, hydrocarbon regulator in the northeast Atlantic and European Arctic. The Arctic Council has addressed numerous offshore oil and gas impacts, but only through lax guidance. In both industries, regulatory gaps persist in geographic coverage, regulatory stringency, and issue coverage.

These findings have concrete policy implications for governing the Arctic as the climate changes. Arctic shipping regulation seems to be on a promising track. Where stringent shipping regulation already exists, implementation is the major concern. Arctic coastal states need to establish sufficient numbers of reception facilities for oil, noxious liquid substances, and other ship-generated wastes (PAME, 2017). Furthermore, flag, coastal, and port states need to collaborate in enforcing regulations through proper surveys, surveillance, and inspections. Where additional protection of the Arctic marine environment is needed, the designation of Particular Sensitive Sea Areas under IMO could be an option (DNV, 2014). Where regulatory gaps persist, additional stimuli from the regional level – for example, from the Arctic Council (Stokke, 2013) – could trigger a second phase of IMO Polar Code negotiations.

The road ahead in Arctic offshore oil and gas regulation is more challenging due to regulatory fragmentation and the institutional weakness of the Arctic Council. Given that a binding Arctic Ocean treaty is unlikely for various reasons (Young, 2011), three alternative strategies merit further exploration: First, IMO could increasingly regulate offshore oil and gas impacts. Second, OSPAR regulations could be gradually extended to larger parts of the Arctic, especially to northwest Russia and western Greenland. Third, the Arctic Council could develop more ambitious and precise standards, and properly monitor and economically incentivize their implementation. Pursuing any of these options, or a combination of them, also needs to consider governance challenges arising from continued fragmentation (Humrich, 2013). While these insights pertain to the realm of operational environmental regulation, they might also be useful for future maritime safety regulation in the Arctic.

Conclusion

The findings highlight a number of general difficulties surrounding international policy responses to climate-induced ocean threats. Climate change provokes changes beyond Arctic shipping and offshore hydrocarbons. Other chapters in this volume point to the accessibility of Antarctic resources (see Chapter 13), fish population changes (see chapters in Part III), ocean renewable energy (see Chapter 24), and geoengineering efforts (see Chapter 26). Many of these changes have environmental implications that call for international policy responses. When should such policies be developed, by whom, on which issues, and how? The findings in this chapter suggest no easy answers. Early action might pave the way for more advanced policies or cement a lax policy trajectory. Those international organizations supplying stringent policies might have inadequate geographic coverages, and vice versa. Moreover, knowledge and technology determine possibilities for regulation. Ocean governance in an era of climate change needs to carefully consider these aspects.

Future research can help to further improve ocean governance. For example, more political science research into drivers and effects of international maritime regulation in the Arctic is needed. Scholars could use the data analysis presented here to investigate the effectiveness of different regulatory designs. They could also link the identified stringency patterns to competing theoretical explanations of regulatory decision making. Together, these lines of research would elucidate under which conditions effective international regulation is likely to emerge – for a warming and more threatened Arctic Ocean, as well as for other sea regions amidst climate-induced changes.

References

Abbott, K. W., Keohane, R. O., Moravcsik, A., Slaughter, A., and Snidal, D. (2000). The concept of legalization. *International Organization*, **54**(3), 401–19.

AMAP. (2010). *Assessment 2007: Oil and Gas Activities in the Arctic – Effects and Potential Effects*, vol. II, Oslo: Arctic Monitoring and Assessment Programme (AMAP).

Andresen, S., Rosendal, K., and Skjærseth, J. B. (2013). Why negotiate a legally binding mercury convention? *International Environmental Agreements: Politics, Law and Economics*, **13**(4), 425–40.

Arctic Council. (2009a). *Arctic Marine Shipping Assessment 2009 Report*, Arctic Council Working Group on the Protection of the Marine Environment (PAME).

Arctic Council. (2009b). *Arctic Offshore Oil and Gas Guidelines 2009*, Arctic Council Working Group on the Protection of the Marine Environment (PAME).

Bambulyak, A., Golubeva, S., Sydnes, M., et al. (2015). Resource management regimes in the Barents Sea. In A. Bourmistrov, F. Mellemvik, A. Bambulyak, et al., eds., *International Arctic Petroleum Cooperation: Barents Sea Scenarios.* London: Routledge, pp. 53–76.

Bernauer, T., Kalbhenn, A., Koubi, V., and Spilker, G. (2013). Is there a "Depth versus Participation" dilemma in international cooperation? *The Review of International Organizations*, **8**(4), 477–97.

Borgerson, S. G. (2008). Arctic meltdown: The economic and security implications of global warming. *Foreign Affairs*, **87**(2), 63–77.

Böhmelt, T., and Pilster, U. H. (2010). International environmental regimes: Legalisation, flexibility and effectiveness. *Australian Journal of Political Science*, **45**(2), 245–60.

Braithwaite, J., and Drahos, P. (2000). *Global Business Regulation.* Cambridge: Cambridge University Press.

Breitmeier, H., Levy, M. A., Young, O. R., and Zürn, M. (1996). *International Regimes Database (IRD) Data Protocol.* Available at www.fernuni-hagen .de/polis/download/lg2/projekte/protocoll_regime_database.pdf (last accessed December 12 2016).

Breitmeier, H., Young, O. R., and Zürn, M. (2006). *International Regimes Database.* Available at www.fernuni-hagen.de/polis/lg2/projekte/InternationalRegimes Database.shtml (last accessed August 6, 2017).

DeSombre, E. R. (2006). *Flagging Standards: Globalization and Environmental, Safety, and Labor Regulations at Sea.* Cambridge, MA: MIT Press.

DNV. (2014). Specially Designated Marine Areas in the Arctic High Seas, report no. 2013–1442, revision 2, Oslo: Det Norske Veritas.

Downs, G. W., Rocke, D. M., and Barsoom, P. N. (1996). Is the good news about compliance good news about cooperation? *International Organization*, **50**(3), 379–406.

Falkner, R. (2008). *Business Power and Conflict in International Environmental Politics*, Basingstoke: Palgrave Macmillan.

Garner, R. (2017). *Sea Ice Extent Sinks to Record Lows at Both Poles.* Available at www.nasa.gov/feature/goddard/2017/sea-ice-extent-sinks-to-record-lows-at-both-poles.

Goertz, G. (2006). *Social Science Concepts: A User's Guide.* Princeton: Princeton University Press.

Guzman, A. T. (2010). *How International Law Works: A Rational Choice Theory.* Oxford: Oxford University Press.

Harsem, Ø., Eide, A., and Heen, K. (2011). Factors influencing future oil and gas prospects in the Arctic. *Energy Policy*, **39**(12), 8037–45.

Humrich, C. (2013). Fragmented international governance of Arctic offshore oil: Governance challenges and institutional improvement. *Global Environmental Politics*, **13**(3), 79–99.

IFC. (2015). *Environmental, Health, and Safety Guidelines for Offshore Oil and Gas Development*, International Finance Corporation, World Bank Group.

IMO. (2015). *International Code for Ships Operating in Polar Waters*, International Maritime Organization, Marine Environmental Protection Committee, MEPC 68/21/Add.1, Annex10.

Jensen, Ø. (2016). The international code for ships operating in polar waters: Finalization, adoption and law of the sea implications. *Arctic Review*, **7**(1), 60–82.

Koivurova, T. (2018). Framing the problem in Arctic offshore hydrocarbon exploitation. In C. Pelaudeix and E. M. Basse, eds., *Governance of Arctic Offshore Oil and Gas*. London: Routledge, pp. 19–30.

Koivurova, T., and Molenaar, E. J. (2009). *International Governance and Regulation of the Marine Arctic*, Oslo: WWF International Arctic Programme.

Koremenos, B., Lipson, C., and Snidal, D. (2001). The rational design of international institutions. *International Organization*, **55**(4), 761–99.

Leeuwen, J. (2010). *Who Greens the Waves? Changing Authority in the Environmental Governance of Shipping and Offshore Oil and Gas Production*. Wageningen: Wageningen Academic Publishers.

Lyons, Y. (2015). Regulation of offshore hydrocarbon exploration and exploitation under international law. In R. Warner and S. Kaye, eds., *Routledge Handbook of Maritime Regulation and Enforcement*. London: Routledge, pp. 193–211

McDorman, T. L. (2014). Canada, the United States and international law of the sea in the Arctic Ocean. In T. Stephens and D. L. VanderZwaag, eds., *Polar Oceans Governance in an Era of Environmental Change*. Cheltenham: Edward Elgar, pp. 253–68.

Mitchell, R. B. (1994). Regime design matters: Intentional oil pollution and treaty compliance. *International Organization*, **48**(3), 425–58.

Mitchell, R. B. (2017). *International Environmental Agreements Database Project*, version 2017.1. Available at http://iea.uoregon.edu/ (last accessed August 6, 2017).

Hansen, C. Ø., Grønsedt, P., Graversen, C. L., and Hendriksen, C. (2016). *Arctic Shipping – Commercial Opportunities and Challenges*. Frederiksberg: CBS Maritime.

OSPAR. (2010). *Quality Status Report 2010*. London: OSPAR Commission.

OSPAR. (2016). *Impacts of Certain Pressures of the Offshore Oil and Gas Industry on the Marine Environment – Stocktaking Report*. London: OSPAR Commission.

OSPAR. (2017). Trends in discharges, spills and emissions from offshore oil and gas installations. *OSPAR Intermediate Assessment 2017*. Available at https://oap .ospar.org/en/ospar-assessments/intermediate-assessment-2017/ (last accessed September 15, 2017).

PAME (2017). *Regional Waste Management Strategies for Arctic Shipping: Regional Reception Facilities Plan (RRFP) and Proposal for IMO Consideration*. Akureyri: Arctic Council Working Group on the Protection of the Marine Environment (PAME).

Raustiala, K. (2005). Form and substance in international agreements. *The American Journal of International Law*, **99**(3), 581–614.

Rayfuse, R. (2014). Coastal state jurisdiction and the Polar Code: A test case for Arctic Ocean governance? In T. Stephens and D. L. VanderZwaag, eds., *Polar Oceans Governance in an Era of Environmental Change*. Cheltenham: Edward Elgar, pp. 235–52.

Reiss, A. J. (1984). Consequences of compliance and deterrence models of law enforcement for the exercise of police discretion. *Law and Contemporary Problems*, **47**(4), 83–122.

Skjærseth, J. B., Stokke, O. S., and Wettestad, J. (2006). Soft law, hard law, and effective implementation of international environmental norms. *Global Environmental Politics*, **6**(3), 104–20.

Soltvedt, I. F. (2017). Soft law, solid implementation? The influence of precision, monitoring and stakeholder involvement on Norwegian implementation of Arctic Council recommendations. *Arctic Review*, **8**, 73–94.

Stokke, O. S. (2013). Regime interplay in Arctic shipping governance: Explaining regional niche selection. *International Environmental Agreements: Politics, Law and Economics*, **13**(1), 65–85.

Tan, A. K. (2005). *Vessel-Source Marine Pollution: The Law and Politics of International Regulation.* Cambridge: Cambridge University Press.

US EPA. (2011). *Oily Bilgewater Separators, Report EPA 800-R-11-007.* Washington, DC: United States Environmental Protection Agency.

Young, O. R. (2009). Whither the Arctic? Conflict or cooperation in the circumpolar north. *Polar Record*, **45**(1), 73–82.

Young, O. R. (2011). If an Arctic Ocean treaty is not the solution, what is the alternative? *Polar Record*, **47**(4): 327–34.

15

The Arctic Ocean's Melting Ice
Institutions and Policies to Manage Black Carbon

THOMAS L. BREWER

Introduction

The role of black carbon (BC) in the melting of Arctic Ocean ice poses distinctive and significant challenges for regional governance. Recent evidence of the increasing magnitude and rate of Arctic Ocean ice melt, as well as model-based forecasts of future changes, have heightened alarm among specialists on Arctic climate change. This chapter focuses specifically on *ocean* ice melt (not land ice melt) as a process that is occurring in the Arctic and also specifically on the contribution of BC emissions to that process. The ice melt occurs in substantial part because of BC's warming effects as an aerosol in the Arctic atmosphere and because of its deposits, which reduce the reflective "albedo" effects of snow and ice. In addition, along with carbon dioxide and other greenhouse gases, BC contributes to warming on a global scale. There is also increasing interest in BC more generally as a significant climate change–forcing agent and as a public health problem and a threat to agricultural production. Although this chapter's focus is quite specific, its implications are global, long-term, and multidimensional.

Because BC is particulate matter (that is, not a gas) it is not directly covered by the United Nations Framework Convention on Climate Change (UNFCCC), and it is not formally on the agendas of UNFCCC conferences of the parties (COPs). However, it is on the agendas of other international institutions, including the Climate and Clean Air Coalition (CCAC) program at the United Nations Environment Programme (UNEP), as well as the International Maritime Organization (IMO). It has also been a subject of interest in the Arctic Council and the Nordic Council and among nongovernmental organizations such as the International Council for Clean Transportation (ICCT) and the International Center for Trade and Sustainable Development (ICTSD).

After a review of facts concerning Arctic Ocean ice melt, this chapter presents basic features of BC that are necessary to understand its role in Arctic climate change processes and associated policy issues. The chapter assesses current governance institutions and policies, particularly at the international level, and it proposes an Arctic black carbon (ABC) agreement that could reduce BC's contribution to Arctic Ocean ice melt and to other elements of the Arctic warming emergency.

Melting of Arctic Ocean Ice

The key trends of Arctic Ocean ice melt to date are well established. They can be summarized as follows (IPCC, 2013: ch. 4, 323–39; US National Snow and Ice Data Center [NSIDC], 2017; US National Aeronautics and Space Administration [NASA], 2017; and US National Oceanic and Atmospheric Administration [NOAA], 2017):

- The extent of ice at its late-summer low has declined about 11 percent per decade over the past four decades.
- The volume of ice at its late-summer low has declined about 17 percent per decade over the past four decades.
- Parallel declines in the maximums at the end of winter have also been observed.
- The rate of decline has increased in recent years, with new records being set (based on the satellite dataset that started in 1979).
- The ice season has been getting shorter and thus the ice-free season longer, for instance, by three *months* for the area extending from the eastern Siberian Sea to the western Bering Sea (IPCC, 2013: 329).
- Observed ice melt rates have been greater than expected, partly because of the inadequacies of the data about black carbon available for modeling ice melt trends.

Extensive graphics depicting these trends, and frequent updates of data, are available at the websites of the US National Snow and Ice Data Center (www.nsidc.org), National Aeronautics and Space Administration (www.nasa.gov), and National Oceanic and Atmospheric Administration (www.noaa.gov) (as of late 2017, pending the outcome of the fiscal year 2018 budget decision process).

What is causing sea-ice melt in the Arctic? The most obvious direct factor is that Arctic temperatures have been increasing at more than twice the rate of the global mean surface temperature. Black carbon particulates in the atmosphere and depositions on ice and snow are major contributing factors. Thus, it is plausible that mitigating only carbon dioxide will not be sufficient to reverse Arctic Ocean ice melt, though, of course, cutting carbon emissions will still be necessary. It will also be necessary to mitigate BC emissions. Even meeting the targets of the 2015 Paris Agreement on climate change for greenhouse gas emissions, including carbon dioxide in particular, will not be adequate to prevent ice-free periods in the Arctic Ocean (and it is unlikely that even the Paris targets will be met).

What are the consequences of Arctic Ocean ice melt? Interest in the impact on changes in the jet stream and thus weather patterns in the Northern Hemisphere has been recently reinforced in the context of Hurricane Harvey in the United States in September 2017. A specific issue is whether the hurricane's northward progression and dissipation were retarded by a stable high-pressure area resulting from deepening and prolongation of the jet stream. The potential contribution of Arctic Ocean ice melt to irreversible permafrost melt is also attracting increasing interest. Melting permafrost has the potential to release enormous quantities of powerful greenhouse gases. In addition, there are oil and gas production issues, international security issues, and local socioeconomic issues, including the implications of increasingly open lanes through the Arctic Ocean as a result of the ice melt (see Chapter 14). Although these are all relevant issues in terms of the impacts of Arctic Ocean ice melt, they are beyond the scope of this chapter. This chapter will focus specifically on the contributions of BC to Arctic Ocean ice melt and on the associated implications for climate change policy and international institutions.

Defining Black Carbon

In brief, BC is carbon *particulate matter* less than 2.5 microns across ($PM_{2.5}$), which results from incomplete combustion of fossil fuels, biofuels, and biomass (European Environment Agency, 2013: 9, Box 1.2). BC's total contribution to global warming has been estimated to be 55 percent of carbon dioxide's, and BC is thus the second leading contributor to global warming (Bond et al., 2013). On a per-ton basis, BC's global warming potential (GWP) index for a 20-year reference period is on the order of thousands of times greater than carbon dioxide's (IPPC, 2013: 740, Table 8.A.6). A distinctive and significant feature of BC emissions is that their effects are *experienced directly at local and regional levels* because they result in particulate matter depositions at the ground level, in addition to them being atmospheric pollutants. In addition to its role as a global climate change forcer, BC is a locally and regionally concentrated pollutant that affects public health and agricultural production. (See the review of BC climate change mitigation issues and other issues by the Science and Technology Advisory Panel of the Global Environment Facility, as reported by Sims, Gorsevski, and Anenberg, 2015.)

Although BC is commonly called "soot," this is not precisely correct because soot contains other elements in addition to BC. Indeed, calculations of BC's climate change effects are complicated by the presence of organic carbon (OC) emissions as copollutants with BC emissions in soot. A key determinant of the *net* warming/cooling impact of emissions is the ratio of BC to OC because OC is a coolant. Importantly, the ratio of BC to OC in diesel emissions, in particular, has been estimated to be as high as 9:1 (Azzara, 2013; 2015), and mobile diesel engines' BC emissions, especially from motor vehicles, have been estimated for the United States to be about 75 percent of soot's total (US EPA, 2017). There is thus now a widespread consensus among emissions specialists that diesel emissions are a main source of BC emissions and a major net contributor to local, regional, and global warming.

Black carbon is sometimes confused with "carbon black," which is a manufactured material used, for instance, to make vehicle tires black. Andreae and Gelencser (2006) provide an extensive discussion of terminology issues related to BC emissions, and Goldberg (1985) has an extensive examination of BC as an atmospheric pollutant. Masiello (2004) discusses BC in marine environments.

Black Carbon in the Arctic

Black carbon contributes to Arctic Ocean ice melt because of its concentration levels as an aerosol in the Arctic atmosphere and because of its deposition directly on Arctic Ocean ice and snow. There is abundant evidence about BC's contribution to Arctic temperature increases and ice melt in a study by the Arctic Monitoring and Assessment Program of the Arctic Council (AMAP, 2015). For instance, the Arctic net surface temperature response from BC in the atmosphere is 0.40°C, and it is 0.22°C from BC depositions on snow (see also Hansen and Nazarenko, 2004; Hegg et al., 2009; US EPA, 2012; 2016). According to the Clean Arctic Alliance (2017), black carbon has an especially potent climate warming effect when emitted at high latitudes; the warming effect is increased by a factor of three in the Arctic region, as compared to emissions over the open oceans.

On the basis of simulation models of BC's direct radiative forcing in the Arctic region, using 2010 data from all regions outside and inside the Arctic, an AMAP study found that Asian countries and Russia were the largest contributors, whereas the Nordic countries contributed the least (AMAP, 2011: 87–88). Gas flaring in Russia is an especially significant sectoral-regional source of BC depositions on snow and ice (see Clean Air Task Force, 2014). Most BC in the Arctic region comes from sources outside the Arctic; much of it is from cooking stoves in Asia. These are important dimensions, respectively, of the geographic context and the sectoral context.

As will be noted latter in the discussion of a proposal for an ABC agreement, the approach presented here is *proactive* and *incremental*. However, this approach need not preclude other approaches with different geographic or sectoral emphases. In fact, a broad array of complementary "solutions" could be effective in their totality. The specific focus here is on two important sectors: maritime shipping and aviation. Their distinctive international institutional and policy circumstances could simulate further parallel analyses from different geographic and sectoral perspectives.

The reality of increasing ship traffic through the Arctic Ocean (see Chapter 14) was highlighted by the cruise of a large luxury ocean liner during August to September 2016. With more than 1,700 people onboard, the *Crystal Serenity* undertook a 32-day tour, leaving from Seward, Alaska, and passing through the Bering Strait and Northwest Passage, before reaching New York City. In 2017, a Russian liquefied natural gas (LNG) carrier, purpose-built to be its own ice breaker, passed through the Northeast Passage without an accompanying ice-breaking vessel. That was the first time for such a voyage. A projection of vessel traffic through the US Arctic, based on a combination of business-as-usual growth and a diversion through the Arctic of 5 percent of the traffic normally passing through the Suez and Panama canals, is 710 to 750 vessels in 2025 (ICCT, 2015). These would be increases from 120 vessels in 2008 and 240 vessels in 2013. Other analyses emphasize a decades-long process of significantly increasing maritime traffic (see Stuards, 2016).

Maritime shipping has been noted by the Arctic Council (2011a) as follows: "Marine shipping in the Arctic region is … potentially high in its impact due to its proximity to Arctic snow and sea ice. Emissions from this sector may increase significantly due to increases in global marine shipping traffic, as well as a lower prevalence in summer sea ice cover." Reports sponsored by the International Council on Clean Transportation (ICCT) projected an increase in black carbon emissions of tons per year by 2025 by a factor of 5 to 120 (Azzara, 2013; Azzara and Rutherford, 2009; 2015; Azarra, Minjares, and Rutherford, 2015). The Norwegian Shipowners Association has similarly warned about increasing traffic in the next ten years (RTCC, 2013; see also Arctic Council, 2009, 2011b).

As for aviation's BC emissions (Fuglestvedt et al., 2010), there is a trans-Atlantic concentration of international traffic just south of Greenland and Iceland – a concentration pattern that contributes to the relatively high warming trends in the Arctic region. There is evidence of underestimates of aviation's BC emissions in general because of the inherent inadequacies of traditional measurement methods. When aviation's BC emissions have been subjected to rigorous analysis in recent years, a key finding is that the commonly used metric of "smoke number" *underestimates* emissions. The smoke number indirectly estimates the mass of BC emissions by optically measuring the visibility of the plume of an aircraft at cruise speed as well as during takeoff and landing. It is not a direct physical chemical measurement. One study found that the smoke number underestimates BC emissions by a factor of 2.5 to 3, thus indicating that current estimates of aviation BC need to be increased by a factor of similar magnitude (Stettler, Eastham, and Barrett, 2013). Another study found that the estimate of global aviation BC emissions should be about three times earlier estimates, or equal to about one-third of the radiative forcing of aviation's carbon dioxide emissions (Stettler et al., 2013). A study of UK airports representing 95 percent of UK passenger traffic in 2005 also found greater BC emissions than those in studies using the smoke number (Stettler, Eastham, and Barrett, 2011); in fact, more direct alternative measurement methods found BC emissions to be eight times greater than smoke number studies. The effects of organic carbon (which is a coolant, as noted earlier) were also found to be higher, but only by a factor of 0.4.

The volume of aviation traffic is expected to increase for at least the next few decades, as it has over the past several decades – typically by more than the rate of increase in world gross domestic product (GDP). Forecasts by the International Civil Aviation Organization (ICAO; 2016) are that total world aviation passenger traffic will increase by an average of 4.6 percent per year over the 20-year period from 2012 to 2032, and 4.5 percent over the 30 years from 2012 to 2042 (ICAO, 2016a: 10, 23; see also Bows-Larkin, 2015, and Air Transport Action Group, 2015). These forecasts compare with 5.2 percent per year on average that was experienced from 1995 to 2012. The parallel forecasts for air freight traffic are averages of 4.6 percent and 4.5 percent per year.

The MIT Joint Program on Science and Policy conducted an extensive modeling exercise of future global aviation emissions, including BC emissions. Their results show BC emissions in 2050 that will be 1.38 to 4.87 times greater than 2006 levels (Brasseur et al., 2016: 567, Table 2). The combination of at least a doubling of air traffic volume, with the greater and more accurate estimates of BC emissions, bodes ill for business-as-usual aviation contributions to global warming. Despite improvements in aviation fuel efficiency, BC emissions will increase substantially. Aviation's contributions to BC in the Arctic are therefore also likely to increase substantially.

International Institutions and Policies

The literature reviews in the volumes of the Fifth Assessment Reports of the Intergovernmental Panel on Climate Change (IPCC, 2013; 2014a; 2014b) include scores of citations of refereed publications about BC. Diplomacy, however, has been less attentive to BC issues. The UNFCCC explicitly focus on *gases* as sources of climate change. Consequently, because it is not a gas, BC has been neglected in formal multilateral climate change negotiations and implementation processes. However, BC has been on the agendas of the Arctic Council, the Nordic Council, the CCAC of UNEP, the Gothenburg Protocol, and the IMO. Such organizations have used UNFCCC COPs as occasions for promoting action on BC emissions and for informing other actors about these organizations' own programs toward that end.

The Arctic Council

The Arctic Council is essentially a discussion forum with working groups that produce studies about environmental and safety issues. Climate change issues have been the focus of several years of work by specialized task forces and expert groups. The council's Task Force on Short-Lived Climate Forcers issued a series of findings and recommendations that included a section on marine shipping (Arctic Council, 2011c; italics added):

Measures to reduce BC from marine shipping in and near the Arctic could include Council-wide adoption of voluntary technical and non-technical measures, adoption of the proposed amendment of MARPOL [International Convention for the Prevention of Pollution from Ships] Annex VI to establish an Energy Efficiency Design Index, and collaboration with IMO on certain other actions. *Marine shipping in the region [was in 2011] a relatively small source of BC, but it is potentially high in impact due to its proximity to snow and ice, and may increase significantly due to projected increases in global ship traffic as well as decreases in summer sea ice cover.*

At its April 2015 Ministerial Meeting, the Council advanced the BC agenda by approving "An Arctic Council Framework for Action" as part of its *Enhanced BC and Methane Emissions Reductions* (Arctic Council, 2015). According to this agenda, each Arctic state was committed to do the following:

- Develop and improve emission inventories and emission projections for BC using, where possible, relevant guidelines from the Convention on Long-Range Transboundary Air Pollution (CLRTAP) and improve the quality and transparency of information related to emissions of BC,
- Enhance expertise on the development of BC inventories, including estimation methodologies and emissions measurements, by working jointly through the Arctic Council and other appropriate bodies.

The Council does not implement its decisions collectively; it has no enforcement authority; and it has no program budget. Implementation and financing of its programs are undertaken by individual member states. The Council is not able to promulgate enforceable mandatory regulations, unless it were to undergo a dramatic transformation in its basic institutional nature, which is unlikely anytime soon. (For additional information on the Arctic Council and climate change issues in the Arctic region, see Yeo, 2015; as well as Arctic Council, 2009, 2011a, 2011b, 2012, 2013, 2014, 2015.)

The Nordic Council

Some of the work of the Nordic Council on Arctic issues has been channeled through the Arctic Council. However, with the withdrawal of the US government from engagement in climate change action internationally, the Nordic Council's work has become more salient and significant. With the advent of the Trump administration in the United States in January 2017, the prospects for action on BC were reduced. Research and policy planning are nevertheless proceeding in many countries, especially the Nordic Council countries: Denmark, Finland, Iceland, Norway, and Sweden. There is much variation among the interests of these countries regarding the Arctic. For instance, Norway's oil and gas reserves are a significant economic and political issue in that country. Denmark's interests are largely derivative from Greenland's position and potential uranium deposits. Yet all of the countries have a strong tradition of environmentalism and a sense of responsibility to indigenous peoples, and thus an unusual sensitivity to Arctic ice melt. Norway, Denmark (through its Greenlandic link), and Iceland all have territory with Arctic Ocean coastlines. (Discussions of Nordic programs and studies are available at www.norden.org.)

The Climate and Clean Air Coalition

The Climate and Clean Air Coalition was created with a mandate to address issues concerning short-lived climate pollutants, including BC, methane, and hydrofluorocarbons. Its membership consists of 39 countries, plus the European Commission. It also has 52 nonstate partners, including the International Council on Clean Transportation, the World Health Organization, and the World Bank. Its relationship with the UNEP provides it with an institutional place within the UN system. Its mandate includes the health effects of BC, as well as the climate effects of it (CCAC, 2014). The CCAC, which is institutionally housed in UNEP, is in a unique position vis-a-vis BC internationally because of its specialized programs in diesel engine emissions (CCAC, 2015, 2016c) and its focus on seaport emissions (CCAC, 2016a, 2016b).

The Gothenburg Protocol

The Gothenburg Protocol to Abate Acidification, Eutrophication and Ground-level Ozone was agreed in 1999 (UNECE, 1999). Revisions to the Protocol in 2012

explicitly included BC as a particulate matter to be reduced for climate change mitigation, as well as for the health benefits (UNECE, 2013, 2015). The Protocol is an
important development in international efforts to address BC issues, and it is an
example of the incremental expansion of the original Protocol, which itself was one
expansion among many to the 1979 Convention on Long-range Transboundary Air
Pollution (UNECE, 1979). Canada, the United States, and 30 European countries,
plus the European Union as a regional entity, have made commitments of varying
percentage reductions in their small-particulate emissions. The focus on stationary
sources of emissions (not mobile sources) limits the scope of relevant issues, of
course, excluding international shipping and aviation, albeit including stationary diesel engines at seaports and airports.

The International Maritime Organization

The IMO has been granted specific sectoral mandates on international trade and climate change issues (see Chapter 23). Its mandate concerning climate change is
embodied in Article 2 of the Kyoto Protocol of the UNFCCC. BC has been on the
agenda of the IMO for several years, but thus far the organization has only agreed
on a definition of the term (which, after two years of consideration, adopted verba-
tim the definition in Bond et al., 2013), though it also has under way an assessment
of BC measurement methods and a reporting protocol. The IMO has been working
on an International Code for Ships Operating in Polar Waters (the Polar Code) to
address safety and environmental issues in the Arctic and Antarctic regions (IMO,
2014b). The code was approved by the Marine Environment Protection Committee
in October 2014 as amendments to the International Convention for the Prevention
of Pollution from Ships (MARPOL), and it was approved by the Maritime Safety
Committee in November 2014 as amendments to the International Convention for
the Safety of Life at Sea, with entry into force on January 1, 2017 (IMO, 2014b).
Although the Code does not include provisions concerning BC per se, its adoption
indicates a willingness to address Arctic-specific environmental issues.

The fuel efficiency regulations of the IMO are also relevant because they can
reduce BC emissions, even though their direct objective is to reduce carbon dioxide
emissions by increasing fuel efficiency (see Chapter 23). The fuel-efficiency regulations are mandatory, tangible, and in force, and will evolve over time. In July 2017,
the Marine Environment Protection Committee (MEPC) agreed on a brief outline
for "an initial IMO strategy on the reduction of GHG emissions from ships" (IMO,
2017), with a view to preparing a draft text by April 2018. The place in the text, if
any, for BC was not apparent as of late 2017, despite recent developments.

Emission Control Areas for Ships

Regional emission control areas (ECAs) regulate ships' emissions of sulfur oxide
(SOx) and nitrogen oxide (NOx) (IMO, 2012a, 2014c; US EPA, 2015). ECAs have

taken the legal form of amendments to Annex VI of MARPOL and are thus officially designated areas subject to IMO regulation. As of late 2018, there were four
such areas in various states of implementation: North America (Canada, United
States, French islands of Saint-Pierre and Miquelon), US Caribbean Sea (Puerto
Rico and the US Virgin Islands), the Baltic Sea, and the North Sea. The latter two,
it should be noted, currently cover only SOx emissions. Possible ECAs have also
been discussed for Norway, Japan, and the Mediterranean, and others have been
proposed for the Arctic. However, the inclusion of BC emissions in ECAs remains
an open question.

International Civil Aviation Organization

The ICAO, which is a sector-specific UN Special Agency like the IMO, has largely
ignored both BC emission issues and Arctic climate change issues. Its outline for a
Carbon Offsetting and Reduction Scheme for International Aviation does not mention BC, nor is there any evidence in ICAO's history to indicate sustained attention
to this type of issue (Rutherford, 2016; ICCT, 2017).

An Arctic Black Carbon Agreement

In light of the inadequate progress to date in addressing the black carbon pollution
in the Arctic (except in the Arctic Council's studies of the problem), more needs to
be done. An international agreement to create a "club"-type partnership is an
appealing option (Brewer 2015a, 2015b, 2016a, 2016b; Aakre, Kallbekkan, Van
Dingenen, and Victor, 2017). An ABC agreement could be a building block or mod-
ule in a network of broader institutionalized arrangements addressing BC-related
aviation and/or maritime shipping issues in the region. Since the multilateral
UNFCCC addresses neither BC, nor aviation, nor shipping, there is a gap in the
multilateral climate change "regime complex" (Keohane and Victor, 2011). This gap
creates an opportunity for a niche agreement with global significance.

   An ABC agreement should be focused on the future; it would surely take time to
formulate and negotiate such an agreement. The objective of the ABC agreement
should be to prevent BC deposition and air pollution in the Arctic from increasing
alongside increases in shipping traffic in the Arctic and other regions from which
shipping emissions are transported to the Arctic. Although the precise target levels
that would be either explicit or implicit within such an agreement would be subject
to further scientific, economic, and diplomatic exercises, the lower the levels, the better. The sooner the issues are on the active diplomatic agenda, the more likely that
an effective and otherwise attractive agreement can be reached. Indeed, the United
States observed in its National Strategy for the Arctic Region (released before the
advent of the Trump administration) that "[u]ncoordinated development – *and the
consequent increase in pollution such as emissions of BC or other substances from fossil fuel combustion* – could have unintended consequences on climate trends, fragile

ecosystems, and Arctic communities. It is imperative [to] *proactively* establish national priorities and objectives for the Arctic region" (US White House, 2013; italics added).

As defined in the political economy literature (Cornes and Sandler 1996; Brewer, Derwent and Blachowicz, 2016), the benefits of "clubs" have two key features: (1) they can be shared among participants, and (2) they can be excluded from nonparticipants. In the context of climate change agreements, the development of clublike international agreements can thus incentivize *participation* and *compliance* (IPCC, 2014b: ch. 13; Victor, 2015a, 2015b, 2017; see DeSombre, 2008; Falkner, 2015; and Weischer, Morgan, and Patel, 2012). As an international governance modality, such an arrangement has the advantages of deterring "free riding" via nonparticipation or noncompliance. Countries or nonstate actors that want to enjoy the benefits of the agreement must participate in it, which would require complying with its rules.

There are two different uses of the club concept that are emerging among climate change specialists. One is based on the restrictive notion adopted earlier, where shareable and excludable benefits are central to the creation of incentives for participation and compliance. The other is based on the number of participants and distinguishes clubs from multilateral arrangements. The distinction between the two notions is important because some multilateral arrangements have the key features of clubs in the restrictive sense, despite their large size. It is possible to combine the two criteria – benefits and size – in order to accommodate both the political economy literature that emphasizes the structure of benefits and much of the climate change discussion to date that emphasizes membership size. Accordingly, one could limit the notion of climate clubs to arrangements that have a relatively small number of participants (see Falkner [2015] for illustrative numbers) *and* have shareable and excludable benefits to encourage participation and compliance. The proposal for an ABC agreement in this chapter would meet both the size and benefits features.

The benefits of *participation* need to be specified. There are many possibilities. An obvious one is the opportunity to operate in Arctic region waters. The agreement would provide that only ships meeting BC-related equipment and operational standards could operate in the Arctic region – for instance, the use of diesel particulate filters (Brewer, 2016b) or low-sulfur distillate fuels (Clean Arctic Alliance, 2017). An international license for Arctic operations by individual ships and ship owner-operators could be issued on the basis of certification that required equipment has been installed and properly maintained, as well as meeting operational standards. This would thus be a public–private sector partnership, in which individual ships, ship owners, ship operators, ship registry governments, all governments participating in the Arctic Council, and other governments in the IMO would all be participants. Participation would be voluntary, but participation would be a precondition for a ship to operate in the Arctic region. A regulatory framework would be established within the IMO by the ABC agreement, in cooperation with other organizations. The details of the division of labor among the organizations would be coordinated by the IMO.

The licensing requirement would be imposed on ships involved in oil or gas exploration or extraction activities, as well as ships engaged in the transport of any goods or people, and thus it would include all types of ships engaged in international commerce. Another benefit that could be shared by participants in the agreement and excluded from nonparticipants would be a technology-transfer agreement, whereby participants would be entitled to assistance in the acquisition of the required technology to meet participation and compliance criteria. The scope and funding levels of the programs would be among the issues to be negotiated. These could be codified and monitored by the IMO with advice and operational support from a variety of organizations.

A *compliance* enforcement system would also be needed. There is already in place a worldwide, satellite-based, real-time tracking system that identifies individual ships, with their position, direction, and speed. Any ship sailing into or through the Arctic region would be required to keep its transponder operating in order to be tracked. Failing to do so would result in a citation of the ship operator, with a substantial fine, embargo of the ship, and cancellation of the operator's right to sail any ships in the Arctic region for a period of years. All licensed ships would be monitored for compliance with the equipment and operational standards.

Participation would be open to nonstate entities such as ship owners and ship operators, as well as to governments. International financial and development institutions, such as the World Bank, regional development banks, UN Industrial Development Organization, and UN Conference on Trade and Development (UNCTAD) could also participate, particularly in technology sharing programs. Initiatives to address the climate change impacts of the international maritime industry should include the IMO as a central forum. In addition, many other activities in numerous organizations already do or could interact with IMO activities in this issue area. The IMO should work directly with the Arctic Council, CCAC, International Organization for Standardization, UN Economic Commission for Europe, World Health Organization, and other organizations to develop an ABC agreement. This initiative should begin as soon as possible. Numerous other international organizations could provide formal or informal advice on a wide range of issues. Among them, the UNFCCC could provide input on international technology transfer issues, including how international maritime programs relate to other UNFCCC technology transfer activities. The Organisation for Economic Co-operation and Development and UNCTAD have relevant expertise on international maritime trade issues.

Several energy-focused institutions might also be able to advise on technical issues related to BC regulation in the Arctic, including those issues associated with exploration and extraction of oil and gas reserves in the region. These institutions include the International Energy Agency (IEA) and Energy Charter Treaty. In addition, other forums, such as the Clean Energy Ministerial and World Energy Council, have expertise that is relevant to international technology transfer relevant to BC regulation in the Arctic. Although each of these institutionalized arrangements has expertise relevant to BC emissions in the Arctic, none has a mandate to regulate international maritime shipping or other activities in the region. Their potential contributions might thus be limited to relatively narrow technical issues, rather than core substantive regulatory issues. Such technical contributions could be obtained from other institutions. For example, among economic development agencies, the World Bank could be particularly important in financing schemes.

Conclusion

Because of its distinctive and especially important role in Arctic climate change generally, and in Arctic Ocean ice melt processes in particular, there is a strong case for a specially tailored international institutional regulatory arrangement along the lines of the ABC introduced earlier. Obvious candidates for regulation are international maritime shipping and aviation. An ABC agreement could slow the rate of Arctic Ocean ice melt if it is designed with an effective mix of incentives to encourage polluters to participate and to comply with the rules of the agreement.

There are many considerations beyond the specific problem of Arctic Ocean ice melt and the role of BC in that melting. There are questions about the local social and economic consequences of Arctic Ocean ice melt and any attempts to address it. There are questions about the global consequences as well, including the possibility of climate change "tipping points" that could radically alter Northern Hemisphere and global weather patterns.

A new Arctic Black Carbon agreement is a conceivable "solution" to potentially catastrophic climate change processes in the Arctic. It is at least a tangible proposal that could stimulate thinking "out of the box" about governance responses to the growing emergency of Arctic warming. These responses can contribute to wider efforts to achieve effective governance of the world's oceans amidst climate change.

References

Aakre, S., Kallbekken, S., Van Dingenen, R., and Victor, D. G. (2017). Incentives for small clubs of Arctic countries to limit black carbon and methane emissions. *Nature Climate Change*, **8**(1), 85–89.

Air Transport Action Group. (2015). *Aviation Climate Solutions*. Geneva: Air Transport Action Group.

Andreae, M. O. and Gelencser, A. (2006). Black carbon or brown carbon? The nature of light-absorbing carbonaceous aerosols. *Atmospheric Chemistry and Physics*, **6**(10), 3131–48.

Arctic Council. (2009). *Arctic Marine Shipping Assessment*. Tromsoe, Norway: Arctic Council [blog post].

Arctic Council. (2011a). *The Arctic Environment and Climate Change*. Tromsoe, Norway: Arctic Council [blog post].

Arctic Council. (2011b). *An Assessment of Emissions and Mitigation Options for Black Carbon for the Arctic Council*. Tromsoe, Norway: Arctic Council [blog post].

Arctic Council. (2011c). *Task Force on Short-Lived Climate Forcers: Progress Report and Recommendations for Ministers.* Tromsoe, Norway: Arctic Council [blog post].

Arctic Council. (2012). *2012 Arctic Report Cards Describe Dramatic Changes in the Arctic.* Arctic Council, December 5 [blog post].

Arctic Council. (2013). *Task Force on Black Carbon and Methane,* 2nd meeting, Stockholm, December 11–13 [blog post].

Arctic Council. (2014). *Addressing Black Carbon and Methane in the Arctic.* Tromsoe, Norway: Arctic Council [blog post].

Arctic Council. (2015). *Enhanced Black Carbon and Methane Emissions Reductions: An Arctic Council Framework for Action.* Tromsoe, Norway: Arctic Council [blog post].

Arctic Monitoring and Assessment Programme of the Arctic Council (AMAP). (2011). *The Impact of BC on Arctic Climate.* Oslo: AMAP.

Arctic Monitoring and Assessment Programme of the Arctic Council (AMAP). (2015). *AMAP Assessment 2015: BC and Ozone as Arctic Climate Forcers.* Oslo, Norway: AMAP.

Asariotis, R. and Benamara, H. eds. (2012) *Marine Transport and the Climate Challenge.* London and New York: Routledge and Earthscan.

Azzara, A. (2013). *Arctic Sea Shipping: Emissions Matter More Than You Might Think, From the ICCT Blogs, 2013.* Washington, DC: International Council on Clean Transportation.

Azzara, A. (2015). *BC Emissions from Shipping: Fact-Checking Conventional Wisdom. From the Blogs.* Washington, DC: International Council on Clean Transportation.

Azzara, A. and Rutherford, D. (2009). *Marine BC Emissions: Identifying Research Gaps.* Washington, DC: International Council on Clean Transportation.

Azzara, A., Minjares R., and Rutherford, D. (2015). *Needs and Opportunities to Reduce BC Emissions from Maritime Shipping.* Washington, DC: International Council on Clean Transportation.

Azzara, A. and Rutherford, D. (2015). *Air Pollution from Marine Vessels in the U.S. High Arctic in 2025.* Washington, DC: International Council on Clean Transportation.

Bond, T. C., Doherty, S. J., Fahey, D. W., et al. (2013). Bounding the role of BC in the climate system: A scientific assessment. *Journal of Geophysical Research: Atmospheres,* 118(11), 5380–552.

Bows-Larkin, A. (2015). All adrift: Aviation, shipping, and climate change policy. *Climate Policy,* 15(6), 681–702.

Brasseur, G. P., Gupta, M., Anderson, B. E., et al. (2016). Impact of aviation on climate, *Bulletin of the American Meterological Society,* 97(4), 561–83.

Brewer, T. L. (2015a). *Arctic Black Carbon from Shipping: A Club Approach to Climate-and-Trade Governance.* Geneva: ICTSD.

Brewer, T. L. (2015b). Arctic black carbon: The challenge for international governance, *BioRes,* 9(5), 16–9.

Brewer, T. L. (2016a). Arctic warming: Context and issues for an Arctic Black Carbon (ABC) agreement, ICTSD Dialogue Process – Background Paper Update.

Brewer, T. L. (2016b). *Black Carbon Problems in Transportation: Technological Solutions and Governmental Policy Solutions.* Working Paper. MIT Center for Energy and Environmental Policy Research. Cambridge, MA: MIT.

Brewer, T. L., Derwent, H., and Blachowicz, A. (2016). Carbon market clubs and the new Paris regime. Discussion Paper for the World Bank Group's Networked Carbon Markets Initiative.

Clean Air Task Force. (2014). *The Last Climate Frontier: Leveraging the Arctic Council to Make Progress on Black Carbon and Methane.* Boston, MA: Clean Air Task Force.

Clean Arctic Alliance. (2017). Overview. Available at www.hfofreearctic.org.

Climate and Clean Air Coalition (CCAC). (2014). *Health: Urban Health Initiative.* Paris: CCAC.

Climate and Clean Air Coalition (CCAC). (2015). *Diesel Factsheet.* Paris: CCAC.

Climate and Clean Air Coalition (CCAC). (2016a). *Ports and Marine Factsheet.* Paris: CCAC.

Climate and Clean Air Coalition (CCAC). (2016b). *Reducing BC from Ports.* Paris: CCAC.

Climate and Clean Air Coalition (CCAC). (2016c). *Reducing BC Emissions from Heavy-Duty Diesel Vehicles and Engines: Results to Date and Strategy for Continued Success.* Paris: CCAC.

Comer, B., Olmer, N., Mao, X., Roy, B., and Rutherford, D. (2017). *Prevalence of Heavy Fuel Oil and BC in Arctic Shipping, 2015–2025.* Washington, DC: ICCT.

Corbett, J., Lack, D., and Winebrake, J. (2010). Arctic shipping emissions inventories and future scenarios. *Atmospheric Chemistry and Physics,* 10(19), 9689–704.

Corbett, J. J. and Winebrake, J. (2008). *The Impacts of Globalization on International Maritime Transport Activity: Past Trends and Future Perspectives.* Paris: OECD.

Corbett, J. J., Winebrake, J. J., Green, E. H., et al. (2007). Mortality from ship emissions: A global assessment. *Environmental Science & Technology,* 41(24), 8512–18.

Cornes, R. and Sandler, T. (1996). *The Theory of Externalities, Public Goods, and Club Goods.* Cambridge, UK: Cambridge University Press.

DeSombre, E. R. (2008). Globalisation, competition, and convergence: Shipping and the race to the middle. *Global Governance,* 14(2), 179–98.

European Environmental Agency (EEA). (2013). Status of BC monitoring in ambient air in Europe. EEA Technical Report No. 18/2013.

Falkner, R. (2015). *A Minilateral Solution for Global Climate Change? On Bargaining Efficiency, Club Benefits and International Legitimacy.* Centre for Climate Change Economics and Policy Working Paper No. 222; Grantham Research Institute on Climate Change and the Environment Working Paper No. 197.

Fuglestvedt, J. S., Shine, K. P., Berntsen, T., et al. (2010). Transport impacts on atmosphere and climate: Metrics, *Atmospheric Environment,* 44(37), 4648–77.

Goldberg, E. D. (1985). *Black Carbon in the Environment: Properties and Distribution.* New York: J. Wiley & Sons.

Hansen, J. and Nazarenko, L. (2004). Soot climate forcing via snow and ice albedos. *Proceedings of the National Academy of Sciences,* 101(2), 423–28.

Hegg, D. A., Warren, S. G., Grenfell, T. C., et al. (2009). Source attribution of BC in Arctic snow. *Environmental Science & Technology,* 43(11), 4016–21.

Hughes, E. (2013). A new chapter for MARPOL Annex VI – requirements for technical and operational measures to improve the energy efficiency of international shipping, Marine Environment Division, IMO.

Intergovernmental Panel on Climate Change (IPCC). (2013). *The Physical Science Basis.* Cambridge, UK: Cambridge University Press.

Intergovernmental Panel on Climate Change (IPCC). (2014a). *Climate Change 2014: Mitigation of Climate Change.* Cambridge, UK: Cambridge University Press.

Intergovernmental Panel on Climate Change (IPCC). (2014b). *Climate Change 2014: Impacts, Adaptation and Vulnerability.* Cambridge, UK: Cambridge University Press.

International Civil Aviation Organization (ICAO). (2016). *Long-Term Traffic Forecasts.* Montreal: ICAO.

International Council on Clean Transportation (ICCT). (2011b). *Reducing Greenhouse Gas Emissions from Ships: Cost Effectiveness of Available Options.* Washington, DC: ICCT.

International Council on Clean Transport (ICCT). (2015). A 10-year projection of maritime activity in the U.S. Arctic region, report prepared for the U.S. Committee on the Maritime Transportation System, January 1.

International Maritime Organization (IMO). (2017). International Maritime Organization moves ahead with oceans and climate change agenda, press briefing 17, July 11.

Keohane, R. O. and Victor, D. G. (2011). The regime complex for climate change, *Perspectives on Politics,* **9**(1), 7–23.

Masiello, C. A. (2004). New directions in black carbon organic geochemistry, *Marine Chemistry,* **92**(1), 201–13.

Rutherford, D. (2016). *Brother, Can You Spare Three Cents (for the Climate)?* Washington, DC: ICCT.

RTCC (Responding to Climate Change). (2013). Norwegians issue shipping warming as Arctic sea lanes open. Available at www.rtcc.org.

Sandler, T. (1997). *Global Challenges: An Approach to Environmental, Political, and Economic Problems.* Cambridge, UK: Cambridge University Press.

Sims, R., Gorsevski, V., and Anenberg, S. (2015). *BC Mitigation and the Role of the Global Environment Facility: A STAP Advisory Document.* Washington, DC: Global Environment Facility.

Stettler, M.E.J., Eastham, S., and Barrett, S.R.H. (2011). Air quality and public health impacts of UK airports. Part I: Emissions. *Atmospheric Environment,* **45**(31), 5415–24.

Stettler, M.E.J, Eastham, S., and Barrett, S.R.H. (2013). Air quality and public health impacts of UK airports. Part II: Impacts and policy assessment. *Atmospheric Environment,* **67**, 184–92.

Stettler, M. E., Boies, A. M., Petzold, A., and Barrett, S. R. (2013). Global civil aviation BC emissions, *Environmental Science & Technology,* **47**(18), 10397–404.

Stuards, M. (2016). Artic shipping passage 'still decades away,' *The Guardian,* February 10.

UN Economic Commission for Europe (UNECE). (1979). *1979 Convention on Long-range Transboundary Air Pollution.* Geneva: UNECE.

UN Economic Commission for Europe (UNECE). (1999). *The 1999 Gothenburg Protocol to Abate Acidification, Eutrophication and Ground-level Ozone.* Geneva: UNECE.

UN Economic Commission for Europe (UNECE). (2013). *1999 Protocol to Abate Acidification, Eutrophication and Ground-level Ozone to the Convention on Long-range Transboundary Air Pollution, as Amended on 4 May 2012.* Geneva: UNECE.

UN Economic Commission for Europe (UNECE). (2015). *Protocol to Abate Acidification, Eutrophication and Ground-level Ozone.* Geneva: UNECE.

US Environmental Protection Agency (EPA). (2012). *Report to Congress on BC.* Washington, DC: US EPA.

US Environmental Protection Agency (EPA). (2015). *Ocean Vessels and Large Ships, 2015*. Washington, DC: US EPA.

US Environmental Protection Agency (EPA) (2016). *Methane and Black Carbon Impacts on the Arctic: Communicating the Science*. Washington, DC: US EPA.

US Environmental Protection Agency (EPA) (2017). *Mitigating BC*. Washington, DC: US EPA.

US National Academies of Sciences, Engineering, and Medicine, Committee on Propulsion and Energy Systems to Reduce Commercial Aviation Carbon Emissions. (2015). *Commercial Aircraft Propulsion and Energy Systems Research: Reducing Global Carbon Emissions*. Washington, DC: The National Academies Press.

US National Aeronautics and Space Administration (NASA). (2017). *End-of-Summer Artic Sea Ice Extent Is Eighth Lowest on Record*. Washington, DC: NASA.

US National Oceanic and Atmospheric Administration (NOAA). (2017). *Arctic Report Card*. Washington, DC: NOAA.

US National Snow and Ice Data Center (NSIDC). (2017). *Arctic Sea Ice News and Analysis*. Boulder, CO: NSIDC.

US White House (2013). *National Strategy for the Arctic Region*. Available at www .whitehouse.gov. [N.B. This document was produced by the Obama administration before the Trump administration entered office in January 2017.]

Victor, D. G. (2015a). *The Case for Climate Clubs*. E15 Initiative. Geneva: International Centre for Trade and Sustainable Development (ICTSD) and World Economic Forum.

Victor, D. G. (2015b). Join the club: Group approaches to tackling climate change. *BioRes*, **9**, 18–19.

Victor, D. (2017). *Three-Dimensional Climate Clubs: Implications for Climate Cooperation and the G20*. Geneva: International Centre for Trade and Sustainable Development (ICTSD).

Weischer L., Morgan, J., and Patel, M. (2012). Climate clubs: Can small groups of countries make a big difference in addressing climate change? *Review of European Community & International Environmental Law*, **21**(3), 177–92.

Yeo, S. (2015). How successfully can the Arctic Council tackle climate change? *The Carbon Brief, blog*, April 28.

Part V

Institutions and Law for Ocean Governance

16

Contested Multilateralism

Toward Aligning Regimes for Ocean and Climate Governance

REUBEN MAKOMERE AND KENNEDY LITI MBEVA

Introduction

Climate change has not only resulted in unprecedented challenges to global well-being but also has created and intensified environmental challenges. Critical to this is the role oceans play and the effects of climate change on oceans. As the links between climate and oceans become more complicated, primarily due to increasing scope and intensity of climate stressors on oceans (Brigg et al., 2013; Mendelsohn et al., 2012), contemporary ocean governance regimes have experienced a noticeable inertia Kim (2012). They have to maintain a status quo as the complexity of the challenges has steadily increased. This is emblematic of the broader challenge of gridlock in global governance, especially regarding "super-wicked" global challenges that exhibit complex interdependence (Keohane and Nye, 2011) and that require unprecedented global cooperation and collective action (Hale et al., 2013). That said, this has begun to change, albeit modestly, in the past few years. The adoption of the 2030 Agenda for Sustainable Development, the Sustainable Development Goals (SDGs), and the Paris Agreement on climate change were significant steps. Climate change and ocean governance were part of the SDGs. It is how these two issues have been interlinked within these and other multilateral frameworks that is the subject of this chapter.

In this chapter, we examine the methods exhibited in efforts to align the ocean governance regime, especially the United Nations Convention on the Law of the Sea (UNCLOS) (see Chapter 2), with the international climate governance regime. We do this in light of the 2015 Paris Agreement. We examine linkage politics as catalysts for coevolving governance. We look at how and why states have introduced ocean governance issues into the international climate regime through their Nationally Determined Contributions (NDCs) under the Paris Agreement. States have increasingly linked climate impacts to ocean challenges, and specifically included this linkage in their NDCs as part of their efforts to adapt governance responses to the complex interactions between climatic and ocean systems. We refer to this approach as "signaling" and argue that states seek to draw attention to the need to adjust the climate and ocean regimes in response to these complex interactions.

Interplay between Regimes for Climate and Ocean Governance

The climate regime has undergone noticeable changes, as illustrated by the 2015 Paris Agreement. That agreement not only signaled a major political milestone but also a significant shift in the legal character of global climate governance responses. In a sense, the agreement was a watershed step toward adopting institutional forms that allow for broad and flexible responses to global climate change challenges. This is because the breadth of climate challenges could not be addressed by a singular approach or by a select set of actors (Keohane and Victor, 2016). The flexibility is exemplified in the agreement's key feature: self-differentiated NDCs. These NDCs signify a more bottom-up approach to climate change governance. The Paris Agreement therefore has been argued to have demonstrated experimentalist forms of governance in an increasingly pluralistic international order, which is an evolution from the more integrated forms of international regimes (Ovodenko, 2014, 2017). The ocean governance regime, in contrast, particularly epitomized by the UNCLOS (see Chapter 2), draws from the early days of international environmental law. It has not significantly changed over time.

Furthermore, the dynamics of interlinkages regarding the ocean and climate regimes remain mostly informal, underexplored, and fragmented (Joyner, 2005; Carlarne, 2010). This has resulted in gaps in treaty application, governance, and enforcement when responding to complex environmental challenges (Rayfuse, 2010). Increasing complexity (Kim, 2013) has highlighted the importance of cross-regime linkages in ways that enhance synergy. Carlarne (2010) articulated this succinctly by noting that "it will be impossible for policy-makers to make well-informed decisions about ocean or climate management without further research into the intimate relationship between the oceans and the climate and improved communication between international ocean and climate regime" (Carlarne, 2010: 278). In the remainder of this chapter we take up this research task by focusing on the Paris Agreement's interaction with ocean issues and the dynamics of associated linkages in attempting to connect the climate regime to the ocean regime, as typified by UNCLOS.

Research Design

Theoretical Framework

We adopt *regime interplay* as the overarching theoretical framework for this chapter. Regimes are "sets of implicit or explicit principles, norms, rules, and decision-making procedures around which actors' expectations converge in a given area of international relations" (Krasner, 1982: 185). Interplay refers to the "full range of interaction across institutional boundaries, whether it concerns substantive international rules, decision-making processes, or related administrative structures (secretariats, other bodies)" (Oberthür and Stokke, 2011: 4). Simply put, regime interplay concerns the interaction of regimes. It is, however, critical to note that as regimes overlap, the majority of regime interplays form continuous relationships (Dunoff, 2016). Different regimes have diverse approaches to governance. As a result, regime overlaps, with various rules for common issues, have emerged from international environmental governance. Additionally, diverse regimes look toward normalizing their rationale in a contest among themselves. This contestation extends to their overriding objectives, goals, and interests (Blome et al., 2016). Blome et al. (2016) posit that these contestations are not only reflected in systems theory but are also recognized in regimes theory, which looks at, among other things, overlaps of different regimes.

It is noteworthy that while these overlaps exist, translation into formal and concrete institutional linkages is yet to be optimized, with the regimes remaining fragmented and having peripheral regime interaction, particularly between the ocean and climate regimes (Joyner, 2005; Carlarne, 2010). However, there have been noticeable attempts to address this and create more intimate relationships between climate and ocean regimes through openings created by frameworks such as the SDGs (UN, 2015). Furthermore, innovations in international environmental governance, including the Paris Agreement's NDCs, have enhanced the space for exploring ways to enhance institutional linkages. We take this theoretical departure point as it examines linkage politics of regime interaction in the context of the Paris Agreement and ocean regime. We consider why and how states have interacted with these regimes in a bid to respond to complex interactions between climate and ocean systems.

Conceptual Framework

Here we adopt "contested multilateralism" as the overarching conceptual framework. Contested multilateralism "occurs when states and/or non-state actors either shift their focus from one existing institution to another or create an alternative multilateral institution to compete with existing ones" (Morse and Keohane, 2014: 387). Contested multilateralism comprises two key features: regime shifting and competitive regime creation. It is defined by three criteria:

- A multilateral institution exists within a defined issue area and with a mission and a set of established rules and institutionalized practices.
- Dissatisfied with the status quo institution, a coalition of actors – whether members of the existing institution or not – shift the focus of their activity to a challenging institution with different rules and practices. This challenging institution can be either preexisting or new.
- The rules and institutionalized practices of the challenging institution conflict with or significantly modify the rules and institutionalized practices of the status quo institution (Morse and Keohane, 2014: 388).

Contested multilateralism has been used for various studies of world politics, such as global constitutionalism (Follesdal, 2016), international development in the energy sector (Hannam, 2016), and explaining China's Asian Infrastructure Investment Bank as a challenge to the postwar multilateral finance and development framework (Chen and Liu, 2017).

We contend that there is a coalition of states that has attempted to catalyze changes in the ocean regime, especially with regard to UNCLOS, that are in line with the interests of this coalition. However, we note that although this may not be the only motivation or coalition geared toward responding to inertia, contestation by this coalition has played an important role in marshaling responses in the ocean regime. The contesting coalition has thus utilized the more bottom-up approach of the Paris Agreement to introduce and give prominence to ocean issues that are in line with its interests. This coalition has also signaled its intention to not only open up avenues that respond to these interests but also to enhance the political profile of these issues and interests. This is especially the case when considering the inertia in the ocean regime.

The coalition that we focus on in this chapter comprises the least developed countries (LDCs) and the small-island developing states (SIDS), which are most vulnerable to climate change and have limited capacity to adapt to climate-induced ocean impacts such as sea-level rise and coral bleaching. It is critical to note that SIDS have been one of the more consistent and prominent *demandeurs* addressing the climate–oceans nexus via the United Nations Framework Convention on Climate Change (UNFCCC), for instance, by pushing for the introduction of a loss-and-damage mechanism (Barnett and Campbell, 2010). Here we are interested in how contested multilateralism has been exhibited as these states have attempted to catalyze changes to UNCLOS in particular and the ocean regime in a broad sense, thereby attempting to enhance interaction and alignment between the climate and ocean regimes.

Toward this end, we seek to answer the following questions:

1) Which states, as a contesting coalition, have strongly highlighted ocean issues into the Paris climate regime?
2) What are the ocean issues being raised by these states?
3) What are the changes in the challenged regime that this contesting coalition is involved in, and how does this relate to the issues they have highlighted in the Paris Agreement?

We analyzed 127 NDCs submitted by states that are party to the UNFCCC. We further linked this analysis to the role that the contesting coalition is playing in marshaling changes within the ocean regime under UNCLOS. This is explored in the context of the realization of the Sustainable Development Goals, particularly SDG 14 (see also Chapter 19).

Hypotheses

We considered a contesting coalition which has signaled its intention to not only open up avenues that respond to interests of the members but also enhance the political profile of these issues and interests. We tested three hypotheses:

(1) The contesting coalition comprises states that are highly vulnerable to ocean-related impacts and have limited adaptation capacity. We considered and tested other explanations as to whether the current ocean regime is sufficient for

addressing the climate-induced ocean challenges. Some works have argued that the current international regime for oceans might be adequate for addressing these challenges (Harrison, 2017). Others have suggested that there is a need for creating a new multilateral regime that responds to climate-induced ocean challenges, such as creating a multilateral environmental agreement on ocean acidification (Kim, 2012).

(2) States that are members of ocean governance treaties and institutions are less likely to demand new regimes. Issues of equity, justice, and fairness are a central feature of many international regimes in world politics, and they have been particularly salient in the climate regime. The primary approach of dealing with these issues – burden sharing in collective action – under the climate regime is through the notion of "common but differentiated responsibilities and respective capabilities." This is one of the pillars of the climate regime. It has been operationalized in practice through various means, such as climate finance mechanisms (Mbeva and Pauw, 2016). We would expect the contesting coalitions to leverage the dynamism of the climate regime to mobilize resources to address the ocean-related climate impacts.

(3) Contesting coalitions leverage the climate regime's funding mechanisms to mobilize resources to address ocean-related climate impacts. (See the appendix to this chapter for details on the methodology used in our analysis.)

Signaling and Contested Multilateralism

There is a contesting coalition of states, comprising LDCs and SIDS, seeking greater links between the ocean and climate regimes, through the inclusion of ocean issues in their NDCs. States with NDCs that include an adaptation component, and a request for support in implementing their NDCs, are very likely to be in this contesting coalition. However, there is no universal process for formalizing interactions between the ocean and climate regimes. This indicates that there is a likelihood of the existence of contestation within parties of both regimes with regard to the politics of formal regime interactions.

A qualitative analysis of the specific NDCs with ocean-related issues highlights what the contesting coalition has sought to advance. This is shown in Figure 16.1.

The three key areas that dominate the ocean-related issues – the marine ecosystem, sea-level rise, and extreme weather/disasters – clearly underscore climate-induced ocean challenges. This result also supports our proposition that the contesting coalition sees the climate regime as increasingly important to addressing these issues. In doing this, the contesting coalition contributes toward highlighting the demand for an avenue, or complex of avenues, which would better to respond to the challenges that are amplified by interactions between climate and ocean systems.

The adoption and ratification of the Paris Agreement provided a much-needed catalyst to overcome the gridlock that has been pervasive in international climate

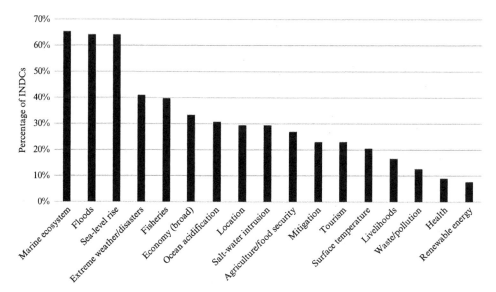

Fig. 16.1.   Percentage of NDCs that include specific ocean and/or marine elements
Source: Authors

governance. It is the innovative "bottom-up" approach in the architecture of the
Paris Agreement, where states submitted their national climate action plans (the
NDCs) that helped expand the window for states dissatisfied with inertia in
the ocean regime (Keohane and Victor, 2016). This window has potentially enabled
states to not only highlight the nexus of ocean and climate issues in governance but
also to raise ocean issues in a regime that might seemingly be more in line with their
strategic interests. The shifting observed here therefore is one of strategy, whereby
the contesting coalitions introduce issues from one type of regime into another that
is better aligned with their aims and interests.

   This aspect of contestation is also manifest because the contesting coalition in
this instance comprises states that have been strongly advocating for the institutionalized linkage of climate and ocean issues most important to them. This has been
through advocating for methods, including the loss-and-damage mechanism in the
Paris Agreement, despite reluctance from other states outside this contesting coalition. This reluctance has been based on, among other things, provisions on legal liability or rights to compensation in this mechanism (Falkner, 2016).

Contesting Coalitions as Catalysts: Signaling in Contested Multilateralism

The question that therefore arises is this: What did the contesting coalition want? An
analysis of the specific contents and contexts of the ocean-related issues that states
within the coalition included in their NDCs indicates that the underlying logic was
that of strategic communication, as shown in Figure 16.2. This is a concept we see as
signaling, akin to the job-market signaling theory developed by Spence (1973).

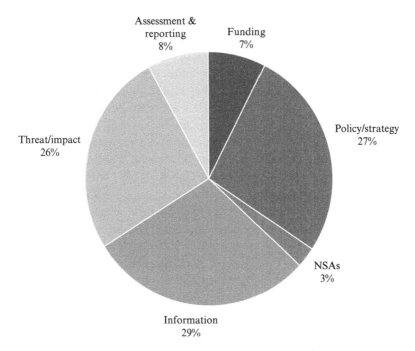

Fig. 16.2.   The components of the "signal" by the contesting coalition
Note: The bulk of the signal is communicating the impacts of climate change on ocean and marine ecosystems. NSAs = Nonstate actors
Source: Authors

The focus was on strategically highlighting the link between climate change and ocean-related issues and how this affects the contesting coalition. Crucially, this made up more than 50 percent of the communication. Accordingly, we argue that the contesting coalition's main motivation for including the ocean-related issues was to highlight the nexus of both the issues addressed by the climate and ocean governance regimes and their response strategies. This is particularly the case because the international climate regime is now in the arena of "high-politics" (Motaal, 2010; Vogler, 2015: Chapter 7). It is striking that only a few members of the contesting coalition's members make an explicit request for financial resources to address the ocean-related issues raised in the contributions.

We believe that the underlying logic for the focus on communicating the climate–oceans nexus is enhancing its political profile in relation to the ocean regime and, in the broader sense, responding to the complex interlinkages between the climate and ocean systems. The three necessary and sufficient conditions for signaling, all of which are evident, are (1) dissatisfaction with one of the regimes, (2) a contesting coalition, and (3) architecture of the "co-opting" regime that allows "shifting" of issues from one regime to another. The signal is also broad enough to outline the interest in the issues being advanced by the contesting coalition, which was done without making specific demands. This was in regard to support or development of specific mechanisms and institutions.

It is imperative to note that the lack of explicit funding requests for ocean-related issues does not imply that the contesting coalition does not intend to seek financial support. Our empirical analysis indicates that states that are likely to request financial support for implementation of their NDCs are more likely to be in this contesting coalition. This is an especially interesting finding, as these states seemingly have broader and higher interests tied to seeking a response to the climate–ocean related issue, not just in funding for implementation of their NDCs. We argue that the diffuse and diverse nature of the international climate regime complex (Keohane and Victor, 2011) makes it conducive to this result because it comprises funding mechanisms and agencies, such as the UNFCCC's financial mechanisms, multilateral funding agencies such as the World Bank, and regional and bilateral funding agencies. Furthermore, the pledge of mobilizing up to US$100 billion per year by 2020 to assist poorer states in the context of the Paris Agreement makes the international climate regime an important entry point for states seeking to optimize these resources. This aspect of multilateralism is part of broader efforts to address challenges exacerbated by the climate and oceans nexus, as the next section outlines (see Harrison, 2017).

This occurred despite diplomatic and political efforts to address the impact of climate change on oceans, as well as broader ocean governance, lagging behind even as significant scientific evidence continues to come to the fore (Henson et al., 2017). This lag has been prevalent against the backdrop of the 2030 Agenda for Sustainable Development, adopted in 2015. The Agenda was a global plan of action by states under the United Nations. It includes 17 Sustainable Development Goals, including SDG 14 – "Life below Water" – which makes a link to climate change in its objective to "Conserve and sustainably use the oceans, seas and marine resources for sustainable development" (UN, 2017a: 1). It notes that "[t]he increasingly adverse impacts of climate change, overfishing, ocean acidification and marine pollution are jeopardizing recent gains in protecting portions of the world's oceans" (UN, 2017a: 1). SDG 14 identifies UNCLOS as the overarching multilateral framework within which to address ocean governance issues (see Chapters 2 and 19).

Perhaps the most revelatory developments in terms of corresponding changes in the ocean governance regime were during the June 2017 United Nations Conference to Support the Implementation of Sustainable Development Goal 14 (Oceans Conference). This was a high-level event whose objective was to support the implementation of SDG 14. The conference had participants from most states, including approximately "4,000 delegates, including 16 Heads of State or Government, two deputy Prime Ministers, 86 Ministers, 16 Vice Ministers, and other government representatives; participants from the UN system, other intergovernmental organizations, international and regional financial institutions, civil society, academic and research institutions, indigenous peoples, local communities, and the private sector" (IISD, 2017: 1). SIDS and LDCs played a significant role in pushing for strong action toward this objective, with Fiji cohosting the meeting, despite its enormous responsibility of presiding over the 2017 UNFCCC climate change negotiations in Bonn, Germany (COP 23).

The Oceans Conference outcomes included an intergovernmental "Call to Action," a Registry of Voluntary Commitments for the implementation of SDG 14, and conclusion of key "Partnership Dialogues." These outcomes further underscored the salience of the climate–oceans nexus in governance responses to challenges brought on by complex interactions between the two major natural systems. States noted that they were "particularly alarmed by the adverse impacts of climate change on the ocean, including the rise in ocean temperatures, ocean and coastal acidification, de-oxygenation, sea level rise, the decrease in polar ice coverage, coastal erosion and extreme weather events" (UN, 2017a: 2). The "Call to Action" made particular linkages to especially the Paris Agreement as being critical in realizing some of the key objectives as set out in SDG 14 (UN, 2017a: Para 4). This was further underscored by strong linkages with other processes, including the Addis Ababa Action agenda on financing development. Further, states committed themselves to the ongoing process of crafting an international, legally binding instrument regarding marine biodiversity in areas beyond national jurisdiction.

The Oceans Conference was indicative of not only an evolving regime complex for ocean governance but also a seemingly conscious effort to align the climate change regime with the ocean governance regime, in the process noting the salience of complex linkages between climate and ocean systems.

Competitive Regime Creation in Contested Multilateralism

Regime creation does not necessarily have to undermine the "challenged" regime but can also seek to alter or modify existing mechanisms. Toward this end, the United Nations Preparatory Committee was tasked with making substantive recommendations toward a new legal instrument under UNCLOS. The committee commenced its work, pursuant to Resolution 69/292 of the General Assembly, in 2016 and submitted its report to the General Assembly in 2017. This was to be done, taking into account the reports of the "Co-Chairs on the work of the Ad Hoc Open-ended Informal Working Group to look into issues relating to the conservation and sustainable use of marine biological diversity beyond areas of national jurisdiction" (UN, 2017b). The committee was also tasked with making decisions on the commencement of an intergovernmental forum, under the auspices of the United Nations, which would look into the committee's recommendations and elaborate elements of the international, legally binding instrument's text. The committee was also to do its work on the basis of it securing, to the furthest extent possible, the broadest approval. This was critical to establishing the legitimacy of the process and the outcome, in addition to incentivize participation. To this end, the committee opened itself to participation from member states of the United Nations, actors from specialized members of the UN, and parties to UNCLOS.

Other actors were invited as observers, in keeping with the traditions of the United Nations (UN, 2017b).

Crucially, the committee's recommendations were divided into two main sections: (1) nonexclusive elements that had the most convergence and (2) those issues that were divergent. Preambular recommendations contained key contextual issues, including the reaffirmation of the status of UNCLOS and the importance of "other existing relevant legal instruments and frameworks" insofar as realizing the mandate of the new legal instrument is concerned. Critically, among the submissions was recognition that low-income developing states (i.e., the LDCs), geographically disadvantaged states, landlocked developing states and SIDS, and coastal African states had an important role to play in the conservation and sustainable use of marine biological diversity in areas beyond national jurisdictions (see Chapter 11). Further, there was the affirmation that a better, more comprehensive global regime was needed in order to address the conservation and sustainable use of marine biodiversity in those areas (UN, 2017b).

The committee recommended that the new legal instrument would look toward fitting into existing relevant regime complexes, thereby promoting coherence and complementarity between the legal instruments, frameworks, and institutions concerned. Toward this end, interpretation and application of the legal instrument was to be done in a way that did not undercut other instruments, frameworks, and bodies (UN, 2017b). Key submissions were also made with regard to the general principles that would guide the legal instrument. Principles included the precautionary principle, resilience to the effects of climate change, and the polluter pays principle, and due consideration of the special requirements of SIDS and least developed states was included (UN, 2017b). Crucially one of the stand-out inclusions was the prevention of transferring, directly or indirectly, a disproportionate and disadvantaged burden of conservation action onto developing states (UN, 2017b). This was an important outcome that is representative of some of the core arguments that these states have always made insofar as international environmental governance is concerned.

The contesting coalition has played a noticeable role in this ongoing process toward the establishment of a new legal order, attempting to align this with their interests. The coalition seeks a mandate that is more consistent with its members' interests. In this regard, we are in consonance with the postulation that contesting coalitions seek to not only respond to the institutional status quo but also attempt to optimize their bargaining leverage in relation to the status quo institution (Morse and Keohane, 2014). This is a striking development since there exists already a myriad of ocean and marine-related international instruments that have been in place since the early days of international environmental governance (Joyner, 2005; Carlarne, 2010). This is further buttressed with the affirmation of the supremacy of UNCLOS as the overarching legal instrument for international oceans governance (see Chapter 19). This affirmation was exhibited by significant attention being devoted to "enhancing the conservation and sustainable use of oceans and their resources by implementing international law as reflected in UNCLOS" (UN, 2017a: Para 11).

There have been numerous situations where a coalition of states has led efforts to alter or create new multilateral institutions that have an influence in the mission or rules of existing institutions (see Morse and Keohane, 2014). This has been critical to catalyzing alignment between not just these institutions and members of these coalitions but also between overlapping institutions. Betts (2009) identifies efforts by coalitions of states to alter existing institutions while examining migration, noting how developed states have tried to respond to obligations on acceptance of migrants as refugees. Toward this end, these states, as a contesting coalition, sought to establish a regime complex that would cap the scope of their obligations, as enshrined in the 1951 UN refugee convention and practices of the UN High Commissioner on Human Rights (Betts, 2009). Along the same wavelength, the 2000 Cartagena Protocol on Biosafety's disruptive effect on World Trade Organization (WTO)–based governance of genetically modified organisms (GMOs) has also been identified as an example of state-led competitive regime creation (Morse and Keohane, 2014). Developing states, as the contesting coalition, sought to respond to the WTO-based governance so as to safeguard their autonomy with regard to regulation of GMOs. The contesting coalition thus took center stage in negotiating the Cartagena Protocol on Biosafety (Oberthur and Gehring, 2006).

Morse and Keohane (2014) have also identified similar manifestations of contested multilateralism in the establishment of the Proliferation Security Initiative, which was designed to disrupt UNCLOS's common interdiction processes and practices so as to identify illicit shipments of ballistic missile technology. The contesting coalition, led by the United States, sought to respond to UNCLOS's mandate pertaining to search and seizure in the high seas. The contesting coalition argued that UNCLOS did not adequately capture and respond to the coalition's interest on search and seizure on the high seas, particularly with regard to weapons of mass destruction (Morse and Keohane, 2014). Kraska (2012) notes that the United States did not try to undermine the challenged regime (UNCLOS). Instead, the contesting states sought to carve out a space within this existing regime that would also protect their interests. This demand was made difficult by UNCLOS's inability to be flexible enough to avoid contestation (Kraska, 2012). The creation of a new operational norm culminated in the formation of the informal Proliferation Security Initiative. This initiative was instrumental in catalyzing changes in the global institutional and legal responses to proliferation of weapons of mass destruction, culminating in Resolution 1540 of the United Nations Security Council in 2004, which contained key elements that essentially moved the existing regime toward the initiative's objectives (UN, 2004; Morse and Keohane, 2014). The intervention thus not only expanded the regime complex on proliferation of weapons of mass destruction but also altered the procedures of interdiction at sea (Morse and Keohane, 2014).

Contested Multilateralism in Responding to Gridlock

Findings and insights from our analysis potentially hold significant promise for understanding linkage and contestation between regimes. Growing complexity from interlinkage of issues has led to gridlock in many areas of world politics (Hale et al., 2013). This is especially the case when the "spillovers" of the regimes are negative, as can happen as a result of the impacts of climate change on oceans. With increased interdependence, Hale et al. (2013) have argued that the demand for international cooperation is at its peak. However, the institutional configuration for state cooperation may not be able to effectively respond to this demand. This exacerbates inertia and the status quo insofar as institutionalization of state cooperation is concerned. In global environmental politics, the role of institutional inertia and multipolarity has been highlighted, especially in terms of contributing to gridlock in environmental governance. Gridlock mechanisms not only forestall multilateral responses but also compound each other, further entrenching the status quo and frustrating cooperation with regard to important environmental issues (Hale et al., 2013). This is especially the case where more complex environmental concerns, including climate change and ocean issues, are involved because they require far deeper degrees of cooperation for alignment.

Toward this end, contesting coalitions thus become critical components in forging responses, thus opening up spaces for untangling cooperation and alignment in global governance. Our analysis of climate change and ocean governance points to the presence of contested multilateralism in linkage and alignment politics. Contested multilateralism has been an avenue through which innovations in one regime (e.g., the climate change regime (Hale, 2017)) catalyze action in a "dormant" regime (e.g., the ocean regime), thus responding to gridlock in global environmental governance and fostering alignment. This is particularly so with regard to the ocean and climate regimes. By challenging seemingly inert or status quo regimes, contesting coalitions catalyze disruptions that enhance responses that are in consonance with the interests of these coalitions – interests that might have been inadequately addressed in the existing "dormant" regime. Further, our case study shows that coalitions have played an important role in signaling the demand for alignment of climate change and ocean issues.

Conclusion

This chapter has explored methods through which global environmental institutions organize in order to govern interactions between the climate and ocean systems. We have looked at pathways through which states structure themselves in order to respond to gridlock in global governance, and thus possibly catalyze alignment. We have discussed some of the dynamics in the processes of alignment, showing that contested multilateralism is present within linkage politics in global governance, specifically climate governance and ocean governance. We show that the two necessary and sufficient conditions for contested multilateralism are (1) regime shifting and (2) competitive regime creation. Low-income developing countries and small-island developing states have formed a contesting coalition which has attempted to optimize the innovative climate regime to further catalyze action in the oceans governance regime through high-level signaling. States in this contesting coalition included ocean issues in their Paris Agreement NDCs as a key component in realizing the conditions for contested multilateralism. We conclude that vulnerability by itself is a necessary, but insufficient, condition for contested multilateralism, but when considered with inadequate adaptive capacity, in this case, the conditions are met.

Importantly, we do not conclude that movements in the status quo ocean regime are solely credited to the LDC–SID coalition, as many other conditions are at play that help generate enough momentum for alignment. However, the contesting coalition has potentially enhanced its influence in marshaling action toward responding to inertia, thus moving toward alignment. This occurs through signaling, regime shifting, and competitive regime creation. This has contributed to enhanced efforts for the creation of a new international legal instrument on marine biodiversity beyond national jurisdictions. We note that important climate change issues, central to the contesting coalition, have indeed been captured within the broad contours of this new legal instrument, including shaping its organizing principles. Noticeably, this new international instrument is envisioned to be under the dominant ocean regime: UNCLOS. This arrangement is designed to avoid undermining the dominant regime because that regime is still of great strategic interest to states involved in the process. This is a strong indicator of the manifestation of contested multilateralism, exemplified by outcomes of the UN Conference to Support the Implementation of Sustainable Development Goal 14 and the Preparatory Committee set up by the General Assembly.

Our analysis of the attempts to align the ocean and climate regimes illustrates promising avenues for future research. These include looking at the response by other institutions, coalitions, and instruments of contestation that can lead up to alignment and cooperation. Another area of research includes looking at how actors and agency dynamics influence the configuration of linkage politics and coalitions, especially in international environmental governance in the context of contested multilateralism. What comes from this is the important message that the apparently separate climate and ocean regimes are nothing but, and that efforts are underway to bring them into alignment for more effective ocean governance amidst climate change.

Acknowledgments

We acknowledge comments and insights from Sebastian Rattansen, Noah Sitati, and Jeffrey McGee on an initial draft of this chapter. The first author acknowledges support from the University of Tasmania, Faculty of Law, and the Research Training Program Scholarship administered by the University of Tasmania and the Commonwealth Department of Education. The second author acknowledges support from the Faculty of Arts, University of Melbourne, through the Melbourne Research Scholarship and the Australian-German Climate and Energy College. The authors take full responsibility for the chapter's contents.

References

Betts, A. (2009). *Forced Migration and Global Politics*, 1st edn, Chichester, U.K.; Malden, MA: Wiley-Blackwell.

Brigg, G. R., Jickells, T. D., Liss, P. S., and Osborn, T. J. (2003). The role of the oceans in climate. *International Journal of Climatology*, **23**(10), 1127–59. Available at http://onlinelibrary.wiley.com/doi/10.1002/joc.926/pdf

Barnett, J. and Campbell, J. (2010). *Climate Change and Small Island States: Power, Knowledge, and the South Pacific*. [online]. London; Washington, DC: Earthscan. Available at https://ezp.lib.unimelb.edu.au/login?url=https://search.ebscohost.com/login.aspx?direct=true&db=cat00006a&AN=melb.b3525449&site=eds-live&scope=site

Blome, K., Fischer-Lescano, A., Franzki, H., Markard, N., and Oeter, S. eds. (2016). *Contested Regime Collisions: Norm Fragmentation in World Society* [online]. Cambridge, UK: Cambridge University Press. Available at www.cambridge.org/core/books/contested-regime-collisions/BF1B4A812026925B08D69E2659342E0A.

Carlarne, C. (2010). Climate change, cultural heritage & the oceans: Rethinking regulatory approaches to climate change. *Southeastern Environmental Law Journal*, **17**(2), 272.

Chen, Z. and Liu, Y. (2017). Granting reassurance while posing challenge: Explaining China's creation of the Asian Infrastructure Investment Bank. Global Economic Governance Working Paper No. 130.

Dunoff, J. L. (2016). How to avoid regime collisions. In: K. Blome, A. Fischer-Lescano, H. Franzki, N. Markard, and S. Oeter, eds., *Contested Regime Collisions: Norm Fragmentation in World Society*. [online]. Cambridge, UK: Cambridge University Press. Available at https://ezp.lib.unimelb.edu.au/login?url=https://search.ebscohost.com/login.aspx?direct=true&db=cat00006a&AN=melb.b6163395&site=eds-live&scope=site

Falkner, R. (2016). The Paris Agreement and the new logic of international climate politics. *International Affairs*, **92**(5), 1107–25.

Follesdal, A. (2016). Implications of contested multilateralism for global constitutionalism. *Global Constitutionalism*, **5**(3), 297–308.

Hale, T. (2017). Climate Change: From Gridlock to Catalyst. In T. Hale and D. Held, eds., *Beyond Gridlock*. Cambridge: Polity Press.

Hale, T., Held, D., and Young, K. (2013). *Gridlock: Why Global Cooperation Is Failing When We Need It Most*. Cambridge, UK and Malden, MA: Polity Press.

Hannam, P. M. (2016). *Contesting Authority. China and the New Landscape of Power Sector Governance in the Developing World*. PhD thesis. [online]. Princeton University. Available at https://media.proquest.com/media/pq/classic/doc/4265622701/fmt/ai/rep/NPDF?_s=OcKkwi5OYJeUDGeKwvaToCUtnxw%3D

Harrison, J. (2017). *Saving the Oceans through Law: The International Legal Framework for the Protection of the Marine Environment*. Oxford: Oxford University Press.

Henson, S. A., Beaulieu, C., Ilyina, T., et al. (2017). Rapid emergence of climate change in environmental drivers of marine ecosystems. *Nature Communications*, **8**, 1–9.

IISD. (2017). Summary of the Ocean Conference: June 5–9, 2017. New York, IISD.

Joyner, C. (2005). *International Law in the 21st Century: Rules for Global Governance*. New York, Oxford: Rowman & Littlefield Publishers.

Keohane, R. O. and Nye, J. S. (2011). *Power and Interdependence* [online]. Longman. Available at https://books.google.com.au/books?id=kt3QtgAACAAJ

Keohane, R. O. and Victor, D. G. (2011). The regime complex for climate change. *Perspectives on Politics*, **9**(1), 7–23.

Keohane, R. O. and Victor, D. G. (2016). Cooperation and discord in global climate policy. *Nature Climate Change*, **6**(6), 570–75. Available at www.nature.com/natureclimatechange

Kim, R. E. (2012). Is a new multilateral environmental agreement on ocean acidification necessary? *Review of European Community & International Environmental Law*, **21**(3), 243–58.

Kim, R. E. (2013). The emergent network structure of the multilateral environmental agreement system. *Global Environmental Change*, **23**(5), 980–91.

Kraska, J. (2012). Using UNCLOS as a force multiplier for American power. *Opinio Juris*. Available at http://opiniojuris.org/2012/06/14/using-unclos-as-a-force-multiplier-for-americanpower/.

Krasner, S. D. (1982). Structural causes and regime consequences: Regimes as intervening variables. *International Organization*, **36**(2), 185–205.

Mbeva, K. and Pauw, P. (2016). Self-differentiation of countries' responsibilities. Addressing climate change through intended nationally determined contributions [online]. *German Development Institute*. Bonn, Germany: Deutsches Institut für Entwicklungspolitik (DIE). Discussion Paper No. 4/2016. Available at www.die-gdi.de/uploads/media/DP_4.2016.pdf.

Mendelsohn, R., Emanuel, K., Chonabayashi, S., and Bakkensen, L. (2012). The impact of climate change on global tropical cyclone damage. *Nature Climate Change*, **2**(3), 205–9.

Morse, J. C. and Keohane, R. O. (2014). Contested multilateralism. *The Review of International Organizations*, **9**(4), 385–412.

Motaal, D. A. (2010). The shift from "low politics" to "high politics" – Climate change. *Environmental Policy and Law*, **40**(2–3), 98–109.

Oberthür, S. and Gehring, T. (2006). *Institutional interaction in global environmental governance: synergy and conflict among international and EU policies*. [online]. Cambridge, Mass.; London: MIT, 2006.

Oberthür, S. and Stokke, O. S. eds. (2011). Institutional interaction in global environmental change. In *Managing Institutional Complexity. [Electronic Resource]: Regime Interplay and Global Environmental Change*. [online]. Cambridge, MA: MIT Press. Available at https://ezp.lib.unimelb.edu.au/login?url=https://search.ebscohost.com/login.aspx?direct=true&db=cat00006a&AN=melb.b5784766&site=eds-live&scope=site

Ovodenko, A. (2014). The global climate regime: Explaining lagging reform. *Review of Policy Research*, **31**(3), 173–98.

Ovodenko, A. (2017). *Regulating the Polluters Markets and Strategies for Protecting the Global Environment*. Oxford: Oxford University Press.

Rayfuse, R. (2010). The anthropocene, autopoiesis and the disingenuousness of the genuine link: Addressing enforcement gaps in the legal regime for areas beyond national jurisdiction. In Elferink, A. G. and Molenaar, E. J., eds. *The Legal Regime of Areas beyond National Jurisdiction: Current Principles and Frameworks and Future Directions*. Leiden; Boston, MA: Martinus Nijhoff, 165–90.

Spence, M. (1973). Job market signaling. *The Quarterly Journal of Economics*, **87**(3), 355–74.

UN. (2017a). *Our Ocean, Our Future Call for Action*. New York: United Nations.

UN. (2017b). *Report of the Preparatory Committee Established by General Assembly Resolution 69/292: Development of an International Legally Binding Instrument under the United Nations Convention on the Law of the Sea on the Conservation and Sustainable Use of Marine Biological Diversity of Areas Beyond National Jurisdiction*. New York: United Nations.

UN General Assembly. (2015). *Resolution 69/292. A/RES/69/292*. New York: United Nations.

UN Security Council. (2004). *Resolution 1540. S/RES/1540*. New York: United Nations.

Vogler, J. (2015). *Climate Change in World Politics*. Hampshire: Palgrave Macmillan UK.

Appendix

Data and Methods

A total of 127 NDCs were analyzed, with the EU considered as a single entity since it has submitted one NDC. Furthermore, only states that have a shoreline were considered, since they are exposed to the ocean and thus the most immediate ocean-related climate impacts. We undertook a binary logistic regression analysis to test our hypotheses, with the model estimated as follows:

$$\text{Signal}_i = \alpha + \beta_1 \, \text{Vulnerability Index}_i + \beta_2 \, \text{Adaptation Capacity}_i$$

$$+ \, \beta_3 \, \text{NDC Adaptation Fundingrequest}_i + \beta_4 \, \text{IMO}_i$$

$$+ \beta_5 \text{UNCLOS}_i + \beta_6 \text{SIDS}_i + \beta_7 \text{DevelopmentLevel}_i$$

$$+ \beta_8 \text{Population}_i + \beta_9 \text{GHG}_i + \beta_{10} \text{GDP}_i + \beta_{11} \text{GDPpercapita}_i + \mu_i \qquad (A16.1)$$

Data for vulnerability index and adaptation capacity were retrieved from the ND-GAIN Index database; the NDC adaptation funding request was coded from NDCs by the authors, retrieved from the UNFCCC online registry; IMO and UNCLOS state membership from the relevant institution's website; SIDS membership and states' levels of development from UNFCCC website; and population, GHG, GDP, and GDP data from the World Bank Development Indicators database. u_i captures unobserved effects in the model. Table A16.1 presents summary statistics of these variables.

Table A16.1. *Summary statistics of variables*

Variable	Observations	Min	Max	Median	Standard Deviation	Data Source
Includes ocean issues in NDC?	127					NDC
Vulnerability to climate change (index)	127	0.2310	0.6800	0.4200	0.1009162	ND-GAIN
Adaptation capacity (index)	127	0.1010	0.9500	0.4800	0.1740752	ND-GAIN
Adaptation funding requested in NDC?	127					NDC
Member of International Maritime Organization	127					IMO Website
Ratified the UN Convention on Law of the Sea	127					UNCLOS Website
Is a Small-Island Developing State?	127					UNFCCC
Level of development (Developed, developing or least developed state)	127					UNFCCC
Has climate policy?	127					GLOBE Database
Population	127	1624	1403500365	7606374	175132715	World Bank Development Indicators
Greenhouse Gas Emissions (MtCO2e)	127	0.005	11735.007	27.097	1218.324	World Bank Development Indicators
Size of economy (GDP in million $)	127	20	18560100	574241	2427629	World Bank Development Indicators
Wealth (GDP per capita)	127	400	127523	11599	20669.22	World Bank Development Indicators

Note: Categorical variables not reported.

Source: The authors.

The dependent variable is whether a state includes ocean governance issues in its NDC. This is a concept we term signaling (Spence, 1973). This binary variable seeks to capture the coalition of states that are significantly pushing for linkage, hence contestation within the climate and ocean governance regimes. As per H1, we would expect the contesting coalition to comprise states that have expressed dissatisfaction with inertia in the ocean regime. Historically, this has been developing and least developed states, and in particular SIDS. We would therefore expect members of the contesting coalition to include ocean governance issues in their NDCs. States engaging in the climate regime have always been split into three main groups – Annex I (industrialized states), least developed countries (LDCs), and other developing and emerging states (developing states). The "development level" variable captures this aspect, thus disaggregating the states so as to test the first hypothesis, since Annex I states have mainly provided support, through obligation and volition, to the other groups. This would be the main explanatory variable for the first hypothesis; thus, the category of LDCs, developing states, and the SIDS variables should correlate positively with signaling.

Membership of UNCLOS and IMO is measured as the "IMO" and "UNCLOS" binary variables, which we test as per the second hypothesis. We also measure the leverage to mobilize resources by the contesting coalition, by analyzing whether members of the contesting coalitions have requested support to implement their NDC. We expect a positive and statistically significant correlation between the dependent variable and the "NDC Adaptation funding request" variable, as per the third hypothesis. We test for other possible explanations that may motivate the contesting coalition. These include whether a state has climate legislation/policy, its vulnerability, and adaptive capacity. We finally control for state-specific factors such as GDP, GDP per capita, greenhouse gas emissions, and population, which were logged before the analysis.

Robustness Tests

Several tests were undertaken to ascertain the robustness of the models used. First, an exploratory analysis was conducted to ensure that the data did not have any distortions that would significantly affect the model and subsequent interpretation of the results. Second, a stepwise regression, including both forward and backward directions – independently and in combination – was undertaken to establish the baseline model with the most significant variables. Fisher scoring iterations were also undertaken to complement the fitness of the models.

Results

The results of the regression analysis are presented in Table A16.2.

Table A16.2. *Regression results*

Variables	1	2	3	4	5	6
Vulnerability		1.5309	1.0580	−5.5544	−4.3093	−3.9432
		(5.1173)	(5.1215)	(5.9990)	(5.7847)	(5.7173)
Adaptation Capacity		0.2472	−0.1991	1.5011	1.0680	0.8161
		(2.9745)	(2.9259)	(3.1805)	(3.1228)	(3.0667)
Wealth (GDP per capita)	**0.9609*****	0.7137	**0.8925***	0.3695	0.5023	0.5001
	(0.3427)	(0.5184)	(0.5395)	(0.5966)	(0.5868)	(0.5765)
Adaptation funding requested in NDC	**0.9915***		**1.0722****			
	(0.5836)		(0.5447)			
Member of International Maritime Organization				1.1740		
				(1.5188)		
Ratified the UN Convention on Law of the Sea				0.8199		
				(0.6649)		
Size of economy (GDP)	**−0.6732***	**−0.7850***	**−0.8679***	−0.3059	−0.4236	−0.4095
	(0.3705)	(0.4589)	(0.4875)	(0.5350)	(0.5421)	(0.5290)
Population		0.2458	0.1172	−0.2917	−0.2312	−0.1806
		(0.4267)	(0.4499)	(0.5065)	(0.5060)	(0.4862)
Greenhouse gas emissions	**0.5956***	0.4168	0.4697	0.3761	0.4237	0.3732
	(0.3551)	(0.3430)	(0.3531)	(0.3467)	(0.3602)	(0.3471)
Developing states as contesting coalition	**1.7267***			**2.8050*****	**2.6886*****	**2.6194*****
	(0.9303)			(1.0374)	(1.0131)	(1.0026)
Least developed states as contesting coalition	**2.6932****			**3.8419*****	**3.7591*****	**3.6507*****
	(1.2395)			(1.4299)	(1.4065)	(1.3874)
Small-island developing state (SIDS) as contesting coalition	**1.2825***	**1.1880***				
	(0.6857)	(0.6511)				
Climate policy exists?					0.2881	
					(0.4686)	

Note: Standard errors are in parentheses and clustered by state. Statistical significance thresholds: * $p < 0.1$, ** $p < 0.05$, *** $p < 0.001$. Model 1: Baseline; Model 2: Membership in SIDS hypothesis; Model 3: Funding request hypothesis; Model 4: Membership in ocean regime hypothesis; Model 5: Climate policy hypothesis; Model 6: Contesting coalitions hypothesis. *Source*: The authors.

17

Climate Change in the Coral Triangle
Enabling Institutional Adaptive Capacity

PEDRO FIDELMAN

Introduction

Large-scale marine governance interventions have become a preferred approach to addressing contemporary pressing issues affecting oceans and coasts worldwide, for example, overexploitation of resources, decline in ecosystem conditions, biodiversity loss, and climate change (Fidelman et al., 2012; Alger and Dauvergne, 2017). Large-scale marine governance includes various approaches, such as those related to seascapes, ecoregions, large-marine ecosystems, and regional seas programs for conservation and fisheries management (see chapters in Parts II and III). Several factors seem to underpin the impetus for large-scale marine governance, including (1) the view that big problems require sizable solutions; (2) the recognition that external drivers of change, including climate change, often limit smaller-scale efforts; (3) a focus on representation and connectivity, given the continuous and incremental pressure on ecosystems; (4) the need to connect and manage ever-increasing numbers of actors in light of globalization; and (5) a need for streamlining resources with the aim of benefiting from economies of scale (Fidelman et al., 2012).

Large-scale marine governance interventions potentially can produce effects (both positive and negative) beyond their primary intended goals. In the context of climate change adaptation, Adger et al. (2011) stress that governance interventions can affect the ability of the systems in which these interventions are embedded to adapt to current and future changes. They demonstrate how several policy interventions, including coastal management in the United Kingdom, affect systems resilience. Fidelman et al. (2017) examine how co-management of coastal resources in Cambodia and Vietnam can facilitate and constrain adaptive capacity. If adaptive capacity to climate change is to be built and enhanced, it is critical to understand the implications of governance interventions.

This chapter examines the implications of large-scale marine governance for adaptive capacity in the context of the Coral Triangle Initiative (CTI). The CTI is an exemplar of the large-scale interventions mentioned earlier. It comprises an intergovernmental agreement involving Malaysia, the Philippines (see Chapter 6), Indonesia (see Chapter 7), Timor Leste, Papua New Guinea, and the Solomon Islands. Its aim is to reverse the degradation of coastal and marine environments and pursue a more sustainable use of coastal and marine resources in the Coral Triangle (CTI Secretariat, 2009). This chapter assesses governance institutions comprising the CTI in relation to their influence on key attributes of adaptive capacity. The chapter draws primarily on research undertaken by the author and his colleagues (see citations to the author's work later). The next section provides an overview of the concept of adaptive capacity and how it may be assessed in the context of governance institutions. It is followed by a description of the CTI and an assessment of its governance institutions. The final two sections discuss the findings and their implications for ocean governance amidst climate change.

Conceptual Background

Adaptation is an important societal response to climate change. It comprises measures to reduce expected adverse impacts or to take advantage of opportunities arising from change (Smit and Wandel, 2006). Critical to adaptation is the notion of adaptive capacity, specifically the preconditions that enable adaptation, which include social and physical resources, and the ability to mobilize such resources in anticipating or responding to change (Nelson et al., 2007; Engle, 2011). Adaptive capacity is therefore a significant property for fostering adaptation; the higher the adaptive capacity of a system, the more likely such a system will adapt (Engle, 2011). Typical determinants of adaptive capacity include information and technology, material resources and infrastructure, organization and social capital, political capital, and wealth and financial capital (see Eaking and Lemos, 2006; Engle and Lemos, 2010; Engle, 2011). Systems of formal rules and social norms that mediate human behavior and interaction, also known as institutions, are regarded as prevailing determinants (Adger et al., 2005; Nelson et al., 2007; Engle, 2011). Institutions regulate social practices, assign roles to participants in such practices, and guide interactions among occupants of relevant roles (Ostrom, 2005; Young, 2005). In this regard, marine governance can be defined in terms of institutions. For instance, marine governance comprises institutions defining the interests represented in such governance (e.g., resource users, government, and industry), their roles, and how they engage with other stakeholders in making decisions (e.g., developing a management plan) to address relevant problems in a given jurisdiction. In this context, institutions play a critical role in how individuals and organizations respond and adapt to climate change (Young, 2002; Ostrom, 2005; Gupta et al., 2010).

As determinants of adaptive capacity, institutions can be described in terms of six attributes. These are the ability of institutions to (1) encourage the involvement of a variety of actors, perspectives, and solutions; (2) enable actors to continuously learn and improve their institutions; (3) allow and motivate stakeholders to self-organize, design, and reform their institutions; (4) mobilize leadership qualities of social actors; (5) mobilize resources for decision making and implementation; and

(6) support principles of fair governance based on legitimacy, equity, responsiveness, and accountability. These attributes comprise the *Adaptive Capacity Wheel* (Gupta et al., 2010), which has been used to examine institutional adaptive capacity in different settings and sectors, for example, water management, coastal protection, agriculture, coastal resources co-management, regional planning and climate vulnerability, and national adaptation strategies (Munaretto and Klostermann, 2011; Termeer et al., 2011; Van den Brink et al., 2011; Bergsma et al., 2012; Grothmann et al., 2013; Grecksch, 2014; Van den Brink et al., 2014; Gupta et al., 2016; Fidelman et al., 2017). These attributes are described in detail in Table 17.1. They will be used to assess the implications of governance institutions pertaining to the CTI for adaptive capacity.

In addition, Fidelman et al. (2017) highlight the importance of the concept of institutional rules in understanding adaptive capacity. Such rules refer to seven categories into which institutions can be classified: position, boundary, choice, aggregation, information, scope, and payoff rules (Ostrom and Crawford, 2005). In the context of this study, these rules are conceptualized as follows: *position* rules specify the participants (who in turn have a combination of resources, perspectives, and preferences) and their roles in marine governance; *boundary* rules specify how participants can take part in such governance; *choice* rules specify the actions participants can take as part of their roles; *aggregation* rules specify decision-making procedures, including arrangements to aggregate the preferences of participants; *information* rules specify the arrangements for information exchange between participants; *payoff* rules specify the incentives and disincentives in terms of resources available to support decision making and action; and *scope* rules specify the functional scope and geographic domain that can be affected by marine governance. These institutional rules, usually in combination, affect the attributes of adaptive capacity (i.e., variety, learning capacity, autonomy, leadership, resources, and fair governance) (see Table 17.1). For example, position, boundary, choice and scope rules determine the participants in marine governance, their actions and potential outcomes; therefore, they may affect variety through inclusive participation, diversity of actions and issues addressed (Fidelman et al., 2017).

The Coral Triangle Initiative

The Coral Triangle is an archipelagic region of approximately 5.7 million km^2 (roughly half the size of the United States), regarded as the global epicenter of marine biodiversity and abundance (see Figure 17.1). Coastal and marine resources provide benefits to all Coral Triangle countries, which depend directly on these resources for economic development, coastal community livelihoods, and food security (CTI Secretariat, 2009). Increasing threats to coastal and marine resources include overfishing, unsustainable fishing practices, land-based sources of marine pollution, coastal habitat conversion, and climate change. Climate change impacts are already observed in the Coral Triangle. For example, because of warmer sea

Table 17.1. *Adaptive capacity attributes and indicators (after Gupta et al., 2010; Termeer et al., 2011; Fidelman et al., 2017).*

Attribute	Definition	Indicators	Relevant rule
Variety	The ability of institutions to encourage the involvement of a variety of actors, perspectives, and solutions. Because environmental change problems are complex and unstructured (lacks agreement on values), embedding diverse interests and perspectives, dealing with such problems requires multiple perspectives and solutions. This includes the participation of relevant stakeholders across different sectors and levels of governance in problem framing and formulation of solutions.	Inclusive participation of relevant actors Diversity of perspectives, and actions and issues addressed	Position, boundary, choice, scope
Learning capacity	Learning is critical for dealing with uncertainty, surprises, and variation that characterize environmental change. There is an ongoing need to revise existing knowledge and understanding to enable adaptation. Learning allows actors to reformulate knowledge and understanding based on experiences. Adaptive institutions are therefore those that enable social actors to continuously learn and experiment to improve their institutions.	Activities and mechanisms that support learning (e.g., meetings, joint decision making, collaborative activities, monitoring and evaluation, etc.)	Information, choice
Autonomy	The ability of social actors to autonomously review and adjust institutions in response to environmental change. Adaptive institutions allow and motivate actors to self-organize, design, and reform their institutions. Authority (legitimate or accepted forms of power) for decision making and implementation is key to autonomy when it is supported (or at least not undermined) by actors and other decision-making entities.	Authority to make and implement decisions Authority is not undermined by other actors/ decision-making entities	Aggregation, choice, payoff

Table 17.1. (*cont.*)

Attribute	Definition	Indicators	Relevant rule
Leadership	Leadership may be regarded as a driver for change when it points to (a) direction(s) and motivates others to follow, facilitates access to resources, and bridges and builds coalitions. Institutions supporting adaptive capacity are those that can mobilize leadership qualities of social actors in the process of (re)designing institutions.	Ability of actors to direct and motivate others to follow, gain access to resources, and bridge and build coalitions	Position, boundary, choice
Resources	Resources are critical in generating incentives and reducing transaction costs for actors to engage in collective decision making and action. Therefore, adaptive institutions have the capacity to mobilize resources (human, financial, technical) for making and implementing decisions (e.g., adaptation measures).	Human, financial, and technical resources	Payoff
Fair governance	Fair governance includes institutions that are accepted and supported by their constituents (legitimacy), considered to be fair (equity), responsive (responsiveness), and/or accountable to social actors (accountability).	Legitimacy, equity, responsiveness, accountability	Boundary, choice, aggregation, information

temperatures, mass coral bleaching events have occurred in the region (Hoegh-Guldberg et al., 2009; McLeod et al., 2010; Burke et al., 2011). Further, climate change is expected to compound many of the other threats (Burke et al., 2011).

In an effort to address the threats to the Coral Triangle outlined earlier and pursue a more sustainable use of the region's resources, the Coral Triangle Initiative on Coral Reefs, Fisheries and Food Security (hereafter CTI or the Initiative) was established. The CTI is an intergovernmental agreement between Malaysia, the Philippines, Indonesia, Timor Leste, Papua New Guinea, and the Solomon Islands, formally adopted in 2009. A Regional Plan of Action is a core component of the CTI; it comprises a legally nonbinding document containing goals and actions over a period of 10 to 15 years. The Regional Plan of Action focuses primarily on (1) designating and effectively managing priority seascapes, (2) applying an

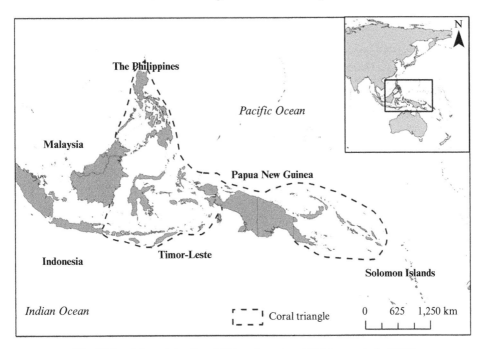

Fig. 17.1.   Map of the Coral Triangle region. (Fidelman et al., 2012)

ecosystem approach to fisheries management and other marine resources, (3) establishing and effectively managing marine protected areas, (4) adopting climate change adaptation measures, and (5) improving the status of threatened species. The regional plan provided an overarching framework for the National Plans of Action developed by each Coral Triangle country (CT country) with support from nongovernmental organizations (NGOs) and other stakeholders.

A regional secretariat, based in Indonesia (see Chapter 7), provides overall coordination to the Initiative. Like other large-scale marine governance efforts, the CTI features institutions for coordinating actions among stakeholders to address resource use and management across jurisdictions. Similar institutional arrangements may also characterize climate change adaptation efforts (Fidelman et al., 2013). In this context, the CTI comprises an appropriate case to examine the implications of large-scale marine governance for adaptive capacity (see also Chapter 11).

Assessing Adaptive Capacity

This section uses the conceptual framework presented in Table 17.1 to assess the influence of governance institutions pertaining to the CTI on key attributes of adaptive capacity, namely variety, learning capacity, leadership, resources, and fair governance.

Table 17.2. *Key official stakeholders of the Coral Triangle Initiative (after Fidelman et al., 2014).*

Stakeholder	Description
CT governments	A lead agency is responsible for coordinating national efforts in each CT country; they include Indonesian Ministry of Marine Affairs and Fisheries; Philippines Bureau of Fisheries and Aquatic Resources and Department of Environment and Natural Resources; Malaysian National Oceanography Directorate of the Ministry of Science, Technology and Innovation; Timor-Leste National Directorate for Fisheries and Aquaculture of the Ministry of Agriculture and Fisheries; Papua New Guinea Department of Environment and Conservation and the National Fisheries Authority; and Solomon Islands Ministry of Environment, Climate Change, Disaster Management and Meteorology, and Ministry of Fisheries and Marine Resources.
Donors	The US Agency for International Development (USAID) led the US Coral Triangle Initiative Support program (USCTI), which also included the US National Oceanic and Atmospheric Administration, the Coral Triangle Support Partnership, and the US Department of State.
	The Global Environmental Facility (GEF) was one of the major contributors of funds to the CTI. The Asian Development Bank (ADB) served as the lead agency in the GEF CTI program, for which planning and implementation involved the CT countries, ADB, FAO, UNDP, UNEP, and the World Bank.
	The Australian government's Department of Sustainability, Environment, Water, Population and Communities led Australia's engagement with the CTI. Assistance from the Australian government included an initial funding commitment and a pledge of long-term support as the CTI evolves.
NGOs	The Coral Triangle Support Partnership (CTSP) was a consortium of international NGOs formed by the World Wildlife Fund (WWF), The Nature Conservancy (TNC), and Conservation International (CI) as part of the USCTI. WWF served as the leading organization for the consortium. The CTSP played a critical role in implementing the USCTI.

Variety

During the period examined in this study, the CTI involved several official stakeholders pertaining to three broad categories: governments, donors, and NGOs (see Table 17.2). Some relevant policy sectors and levels of governance, such as fishing and tourism interests, were not directly represented. The CTI Regional Plan of Action (RPOA) was developed in a top-down fashion, involving delegates from CT countries and representatives of international NGOs (Rosen and Olsson, 2013). Despite addressing diverse issues and actions, problem framing and associated solutions proposed in the RPOA seem to draw predominantly on Western intellectual frameworks, rather than on national and local contexts (Clifton and Foale, 2017). The use of English as the negotiating language prevented stakeholders (other than international NGOs) from effectively contributing to the RPOA. As a result, the ability to consider alternative views to the planning process may have been constrained (Von Heland et al., 2014). This may, in turn, have contributed to the strong focus of the CTI on the ecological aspects of marine governance, rather than development goals, such as alleviating poverty and achieving food security (Foale et al., 2013).

Nevertheless, the Leaders Declaration (2009) states that the CTI aims to achieve its regional goals "through accelerated and collaborative action, taking into consideration *multi-stakeholder participation*" (emphasis added). Further, guiding principle No. 8 of the RPOA states that "Multiple stakeholder groups should be actively engaged in the CTI, including other national governments, local governments, NGOs, private sector companies, bilateral donor agencies, multilateral agencies, indigenous and local communities, coastal communities, and the academic and research sector" (CTI Secretariat, 2009: 9).

Learning Capacity

The planning and decision-making entities of the CTI (see Table 17.3), given their collaborative nature, would entail interaction and information exchange and, ultimately, learning among stakeholders. However, constraints to *variety* (e.g., limited diversity of perspectives) have significant implications for learning. These include limited consideration of information about the human dimensions of marine governance (Foale et al., 2013), particularly in light of the diverse national and local socioeconomic, cultural, and political contexts of the Coral Triangle (Fidelman et al., 2012). As noted earlier, the use of English was a challenge to communication, preventing CT government representatives from fully contributing their views in negotiations (Von Heland et al., 2014). Further, understanding of core concepts underpinning the RPOA, such as ecosystem approach to fisheries management, marine protected areas, and climate change adaptation and/or how to implement them varied among stakeholders (Fidelman et al., 2014). The midterm report of the US Coral Triangle Initiative Support program (USCTI) identified inconsistent coordination and communication between stakeholders, and confusion among and within CT governments, partners, and other stakeholders about the CTI vision (The WorldFish Centre/USAID, 2010). Some stakeholders reported that information about funding, political objectives, values, and implementing activities was not easily accessible (Von Heland et al., 2014).

The RPOA includes several propositions about learning from existing and past practices and programs and information, data, and technology sharing. These would be facilitated by seminars, workshops, exchange visits, Internet list servers, lessons

Table 17.3. *Key planning and decision-making entities of the Coral Triangle Initiative (after Rosen and Olsson, 2013).*

Entity	Description
CTI Secretariat	Administrative office responsible for regional coordination.
National Coordination Committees (NCCs)	Comprise national delegates from CT countries and representatives from NGOs (WWF, TNC, CI); their responsibilities include national coordination, developing national plans of action, and negotiating national interests in the CTI.
CTI Coordination Committee (CCC)	Consists of national delegates from CT countries and technical advisors (mostly NGO representatives); it was responsible for developing the Regional Plan of Action.
Senior Officials Meetings (SOM)	Comprise representatives from NCCs responsible for reviewing the work of CCC and reporting to MM. NGO and donor representatives may assist in convening and facilitating the meetings.
Ministerial Meetings (MM)	Consist of ministers from CT countries responsible for formalizing the CTI.

learned documents, and learning networks. For example, in the context of the fisheries management, the Plan states that "[i]nformation on country-level legislative, policy and regulatory reform efforts will be actively shared across the CT6 countries, to help promote harmonization and effective action" (CTI Secretariat, 2009: 8). Another learning indicator proposed in the Regional Plan of Action refers to a monitoring and evaluation program to track the progress in achieving the targets of the Regional Plan of Action.

Autonomy

The primary formal source of authority for the CTI is the Leaders' Declaration signed by the CT heads of state in 2009. This declaration formalized the adoption of the CTI and its RPOA at the highest political level (Leaders Declaration, 2009). The backing of international NGOs and donors helped gather further support from stakeholders within and outside the region (Fidelman et al., 2014). Further, the core objectives of the CTI align with those of international arrangements relating to the region's governance (Fidelman and Ekstrom, 2012). In sum, despite not being legally binding, the CTI enjoyed considerable authority in terms of support from prominent stakeholders and alignment with existing international institutions.

However, power imbalance among stakeholders may have undermined the Initiative's authority. For instance, the key NGOs involved (WWF, TNC, and CI) had a disproportionate influence in the development of the CTI (Fidelman et al., 2014; Von Heland et al., 2014). In fact, several stakeholders suggested that the goals and targets of Plan directly reflect the specific agenda of these NGOs. Similarly, among the CT countries, the government of Indonesia was perceived as a dominant force attempting to drive its own agenda (Von Heland et al., 2014).

Tensions between regional and national priorities illustrate additional challenges to the CTI authority. For example, community-based management and institutions comprise the main approach to marine management in the Solomon Islands and Papua New Guinea, rather than no-take marine protected areas proposed in the Regional Plan of Action (Fidelman et al., 2014). Ultimately, these challenges may affect the willingness and ability of CT governments to prioritize and implement the RPOA. In this regard, key government agencies (beyond those responsible for conservation and fisheries management) did not relate to the CTI, and the Initiative's priorities were yet to be integrated into ministry-level policy (Von Heland et al., 2014).

Leadership

The role of leadership in mobilizing support for the CTI cannot be overestimated. The CTI was initially proposed by President Yudhoyono of Indonesia in 2006, during the Eighth Conference of the Parties to the Convention on Biological Diversity held in Brazil. In the following year, President Yudhoyono's proposition was welcome by 21 heads of state at the Asian-Pacific Economic Cooperation Meeting in Australia and gained support from other heads of state in the Coral Triangle.

Interestingly, the Initiative was de facto orchestrated to a considerable extent by WWF, TNC, and CI. Rosen and Olsson (2013) refer to them as *institutional entrepreneurs* who worked behind the scenes to leverage technical, financial, and political resources. This included strategically using informal networks and lobbying to gather support from the president of Indonesia around the CTI concept, as well as identifying and mobilizing partners and technical and political support (Rosen and Olsson, 2013). The USCTI played a critical role in providing leadership and facilitating coordination with other CT countries, bearing much of the transaction costs in moving the Initiative forward. On the other hand, the leadership provided by international NGOs and donors may have been seen by national stakeholders as a disproportionate foreign influence in regional affairs, ultimately limiting their sense of ownership of the CTI (Fidelman et al., 2014; Von Heland et al., 2014).

Resources

International donors and NGOs provided most of the human, technical, and financial resources to the CTI. The Global Environmental Facility (GEF) was one of the major contributors. In addition to investments of US$63 million, it catalyzed co-financing of over US$300 million. The USAID provided US$42 million as part of the five-year USCTI, and assistance from the Australian government included an initial commitment of A$2 million (approximately US$2 million at the time) and a pledge of long-term support as the CTI evolved. In addition, the consortium of international NGOs provided human and technical resources. The availability of external funding possibly mitigated the transaction costs that would otherwise place a higher burden on regional stakeholders. As a result, at the outset, the Initiative gathered considerable momentum (Fidelman et al., 2014).

Overall, human and financial capacity of the CT countries was limited, particularly considering the magnitude and complexity of the CTI. The amount of funds available was far from sufficient to implement the RPOA over the following decade. Clifton (2009) estimates that the cost of implementing the network of marine protected areas alone was approximately US$400 million. Unclear funding arrangements, absence of representation from CT governments in funding allocation, and lack of long-term financing strategies were compounding issues (Fidelman et al., 2014). Further, the reliance on external resources raised questions relating to the introduction of foreign interests in regional and national policy making and a foreign constituency to which the CTI needed to respond and remain relevant (Von Heland et al., 2014).

Fair Governance

As seen earlier, the CTI has not been largely inclusive of stakeholders, sectors, and levels of governance, which has obvious implications for legitimacy and accountability. Further, the CTI was far from being a level playing field. For example, because CTI meetings were usually conducted in English, international NGOs and donors – which were generally Anglophone – had a considerable advantage in influencing planning and decision making. Further, they had relatively well-developed capacity to access and distribute information and master sophisticated concepts underpinning the CTI goals, as well as mobilize human, technical, and financial resources (Von Heland et al., 2014). As a result, national stakeholders, including government representatives, perceived the CTI as a process dominated by international NGOs and donors, whose agenda was not necessarily aligned with national and local priorities (Foale et al., 2013; Fidelman et al., 2014). National stakeholders also questioned the transparency in relation to objectives, investments, roles, and time frames of external stakeholders (Von Heland et al., 2014).

Institutions and Adaptive Capacity

The influence of large-scale marine governance institutions on adaptive capacity in the context of the CTI has been examined. The institutions examined enabled multiple stakeholders to collaborate to address pressing issues affecting the Coral Triangle; mobilize resources, leadership, and authority; and contribute to improved regional governance. Ultimately, they contributed to building and mobilizing adaptive capacity. On the other hand, these institutions were associated with disabling conditions characterized by a top-down approach to governance and power asymmetry among stakeholders. As mentioned previously, institutions can both enable and disable adaptive capacity (Engle and Lemos, 2010), given their inherent capacity to expand and limit decision making and action (Ostrom, 2005). Further, given the interdependent nature of the attributes of adaptive capacity, these attributes can reinforce and/or undermine each other (Gupta et al., 2016; Fidelman et al., 2017). Last, the context in which the CTI takes place is remarkably complex and, therefore, can both enable and constrain governance efforts (Fidelman et al., 2012).

Power relations emerge as a critical factor underpinning institutional constraints to adaptive capacity. More specifically, it appears that such relations-enabled international NGOs to ensure that institutional design served their interests. The ability of international NGOs to access legal authority, public opinion, information, supporters, financial resources, and skillful leadership gave them disproportionate influence in relation to other stakeholders. They used such ability to forge the CTI and shape its development (see Rosen and Olsson, 2013). This includes identifying and mobilizing the organizations involved in the CTI, therefore affecting *variety* in terms of inclusive (or otherwise) participation. The limited representation of relevant stakeholders, in turn, affected the diversity of perspectives and issues addressed. Power relations also affected *learning capacity*, given the ability of international NGOs to access and distribute information and make their knowledge become common ground (Von Heland et al., 2014). This, in turn, contributed to the strong focus on ecological aspects of marine governance underpinning the CTI. While international NGOs provided important *leadership*, including the mobilization of human, technical, and financial *resources*, such influence may have limited the sense of ownership of national stakeholders (The WorldFish Centre/USAID, 2010) and raised questions about *fair governance* in terms of equity, legitimacy, and accountability. These institutional constraints associated with power relations are not exclusive to the CTI. Brosius and Russell (2003) note that large-scale, ecoregional initiatives such as the CTI have the capacity to reinforce power asymmetry by favoring transnational conservation organizations. Particularly in less developed countries, international NGOs have been influential in shaping marine governance (Rodríguez et al., 2007). Many of the issues associated with international NGOs highlighted here have also been recognized elsewhere (Foale, 2001; Agardy, 2005; Pajaro et al., 2010).

Power relations are inherently embedded in governance institutions (Nightingale, 2017), which may explain the constrains to adaptive capacity highlighted earlier. On one hand, institutions affect power distribution, which, in turn, determine who is involved in adaptation efforts and in which manner. On the other hand, power distribution directly affects institutional design and change (Clement, 2012; Kashwan, 2016; Nightingale, 2017). In this context, addressing power asymmetry in the context of the CTI would involve, to some extent, reviewing those rules that may contribute to creating and reinforcing such asymmetry – for example, choice rules, which limit or expand stakeholders' action, and aggregation rules, which limit or expand decision making of these stakeholders as they interact. In this case, institutional change would aim to enhance the capacity of relevant stakeholders in terms of decision making and action. Such capacity would be further enhanced by changes in payoff rules aiming to provide more adequate levels of resources to those stakeholders. In addition, changes in prevailing discourses that legitimate institutions would also be needed (Clement, 2010). This would involve reviewing the global conservation narrative that underpins the CTI, so that such narrative does not override local and national priorities (Fidelman et al., 2012). Last, institutional change aiming to enhance adaptive capacity – for example, changing position and boundary rules to increase the variety of interests represented in the CTI – needs to pay attention to how the different interests mediate power relations. This includes understanding struggles over authority and desires for social and political recognition by participants in the CTI (Nightingale, 2017). This is particularly important if one assumes that the different social, economic, and political contexts within and across the CT countries (see Fidelman et al., 2012) are underscored by power relations.

Conclusion

Large-scale marine governance interventions such as the Coral Triangle Initiative may affect adaptive capacity to climate change. Most significantly, the case of the CTI underscores how the interplay between institutions and power may constrain key attributes of adaptive capacity (i.e., variety, learning capacity, leadership, resources, and fair governance). In this context, enhancing the capacity of CTI countries to adapt to climate change capacity will require, in addition to reforming governance institutions, considering power relations that shape and are shaped by institutions. With very few exceptions (e.g., Nightingale, 2017), the literature on climate change adaptation has yet to address the role of power relations in adaptation efforts. Assessments of institutional adaptive capacity, in particular, are silent in relation to such a role. In the context of institutional analysis, despite positive developments (Clement, 2010, 2012; Kashwan, 2016), works addressing the interplay between institutions and power are still rather limited (Agrawal, 2003). The discussion in this chapter therefore makes empirical and intellectual contributions to the scholarship on marine governance, adaptation to climate change, and institutional analysis.

Future research seeking to incorporate the notion of power in institutional analysis would further expand the scope of institutional adaptive capacity assessments. In this regard, developing an institutional analysis framework that enables mapping specific institutional rules that shape, and are shaped by, power would be particularly beneficial. The framework articulated in Fidelman et al. (2017) may provide a promising starting point (see also Clement, 2010). Adaptation, institutions, and power manifest at multiple governance levels. Future research exploring institutional and power interplay across these different levels, and how interplay affects multilevel adaptation, would also be beneficial. In this regard, the works of Fidelman et al. (2013), Clement (2010), and Kashwan (2016) may provide useful insights into how to make ocean governance more effective as the impacts of climate change increase.

References

Adger, N. W., Arnell, N. W., and Tompkins, E. L. (2005). Successful adaptation to climate change across scales. *Global Environmental Change*, **15**(2), 77–86.

Adger, N. W., Brown, K., Nelson, D. R., et al. (2011). Resilience implications of policy responses to climate change. *Wiley Interdisciplinary Reviews Climate Change*, **2**(5), 757–66.

Agardy, T. (2005). Global marine conservation policy versus site-level implementation: The mismatch of scale and its implications. *Marine Ecology Progress Series*, **300**, 242–48.

Agrawal, A. (2003). Sustainable governance of common pool resources: Context, methods, and politics. *Annual Review of Anthropology*, **32**(1), 243–62.

Alger, J. and Dauvergne, P. (2017). The global norm of large marine protected areas: Explaining variable adoption and implementation. *Environmental Policy and Governance*, **27**(4), 298–310.

Bergsma, E., Gupta, J., and Jong, P. (2012). Does individual responsibility increase the adaptive capacity of society? The case of local water management in the Netherlands. *Resources, Conservation and Recycling*, **64**, 13–22.

Brosius, J. P. and Russell, D. (2003). Conservation from above: An anthropological perspective on transboundary protected areas and ecoregional planning. *Journal of Sustainable Forestry*, **17**(1), 39–65.

Burke, L., Reytar, K., Spalding, M., and Perry, A. (2011). *Reefs at Risk Revisited*. Washington, DC: World Resources Institute, p. 114.

Clement, F. (2010). Analysing decentralised natural resource governance: Proposition for a "politicised" institutional analysis and development framework. *Policy Sciences*, **43**(2), 129–56.

Clement, F. (2012). For critical social-ecological system studies: Integrating power and discourses to move beyond the right institutional fit. *Environmental Conservation*, **40**(1), 1–4.

Clifton, J. (2009). Science, funding and participation: Key issues for marine protected area networks and the Coral Triangle Initiative. *Environmental Conservation*, **36**(2), 1–6.

Clifton, J. and Foale, S. (2017). Extracting ideology from policy: Analysing the social construction of conservation priorities in the Coral Triangle region. *Marine Policy*, **82**, 189–96.

CTI Secretariat. (2009). *Regional Plan of Action*. Jakarta: Interim Regional CTI Secretariat. p. 87.

Eaking, H. and Lemos, M. C. (2006). Adaptation and the state: Latin America and the challenge of capacity-building under globalization. *Global Environmental Change*, **16**(1), 7–18.

Engle, N. L. (2011). Adaptive capacity and its assessment. *Global Environmental Change*, **21**(2), 647–56.

Engle, N. L. and Lemos, M. C. (2010). Unpacking governance: Building adaptive capacity to climate change of river basins in Brazil. *Global Environmental Change*, **20**(1), 4–13.

Fidelman, P. and Ekstrom, J. A. (2012). Mapping seascapes of international environmental arrangements in the Coral Triangle. *Marine Policy*, **36**(5), 993–1004.

Fidelman, P., Evans, L., Fabinyi, M., et al. (2012). Governing large-scale marine commons: Contextual challenges in the Coral Triangle. *Marine Policy*, **36**(1), 42–53. doi:10.1016/j.marpol.2011.03.007

Fidelman, P., Evans, L., Foale, S., et al. (2014). Coalition cohesion for regional marine governance: A stakeholder analysis of the Coral Triangle Initiative. *Ocean and Coastal Management*, **95**, 117–128.

Fidelman, P., Leitch, A., and Nelson, D. R. (2013). Unpacking multilevel adaptation in the Great Barrier Reef. *Global Environmental Change*, **23**(4), 800–12.

Fidelman, P., Truong Van, T., Nong, K., and Nursey-Bray, M. (2017). The institutions-adaptive capacity nexus: Insights from coastal resources co-management in Cambodia and Vietnam. *Environmental Science and Policy*, **76**, 103–12.

Foale, S. J. (2001). 'Where's our development?' Landowner aspirations and environmentalist agendas in Western Solomon Islands. *The Asia Pacific Journal of Anthropology*, **2**(2), 44–66.

Foale, S., Adhuri, D., Alino, P., et al. (2013). Food security and the Coral Triangle Initiative. *Marine Policy*, **38**, 174–83.

Grecksch, K. (2014). Adaptive capacity and regional water governance in north-western Germany. *Water Policy*, **15**(5), 794–815.

Grothmann, T., Grecksch, K., Winges, M., and Siebenhuner, B. (2013). Assessing institutional capacities to adapt to climate change: Integrating psychological dimensions in the Adaptive Capacity Wheel. *Natural Hazards and Earth System Sciences*, **13**(12), 3369–84.

Gupta, J., Bergsma, E., Termeer, C., et al. (2016). The adaptive capacity of institutions in the spatial planning, water, agriculture and nature sectors in the Netherlands. *Mitigation and Adaptation Strategies for Climate Change*, **21**(6), 883–903. doi:10.1007/s11027-014-9630-z

Gupta, J., Termeer, C., Klostermann, J., et al. (2010). The Adaptive Capacity Wheel: A method to assess the inherent characteristics of institutions to enable the adaptive capacity of society. *Environmental Science and Policy*, **13**(6), 459–71.

Hoegh-Guldberg, O., Hoegh-Guldberg, H., Veron, J. E. N., et al. (2009). *The Coral Triangle and Climate Change: Ecosystems, People and Societies at Risk*. Brisbane: World Wild Fund Australia.

Kashwan, P. (2016). Integrating power in institutional analysis: A micro-foundation perspective. *Journal of Theoretical Politics*, **28**(1), 5–26.

Leaders Declaration. *Coral Triangle Initiative Leaders' Declaration on Coral Reefs, Fisheries and Food Security*. Signed May 15, 2009, Manado.

McLeod, E., Moffitt, R., Timmermann, A., et al. (2010). Warming seas in the Coral Triangle: Coral reef vulnerability and management implications. *Coastal Management*, **38**(5), 518–39.

Munaretto, S. and Klostermann, J. (2011). Assessing adaptive capacity of institutions to climate change: A comparative case study of the Dutch Wadden Sea and the Venice Lagoon. *Climate Law*, **2**(2), 219–50.

Nelson, D. R., Adger, N. W., and Brown, K. (2007). Adaptation to environmental change: Contributions of a resilience framework. *Annual Review of Environment and Resources*, **32**, 395–419.

Nightingale, A. (2017). Power and politics in climate change adaptation efforts: Struggles over authority and recognition in the context of political instability. *Geoforum*, **84**, 11–20.

Ostrom, E. (2005). *Understanding Institutional Diversity*. Princeton: Princeton University Press.

Ostrom, E. and Crawford, S. (2005). Classifying rules. In E. Ostrom, ed., *Understanding Institutional Diversity*. Princeton: Princeton University Press, pp. 187–215.

Pajaro, M. G., Mulrennan, M. E., and Vincent, A. C. J. (2010). Toward an integrated marine protected areas policy: Connecting the global to the local. *Environment, Development and Sustainability*, **12**(6), 945–65.

Rodríguez, J. P., Taber, A. B., Daszak, P., et al. (2007). Globalization of conservation: A view from the south. *Science*, *317*(5839), 755–56.

Rosen, F. and Olsson, P. (2013). Institutional entrepreneurs, global networks, and the emergence of international institutions for ecosystem-based management: The Coral Triangle Initiative. *Marine Policy*, **38**, 195–204.

Smit, B. and Wandel, J. (2006). Adaptation, adaptive capacity and vulnerability. *Global Environmental Change*, **16**(3), 282–92.

Termeer, C., Biesbroek, R., and Van Den Brink, M. (2011). Institutions for adaptation to climate change: Comparing national adaptation strategies in Europe. *European Political Science*, **1**, 1–13.

The WorldFish Centre/USAID. (2010). *The US Coral Triangle Initiative (CTI) Support Program: Midterm Program Performance Evaluation Report*. Penang: The WorldFish Center (WorldFish) and the United States Agency for International Development (USAID), p. 54.

Van den Brink, M., Meijerink, S., Termeer, C., and Gupta, J. (2014). Climate-proof planning for flood-prone areas: Assessing adaptive capacity of planning institutions in the Netherlands. *Regional Environmental Change*, **14**(3), 981–95.

Van den Brink, M., Termeer, C., and Meijerink, S. (2011). Are Dutch water safety institutions prepared for climate change? *Journal of Water and Climate Change*, **2**(4), 272–87.

Von Heland, F., Crona, B., and Fidelman, P. (2014). Mediating science and action across multiple boundaries in the Coral Triangle. *Global Environmental Change*, **29**, 53–64.

Young, O. R. (2002). *The Institutional Dimensions of Global Environmental Change: Fit, Interplay and Scale*. Cambridge, MA: MIT Press.

Young, O. R. (2005). *Science Plan – Institutional Dimensions of Global Environmental Change*. Bonn: International Human Dimensions Programme on Global Environmental Change, Institutional Dimensions of Global Environmental Change, p. 15.

18

Nonterritorial Exclusive Economic Zones
Future Rights of Small-Island States

ORI SHARON

Introduction

To small-island developing states (SIDS), climate change-induced sea-level rise poses an existential threat. Many of these low-lying island territories will become uninhabitable or even completely submerged by sea-level rise (IPCC, 2014: 1618). The threat of disappearance has contributed to SIDS' image as the poster child or "canary in the coal mine" of climate change's imminent threat to human society (McAdam, 2010: 107) (see Chapters 3 and 6–8). One issue that is captivating law scholars is the question of what is to happen to the most valuable national asset of SIDS, their resource-rich exclusive economic zones (EEZs). According to conventional legal thinking, SIDS' rights to marine resources will be extinguished as their territory submerges into the ocean. Under the Montevideo Convention, territory is a criterion for statehood (Montevideo Convention, 1933). No territory, scholars argue, means no state, and therefore no right to an EEZ (Rayfuse, 2009: 6–7).

For SIDS, the issue of losing EEZ rights is intertwined with the question of political and communal survivability. Focusing on communal and cultural survival, several SIDS have been trying to secure territory in other countries to which they can safely relocate their nation and avoid piecemeal migration (McAdam, 2010: 122). Maintenance of EEZ rights post-relocation will greatly increase SIDS' prospects to secure alternative territory; it may also serve as an economic foundation to the legal entity that might succeed the state (McAdam, 2010: 122).

This chapter argues that the loss of national territory due to climate change–induced sea-level rise is an event so unique and unprecedented in international law that it cannot be addressed through strict adherence to the statehood criteria of the Montevideo Convention. It surveys legal doctrines of rights survivability in private and international law to answer the question of the legal fate of SIDS' EEZs in a post–climate change world. The comparative review reveals that loss of legal status does not necessarily entail loss of rights. To identify whether a right is extinguished with the loss of legal status, one must answer a series of questions pertaining to the nature of the right, the character of the rights holder, the legal relationship that established the right, and the circumstances that led to the loss of legal status. The chapter applies two legal frameworks to generate answers to these questions: a communitarian property theory of national resources and a contractarian theory of international treaties. The analysis suggests that EEZs are communitarian rather than national rights. As such, they belong to the people, not the sovereign. This conclusion opens the way for institutional and legal remedies unavailable under a Montevideo-based analysis.

Sovereignty and Territory

The question of state continuity, or when exactly statehood is lost, is not easy to answer. According to customary international law, for an entity to be recognized as a state, it must meet several requirements known as the "Montevideo Criteria" (Shaw, 2014: 198). The four requirements are permanent population, defined territory, effective government, and capacity to enter into relations with other states. Considering the Montevideo criteria in the abstract, one might conclude that SIDS will lose statehood once any one of the four indicia of statehood is lost. This view, which dominates global political discourse on climate change and SIDS, has also been advanced by several legal scholars (Rayfuse, 2009: 6). However, the issue of statehood is more complex than a simple reading of the Montevideo Convention suggests. First, the norms of international law that determine the status of states were developed in the context of creating new states, not as mechanisms for determining the continuity of already recognized states (Costi and Ross, 2017: 101–2). Unlike the question of establishing statehood, the issue of "[c]ontinuity of already-established statehood is a separate legal question" (Costi and Ross, 2017: 111). The problem with trying to answer this question is that the issue of involuntary state "extinction" is unknown to modern international law.

Since the establishment of the UN in 1945, there has not been even one case of involuntary state extinction (McAdam, 2010: 110). There is no "international law of state extinction." There are no precedents, no rules, no authority, and no custom to support the view of climate-induced loss of statehood (Costi and Ross, 2017: 102–3, 114). Thus, not only is there no legal basis for the assumption of state disappearance but "to assume that sovereignty is lost with state disappearance retards the potential for creative solutions to an entirely novel problem" (Costi and Ross, 2017: 103, 113). If anything, one fundamental tenet of international law negates the possibility of state elimination. International law is premised on the principle of state continuity: once a state is recognized in international law, it continues to exist indefinitely (Crawford, 2007: 667–68). The fundamental principle of state continuity works against an *a priori* assumption of climate change–induced state extinction. It is the reason many states have continued to exist in international law despite drastic changes in their government, territory, population, or international relations (Crawford, 2007: 667–68). In fact, the history of international law is replete with examples of recognized states that at one time, or even continually, have failed to meet one or more of the Montevideo Criteria (Rayfuse, 2010: 10–11).

Based on these precedents, Stoutenburg suggests that SIDS could maintain legal status as "authorities in exile," and Costi and Rossi argued for a nonstate sovereign entity of international law (Stoutenburg, 2013: 70; Costi and Ross, 2017: 123–25).

Burkett and Rayfuse provide legal infrastructure for SIDs permanence as "deterritorialized states," an evolutionary legal entity of international law (Rayfuse, 2010: 10; Burkett, 2011). More conservative solutions include establishing a confederacy with another state or suggesting that other states cede territory to a threatened island for its continued existence. In these cases, "pre-existing maritime zones would continue to remain effective" (Rayfuse, 2010: 8–9).

Legal Status and Rights

In law, loss of legal status does not necessarily entail loss of rights associated with the status. Testate and intestate succession are two common universal examples of survival of legal rights when the legal entity who held the rights has ceased to exist. When a person dies, some rights are terminated upon death, whereas other rights survive the deceased. In business law, a "dead" corporation may be revived years after its complete and final termination and immediately regain its rights, as if it never ceased to exist (*Reliable Life Insurance* v. *Ingle et al.*, 2009: 28225). Similarly, in international law, a state could disappear and reemerge with the same rights it held before "termination." Under the 1938 Anschluss, Austria ceased to exist as a state, but it reemerged in 1946 reasserting the rights it held as a "person" in international law before its "termination" (Crawford, 2007: 669–70). When a state is succeeded by another state, the new state "inherits" the rights and duties of the extinct state.

What these examples tell us is that loss of status or legal capacity does not automatically entail loss of rights. If we follow standard analysis, the question of whether a right survives loss of status is often contingent on the classification of the right. In private law, a personal right extinguishes with death, a nonpersonal right does not (*Hebrew Univ. of Jerusalem* v. *Gen. Motors LLC,* 2012: 936). A similar distinction cannot be found in international law, because states do not have personal rights. But international law does recognize a distinction between contractual and communal rights. The Vienna Convention on Succession of States governs the transfer of treaty-based rights from predecessor to successor states. "Natural wealth" rights, however, are not affected by the provisions of the Convention (Vienna Convention, 1978).

Rights to marine resources are "natural wealth" rights. To determine the fate of these rights, we must apply the legal principles underlying rights to natural wealth. We cannot simply assume what needs instead to be proven – that if there is no state, there are no rights. To answer the question of EEZ survivability, we must perform a substantive analysis that starts by understanding the nature of the rights and the entity to which they are attached.

The Nature of the National Right to Marine Resources

The Proprietary View

The rationales underlying traditional maritime jurisdiction doctrines are proprietary (Barnes, 2009: 252). Hugo Grotius, the father of modern international maritime law, asserted that the ocean is a form of property and that what distinguishes it from other forms of property is that it cannot be occupied, a necessary condition for nations to claim title in ocean spaces (Churchill and Lowe, 1999: 71). The counter-doctrine of *mare clausum* (the closed sea), as advanced by Welwood and Selden, was also proprietary in nature; like Grotius, these authors used the language of exclusion as a justification for sovereignty over the ocean (Leary, 2007: 81).

The scholars who constructed the framework of maritime entitlements during the seventeenth to nineteenth centuries used the proprietary doctrine of adverse posses-sion (Friedheim, 1993: 12). The arguments advanced under this doctrine were fac-tual, revolving around a nation's *ability* to exclude others from parts of the ocean. Grotius's theory maintained that the sea is free from state ownership because no country could effectively exclude others from the sea. Grotius's opponents countered that "countries could control as much sea territory as they could dominate milita-rily" (Leary, 2007: 81). The legal rationale underlying these debates is that exclusion entails ownership.

The use of property theory to justify sovereignty over marine resources is not sur-prising when taken for what it is – a discussion about exclusive control over resources. When seen in this light, the parallel to property is immediate, because exclusion is the "most essential stick" in the bundle of rights we call property (Lucas, 1992: 1044). To the scholars who laid the foundations for the law of the sea, sovereignty (*imperium*) and ownership (*dominium*) were not separate, but rather existed in conjunction: "so the power to rule and to legislate, which is the power of imperium, could extend so far as the ruler and legislator possessed dominium, or the rights of an owner" (Shearer, 2013: 52). It is not surprising that during the 250 years that this juridical approach dominated international maritime law, the evolution of marine entitlements followed standard theory regarding the emergence of property rights in natural resources. In historical perspective, the extent of marine entitle-ments correlates with changes in fishing techniques, growth of global demand for fish, and most importantly, the development of technology for catching fish (Krueger, 1968: 1) (on fisheries issues more generally, see the chapters in Part III). The further a nation could exert its influence, the more ocean territory it claimed (Dupuy and Vignes, 1991: 262–63).

To most international law scholars, the thesis of national ownership of marine resources is outdated. However, not everyone agrees with this view. Some scholars, and not a few developing coastal states, continue to ascribe to the proprietary posi-tion. Those who maintain the proprietary doctrine do so based on the observation that the extension of sovereignty to the 200-nautical-mile line under the United Nations Convention on the Law of the Sea (UNCLOS) was done in the service of an economic goal: the extraction of resources. (On UNCLOs, see Chapter 2 and other chapters in Part V.) As such, while EEZ does have some characteristics of sovereignty, it cannot be said to reflect sovereignty in the same way it is reflected in the police power of the state (Dupuy and Vignes, 1991: 254). Indeed, "the coastal state does not have sovereignty over the sea included in the EEZ but only 'sovereign rights' for a specific purpose – the management of natural resources and other economic activities" (Restatement [Third] of The Foreign Relations Law of the United States, 1987: § 57). If the global regime of ocean governance is one that is wholly justified by the efficient and effective management of resources, then it is more accurate to frame it in proprietary terms, and necessary to do (Barnes, 2009: 252).

The Contractarian Approach

The contractarian approach views the entry into force of UNCLOS as a constitutive moment in international maritime law. Under this view, the signing of UNCLOS and the granting of equal EEZ rights to all coastal states was a juridical process in which states, the "persons" of international law, mutually agreed to depart from the previous control-based justification in favor of a more egalitarian approach to ocean governance. The traditional proprietary view evolved at a time when few maritime nations dominated the seas. It reflected political (and technological) realities of a colonial era, and it mostly benefited the strong and the quick to assert claims (Brilmayer and Klein, 2001: 712).

The dramatic increase in the number of state actors in the postcolonial era changed the balance of power in global politics and consequently in international law (Dupuy and Vignes, 1991: 281). During the second half of the twentieth century, a new discourse emerged, one that was focused more on equity and conservation rather than on the use of power (Brilmayer and Klein, 2001: 712). The legal regime this discourse facilitated views maritime entitlements as an inherent sovereign right of any coastal state, regardless of occupancy or ownership (Osherenko, 2006: 334). Under the new postcolonial doctrine, rights of nations over marine resources "extend not from proprietorship but from sovereignty," that is from the inherent "right of a nation to exercise power over its territory" (Turnipseed et al., 2009: 37).

Sovereignty, as the notion came to be understood in the postcolonial era, is embodied in the state (Turnipseed et al., 2009: 32). It is the inherent right of a state to "lawful control over its territory generally to the exclusion of other states, authority to govern in that territory, and authority to apply law there" (Restatement, 1987: § 206 cmt. b). Under the modern view of sovereignty, all nations enjoy permanent sovereignty over their natural resources (UNGA, 1963, 1973), "a basic constituent of the right [of nations] to self-determination, and, therefore, essential to a nation's economic sovereignty and development" (Turnipseed et al., 2009: 32).

The agreement reached among nations in UNCLOS extended a coastal state's sovereignty rights to specific maritime resources. Under the contractarian view, UNCLOS is viewed as a constitutive document of international law, a rights-establishing instrument that rearranged, reclassified, and recodified the array of ocean-related claims that existed before it came into force (Shearer, 2013: 57). What UNCLOS did not do is grant property rights in maritime resources (Osherenko, 2006: 334). In accordance with the views of its signatories, UNCLOS refrains from using proprietary terms (Osherenko, 2006: 333), signifying the deliberate choice to depart from occupancy-based regimes (Brilmayer and Klein, 2001: 710).

The Character of Rights to Exclusive Economic Zones

International law expert Peter H. Sand classifies sovereignty rights in the EEZ as fiduciary obligations (Sand, 2004: 48, 55). According to Sand, given the array of restrictions imposed on the coastal state in its management of the EEZ, "the analogy to 'ownership' rights becomes so diluted as to evoke a different legal analogy altogether, that is, the role of the nation state becomes more akin to a kind of public trusteeship" (Sand, 2004: 48). For a property scholar, Sand's observation may seem peculiar, as the legal concept of "public trust" is firmly grounded in property law (Archer et al., 1994: 21). Under the doctrine of public trust, which evolved from ancient Roman law (Frey, 1974: 220–24), the sovereign holds certain commonly owned natural resources in trust in perpetuity for the public interest (Frey and Mutz, 2007: 918). Sand acknowledges the proprietary origins of the public trust doctrine but argues that the doctrine has evolved beyond the contours of property law to occupy a separate field of law, which Sand identifies as environmental trusteeship (Sand, 2004: 49–54).

To support his observation, Sand identifies forms of environmental trusts in comparative as well as international environmental law. An environmental trust, according to Sand, is a legal structure that imposes fiduciary responsibilities on the government in its management of designated natural resources (Sand, 2004: 49). What used to be a common law property-based doctrine, explains Sand, is now a universal construct of environmental law. Since its rediscovery by Sax in 1969 (Sax, 1969: 471), the public trust doctrine has expanded to many jurisdictions and is enjoying growing international recognition, with many countries and international treaties applying environmental trusteeship to a constantly increasing pool of natural resources (Turnipseed et al., 2012: 2–3).

As an ancient Roman law doctrine, the notion of public trust percolated through various legal regimes, including the common law, and the civil law of France and Spain (Wilkins and Wascom, 1992: 863). Many scholars regard it as a natural law construct, basing this classification on the fact that the view of the sovereign as a protector of natural resources for the people has emerged independently in many disparate cultures and legal regimes (Turnipseed et al., 2009: 10). Even according to the Institutes of Justinian, the Roman legal code associated with the introduction of the public trust in the West, the doctrine is rooted in natural law (Frey, 1974: 221).

Two processes support Sand's separation of the public trust doctrine from its proprietary origins. The first is the movement of international maritime law away from concepts of property. The second is the emergence of nonproprietary environmental trusts in international and domestic law. However, it is important to note that while sovereign EEZ rights could be classified as a modern nonproprietary environmental trust, there is also a strong case for their classification as a classic property-based trust.

The proprietary framework underlying modern international maritime law (Turnipseed et al., 2009: 34–35) views the high seas as the common property of mankind (Osherenko, 2006: 328). Article 86 of UNCLOS makes clear that the EEZ is not included in the parts of the ocean that constitute the high seas. Appropriation of the EEZ by a coastal state did not change the classification of EEZ space from common property to private property – it merely contracted the pool of "owners." UNCLOS "reduced the common owners from a global community to citizens of particular states, but ... did not change the fundamental nature of [common property] ownership" (Osherenko, 2006: 331). The regime established under UNCLOS provides coastal states with the right "to manage the EEZ on behalf of the people, the common property owners" (Osherenko, 2006: 340). Under this view, the duties of the state to the people are revealed as a classic proprietary public trust.

A US Supreme Court case from 1979 illustrates the difference between a traditional property-based public trust and Sand's modern environmental trusteeship. In *Hughes* v. *Oklahoma* (1979: 322), the Supreme Court struck down an Oklahoma law prohibiting the export of fish caught in Oklahoma as violating the Constitution's Commerce Clause. Oklahoma's defense of the law was rooted in the common law notion of state ownership of wildlife. Common law up to the *Hughes* decision held that "the wild animals and fish within a state's border are ... owned by the state in its sovereign capacity for the common benefit of all its people" (*Hughes* v. *Oklahoma*, 1979: 324–25). The state ownership of wildlife, argued Oklahoma, conveys with it a right and duty to regulate the taking of wildlife (*Hughes* v. *Oklahoma*, 1979: 325). The Supreme Court rejected Oklahoma's argument. In a landmark decision that overturned centuries of settled common law, the Court severed state regulation of wildlife from its classic proprietary justification. According to the Court, the "ownership language" used in cases leading to *Hughes* should be understood as legal fiction expressing "the importance to its people that a State have power to preserve and regulate the exploitation of an important resource" (*Hughes* v. *Oklahoma*, 1979: 335). Modern legal analysis, opined the Court, does not require the continued use of fictional ownership constructs, as it is widely accepted that the state has police power to regulate its own resources irrespective of communal ownership rights of the people (*Hughes* v. *Oklahoma*, 1979).

In dissent, Justice Rehnquist, joined by Chief Justice Burger, criticized the Court's construction of state ownership rights as misrepresenting the relationship between the state as sovereign, the people, and the state's resources. The traditional ownership-based doctrine, explained the dissenters, is not a legal fiction, but rather a public trust. Under this public trust, "the wild fish and game located within the territorial limits of a State are the common property of its citizens and that the State, as a kind of trustee, may exercise this common 'ownership' for the benefit of its citizens" (*Hughes* v. *Oklahoma*, 1979: 341). The state is therefore not an "owner" in the conventional sense, but rather a trustee with a "substantial interest in preserving and regulating the exploitation of the fish and game and other natural resources within its boundaries for the benefit of its citizens" (*Hughes* v. *Oklahoma*, 1979: 342). The

Supreme Court's departure from wildlife ownership constructs did not absolve the state from its fiduciary responsibilities to the people. As the California Court of Appeals subsequently explained: "while the fiction of state ownership of wildlife is consigned to history, the state's responsibility to preserve the public's interest through preservation and wise use of natural resources is a current imperative" (*Ctr. for Biological Diversity*, 2008: 599).

In many ways, the *Hughes* holding parallels the movement from property to sovereignty in international maritime law. The Court in *Hughes* disconnected ownership from a state's police power, but separation of the two did not vacate the state's fiduciary obligations – it merely changed the legal justification for the trusteeship. Justice Rehnquist's dissent in *Hughes* highlights that while the fiduciary obligation could be attributed to the state's sovereign interest in protecting its common resources, the proprietary analysis would reach the same result, albeit by a different legal path.

How is it that both views, the proprietary and the contractarian, lead to the same result? First, while many have tried, complete severance of sovereignty from property is a challenge yet to be met by legal scholars (Turnipseed et al., 2009: 36). The classification of EEZ sovereign rights as fiduciary relationships does not really solve the problem because, at its core, a trust "is a fiduciary relationship with respect to property" (Restatement, 2008: § 2). Crawford's observation that "the substrate of the State is not property, it is the people of the State seen as a collective" (Crawford, 2006: 717) sheds some light on the tripartite community–sovereign–resources relationship underlying the trusteeship. It emphasizes that regardless of the legal path we choose to take, proprietary or contractarian, the state remains an agent of the collective (Archer et al., 1994: 31). The relationship between sovereignty and property, as explained by the Supreme Court, is one of subordination. Under the modern theory of sovereignty, property interests are not extinguished, but rather follow sovereign authority and are subordinated to it (*United States* v. *State of Texas,* 1950: 719).

According to the contractarian view, sovereignty over natural resources emanates from the right of nations to self-determination, which, by definition, is a communal right (Koskenniemi, 1994: 246). It is the community's distinctive history, language, ethnicity, national identity, culture, etc., that comprise its right to self-determination (McVay, 2012: 37–40). Similarly, it is the community's heritage that stands at the basis of the intimate connection between the people and their territory. The community–territory relationship is what supports the community's right to exercise sovereignty over its territory. But it is the same relationship that underlies the notion that certain natural resources are common to the nation's citizens.

The Legal Significance of the Fiduciary Relationship

Under a fiduciary regime, the state acts as trustee of the natural resource for the benefit of the people (Sand, 2004: 55). Mary Christina Wood explains the legal nature of the relationship: "At its core, the doctrine declares public property rights originally and inherently reserved through the people's social contract with their sovereign governments. The trust remains an attribute of sovereignty that cannot be alienated by any legislature. This principle designates government as trustee of crucial natural resources and obligates it to act in a fiduciary capacity to protect such assets for the beneficiaries of the trust, which includes both present and future generations of citizens" (Wood, 2014: 336–37).

What Makes States Entitled to Exclusive Use of Marine Resources?

The permanent right to sovereignty over marine resources is "a basic constituent of the right to self-determination" because all peoples have a fundamental right to exclusive use and enjoyment of their natural environment (Turnipseed et al., 2009: 32). Thus, once the right to self-determination of a people has been recognized and established in international law, it immediately conveys with it the inherent right to permanent sovereignty over their natural resources. Alternatively, the extension of permanent sovereignty to marine resources could be viewed as a natural progression of communal property rights, that it emanates from common ownership by the people.

What Governance Regime Does the Character of EEZ Rights Warrant in the Context of Climate Change?

The identification of the EEZ regime as a fiduciary relationship is not only analytically accurate, it is also extremely useful. Legal challenges that arise in the context of EEZ rights and obligations "can be elucidated by comparison to the well-developed body of law regarding private and charitable trusts" (Turnipseed et al., 2009: 32). The fiduciary relationship underlying EEZ rights highlights that the focus of legal scholars on statehood is misguided. The state in this relationship is an *agent* of the beneficiaries entrusted with responsibilities to manage the EEZ on their *behalf*. The relevant legal entity for determining the rights in this relationship is not the state-trustee but the people-beneficiaries. The beneficiaries are the core of the fiduciary relationship, they are the "holders" of the equitable title (Sand, 2004: 56).

State disappearance due to climate change–induced sea-level rise may have detrimental effects on the rights of the beneficiaries. For instance, it could be determined that, in the absence of a competent state-trustee, the fiduciary relationship cannot be maintained, but that is not equivalent to a disappearance of marine entitlements. The entitlements do not disappear; rather, the social contract between the government and the people-beneficiaries is voided. In such a case, the people-beneficiaries may be entitled to remedies, but the conclusion that their marine entitlements somehow disappear is simply unwarranted. Since the rights in the EEZ belong to the "people" (Cambou and Smis, 2013: 359–60), only a disappearance of the "people" could entail loss of rights. For the disappearance of the state to extinguish the rights, one must demonstrate that "statehood" is a constitutive criterion of "people."

Simply put, for the formula "no state equals no EEZ" to be correct, one must first establish that "no state equals no 'people.'"

The inhabitants of SIDS, the people of low-lying island nations, have distinct ethnicities, unique cultural traditions, histories, languages, intimate connections to their homelands, and political representation (see Chapter 8). These attributes comprise SIDS people's national identity and are the core of their right to self-determination (Koskenniemi, 1994: 246). The "people" are the holders of the right to self-determination (Raic, 2002: 172), which only a "people" could have (ICJ, 1975: 25). The right to self-determination safeguards "the cultural, ethnic and/or historical identity or individuality (the 'self') of a collectivity, that is, of a 'people'" (Raic, 2002: 223). Statehood is not a factor in recognizing self-determination, it is a consequence of the recognition (Raic, 2002: 172). Once a state has been established, the principle of self-determination becomes dormant, "enclosed within sovereignty." However, in times of crisis, when state existence is threatened, self-determination springs back to the fore and is once again relevant "to reconstitute the political normality of statehood" (Koskenniemi, 1994: 246).

Although the features of the "self" in "self-determination" are the subject of controversy (McVay, 2012: 39), once a "people" have been recognized as a "self," a nation among nations, the loss of statehood, by itself, cannot be said to strip the "people" of their "self." It is therefore clear that the loss of statehood has no bearing on the right to self-determination or the identification of the collective as a "people." If anything, it strengthens it. Statehood is therefore not a legally relevant factor for determining the EEZ rights of the people-beneficiaries. In analogy to the law of trusts, the incapacity of the trustee cannot be said to extinguish the rights of the beneficiaries.

The existence of a territory, however, *is* a relevant factor in the preservation of the "self." At least for purposes of recognizing self-determination, a connection to a territory is a prerequisite (Summers, 2014: 122). This raises the question whether a nation that has been recognized as a "people" with territory could be stripped of the recognition once the territory disappears. It is worth noting that for recognizing a community as deserving of self-determination, the territorial criterion could be satisfied by "real, imagined, past, present, and future" territory (Ohlin, 2016: 80). This is especially true in cases where territorial integrity is compromised. In such instances, historical ties to territory could prevail over territorial integrity (Summers, 2014: 122). For that reason, McVay (2012) concludes that refugees residing *as a collective* in a host country are still deserving of the right to self-determination. As long as they "continue to possess a separate identity," their recognition as a "people" cannot be revoked (McVay, 2012: 46). The key is the character of the people as a *collective* (McVay, 2012). While a link to a territory is fundamental for the identification of a national self, a territory could be lost, be replaced, or even be imaginary. As long as the people maintain their separate identity as a collective, they maintain their identification as a "people."

The identification as a "people" entitled to self-determination is therefore detached from the existence of a territory. The question of territory might be important for the *exercise* of certain rights associated with the right to self-determination, sovereignty prominent among them. But the loss of territory is not an event that undermines the integrity of the "self." In the context of climate change–induced sea-level rise, this conclusion indicates that populations of SIDS states that relocate *as a community* and maintain their separate collective identity will continue to be recognized in international law as a "people."

The maintenance of legal status as a "people" is significant to EEZ rights because under both theories of EEZ, the proprietary and the contractarian, the rights in the EEZ attach to the "people" (Cambou and Smis, 2013: 358). Loss of territory is therefore significant not for the definition of the "self," but rather for the *validity* of the people's claim to the EEZ. While marine entitlements originate in the "people," an entitlement to a *specific* EEZ flows *ab initio* (i.e., from the beginning) from the territory to which the EEZ is attached (Brilmayer and Klein, 2001: 710). It therefore follows that the claim of a relocated climate change refugee community to EEZ rights in its former territory is contingent on the community's ability to demonstrate meaningful connection to that territory. Here, time is of essence. The more time elapses from the date of departure, the harder it is for relocated refugees to demonstrate a lasting and meaningful connection to a previous territory.

International law, however, provides a "safe harbor" for temporary situations that could be used by SIDS communities. As previously noted, in not a few cases, states that have lost one or more of the indicia of statehood enjoyed continued recognition in international law as sovereign entities. The common thread among these precedents is temporariness, the belief that the state of affairs undermining the integrity of the state is provisional. A similar argument could be made by peoples of SIDS. At least in the short term, the argument that dislocation is temporary provides a strong background for maintaining legal connection with the territory. As long as there is a possibility that efforts to combat climate change yield positive results in terms of reducing sea-level rise to the level it was before inundation, a relocated community may still have a valid argument that the lost territory is "theirs."

The Question of Remedies

The identification of EEZ rights as connected to the "people" allows legal scholars to break the logjam that hindered discussions about EEZ rights in a post–climate change world. Moving from abstract discourses to analyses of particular rights allows new thinking about the legal environments surrounding these rights. This movement invites identification and development of institutional arrangements for safeguarding rights and appropriate remedies for potential violations (Sand, 2014: 57). Such arrangements could come in the form of designated guardians with legal powers to represent the people-beneficiaries or the establishment of international institutions to ensure enforcement of the trust (Sand, 2014). The issue of remedies is especially important for a discussion of equity. A trust is a legal instrument of equity. Equity is one of the general principles of law recognized by civilized nations (Sofroniou, 2017: 71), and as such serves as a source of international law (McIntyre, 2007: 122).

The major legal challenges underlying the issue of "sinking" states have no specified recourse in international maritime law. Existing legal frameworks of ocean governance were not designed to address such an unthinkable eventuality. Indeed, "the physical undermining of whole states represents an extreme failure of the international community and constitutes an entirely novel problem for public international law" (Costi and Ross, 2017: 106). However, the fact that there are no specifically designed solutions in law is no reason for allowing injuries to occur. One of the fundamental principles of equity is that "equity will not suffer a wrong to be without a remedy" (McIntyre, 2007: 126).

Three remedies in equity are recognized in international law: equity *infra legem* (adaptation of the law to the circumstances of a particular case to avoid unjust results), equity *praeter legem* (the use of equity to fill gaps in the law), and equity *contra legem* (refusing to apply the law to prevent unjust results) (McIntyre, 2007: 129). Any of these legally recognized remedies could be applied by a competent international tribunal to safeguard the people-beneficiaries' rights under the EEZ trusteeship.

Moreover, specific trust-related remedies in equity are worthy of exploration in the context of SIDS' people's EEZ rights. The first that has precedent in international law is the international equitable trust. An international equitable trusteeship for safeguarding the rights of incapacitated peoples is not a new invention. In 1945 the UN established a trusteeship system that placed vast parts of the planet in a "transitory" trust system designed to support and guide collectives as they transition from incapacity to sovereignty (Head, 2017: 316). This system "reflected the main contours of an equitable trust" (Head, 2017) and may serve as an international precedent for outlining an equitable EEZ trusteeship.

A second trust-related equitable remedy is the constructive trust. A constructive trust applies when certain property "has been acquired in such circumstances that the holder of the legal title may not in good conscience retain the beneficial interest" (*Beatty* v. *Guggenheim Expl. Co.,* 1919: 378). In these instances, equity will vest title in the wronged party and hold the holder of the property as a trustee (*Beatty* v. *Guggenheim Expl. Co.,* 1919). Thus, if no institutional arrangement is advanced to safeguard the rights of SIDS people-beneficiaries, relocated SIDS communities could sue vessels that violate their EEZ rights post-submergence by advancing an equitable trust-based argument. Constructive trusts belong to a class of remedies known as restitution (Pound, 1920: 421). Restitution is a remedy devised to prevent unjust enrichment (Yzenbaard et al., 2017: § 471). Both unjust enrichment and restitution are recognized in international law (*U.S.* v. *Mexico,* 1942). Moreover, unjust enrichment is viewed as a general principle of international law (*Saluka Investment BV (The Netherlands)* v. *The Czech Republic,* 2006: 92) that "stands on equal footing with the general principle of equity" (*Saluka Investment BV (The Netherlands)* v. *The Czech Republic,* 2006).

Lastly, an international tribunal could use equity *infra legem* or equity *praeter legem* (depending on the circumstances of the case) to incorporate into international law the private law equity doctrine of *Cy pres*. The doctrine of *Cy pres* is an ancient equity-based principle of charitable trust law (Chasin, 2015: 1465). It is applied to maintain the operation of charitable trusts whose particular original purpose had become impossible or impracticable to carry out (Restatement (2d) of Trusts, 1959: § 399). Under *Cy pres*, instead of allowing the trust to fail, a court will redesignate the purpose of the trust to some other purpose "that reasonably approximates the designated purpose" (Restatement (Third) of Trusts, 2003: § 67).

Cy pres is not restricted to trust law. Courts and legislators have expanded its operation to the fields of gift law (Sullivan, 2017: 107–108), class action law (Chasin, 2015: 1470), and estate law (*In re Estate of Lamb,* 1971: 49). As an equitable doctrine designed to ease the tension between explicit intent and changing circumstances (Fisch, 1952: 384), it is especially fitting to resource management regimes. *Cy pres* could be applied as a remedy in international law to address the plight of SIDS communities. Using *Cy pres*, the trust regime of the EEZ would be adjusted to maintain its original purposes, that is, benefitting the people of SIDS while maintaining equity, sustainability, and subsistence. For instance, such a regime could impose trusteeship responsibilities on third parties vis-à-vis SIDS communities for a designated period.

Conclusion

Sea-level rise poses unprecedented challenges to existing ocean governance regimes. The legal frameworks governing rights and obligations of states in ocean resources were not designed for, nor do they contemplate, the eventualities brought about by anthropocentric climate change. Strict adherence to existing inflexible frameworks is not only guaranteed to lead to unjust results, it will surely introduce disincentives for some states to "game" the system at the expense of others. The advance of sea-level rise threatens the modern legal regime of ocean governance that is based on the three fundamental principles of peace, equity, and sustainability. To meet the challenges created by climate change, we must search for legal solutions that are adaptive to changing circumstances. The legal doctrine of equity provides competent international tribunals with much-needed flexibility to fill gaps in legal frameworks while maintaining stability. Using equity, courts could adhere to fundamental principles underlying existing governance regimes while tailoring specific solutions that meet the needs of the hour.

Specifically with regard to SIDS, most of the deliberation about their legal options in a post–climate change world has been taking place in the abstract, leading law scholars to conclusions that are in conflict with accepted legal theory and the concrete rights SIDS enjoy under international law (McAdams, 2010: 105). As this chapter has argued, to identify whether a right is extinguished with the loss of legal status, one must answer a series of questions pertaining to the nature of the right, the character of the right's holder, the legal relationship that established the right, and the circumstances that led to the loss of legal status. When this analysis is applied to the case of SIDS, we can see that the right to marine resources is attached to the "people" who are protected under a fiduciary structure designating the state as a trustee. The identification of EEZ rights as belonging to the "people" under a system of trusteeship creates new opportunities for thinking creatively about legal remedies, a path inaccessible when one starts the analysis by assuming that EEZ rights are terminated upon submergence of national territory.

Taking the view of EEZ rights as attached to the "people," potential legal pathways that SIDS communities could pursue to safeguard their EEZ rights become evident. The resulting legal solutions are by no means exhaustive. Moving from abstract discourses to analyses of particular rights allows new thinking about the legal environments surrounding these rights. This movement invites identification and development of legal and institutional arrangements for safeguarding these rights. The paths delineated in this chapter are examples of legal options that become available with the identification of EEZ rights as attached to the "people." These solutions serve as a thought experiment for envisioning legal roads not taken. They may be a grain of hope for landless people and possible ways of thinking about ocean governance to protect the rights and interests of other people threatened by climate change.

References

American Law Institute. (1959). *Restatement (2d) of Trusts § 399.*
American Law Institute. (1987). *Restatement (Third) of The Foreign Relations Law of the United States.*
American Law Institute. (2003). *Restatement (Third) of Trusts § 67.*
American Law Institute. (2008). *Restatement (Third) of Trusts § 2.*
Anderson, T. L. and Leal, D. R. (2001). *Free Market Environmentalism, rev. edn.,* New York: Palgrave Macmillan.
Archer J. H., Connors, D. L., Laurence, K., and Bowen, R. (1994). *The Public Trust Doctrine and the Management of America's Coasts,* Amherst: University of Massachusetts Press.
Barnes R. (2009). *Property Rights and Natural Resources,* Oxford: Hart.
Beatty v. *Guggenheim Expl. Co.* (1919). 225 N.Y. 380, 386.
Brilmayer, L. and Klein, N. (2001). Land and sea: Two sovereignty regimes in search of a common denominator. *NYU Journal of International Law & Policy,* **33**, 703–68.
Burkett, M. (2011). The nation ex-situ: On climate change, deterritorialized nationhood and the post-climate era. *Climate Law,* **2**, 345–74.
Cambou, D. and Smis, S. (2013). Permanent sovereignty over natural resources from a human rights perspective: Natural resources exploitation and indigenous peoples' rights in the Arctic. *Michigan State International Law Review,* **22**(1), 347–76.
Chasin, C. J. (2015). Modernizing class action Cy Pres through democratic inputs: A return to Cy Pres Comme possible. *University of Pennsylvania Law Review,* **163**, 1463–95.
Churchill, A. and Lowe, A. V. (1999). *The Law of the Sea,* Manchester: Manchester University Press.

Costi, A. and Ross, N. J. (2017). The ongoing legal status of low-lying states in the climate change future. In P. Butler and C. Morris, eds., *Small States in a Legal World*. Vol. I. Cham: Springer, pp. 101–38.

Crawford, J. R. (2006). *The Creation of States in International Law*, 2nd edn, Oxford: Oxford University Press.

Crawford, J. R. (2007). *The Creation of States in International Law*, Oxford: Oxford University Press.

Ctr. for Biological Diversity, Inc. v. *FPL Group, Inc.* (2008). 83 Cal. Rptr. 3d 588, 599.

Dupuy R. J. and Vignes D. (1991). *Handbook on the New Law of the Sea*. Vol. I. Leiden; Boston, MA: Martinus Nijhoff Publishers.

Fisch, E. L. (1952). Cy Pres doctrine and changing philosophies. *Michigan Law Review*, **51**, 375–88.

Frey, B. C. (1974). The public trust in public waterways. *Urban Law Annual*, **7**, 219–46.

Frey, B. C. and Mutz, A. (2007). The public trust in surface waterways and submerged lands of the Great Lakes states. *University of Michigan Journal of Law Reform*, **40**(4), 907–94.

Friedheim, R. L. (1993). *Negotiating the New Ocean Regime*, Columbia: University of South Carolina Press.

Head, J. W. (2017). *International Law and Agroecological Husbandry: Building Legal Foundations for a New Agriculture*, London; New York: Routledge.

Hebrew Univ. of Jerusalem v. *Gen. Motors LLC*. (2012). 903 F. Supp. 2d 932, 936.

Hughes v. *Oklahoma*. (1979). 441 U.S. 322.

In re Estate of Lamb. (1971). 97 Cal. Rptr. 46, 49.

Intergovernmental Panel on Climate Change, Working Group II. (2014). *Fifth Assessment Report, Impacts, Adaptation, and Vulnerability, Part B Regional Aspects*.

International Court of Justice Western Sahara Advisory Opinion. (1975). ICJ GL No61, [1975] ICJ Rep 12.

Koskenniemi, M. (1994). National self-determination today: Problems of legal theory and practice. *The International and Comparative Law Quarterly*, **43**(2), 241–69.

Krueger, R. B. (1968). The convention of the continental shelf and the need for its revision and some comments regarding the regime for the lands beyond. *Natural Resources Lawyer*, **1**(3), 1–18.

Leary, D. K. (2007). *International Law and the Genetic Resources of the Deep Sea*, Leiden; Boston, MA: Martinus Nijhoff Publishers.

Lucas v. South Carolina Coastal Council. (1992). 505 U.S. 1003, 1044.

McAdam, J. (2010). Disappearing states, statelessness and the boundaries of international law. In J. McAdam, ed., *Climate Change and Displacement: Multidisciplinary Perspectives*. Oxford: Hart Publishing, pp. 105–28.

McIntyre, O. (2007). *Environmental Protection of International Watercourses Under International Law*, Hampshire: Ashgate Publishing.

McVay, K. (2012). Self-determination in new contexts: The self-determination of refugees and forced migrants in international law. *Utrecht Journal of European and International Law*, **28**(75), 36–52.

Montevideo Convention on the Rights and Duties of States. (1933). 165 L.N.T.S. 19.

Ohlin, J. D. (2016). The right to exist and the right to resist. In F. R. Tesón, ed., *The Theory of Self-Determination*. Cambridge: Cambridge University Press, pp. 70–93.

Osherenko G. (2006). New discourses on ocean governance: Understanding property rights and the public trust. *Journal of Environmental Law and Litigation*, **21**(2), 317–82.

Pound, R. (1920). The progress of the law, 1918–1919: Equity. *Harvard Law Review*, **33**(3), 420–41.

Raic, D. (2002). *Statehood and the Law of Self-Determination*, The Hague; London; New York: Kluwer Law International.

Rayfuse, R. R. (2009). W(h)ither Tuvalu? International law and disappearing states. *UNSW Law Research Paper No. 2009-9*, 1–13.

Rayfuse, R. R. (2010). International law and disappearing states: Utilising maritime entitlements to overcome the statehood dilemma. *UNSW Law Research Paper No. 2010-52*, 1–13.

Reliable Life Insurance v. *Ingle et al.* (2009). CanLII 28225.

Saluka Investment BV (The Netherlands) v. The Czech Republic. (2006). ICGJ 368 (PCA).

Sand, P. H. (2004). Sovereignty bounded: Public trusteeship for common pool resources? *Global Environmental Policy*, **4**(1) 47–71.

Sax, J. L. (1969). The public trust doctrine in natural resource law: Effective judicial intervention. *Michigan Law Review*, **68**, 471–566.

Shaw, M. N. (2014). *International Law*, 7th edn., Cambridge: Cambridge University Press.

Shearer, I. (2013). The limits of maritime jurisdiction. In C. H. Schofield, S. Lee, and M. Kwon, eds., *The Limits of Maritime Jurisdiction*. Leiden; Boston, MA: Martinus Nijhoff Publishers, pp. 51–64.

Sofroniou, A. (2017). *International Law, Global Relations*, World Powers, Andreas Sofroniou.

Stoutenburg, J. G. (2013). When do states disappear? In M. B. Gerrard and G. E. Wannier, eds., *Threatened Island Nations: Legal Implications of Rising Seas and a Changing Climate*. Cambridge: Cambridge University Press, pp. 57–88.

Sullivan, W. P. (2017). The restricted charitable gift as third-party-beneficiary contract. *Real Property, Trust and Estate Law Journal*, **52**, 79–119.

Summers, J. (2014). *Peoples and International Law*, 2nd rev. edn., Leiden; Boston, MA: Martinus Nijhoff Publishers.

The Vienna Convention on Succession of States in Respect of Treaties. (1946). 17 ILM 1488 (1978).

Turnipseed, M., Berkman, J., Blumm, M., et al. (2012). The public trust doctrine and Rio+20. *Third Nobel Laureate Symposium on Global Sustainability*, 1–7.

Turnipseed, M., Roady, S. E., Sagarin, R., and Crowder, L. B. (2009). The silver anniversary of the United States' exclusive economic zone: Twenty-five years of ocean use and abuse, and the possibility of a blue water public trust doctrine. *Ecology Law Quarterly*, **36**, 1–70.

United Nations Convention on the Law of the Sea. (1982). 1833 U.N.T.S. 397.

United Nations General Assembly. (1963). Resolution 1803 Permanent sovereignty over natural resources. U.N. Doc. A/RES/1803.

United Nations General Assembly. (1973). *Resolution 3016 Permanent Sovereignty Over Natural Resources of Developing Countries.* U.N. Doc. A/RES/3016.

United States v. *State of Texas.* (1950). 339 U.S. 707.

United States v. *Mexico.* (1942). US Dept. of State, Publication 2859, Arbitration Series 9,546.

Wilkins, J. G. and Wascom M. (1992). The public trust doctrine in Louisiana. *Louisiana Law Review*, **52**(4), 861–905.

Wood, M. C. (2014). *Nature's Trust: Environmental Law for a New Ecological Age*, New York: Cambridge University Press.

Yzenbaard, C. A., Hess, A. M., Bogert, G. G., et al. (2017). *Bogert's Trusts and Trustees, dig. ed.,* St. Paul, MN: West Academic Publishing.

19

International Law and Marine Ecosystem Governance

The Climate Change Nexus

FREEDOM-KAI PHILLIPS AND KONSTANTIA KOUTOUKI

Introduction

The governance of marine ecosystems under international law sits at a crucial nexus for climate change. This chapter explores legal measures related to climate change and the governance of aquatic ecosystems, including watercourses, wetlands, and biodiversity, to highlight interactions and identify areas of mutual supportive implementation. First, a summary is provided of key legal provisions of the United Nations Framework Convention on Climate Change (UNFCCC) and the Paris Agreement on climate change. Second, an overview is provided of the substantive and procedural obligations found in international instruments governing watercourses, wetlands, and biodiversity. Specifically, emphasis is on provisions of the United Nations Convention on the Law of the Non-Navigational Uses of International Watercourses (1997) (the New York Convention), the Convention on the Protection and Use of Transboundary Watercourses and International Lakes (1992) (Helsinki Convention), the Convention on Wetlands of International Importance especially as Waterfowl Habitat (1971) (the Ramsar Convention), and the Convention on Biological Diversity (CBD, 1992). Third, key synergies across regimes are identified, thereby illustrating legal and policy intersections. Finally, modalities to further strengthen marine governance are discussed in the climate context to illustrate mutually supportive legal measures which assist in the achievement of Sustainable Development Goals (SDGs) related to water (SDG 6), oceans (SDG 14), climate (SDG 13), and biodiversity (SDG 15). Holistic governance of marine ecosystems allows for informed policy trade-offs, supports carbon sequestration and climate resilience, and facilitates effective mitigation and adaptation practices.

Legal Measures Under the UNFCCC and the Paris Agreement

International efforts to address climate change were initially catalyzed with the passage in 1992 of the UNFCCC at the United Nations Conference on Environment and Development in Rio de Janeiro (the so-called Earth Summit). Marine ecosystems were recognized under the UNFCCC both as valuable sinks for greenhouse gas (GHG) emissions and as particularly exposed to changes in global climate

(UNFCCC, 1992: Preamble). Article 4.1(d) and (e) recognize the importance of promoting sustainable management of both terrestrial and marine sinks and fostering cooperation in strategic planning relating to climate change adaptation, preparation of coastal zone and water management, and ecosystem restoration (UNFCCC, 1992, Article 4.1[d–e]). Parties are obliged to protect GHG sinks, as well as to monitor and report emission reductions as a result of national sinks and reservoirs (UNFCCC, 1992: Article 4.2[a–c], 12.1–2). Oceans and coastal zones are estimated to have sequestered 40,000 gigatons ($GtCO_2$), or 93 percent of total historical emissions (Nellemann et al., 2009: 11). Aquatic ecosystems have been shown to be significantly more efficient as carbon sinks than terrestrial ecosystems (Duarte et al., 2010; Lavery et al. 2013; Chmura 2014: 25–31), with wetlands, tidal salt marshes, mangroves, and seagrass meadows covering only 9 percent of comparable territory yet providing, at minimum, 27 to 100 times more sequestration potential (Laffoley and Grimsditch, 2009: 48).

Like the UNFCCC, the 1997 Kyoto Protocol emphasized the importance of preservation and enhancement of GHG sinks in line with obligations under relevant international instruments (Kyoto Protocol, 1997: Article 2.1[a][ii]). The carbon sinks resulting from direct human-induced land-use changes and forestry activities that were identified as needing verification in national-emissions stock taking were afforestation, reforestation, and deforestation (Kyoto Protocol, 1997: Article 3.3). Parties to the Kyoto Protocol are required to implement a system for the estimation of emissions and to report annually, including on removal of emissions by sinks (Kyoto Protocol, 1997: Article 5.1, 7.1). Those parties listed in Annex I (i.e., developed states) may use the Clean Development Mechanism (CDM) to transfer credits generated from the restoration of carbon sinks in non-Annex I jurisdictions to offset or reduce their national emissions (Kyoto Protocol, 1997, Article 6.1(b), 12.3). Regardless of the strides made by the Kyoto Protocol, in 2015 global emissions exceeded 36.2 $GtCO_2$, with the Intergovernmental Panel on Climate Change (IPCC) projecting continued increases in sea level, surface temperature, and ocean acidification (IPCC, 2014: 10–13).

The Paris Agreement, which was adopted in 2015 and entered into force on November 4, 2016, acknowledges in its preamble that climate change is a common concern of mankind, emphasizing the importance of protecting biodiversity and maintaining ecosystem integrity in particular oceans (Paris Agreement, 2015: Preamble). Key objectives are established, including holding global temperature rise well below 2°C in pursuit of only a 1.5°C rise (Paris Agreement, 2015: Article 2.1[a]), fostering resilience to climate-related impacts, and mobilizing finance flows as a pathway to low emission and climate resilient development (Paris Agreement, 2015: Article 2.1[b–c]). Enhancing adaptive capacity is identified as a global goal, with parties acknowledging that adaptation should follow a country-driven, gender-responsive, and participatory approach that considers vulnerable groups, communities, and ecosystems, and is grounded in the best available scientific information and, where appropriate, traditional knowledge (Paris Agreement, 2015: Article 7.5). National commitments are made under the Paris Agreement by way of nonbinding, nationally determined contributions (NDCs) (Paris Agreement, Article 4.3).

Reviewed every five years, a total of 157 NDCs have been communicated to the secretariat from 197 jurisdictions (Paris Agreement, 2015: Article 4.9; NDC Registry, 2017).

Measures to achieve the Paris goals are enshrined in Article 6 and are aimed at promoting adaptation and mitigation actions, fostering sustainable development, and generating transferable mitigation outcomes toward domestic NDCs (Paris Agreement, 2015: Article 6.1–3). Article 6.4 provides a market-based mechanism to support sustainable development by positively incentivizing emission reductions, while Article 6.8 offers an alternative holistic and balanced nonmarket mechanism aimed at fostering mitigation and adaptation actions, enhancing public–private partnerships, and enabling institutional coordination across relevant international instruments (Paris Agreement, 2015, Article 6.4, 6.8). A commitment is made by the parties to strengthen cooperation to enhance technology development and transfer, capacity building, climate change education, and public participation, with implementation of the agreement to be conducted in accordance with the principle of common but differentiated responsibilities (Paris Agreement, 2015, Article 2.2, 10–12).

Parties to the agreement further emphasize the need to cooperate to avert, minimize, and address climate change loss and damage, including actions relating to early warning systems; risk assessments; and resilience of communities, livelihoods, and ecosystems (Paris Agreement, 2015: Article 8.4[a], 8.4[e], 8.4[h]). An enhanced transparency and accountability reporting framework is established to provide flexibility domestically and internationally a clear picture of climate change actions which contribute to achievement of the NDCs (Paris Agreement, 2015: Article 13). The significance of the marine/climate nexus is exemplified by specific references to wetland and coastal ecosystems across 28 and 59 NDCs, respectively, without taking into consideration references to the broader Land Use and Land Use Changes and Forestry (LULUCF) (Herr and Landis, 2016: 8–9; Martin et al., 2016). Global climate efforts are encapsulated in SDG 13 on climate change, which encourages action under the UNFCCC process.

Legal Measures Governing Watercourses, Wetlands, and Biodiversity

United Nations Convention on the Law of the Non-Navigational Uses of International Watercourses (New York Convention)

The New York Convention, adopted May 21, 1997, and entered into force August 17, 2014, currently has 36 parties (UNTS No. 52106). The convention principally aims to harmonize key terms and provisions of watercourse agreements; promote cooperation among riparian states in support of conservation, protection, and management efforts; and provide a constructive avenue for dispute settlement (New York Convention, 1997: Preamble, Article 3.2, 6.2, 33). The convention has an operative scope inclusive of international watercourses for the purposes of protection, preservation, and management, with "watercourse" broadly defined to include the water system as well as surface and groundwater as a unitary whole (New York Convention, 1997: Article 1–2). Using an ecosystem approach to governance, general principles include (1) equitable and reasonable utilization, (2) prevention of significant harm being caused to another state, (3) good faith cooperation through joint commissions or mechanisms, and (4) regular sharing of data (hydrological, meteorological, hydrogeological, and ecological) with other watercourse states (New York Convention, 1997: Article 5, 7–9).

Aiming to foster environmental protection of the watercourse, states are individually or jointly obliged to preserve and protect ecosystems in multiple ways. These include through prevention and control of pollution (New York Convention, 1997: Article 20, 21.2), prevention of alien species introduction (New York Convention, 1997: Article 22), preservation of the marine environment as a contiguous whole and inclusive of estuaries (New York Convention, 1997: Article 23), utilization of consultations relating to ecosystem management (New York Convention, 1997: Article 24–25), and utilization of best available efforts to maintain the integrity of installations on the watercourse (New York Convention, 1997: Article 26.1). Parties must take reasonable efforts to prevent and mitigate conditions harmful to the watercourse from both human and natural causes, with a duty of consultation in place where there is a reasonable belief that a work may result in significant adverse effects (New York Convention, 1997: Article 11–12, 26.2, 27). Disputes arising under the convention may be resolved through special arbitration procedures, or if the parties elect otherwise they may utilize third-party arbitration or submit the dispute to the International Court of Justice (ICJ) or other applicable arbitral forums (New York Convention, 1997: Article 33, Annex).

Convention on the Protection and Use of Transboundary Watercourses and International Lakes (Helsinki Convention)

The Helsinki Convention, adopted on March 17, 1992, and entered into force October 6, 1996, currently has 41 parties (UNTS No. 33207). Originally agreed by the senior advisors to the Economic Commission for European Governments on Environmental and Water Problems as a regional instrument, on February 6 2013, the convention was opened for accession by all UN member states, thus making the convention a global water governance instrument (UNECE, 2004). The convention aims to foster cooperation for the prevention and mitigation of transboundary pollution of watercourses, to provide for ecologically sound and rational utilization, and to encourage conservation and restoration actions (Helsinki Convention, 1992: Article 2.1–2). Currently, the convention encompasses a variety of transboundary bodies of water, including 140 rivers, 25 lakes, 200 groundwater aquifers, and 25 wetlands of international importance identified list as "Ramsar sites" under the Ramsar Convention (on Ramsar, see later) (UNECE 2011). Two key definitions are provided by the convention: first, "transboundary waters" are defined to include both surface and groundwaters that mark, cross, or are located on boundaries with at least one other jurisdiction; and second, "transboundary impact" is defined as a significant adverse environmental effect caused by human activity that changes the conditions of the transboundary waters and has impacts on human health, biodiversity, climate, historical and cultural structures, cultural heritage, and socioeconomic conditions (Helsinki Convention, 1992: Article 1.1–2).

The convention is grounded in the precautionary principle, the polluter pays principle, and the principle that water resources should be managed for both present and future generations (Helsinki Convention, 1992: Article 2.5). Measures are established to prevent, control, and reduce the introduction of pollution into transboundary waters, including (1) utilization of low and nonwaste technology; (2) prior licensing and discharge limits for point-of-source wastewater discharge; (3) strict requirements or prohibition where ecosystem water quality is declining; (4) at a minimum, biological treatment for municipal wastewater; (5) application of best available technology to reduce nutrient concentrations in discharges from industrial, municipal, and agricultural sources; (6) application of environmental impact assessments and sustainable water resource management grounded in the ecosystem approach; (7) development of contingency planning; (8) implementation of measures to prevent pollution in groundwater; and (9) use of precautionary measures to minimize risks associated with accidental pollution (Helsinki Convention, 1992: Article 3.1). Emission limits are to be established at the domestic level to prevent, control, and reduce pollution discharge from industrial sectors based on defined water quality limits (Helsinki Convention, 1992: Article 3.2). Development of legal and policy measures on a domestic or multilateral basis is encouraged to facilitate pollution prevention, research, and development based on available technologies and techniques, data evaluation and joint monitoring, and exchange of information relating to changes in ecological character among riparian parties (Helsinki Convention, 1992: Article 5–6, 9, 11, 13). Settlement of disputes is provided for through submission to the ICJ or arbitration in accordance with procedures established in a specialized annex (Helsinki Convention, 1992: Article 22, Annex IV).

Convention on Wetlands of International Importance Especially as Waterfowl Habitat (Ramsar Convention)

An early multilateral environmental agreement, the Ramsar Convention was adopted February 2, 1971, entered into force February 17, 1976, and has 169 Parties (UNTS No 14583; Ramsar Convention, 2013). The convention aims to promote conservation, management, and "wise use" of wetlands, defining "wetland" broadly to include natural, artificial, permanent, or temporary marsh, fen, or peatlands, as well as marine water not exceeding six meters in depth at low tide (Ramsar Convention, 1971: Article 1.1, 3.1). Wetlands of ecological significance are selected for inclusion on the List of Wetlands of International Importance, with parties required to establish domestic planning measures to promote the conservation and wise use of wetlands, monitor for likely or existing human-induced changes in ecological character, and provide notification to the secretariat where changes are observed (Ramsar Convention, 1971: Article 3.2, 8.2). Key terms under Article 3.2, notably "ecological character" and "change in ecological character," were defined by the Conference of the Parties (COP) at COP VI (1996), with the former term referring to "the structure and interrelationships between the biological, chemical, and physical components of the wetland [which] derive from the interactions of individual processes, functions, attributes, and values of the ecosystem(s)," and the latter term defined as "the impairment or imbalance in any of those processes and functions which maintain the wetland and its products, attributes and values" (Ramsar COP Resolution VI.1: para 1.1–1.2).

Article 8.2 establishes the organizational modalities of the secretariat in relation to party reporting obligations under the convention. The responsibilities of the Ramsar secretariat comprise the organization of the meetings of the COP (Ramsar Convention, 1971: Article 8.2[a]); maintenance of "the list," including extensions, deletions, or restrictions made by parties (Ramsar Convention, 1971: Article 8.2[b]); and acting as a focal point for contact from parties, including receiving notifications regarding changes in ecological character (Ramsar Convention, 1971: Article 8.2[c]). The secretariat, upon notification by a party or a third party (party, nonparty, or civil society organization), subsequently informs the COP of all notifications received relating to alterations to the list and changes in ecological character (Ramsar Convention, 1971: Article 8.2[d]) to establish recommendations regarding the sites (Ramsar Convention, 1971: Article 8.2[e]). Options include mobilization of a Ramsar Advisory Mission (RAM) to bring international expertise to assist in identifying appropriate actions (Ramsar COP Resolution VIII.8: para 19) and inclusion of the site on the Montreux Record indicating significant adverse change "has occurred, is occurring or is likely to occur," so as to guide implementation of available technical and financial resources (Ramsar COP Resolution VI:1: para 3.1; Ramsar COP Resolution V.4).

Convention on Biological Diversity

A second of the three "Rio Treaties" put forward at the Rio Earth Summit, the CBD is the principal international instrument governing biodiversity (Cordonier Segger and Khalfan, 2004: 15–18). Entering into force December 29, 1993, and currently with 196 parties, the CBD objectives are the conservation and sustainable use of biodiversity and the fair and equitable sharing of benefits resulting from its utilization (CBD, 1992: Article 1). Two key definitions are of note, "biological diversity," which includes all living organisms, as well as terrestrial, marine, and other aquatic ecosystems, and "ecosystem" inclusive of the dynamic interrelationships of plant, animal, and microorganism communities as a functional unit (CBD, 1992: Article 2). While holding sovereign rights over exploitation of biological resources, parties are obliged to ensure activities under their jurisdiction do not cause adverse environmental damage extraterritorially or to areas beyond national jurisdiction (CBD, 1992: Article 3). Additionally, parties are to develop National Biodiversity Strategies and Action Plans (NBSAP) as a mechanism to progressively integrate and monitor biodiversity indicators in policy planning (CBD, 1992: Article 7).

Under the CBD, in situ conservation is to be fostered through (1) establishment of a system of protected areas; (2) development of management guidelines; (3) conservation and sustainable use of biodiversity both inside and outside of protected areas; (4) promotion of protective measures for species, ecosystems, and natural habitats; (5) environmentally sound and sustainable development adjacent to protected areas; (6) ecosystem restoration and rehabilitation; (7) control of risks related to the use of living modified organisms; (8) prevention of the introduction of alien invasive species; (9) nurturing of conditions conducive to conservation and sustainable use; (10) preservation, protection, and respect for traditional knowledge (TK) relating to biodiversity; (11) development of legislation for the protection of threatened species; (12) regulation of relevant activities that cause significant adverse effect on biodiversity; and (13) cooperation relating to financial and technical support (CBD, 1992: Article 8).

Additionally, parties are required to establish measures that promote and incentivize conservation and sustainable use of biodiversity ex situ, delivery of biodiversity-related education and public mainstreaming, development of appropriate environmental impact assessment mechanisms, access and benefit sharing (ABS) relating to the utilization of genetic resources (GRs) and TK based on fair and equitable terms, and fostering of technology transfer and information relating to biodiversity conservation (CBD, 1992: Article 9–17). The ABS provisions of the CBD were further developed in the Nagoya Protocol, with the core components of prior informed consent (PIC) and mutually agreed terms (MAT) acting as the bedrock for engagement with Indigenous Peoples and Local Communities (ILCs) pertaining to the utilization of biological resources (Nagoya Protocol, 2010: Article 1, 3, 5–7, 12).

The Marine and Climate Change Nexus: Legal and Policy Intersections

International legal obligations relating to watercourses, wetlands, and biodiversity have significant legal and policy intersections at the nexus of the United Nations Convention on the Law of the Sea (UNCLOS) and the global climate framework. First, the scope of environmental protection obligations under UNCLOS is influenced and informed by obligations under other multilateral environment agreements (MEAs). In the South China Sea dispute, the tribunal noted that the general obligation for environmental protection found in Article 192 of UNCLOS was informed through reference to specific obligations relating to prevention of pollution found in Article 194(1) and protection of fragile ecosystems in Article 194(5) (UNCLOS, 1982: Article 192, 194[1], 194[5]), legal obligations established in relevant international conventions, and the "general corpus of international law" (South-China Sea PCA Case No 2013–19, Award: para 914–44, 959). Such a broad interpretation by the tribunal has implications for the climate regime as it recognizes the practical implications of ecosystem interconnection, reinforces the integrity of obligations found under the parallel regimes, and demonstrates the close relationship of fundamental international legal norms (Phillips and Maciunas, 2017). Obligations relating to "rational use" under international law, "wise use" under the Ramsar Convention, "optimal utilization" and "rational management" under UNCLOS, "sustainable use" under the CBD, "ecologically sound and rational water management" under the Helsinki Convention, "protection, preservation and management" under the New York Convention, and "conservation and enhancement" of GHG sinks under the Paris Agreement work synergistically, rather than in isolation, to inform the marine climate nexus. Obligations relating to upstream or transboundary pollution prevention, conservation and sustainable use of biodiversity, and protection of wetlands are all functionally encompassed under the core environmental provisions of UNCLOS.

Second, prevention of anthropogenic emissions, while the principal subject matter of the UNFCCC, is nonetheless substantively referenced under UNCLOS. Implicitly included under the scope of Article 192, prevention of atmospheric emissions is explicitly addressed in Article 212, which requires parties to prevent, reduce, and control pollution of the marine environment from or through the atmosphere and work through competent international organizations to establish global rules, standards, and practices (UNCLOS, 1982; Article 212). Additionally, Article 237 requires parties to implement obligations developed under special instruments relating to the protection and preservation of the marine environment in a consistent and mutually supportive manner with the obligations under UNCLOS (UNCLOS, 1982; Article 237). Together, these obligations create an interconnected system that reinforces the importance of applying legal and policy tools, both domestically and internationally. Forums such as the COP and intersessional meetings of each respective instrument, and the United Nations Environment Assembly (UNEA) of the United Nations Environment Programme (UNEP) provide valuable modalities for harmonization of broad climate priorities to achieve domestic NDCs in accordance with obligations under parallel instruments. Strategic planning tools, such as the NBSAP under the CBD, or technical assistance provided through RAM under the Ramsar Convention, inform domestic management of marine and climate drivers. Furthermore, funding mechanisms such as the Green Climate Fund (GCF) and the Global Environment Facility (GEF) support establishment of appropriate legal and governance frameworks at the domestic level. Ecosystem-based management practices support informed trade-offs in pursuance of sustainable development, mutually supportive implementations of international legal obligations related to the aquatic environment, and achievement of domestic NDCs.

Third, the responsibility and liability standard under UNCLOS has potential implications influencing global climate efforts. Article 235 of UNCLOS provides that states are responsible for the fulfilment of international obligations to preserve and protect the marine environment and subsequently hold liability under international law (UNCLOS, 1982; Article 235[1]). States are obliged under international law to exercise due diligence, including "use all means at [their] disposal" to ensure domestic activities under their authority do not cause transboundary environmental damage (*Pulp Mills on the River Uruguay,* 2010 ICJ Reports 14: para 101). The 2011 Advisory Opinion on Activities in the Area delivered by the Seabed Dispute Chamber of the International Tribunal for the Law of the Sea (ITLOS) noted and elaborated upon this due diligence standard (*Activities in the Area,* 2011 Case No. 17, ITLOS Dispute Chamber: para 111, 114–115). Recognizing that the standard is variable and must be proportionate to the associated risk, the tribunal notes that due diligence encompasses both procedural and direct obligations (*Activities in the Area,* 2011: para 117, 121; Freestone and Phillips, 2017: 328–30). Due diligence is identified as placing on states requirements of both a conduct-based approach – "a duty to ensure" – and direct responsibilities to implement the precautionary approach, application of best environmental practices, availability of specialized protective measures in case of emergency, utilization of environmental impact assessment (EIA), and availability of resources to compensate for environmental pollution (*Activities in the Area,* 2011: para 117, 121–23, 127–35).

In 2015, ITLOS again interpreted and applied the due diligence standard to obligations of marine environmental protection in their decision Request for an Advisory Opinion Submitted by the Sub-regional Fisheries Commission (SRFC) (*Sub-regional Fisheries,* 2015 Case No.21 ITLOS: para 120, 125–33, 146; Freestone and Phillips, 2017: 333–34). Referencing the standard defined in Activities in the Area, the tribunal recognizes that a state would only be freed of liability where it has taken "all necessary and appropriate measures" to effectively discharge the due diligence obligation (*Sub-regional Fisheries,* 2015: para 148–49, 168). Additionally, the tribunal notes that obligations to cooperate, investigate impacts, and take remedial action as necessary are aspects of due diligence (*Sub-regional Fisheries,* 2015: 61). The conduct-based elements of the due diligence standard are effectively supported through the cooperative, collaborative, and reporting mechanisms found in the CBD and the New York, Helsinki, and Ramsar Conventions. Esteemed legal scholar Philippe Sands suggests that the international landscape is ripe for an advisory opinion relating to the scope of marine ecosystem obligations in light of continued anthropogenic pollution, with ITLOS being the ideal forum (Sands, 2016: 19–35). Where a state fails to use best available efforts to prevent continued emissions growth through achievement of their domestic NDC, it is *prima facie* plausible that this could be considered a breach of the environmental protection obligations under UNCLOS, resulting in exposure to potential liability for emissions damage.

Lastly, marine ecosystem protection in light of continued climate change implicates a range of intersections across the SDGs, including food (SDG 2), water (SDG 6), sustainable consumption and production (SDG 12), climate (SDG 13), oceans (SDG 14), and biodiversity (SDG 15). Relating to oceans, SDG 14.1 on prevention of marine pollution from land-based sources has broad implications on SDG 2.3 and 2.4 related to increasing agricultural productivity and food production, SDG 6.3 on pollution reduction in water, SDG 8.4 related to resources efficiency in consumption and production, and SDG 12.4 addressing environmentally sound management of chemicals (2030 Agenda, 2015: SDG 2.3–2.4, 6.3, 8.4, 12.4). SDG 14 broadly, specifically SDG 14.2 on marine ecosystem protection, SDG 14.3 on halting ocean acidification, SDG 14.5 on preservation of coastal areas, and tangentially SDG 14.7 on payment for ecosystem services all have intersections with SDG 6.6 on the protection of water-related ecosystems (2030 Agenda, 2015: SDG 14, 6.6). SDG 15 on terrestrial biodiversity, specifically SDG 15.1 on conservation and sustainable use of terrestrial and inland freshwater ecosystems, SDG 15.6 on promotion of fair and equitable benefit sharing, in particular relating to marine genetic resources, SDG 15.9 on integration of ecosystem values into strategic planning and decision-making processes have direct impact on the achievement of SDG 14 on oceans and SDG 13 on addressing climate change.

Relating to climate change, SDG 13.1 on strengthening climate resilience, SDG 13.2 on integration of climate measures into national planning, and SDG 13.3 on mainstreaming and capacity building are crucial for the achievement of SDG 14 on oceans and SDG 15 on biodiversity. Modalities for payment for ecosystem services, creation of marine protected areas, broadening of access and benefit sharing, and strategic planning through an NBSAP supports synergistic achievement of SDG 13 on climate and SDG 14 on oceans (2030 Agenda, 2015: SDG 15. 13–14).

Areas of Focus for Strengthening Marine Ecosystem Governance

There are many opportunities for strengthening marine ecosystem governance, for example, through "source-to-sea" governance, incentives to promote conservation and sustainable use of resources, equitable use of traditional knowledge, and greater regulation of ocean areas outside national jurisdictions.

Adoption of Source-to-Sea Governance

Concurrent achievement of enhanced climate reliance, protection of marine resources, and economic growth in accordance with sustainable development principles requires comprehensive national resource governance that fosters synergistic and mutually supportive objectives. Evolving from the "ridge-to-reef concept" (UNEP, 1995), adoption of source-to-sea governance models has contributed to achieving integrated water resource management. Grounded in the ecosystem approach, source-to-sea conceptualizes governance of aquatic environments as a contiguous whole based on system flows (water, sediment, pollution, biota, material, and ecosystem services) and interconnected ecological systems (land, freshwater, estuaries, deltas, coastline, sea self, ocean) (Granit et al., 2017: 5–8). Key areas of focus are wetlands, in particular mangrove forests, which are vital for carbon sequestration but also enchased climate resilience for small-island developing states (SIDS).

Achievement of the goals established under the Paris Agreement, both in terms of reduction of emissions and preservation of carbon sinks, requires identification of domestic drivers and development of localized strategic initiatives harmonized through a source-to-sea approach. For example, Finland has established protection plans for water management broadly, groundwater aquifers, river basins, and marine areas grounded in harmonized policy priorities established with regional partners for transboundary continuity (Laki vesienhoidon ja merenhoidon järjestämisestä, 2004: §1 §13, §10, §11, §26). All domestic climate action must be implemented in accordance with the water management framework (Laki tulvariskien hallinnasta, 2010: §2, §7–10, §12; Ilmastolaki, 2015: §2–3). Source-to-sea governance allows for informed trade-offs and comprehensive management of marine ecosystems, while also fostering mitigation and adaptation efforts that enhance climate resilience.

Positive Incentives for Conservation and Sustainable Use

Promotion of conservation and sustainable use practices through establishment of domestic incentive schemes, localized governance models, and multistakeholder participation support concurrent achievement of marine and climate priorities (TEEB, 2010: 12–18,). Since 1998, Costa Rica has employed a payment for ecosystem services (PES) program to compensate local communities or farmers for protection of biodiversity-rich regions, positively incentivizing conservation and sustainable use of biodiversity (Ley N° 7788 1998: Article 37; Cabrera et al., 2014: 18–19). Modalities are established to remunerate participants of the PES program for conservation of national priority areas, voluntary inclusion of private property as a conservation area, management of national protected areas, and preferential tax incentives (Ley N° 7788 1998: Article 37, 100). Similarly, China has developed a PES program under the Ecological Conservation Compensation (ECC) system to provide positive incentives for conservation of forests, wetlands, sea, water, desert, and grasslands through the mobilization of capital (Environmental Protection Law, 2014: Article 31; Opinions of the General Office of the State Council, 2016).

The Democratic Republic of the Congo (DRC) allows for the granting of forest concessions to eligible ILCs for sustainable management and economic exploitation (Loi n° 011/2002: Article 82, 84, 96–97, 99–100). Classified as "local community forests," specialized procedures are established for the granting of forest concessions to ILCs, including inquiry by the local government, issuance of public notice for comment, a maximum scale of 50,000 hectares (ha), development of a map of the area in consultation with neighboring communities, and the ability to structure the concession as private entity (for profit or non-for-profit), or license the concession to a third party to manage (Décret n° 14/018: Article 4–6, 8–12, 18–19, 20–21). Climate-related shifts disproportionately affect ILCs, particularly in LDCs and SIDS. Positive incentivization and localization of governance assist in operationalizing the marine climate nexus and empowering ILCs to engage in mitigation and adaptation efforts.

Equitable Use of Traditional Knowledge to Support Sustainable Development

Development of mechanisms that recognize, protect, and reinforce rights of ILCs to TK and facilitate utilization based on principles of ABS promotes sustainable development (Phillips, 2016). Region-specific environmental techniques and expertise inherent in TK provide a gateway to sustainability practices applied by ILCs for generations (Cordonier Segger and Phillips, 2015). South Africa established a general duty of care for the environment under a system of coastal protection zones, along with modalities for cooperative management with the tribal councils of local communities (National Environmental Management Act, 1998: Article 28, 34–35; Integrated Coastal Management Act, 2008: Article 28, 34–35). This localization of governance over both marine and terrestrial protected areas is coupled with a permit regime for access and utilization of TK based on PIC and establishment of MAT (Biodiversity Act, 2004: Article 81–84; Regulations on Bioprospecting and Access and Benefit, 2008, Article 8[1]) and a requirement for users to disclose in the patent process where the invention was based on TK (Act No. 20 of 2005: Section 2), along with robust guidelines informing users of their obligations (Guidelines for Providers, Users and Regulators, 2012; Phillips, 2016: 301–02).

Global Landscape Funds (GLFs) privately develop funds that provide capital investment or risk assurance to landscape-based initiatives, while providing a medium for the application of TK aimed at providing tangible returns on investments made in developing or emerging markets. Returns may be monetary in nature or through conservation, ecosystem restoration, generation of carbon credits, or development of sustainable livelihoods for local communities. One of the many GLFs in operation, the Livelihoods Fund, developed out of a partnership with Danone, the Ramsar Convention, and IUCN. The Livelihoods Fund provided funding to the Oceanium project supporting reforestation of over 12,000 ha of mangroves in Senegal, with the project leader, Director Jean-Christophe Henry, winning a Ramsar Award in 2015 (Livelihoods Fund; Ramsar, Award for Innovation 2015). Other GLFs include the Moringa Partnership "Moringa Fund," the Terra Global Capital "Terra Bella Fund," the Conservation International Critical Ecosystem Partnership Fund, the African Development Bank "Congo Basin Forest Fund," and the African Agricultural and Trade Investment Fund. By making capital available for climate-focused, ecosystem-related initiatives that are grounded in a respect for the rights of ILCs, ecosystem-related innovation such as techniques for conservation and restoration of carbon sinks can be fostered in a fair, equitable, and inclusive manner.

Governance of Areas Beyond National Jurisdiction

Governance of marine ecosystems on the high seas, classified as Areas Beyond National Jurisdiction (ABNJ), is a crucial gap in implementation of SDG 13 and SDG 14. Under UNCLOS, ABNJ are defined as the water column adjacent to the exclusive economic zone (EEZ), including the seabed beyond the continental shelf, with the deep seabed uniquely classified as "common heritage of mankind." These account for 40 percent of the global surface area, 64 percent of the oceans' surface, and 95 percent of total ocean volume (UNCLOS, 1982: Article 86, 125, 136; FAO, 2016: 4–5). In July 2015 the UN General Assembly (UNGA), recognizing a call to foster conservation and sustainable use of marine biodiversity beyond national jurisdictions (BBNJ) at Rio +20, decided to establish a legally binding instrument under UNCLOS (A/RES/69/292 2015: Preamble). Meetings of the Preparatory Committee (PrepCom) were held in both 2016 and 2017 (UN Division for Ocean Affairs and the Law of the Sea, 2017, Res. 69/202). Early negotiation drafts provide a broad set of inclusive definitions, a geographic and material scope, and core objectives of conservation and sustainable use of BBNJ, with implementation intended to be done in a manner mutually supportive of UNCLOS, UNFSA, and the CBD (UN Chair's Non-Paper, 2017; 5–15). In July 2017, following PrepCom 4, the committee approved areas of convergence for an international legally binding instrument and recommended that the UNGA converge an intergovernmental conference. The first session of Intergovernmental Conference on an international legally binding instrument under the UN Convention on the Law of the Sea (UNCLOS) on the conservation and sustainable use of marine biological diversity of areas beyond national jurisdiction (BBNJ) was held in September 2018 with negotiations expected to be ongoing throughout 2019–2020. Progress, while slow, remains positive to further clarify obligations of states to conserve, protect, and sustainably use the global oceans.

Conclusion

Attainment of the global goal of holding temperature rise below 2°C as per the Paris Agreement and SDG 13 requires extensive and inclusive social, legal, and economic reform. Increased emphasis must be placed on the marine–climate nexus to promote increased climate reliance, enhanced carbon sequestration, and expanded application of source-to-sea principles. Harmonization of polices across jurisdictions allows for informed upstream/downstream developmental trade-offs, supports identification of climate-induced tipping points, and contributes to long-term strategic planning. Legal obligations under the UNFCCC are deeply interconnected with the environmental protection components of UNCLOS, the CBD, and the New York, Helsinki, and Ramsar Conventions. Under each of these instruments, modalities can be found for strategic planning, data sharing, consultation, cooperation, and mobilization of financial and technical capacity to provide a network of forums, tools, and initiatives. These modalities can help to comprehensively govern the marine environment in a way that supports continued climate change resilience and ongoing emissions reductions.

Legal and policy approaches that promote source-to-sea governance, as seen in Finland; positive incentivization of conservation efforts through payment for ecosystem services or local forest concessions, as seen in Costa Rica and the DRC, respectively; and localized governance allowing for equitable application of traditional knowledge, as seen in South Africa synergistically support protection of the marine environment and enhanced climate resilience through conservation of carbon sinks, operationalizing Sustainable Development Goals on climate and oceans. Further work to solidify an international legally binding instrument on marine biodiversity beyond national jurisdictions is needed to clarify international obligations, create institutional capacity, and mobilize technical and financial resources to assist in the preservation of oceans globally. Achievement of the goals of the Paris Agreement requires a further strengthening of the marine–climate nexus to ensure global climate efforts are not undermined by inadequate attention to comprehensive marine ecosystem preservation.

References

Breithaupt, J. L., Smoak, J. M., Smith, T. J., Sanders, C. J., and Hoare, A. (2012). Organic carbon burial rates in mangrove sediments: Strengthening the global budget. *Global Biochemical Cycles*, **26**(3). Available at onlinelibrary.wiley.com/doi/10.1029/2012GB004375/full.

Cabrera, J., Phillips, F. K., and Perron-Welch, F. (2014). *Biodiversity Legislation Study: A Review of Biodiversity Legislation in 8 Countries*. Hamburg, Germany: World Future Council. Available at www.worldfuturecouncil.org/fileadmin/user_upload/PDF/Final_Biodiversity_Legislation_Study.pdf.

Case Concerning Pulp Mills on the River Uruguay. Argentina v. Uruguay, 2010 ICJ Reports 14. Available at www.icj-cij.org/docket/files/135/15877.pdf.

China. (2014). Environmental Protection Law of the People's Republic of China, *Chinadialogue*, April 24, 2014. Available at www.chinadialogue.net/Environmental-Protection-Law-2014-eversion.pdf.

China. (2016). Opinions of the General Office of the State Council on Improving the Compensation Mechanism for Ecological Protection (April 28, 2016), State Council Issue No 31. Available at www.gov.cn/zhengce/content/2016-05/13/content_5073049.htm.

Chmura, G. L. (2014). What do we need to assess the sustainability of the tidal salt marsh carbon sink? *Ocean and Coastal Management*, **83**, 25–31. Available at sites.duke.edu/bluecarbonmastersproject/files/2012/03/Chmura-2011-sequestration-.pdf.

Convention on Biological Diversity. (1992). June 5, 31 ILM 822 (entered into force December 29, 1993).

Convention on the Law of the Non-Navigational Uses of International Watercourses. (1997). May 21, annexed to GA Res 51/229, UNTS 52106 (entered into force August 17, 2014).

Convention on the Protection and Use of Transboundary Watercourses and International Lakes. (1992). March 17, 1936 UNTS 269 (entered into force October 6, 1996).

Convention on Wetlands of International Importance Especially as Waterfowl Habitat. (1971). February 2, 1996 UNTS 245 (entered into force December 21, 1975).

Costa Rica. (2008). Ley N° 7788 del 30/04/1998 de Biodiversidad (modificada por última vez por la Ley N° 8686 del 21 de noviembre de2008). Available at www.wipo.int/edocs/lexdocs/laws/es/cr/cr082es.pdf.

Democratic Republic of Congo. (2002). Loi n° 011/2002 du 29 Aout 2002 portant code forestier. Available at siteresources.worldbank.org/EXTINSPECTIONPANEL/Resources/Annex16.pdf.

Democratic Republic of Congo. (2014). Décret n° 14/018 du 02 août 2014 fixant les modalités d'attribution des concessions forestières aux communautés locales. Available at faolex.fao.org/docs/pdf/cng140362.pdf.

Duarte, C. M., Marbà, N., Gacia, E., et al. (2010). Seagrass community metabolism: Assessing the carbon sink capacity of seagrass meadows. *Global Biochemical Cycles* **24**, GB4032. Available at digital.csic.es/bitstream/10261/46309/1/SeagrassCommunity.pdf.

FAO. (2016). *Vulnerable Marine Ecosystems: Processes and Practices in the High Seas.* FAO Technical Paper 595, Rome: FAO. Available at www.fao.org/3/a-i5952e.pdf.

Finland. (2004). *Laki vesienhoidon ja merenhoidon järjestämisestä* (30.12.2004/1299). Available at http://www.finlex.fi/fi/laki/ajantasa/2004/20041299?search%5Btype%5D=pika&search%5Bpika%5D=30.12.2004%2F1299%20http://www.finlex.fi/fi/laki/alkup/2017/20170094 http://www.finlex.fi/fi/laki/alkup/2017/20170094.

Finland. (2010). *Laki tulvariskien hallinnasta* (24.6.2010/620). Available at www.finlex.fi/fi/laki/ajantasa/2010/20100620.

Finland. (2015). *Ilmastolaki* (22.5.2015/609). Available at www.finlex.fi/fi/laki/ajantasa/2015/20150609?search%5Btype%5D=pikaandsearch%5Bpika%5D=82%2F2014#P.

Freestone, D. and Phillips, F. (2017). The United Nations Convention on the Law of the Sea (LOSC) and Sustainable Development Jurisprudence. In Marie Claire Cordonier Segger and C. G. Weeramantry eds., *Sustainable Development in International Courts and Tribunals.* London: Routledge, 319–38.

Granit, J., Lymer, B. L., Olsen, S., et al. (2017). A conceptual framework for governing and managing key flows in a source-to-sea continuum. *Water Policy Uncorrected Proof,* 1–19. Available at wp.iwaponline.com/content/ppiwawaterpol/early/2017/04/19/wp.2017.126.full.pdf.

Herr, D. and Landis, E. (2016). *Coastal Blue Carbon Ecosystems: Opportunities for Nationally Determined Contributions.* Policy Brief, Gland, Switzerland: IUCN.

IPCC. (2014). Climate Change 2014: Synthesis Report Summary for Policy Makers. Available at www.ipcc.ch/pdf/assessment-report/ar5/syr/AR5_SYR_FINAL_SPM.pdf.

Koutouki, K. and Phillips, F. (2016). SDG 14 on Ensuring Conservation and Sustainable Use of Oceans and Marine Resources. Issue Brief, Montreal, CISDL/UNEP. Available at http://www.cisdl.org/wp-content/uploads/2018/03/SDG_14_Marine_Resources_-_Issue_Brief_-_07.09.2016_-_Final_UNEP-1-1.pdf.

Kyoto Protocol to the United Nations Framework Convention on Climate Change. (1997). December 11, 2303 UNTS 148, 37 ILM 22 (entered into force February 16, 2005).

Laffoley, D. and Grimsditch, G., eds. (2009). *The Management of Natural Coastal Carbon Sinks,* Gland, Switzerland: IUCN. Available at cmsdata.iucn.org/downloads/carbon_managment_report_final_printed_version_1.pdf.

Lavery, P. S., Mateo, M. Á., Serrano, O., and Rozaimi, M. (2013). Variability in the carbon storage of seagrass habitats and its implications for global estimates of blue carbon ecosystem service. *PLoS One* 8, e73748. Available at https://journals.plos.org/plosone/article?id=10.1371/journal.pone.0073748.

Livelihoods Fund. *Senegal: The Largest Mangrove Restoration Programme in the World.* Available at: www.livelihoods.eu/projects/oceanium-senegal/

Martin, A., Landis, E., Bryson, C., et al. (2016). *Nationally Determined Contributions Inventory. Appendix to: Coastal Blue Carbon Ecosystems. Opportunities for Nationally Determined Contributions.* Arendal, Norway: GRID. Available at bluecsolutions.org/dev/wp-content/uploads/Blue-Carbon-NDC-Appendix.pdf.

Nagoya Protocol on Access to Genetic Resources and the Fair and Equitable Sharing of Benefits Arising from their Utilization to the Convention on Biological Diversity. (2010). October 29, UNEP/CBD/COP/DEC/X/1, UNTS No 30619 (entered into force October 12, 2014).

Nellemann, C., Corcoran, E., Duarte, C. M., et al. (2009). *Blue Carbon: A Rapid Response Assessment*. United Nations Environment Programme GRID-Arendal. Available at gridarendal-website.s3.amazonaws.com/production/documents/:s_document/83/original/BlueCarbon_screen.pdf?1483646492.

Paris Agreement to the United Nations Framework Convention on Climate Change. (2015). December 12, Decision CP.21, FCCC/CP/2015/L.9, UNTS No. 54113 (entered into force November 4, 2016).

Phillips, F. K. (2016a). Intellectual property rights in traditional knowledge: Enabler of sustainable development. *Utrecht Journal of International and European Law*, **32**(83), 1–18

Phillips, F. K. (2016b). Sustainable bio-based supply chains in light of the Nagoya Protocol. In L. Bals and W. Tate., eds., *Implementing Triple Bottom Line Sustainability into Global Supply Chains*. London: Greenleaf Publishing, pp. 294–316.

Phillips, F. K. and Maciunas, S. (2017). How a fight over South China islands led to climate win. CIGI Commentary. Available at www.cigionline.org/articles/how-fight-over-south-china-islands-led-climate-win.

Phillips, F. K., Miles, C. A., Khalfran, A., and Reynal, M. L. (2016). *SDG 6 on Ensuring Water and Sanitation for All: Contributions of International Law, Policy and Governance*. Issue Brief, Montreal: CISDL/UNEP. Available at http://www.cisdl.org/wp-content/uploads/2018/03/SDG_6_Water_-_Issue_Brief_-_06.09.2016_-_Final_-_UNEP-1.pdf.

Ramsar Convention. (1993). COP Resolution V.4: The record of Ramsar sites where changes in ecological character have occurred, are occurring, or are likely to occur ("Montreux Record). COP V, June 9–16, 1993. Available at www.ramsar.org/sites/default/files/documents/pdf/res/key_res_5.4e.pdf.

Ramsar Convention. (1996). COP Resolution VI.1: Working definitions of ecological character, guidelines for describing and maintaining the ecological character of listed sites, and guidelines for operation of the Montreux Record. COP VI, March 19–27, 1996. Available at www.ramsar.org/sites/default/files/documents/pdf/res/key_res_vi.01e.pdf.

Ramsar Convention. (2002). COP Resolution VIII.8: Assessing and reporting the status and trends of wetlands, and the implementation of Article 3.2 of the Convention. COP VIII, November 18–26, 2002. Available at www.ramsar.org/sites/default/files/documents/pdf/res/key_res_viii_08_e.pdf.

Ramsar Convention. (2015). Award for Innovation 2015: Oceanium. Available at www.ramsar.org/activities/award-for-innovation-2015.

Request for an Advisory Opinion Submitted by the Sub-regional Fisheries Commission (SRFC). Case No. 21, Advisory Opinion (ITLOS, April 2, 2015). Available at www.itlos.org/fileadmin/itlos/documents/cases/case_no.21/advisory_opinion/C21_AdvOp_02.04.pdf.

Responsibilities and Obligations of States Sponsoring Persons and Entities with Respect to Activities in the Area. Case No. 17, Advisory Opinion (ITLOS Seabed Disputes Chamber Feb. 1, 2011), 50 ILM (2011). Available at www.itlos.org/fileadmin/itlos/documents/cases/case_no_17/adv_op_010211.pdf.

Sands, P. (2016). Climate change and the rule of law: Adjudicating the future in international law. *Journal of Environmental Law*, **28**, 19–35.

Segger, M. C. C. and Khalfan, A. (2004). *Sustainable Development Law: Principles, Practices and Prospects*. Oxford: Oxford University Press.

Segger, M. C. C. and Phillips, F. K. (2015). Indigenous traditional knowledge for sustainable development: The biodiversity convention and plant treaty regimes. *Journal of Forest Research*, **20**(5), 430–7.

South Africa. (1998). *National Environmental Management Act 1998* (Act No 107 of 1998), Government Gazette vol. 401, No 19519 (Cape Town, November 27). Available at www.environment.gov.za/sites/default/files/legislations/nema_amendment_act107.pdf.

South Africa. (2004). *National Environmental Management: Biodiversity Act 2004* (Act No.10 2004). Available at faolex.fao.org/docs/pdf/saf45083.pdf.

South Africa. (2005). *Act No.20 of 2005: Patents Amendment Act, 2005*, Government Gazette Vol. 486, No. 28319 (December 9). Available at www .wipo.int/edocs/lexdocs/laws/en/za/za029en.pdf.

South Africa. (2008). *National Environmental Management: Biodiversity Act 2004* (Act No.10 2004) *Regulations on Bioprospecting and Access and Benefit Sharing 2008*. Available at www.wipo.int/wipolex/en/text.jsp?file_id=179663.

South Africa. (2008). National Environmental Management: Integrated Coastal Management Act 2008 (Act No 24 of 2008), Government Gazette vol. 167, No 40638. Available at www.environment.gov.za/sites/default/files/legislations/ nemicma24of2008_reclamation_regulations.pdf.

South Africa. (2012). *Bioprospecting, Access and Benefit-Sharing Regulatory Framework: Guidelines for Providers, Users and Regulators.* Available at www .environment.gov.za/sites/default/files/legislations/bioprospecting_regulatory_ framework_guideline.pdf.

South China Sea Arbitration. (2016). Philippines v China, Award, PCA Case No 2013–19, ICGJ 495 (PCA 2016), July 12, Permanent Court of Arbitration. Available at www.pcacases.com/web/sendAttach/2086.

TEEB. (2010). *The Economics of Ecosystems and Biodiversity for Local and Regional Policy Makers*. Brussels: UNEP. Available at www.teebweb.org/media/2010/09/ TEEB_D2_Local_Policy-Makers_Report-Eng.pdf.

UN. (1997). General Assembly Resolution 51/229: Convention on the law of the non-navigational uses of international watercourses A/RES/51/229 (July 8). Available at www.un.org/documents/ga/res/51/ares51-229.htm.

UN. (2012). Reference: C.N.639.2012.TREATIES-XXVII.5.b (depositary notification), amendments to Articles 25 and 26 (November 14). Available at treaties. un.org/doc/Publication/CN/2012/CN.639.2012-Eng.pdf.

UN. (2015). Resolution adopted by the General Assembly on 19 July 2015, "Development of an international legally binding instrument under the United Nations Convention on the Law of the Sea on the conservation and sustainable use of marine biological diversity of areas beyond national jurisdiction" A/ RES/69/292 (July 6). Available at documents-dds-ny.un.org/doc/UNDOC/ GEN/N15/187/55/PDF/N1518755.pdf.

UN. (2015). Transforming Our World: The 2030 Agenda for Sustainable Development, UN Doc A/RES/70/1. Available at www.un.org/ga/search/view_ doc.asp?symbol=A/RES/70/1andLang=E.

UN. (2017). Chair's non-paper on elements of a draft text of an international legally-binding instrument under the United Nations Convention on the Law of the Sea on the conservation and sustainable use of marine biological diversity of areas beyond national jurisdiction (February 28, 2017). Third Session of the Prep Com (March 27–April 7, 2017). Available at www.un.org/depts/los/ biodiversity/prepcom_files/Chair_non_paper.pdf.

UN Division for Ocean Affairs and the Law of the Sea. (2017). Report of the Preparatory Committee established by General Assembly resolution 69/292: Development of an international legally binding instrument under the United Nations Convention on the Law of the Sea on the conservation and sustainable use of marine biological diversity of areas beyond national jurisdiction. A/AC.287/2017/PC.4/2 (31 July 2017). Available at http://www.un.org/ga/search/view_doc.asp?symbol=A/AC.287/2017/PC.4/2.

UN Economic Commission for Europe. (2004). Meeting of the Parties to the Convention on the Protection and Use of Transboundary Watercourses and International Lakes: Amendment to Articles 25 and 26 of the Convention, ECE/MP.WAT/14 (January 12). Available at unece.org/fileadmin/DAM/env/documents/2004/wat/ece.mp.wat.14.e.pdf.

UN Economic Commission for Europe. (2011). Convention on the Protection and Use of Transboundary Watercourses and International Lakes: Second Assessment on Transboundary Watercourses and International Lakes. Geneva: United Nations. Available at unece.org/fileadmin/DAM/env/water/publications/assessment/English/ECE_Second_Assessment_En.pdf.

UNEP. (1995). Global Program of Action for the Protection of the Marine Environment from Land-Based Activities. 3 November 1995, UNEP(OCA)/LBA/IG.2/7 (December 5). Available at www.ucc.ie/research/crc/pages/projects/ICZM/documents/policy/international/GlobalPAMarine.pdf.

UNFCCC. NDC Registry (interim). Available at unfccc.int/focus/ndc_registry/items/9433.php.

United Nations Convention on the Law of the Sea. (1982). December 10, 1833 UNTS 3, 21 ILM 1261 (entered into force November 16, 1994).

United Nations Framework Convention on Climate Change. (1992). May 9, 31 ILM 849 (entered into force March 21, 1994).

UN Treaty Collection. (1976). Convention on wetlands of international importance especially as waterfowl habitat (Ramsar, February 2, 1971), February 17, No 14583. Available at treaties.un.org/doc/Publication/UNTS/Volume%20996/volume-996-I-14583-English.pdf.

UN Treaty Collection. (1996). Chapter XXVII Environment: Convention on the Protection and Use of Transboundary Watercourses and International Lakes, (Helsinki, March 17, 1992), October 6, No.33207. Available at treaties.un.org/doc/Publication/MTDSG/Volume%20II/Chapter%20XXVII/XXVII-5.en.pdf.

UN Treaty Collection. (2014). Chapter XXVII Environment: Convention on the Law of the Non-Navigational Uses of International Watercourses (New York, May 21, 1997), August 17, No. 52106. Available at treaties.un.org/doc/Publication/MTDSG/Volume%20II/Chapter%20XXVII/XXVII-12.en.pdf.

20

Managing Marine Resources

Can the Law of the Sea Treaty Adapt to Climate Change?

ANASTASIA TELESETSKY

Introduction

As the "constitution of the oceans," with 167 states and the European Union as parties, the United Nations Convention on the Law of the Sea (the Convention) influences global ocean policy and marine governance (see especially Chapters 2, 16, 18, and 19). Achievement of global climate change mitigation and adaption will require interpreting the Convention as a living document that is capable of meeting new challenges. However, it is not clear that the Convention's negotiators were thinking about global climate change at all during the almost decade of treaty negotiations between 1973 and 1982. The impact of greenhouse gases (GHGs) on the oceans was not an unknown phenomenon at that time. In 1957, US oceanographer Roger Revelle and chemist Hans Suess observed that seawater could not absorb all the additional carbon dioxide that was entering the atmosphere and, in 1975, the term "global warming" entered the scientific lexicon (Revelle and Suess, 1957: 18–27; Broecker, 1975: 460–63). Although scientific consensus was emerging in the 1970s and 1980s regarding the drivers and impacts of climate change, the topic was not central enough to the diplomatic negotiations to merit an explicit mention in the Convention. Nevertheless, the drafters of the Convention anticipated global changes that would require coordinated responses from states and their commitment to protecting marine waters (UNCLOS, 1982: Preamble).

This chapter examines how the fulfilment of certain codified state obligations under the Convention might make a substantial contribution to systematically addressing impacts of global climate change on marine resources. The Convention should be interpreted progressively by state parties to bolster individual and regional state efforts to reduce GHGs, secure legal maritime jurisdictional interests, and effectively manage shared marine resources for the future. Implementation of the Convention offers at least three mitigation and adaptation strategies: pollution prevention and reduction, revised fisheries resource management, and securing present boundaries, in spite of physical oceanographic changes resulting from climate change, through the "freezing" of coastal "low-water lines." This chapter looks at these strategies in turn. It concludes with a brief reflection on whether the continental shelf regime identified

in the Convention needs to be reimagined in light of a "leave it in the ground" movement to reduce fossil fuel extraction.

Pollution Reduction: Mitigating Greenhouse Gases

One strategy for addressing climate change impacts on the oceans is better implementation of the Convention's obligations to prevent pollution (see Chapter 19).
Article 1 of the Convention defines "pollution of the marine environment" as:

[T]he introduction by man, directly or indirectly, of substances or energy into the marine
environment, including estuaries, which results or is likely to result in such deleterious effects
as harm to living resources and marine life, hazards to human health, hindrance of marine
activities, including fishing and other legitimate uses of the sea, impairment of quality for use
of sea water and reduction of amenities. (UNCLOS, 1982)

In order to fall within this definition, there must be (1) a direct or indirect introduction by humans of substances or energy into the marine environment and (2) the
substance or energy must cause or be likely to cause certain types of harms to
humans and ecosystems.

In terms of the first element, it is a reasonable textual interpretation that carbon
dioxide and other GHGs qualify as marine pollutants because they are substances
being directly introduced by humans into the marine environment, even if the introduction is unintentional. Legal systems have recognized GHGs as air pollutants.
For example, on April 2, 2007, the United States Supreme Court in *Massachusetts
v. the Environmental Protection Agency* identified GHGs, including carbon dioxide,
as pollutants (549 U.S. 497). Similar logic applies to identifying GHGs as ocean
pollutants.

In addition to the direct introduction of GHGs into ocean waters, the ongoing
emissions of GHGs are indirectly introducing new sources of thermal energy into
the oceans. As various GHGs interact with the atmosphere, the atmosphere traps
heat that is directed back toward the oceans. By warming a narrow surface area of
the ocean (0.1 to 1 mm thick), the heat in the atmosphere interrupts the heat
exchange cycles between the ocean surface layer and the atmosphere. As the ocean
surface layer continues to warm, the flow of heat out of the ocean into the atmosphere slows such that more heat remains trapped in the ocean. The introduction of
additional small amounts of energy into the surface layer of the ocean (its "cool skin
layer") may thus lead to significant consequences for the entire ocean system (Fairall
et al., 1996: 1295–1308).

The Intergovernmental Panel on Climate Change (IPCC) has identified numerous
"deleterious effects" on marine living resources arising directly from atmospheric
GHGs and indirectly from the addition of new thermal energy. The IPCC has high
confidence that more than 90 percent of the energy accumulated from the climate
system between 1971 and 2010 is in the oceans (IPCC "A", 2014: 40). Likewise,
ocean acidification triggered by human introductions of GHGs is affecting the

formation of marine organism shells (IPCC "A", 2014: 51). These harms directly affect ecosystem health by harming coral reefs (see Chapter 22) and indirectly affect human health as marine species that people depend upon decline in numbers or size (Cheung et al., 2013: 254–58) (see chapters in Part III).

States have numerous obligations to protect and preserve the marine environment by taking measures to prevent, reduce, and control pollution (see Chapter 19). The combined language in both Article 194 and Article 196 of the Convention implicitly oblige states to undertake GHG mitigation efforts. Because GHGs qualify as pollution of the marine environment, states are expected to "take *all* measures consistent with this Convention that are necessary to prevent, reduce, and control pollution" (UNCLOS, 1982: Art. 194[1]); "take *all* measures necessary to ensure that activities under their jurisdiction or control are so conducted as not to cause damage by pollution to other states and their environment" (UNCLOS, 1982: Art. 194[2]); and "take *all* measures necessary to prevent, reduce, and control pollution of the marine environment resulting from the use of technologies under their jurisdiction or control" (UNCLOS, 1982: Art. 196[1]).

Translating this expansive obligation into practical measures for states is not straightforward. One implementation mechanism that would distribute state responsibility might include a global calculation of the total quantity of GHGs anticipated as being absorbed into the oceans on an annual basis. Based on the annual GHG emissions generated by each state, each state might commit to remove from the atmosphere a certain percentage of the sum of GHGs calculated to be entering the oceans. To at least minimally meet Article 194 and Article 196 obligations, state parties should be able to point to a set of public domestic actions that have reduced the production of GHG emissions. Implementing a system of mitigation responsibility would be supported by Convention articles requiring states to cooperate in exchanging "information and data acquired about pollution of the marine environment" and "establishing appropriate scientific criteria for the formulation and elaboration of rules ... for the prevention, reduction and control of pollution of the marine environment" (UNCLOS, 1982: Arts. 200 and 201). This approach has a certain attraction because it would ensure that all Convention members undertake some degree of mitigation effort for the sake of the oceans. In addition to nationally determined contributions (NDCs) under the 2015 Paris Agreement on climate change, assigning GHG targets as part of their Convention obligations would fulfil in part the requirement under Article 235 for states to take responsibility and submit to liability "in accordance with international law." As of the end of 2017, while many states discuss coastal and marine resources in their NDC, (see Chapter 16) no state has identified GHGs as "marine pollutants" in its NDC.

There would be, however, several hurdles in attempting to apply a fair and proportional approach to marine pollution reduction from GHGs. Self-reporting may be unreliable and compliance might be administratively burdensome if states rely on a third-party verification system. Perhaps more frustrating from a mitigation perspective, elimination of a certain quantity of carbon and carbon equivalents from sources may not actually make a measurable difference in marine pollution levels as long as certain thresholds of carbon dioxide and other GHGs still remain in the atmosphere and interact with ocean waters. There can also be a problem with "free riders" like the United States, which is not formally a party to the Convention and may be unwilling to accept that Article 194 and Article 196 bind it as customary international law.

Article 212 of the Convention deserves special attention. As mentioned previously, the term "climate change" does not appear in the Convention's text, but Article 212 calls upon all parties to "adopt laws and regulations to prevent, reduce and control pollution of the marine environment from or through the atmosphere" associated with a state's airspace and a state's vessels and aircraft. "Airspace" is understood as that region above a state's territory, including "territorial waters" up to the boundary with "outer space" (ICAO, 1944: Arts. 1 and 2). This Article is prima facie not limited to emission regulation of state-registered vessels and aircraft but also encompasses other atmospheric pollution sources that emit into a state's airspace. States are further required to "take other measures as may be necessary to prevent, reduce and control such pollution." States are obliged under Article 222 to enforce their laws and rules adopted under Article 212.

Article 212 reflects a clear synergy with the goals of the United Nations Framework Convention on Climate Change (UNFCCC) and its objective of "stabilization of greenhouse gas concentrations in the atmosphere at a level that would prevent dangerous anthropogenic interference with the climate system" (UNFCCC, 1992: Art. 2). Article 212 requires more of states than they might otherwise be required to do under the UNFCCC. The UNFCCC only requires Annex I (developed/industrialized) countries to "adopt national policies and take corresponding measures on the mitigation of climate change" (UNFCCC, 1992: Art. 4[2][a]). UNCLOS goes beyond "soft law" mechanisms like policies and requires instead the adoption and enforcement of "laws and regulations" (UNCLOS, 1982: Arts. 212[2] and 222). In practice, states where parties have been harmed by marine pollution such as fishermen whose stocks have been affected by ocean acidification must "ensure that recourse is available in accordance with their legal systems for prompt and adequate compensation or other relief in respect of damage caused by pollution of the marine environment by natural or juridical persons under their jurisdiction" (UNCLOS, 1982: Art. 235[3]).

In July 2017, the Carbon Majors Project released a report detailing the 100 companies that were responsible for 52 percent of the GHG emissions since the Industrial Revolution as direct producers of fossil fuels, investor-owned companies, or major users of fossil fuel (Griffin, 2017: 5) The top ten emitters in 2015 were Saudi Arabian Oil Company-Aramco (4.6 percent global GHG emissions), Gazprom (2.7 percent), National Iranian Oil Co. (2.4 percent), Coal India (2.4 percent), Shenhua Group Corp Ltd. (2.4 percent), Rosneft (1.8 percent), China National Petroleum Company (1.5 percent), Abu Dhabi National Oil Co. (1.4 percent) and Exxon Mobil (1.4 percent), and Petroleos Mexicanos (1.3 percent) (Griffin, 2017: 14). The states of incorporation for a number of these companies are members of the Convention. Applying state obligations under Article 235, legal recourse should be available within these states, including, for example, Russia and Mexico, for harms to the marine environment from atmospheric pollution. There are, however, obvious hurdles in accessing justice, given that many of the Carbon Majors companies are state-owned and some courts would probably allow sovereign immunity defenses. In spite of the legal requirement for accountability mechanisms under the Convention, there is no obvious practical recourse "for prompt and adequate compensation" from the world's largest carbon emitters for pollution of the marine environment.

For more extensive GHG mitigation, some states may hope to apply geo-engineering techniques to enhance the "carbon sink" qualities of the marine environment (see Chapter 26). Some of these techniques involve the addition of substances such as iron into ocean water. While in theory the introduction of these substances would be intended to ameliorate harm to ecosystems and human health caused by excess carbon, there may still be deleterious effects on ocean ecosystems. An attempt at mitigation of one pollutant (carbon) may lead to the generation of a different source of pollution if geo-engineering attempts fail to capture carbon and lead to harmful algal blooms (Williamson et al., 2012: 475–88). Article 195 of the Convention prohibits geo-engineering efforts that would "transform one type of pollution into another." The Convention also prohibits pollution from "dumping," but "dumping" does not prohibit the "placement of matter for a purpose other than the mere disposal thereof," unless the placement is in contravention of the Convention (UNCLOS, 1982: Arts. 201 and 1[5][b][ii]). The potential for geo-engineering interventions to address marine pollution from carbon dioxide is primarily addressed under the 1971 London Convention on Dumping and its 1996 Protocol, rather than UNCLOS (Resolution LP.4[8], 2013).

Reconceiving Fisheries Management: Adapting to Warmer and More Acidic Oceans

A second strategy for addressing climate change impacts on the oceans is better management of fishery stocks under the Convention that takes into consideration changing physical conditions of the ocean (see chapters in Part III and Chapter 13). The movement of fish due to ocean warming and the decline in biomass due to the destruction of key reef spawning habitat is widely discussed in the fisheries research literature. It is unclear how parties to the Convention will respond through existing institutions such as the Regional Fisheries Management Organizations to the possibilities of large fish migrations, loss of biomass, and the destruction of certain sensitive fisheries habitat due to climate impacts (IPCC "A", 2014: 493). Fostering a future-oriented cooperative approach that protects the underlying marine resources as it crosses maritime borders will be critical in light of the post–World War II excesses of industrial fishing that have already weakened the genetic stocks of many commercial species.

The large-scale movement of fish across maritime borders due to warming happened long before the negotiation of the Convention. In the early part of the twentieth century, the Atlanto-Scandian herring existed in cold waters off the coasts of Norway and in the region of the Barents Sea (Viljhalmsson, 2007: 225–70). When oceans began warming in the 1920s, the herring stocks soared as their feedstock of plankton multiplied to become the largest herring stock in the world, covering a wide ocean range between Norway and Iceland (Viljhalmsson, 2007: 251). The stocks contracted again following ocean cooling events from 1964 to 1970 accompanied by heavy fishing pressure (Viljhalmsson, 2007: 253–54). More recently with warming of the waters, the stocks are on the move again, including into Icelandic and Faroese waters (Viljhalmsson, 2007: 255). These types of stock "expansion and contraction" events connected to ocean temperatures might have been in the minds of at least some negotiators during the original Convention negotiations.

The understanding that changes in environmental conditions can have major implications for marine ecosystems, both at the level of primary producers (planktons) and top-level predators, is captured in the text of the Convention covering the conservation and management of species. For example, Article 61 gives coastal states the power to determine the total allowable catch, but in return the states are expected to use "the best scientific evidence available to it" to ensure that stocks are not overexploited. In setting a maximum sustainable yield, states must qualify the level of harvest with "relevant environmental ... factors." Article 119 regarding the management of high-seas living resources requires that high-seas states also consider "relevant environmental ... factors." While the term "relevant environmental factors" is never defined, climate impacts, including warming oceans and ocean acidification, are the types of factors that states will need to re-evaluate in setting total allowable catches. In some instances warming oceans, as with the case of the 1940s Atlanto-Scandian herring, might improve stock numbers in a particular locale. In many other cases, warming oceans are likely to lead to an exodus of fish or the decimation of sedentary or semisedentary species.

Article 61 and Article 119 of the Convention require states to "take into consideration the effects on species associated with or dependent upon harvested species with a view to maintaining or restoring populations of such associated or dependent species above levels at which their reproduction may become seriously threatened" (UNCLOS, 1982: Arts. 62[4] and 119[2]). This language is intended to catalyze states into adopting an ecosystem approach to fisheries management. This obligation is particularly important, given the connection between fish nurseries and vulnerable habitats, including tropical coral reefs and coastal wetlands. UNCLOS requires active state interventions to conserve populations of "associated or dependent species" and engage in ecological restoration.

The Convention does not dictate a particular strategy for achieving these outcomes, and states can pursue a variety of approaches, including establishing marine protected areas or funding community-based restoration efforts. States have invested in regional efforts to jointly improve shared habitats in the face of climate change such as the Coral Triangle Initiative (CTI) (see Chapter 17) (Coral Triangle Initiative, 2009). The CTI Regional Plan of Action recognizes the need for coordination on transboundary issues and threats, and the CTI states have committed themselves to implementing an ecosystem approach to the management of fisheries and achieving climate change adaptation through the development of an "early action plan for climate change adaptation for the near-shore marine and coastal environment and small island ecosystems."

With biological changes arising from physical changes in the oceans, joint management is necessary for shared and migratory stocks. Article 63 of the Convention provides that states that share stocks across exclusive economic zones (EEZs), including "associated species," must "seek, either directly or through appropriate sub-regional or regional organizations, to agree upon the measures necessary to coordinate and ensure the conservation and development of such stocks." Article 64 of the Convention provides that states that participate in the fishing of highly migratory species (including tuna, marlins, and swordfish) must "cooperate directly or through appropriate international organizations with a view to ensuring conservation and promoting the objective of optimum utilization of such species."

States who share waters frequented by anadromous and catadromous stocks such as salmon and eels will need to pay special attention to the interface between freshwater and marine ecosystems. Higher temperatures in freshwater systems may lead to decreasing nutrient circulation and potentially higher algal production with uncertain impacts on fisheries resources (Segel et al., 2016: 7). Changes in regional governance schemes are already proposed to manage impacts on salmon fisheries in the Columbia River Basin in the Western United States resulting from increased sea surface temperatures plus changes in precipitation and snowpack that will affect dam releases (McIlgorm et al., 2010: 175) (see Chapter 10).

States that have already negotiated bilateral or multilateral shared stocks agreements may need to have further negotiations due to climate impacts on fisheries. Efforts to create quota-sharing agreements in areas where fishing stocks are migrating may lead to tensions, as states that are likely to gain fishery resources due to warming are reluctant to enter into new allocation agreements unless they benefit. Already, the EU, Faeroe Islands, Iceland, Norway, and Greenland are in disagreement about mackerel management, as mackerel schools have shifted toward the Faroe Islands and Iceland (EU Directorate General, 2015: 84). The Faroe Islands, the EU, and Norway have reached an agreement on quota shares, but Iceland and Greenland (which has autonomy under the EU Common Fisheries Policy) have not agreed to be part of the arrangement. Conflicts are likely to increase as fish continue to change migration routes and enter new territorial waters (see also Chapter 12).

States have focused increasing attention on the Arctic as a place where climate impacts are rapidly manifesting (see Chapters 12, 14, and 15). While the Arctic is losing sea ice, it is also expected to be a net recipient of migratory fish, with predictions of increases in total fisheries catch and revenue of 14 percent to 59 percent by 2050 from a baseline of 2000 (Lam et al., 2016: 335–57). Migration of fish from warmer waters to cooler waters may, however, have an unintended consequence of intensifying competition among species. Arctic species are generally long lived, slow growing, and late maturing and may find themselves competing with shorter-lived, faster-growing, and opportunistic species (EU Directorate-General, 2015: 99). This could lead to additional challenges for states fishing in the Arctic regions in implementing an ecosystem approach that is protective of the marine environment.

Of greater concern from a food security perspective are the biogeographical shifts occurring in the tropical seas where warming temperatures and hypoxic events are expected to lead to both the migration of fish from coastal areas dependent on those fishery resources and to smaller body sizes for fish. Large pelagic fish are expected to move several hundred miles east in the Pacific in response to warming oceans (Segel et al., 2016: 15) (see Chapter 11). Food sources for fish such as zooplankton are also migrating. The IPCC predicts a possible decrease of up to 40 percent of marine catch by 2055 on the basis of 2005 baselines (Segel et al., 2016: 11). Some decline in fisheries may arise from the spread of pathogens that thrive in warmer waters.

Regional fisheries management organizations will need to be nimble in responding to climate impacts (see Chapters 9 and 10). As with the subject of maritime boundaries discussed later, the extent of climate impacts on living marine resources will depend on the location of a fishery. Some tropical regions, such as the subtropical gyre regions, are expected to have reduced oxygen availability, leading to long-term impacts on habitat and species. These biophysical impacts may pose sustenance threats to human communities in places such as East Africa where states such as Mozambique, Tanzania, Kenya, and Somalia are already experiencing ocean warming, increased acidification, and decreased oxygen (Segel et al., 2016: 27). Asia is likely to experience large impacts to coral reef economies that will affect millions of people (see Chapters 7 and 17). Adapting to the biophysical changes and reducing the socioeconomic impact on many states will require institutional responses, including better implementation of existing laws and potentially the development of new laws. As oceanographers continue to refine scientific understandings of the linkages between climate change and living marine resources, there may need to be more flexible management of fisheries that might include, for example, reducing the landings of fish that have been historically captured and increasing the landing of new species, especially in those areas such as the Mediterranean, where recently one new fish species has been arriving every month from tropical waters (IPCC "B", 2014: 1295).

Baselines and Maritime Jurisdiction: "Freezing" Baselines

A third strategy for addressing climate change impacts on maritime affairs is freezing baselines based on charts submitted by states to the United Nations. One of the major contributions of the Convention was to provide a legal regime that defined maritime jurisdictions for coastal states. This was necessary to offer states some degree of predictability for planning conservation and management efforts and to create a basis for stable relations with neighboring states. Until the maritime jurisdictions provided for in the Convention were unanimously adopted by all states, there was uncertainty about how far a state's legislative jurisdiction might extend seaward. The phenomenon of sea-level rise has raised existential questions about what reasonable expectations states can hold regarding the stability of their maritime jurisdictions and the security of their maritime boundaries (see Chapter 18). While climate models have greatly improved in recent decades, there remains uncertainty about what level of sea-level rise states should be planning for in the years to come. Depending on the assumptions behind a given climate model, the proposals for the global rise in sea levels have ranged from 0.5 meters to 2 meters by 2100 if the global temperatures increase by 4°C (Nicholls et al., 2011:1). The most recent reports indicate a range of 0.26 meters to 0.98 meters by 2100, with much of this rise taking place between 2081 and 2100 (Church et al., 2013, 1180). Of particular concern among scientists is not just the amount of sea-level rise but also the rate of sea-level rise, given the observation that the rate of sea-level rise appears to have accelerated since 1971 with an increase of 3.2 mm/year of sea-level rise between 1993 and 2000 as compared to 1.7 mm/year over the entire twentieth century (Church et al., 2013: 1146, 1150).

Projected increases in sea level will have impacts on territorial integrity, including the physical integrity of low-lying states and island states. The Convention does not provide explicit tools for incorporating sea-level rise as a consideration for delineation. The unprecedented rate of sea-level rise was not anticipated by Convention negotiators in the 1970s. With the exception of Article 7(2) of the Convention that contemplates the jurisdictional fixing of a dynamic coastline due to "the presence of a delta and other natural conditions," the possibility of sea-level rise or sea-level retreat disturbing state expectations regarding the delineation of maritime jurisdictions was never acknowledged in the Convention text. Rather, the text seems to permit variability where baselines might migrate depending on the location of the "low-water line along the coast" (Caron, 1990: 621–53). Articles 5 and 6 of the Convention also include reference to a "low-water line" but do not acknowledge the same potential variability due to "other natural conditions."

With sea-level rise, "low-water lines" are shifting landward or will be shifting landward over the course of the century. Because the land dominates the sea in maritime delimitations, states may raise questions about other states' assertions of a status quo maritime jurisdiction where a low-water line has migrated landward. It is possible that a coastal state can act unilaterally and refuse to legally recognize the migration of coastlines since Article 5 of the Convention indicates that "the breadth of the territorial sea is the low-water line along the coast as marked on large-scale charts officially recognized by the coastal State." Likewise, in the case of reefs, UNCLOS Article 6 indicates that the territorial sea is measured from the "seaward low-water line of the reef, as shown by the appropriate symbol on charts officially recognized by the Coastal State." In each of these cases, the baseline depends on the identification of a "low-water line" which may move landward if a region experiences a rise in sea level.

States have an obligation to put other states on notice of their jurisdictional claims over a territorial sea (UNCLOS, 1982: Art. 16). The clauses in this Article "as marked on large-scale charts officially recognized by the coastal State" and "on charts officially recognized by the State" may be read to provide some flexibility for a coastal state. Assuming that a state has not submitted a chart under Article 16, the state might assert that charts in use defining the state's coastline have not been "officially recognized" for purposes of defining the "low-water mark" or the "low-water line," even though they may be relied upon for navigation purposes. Given that coastal geography is likely to change for some states due to sea-level rise, the use of a chart to fix jurisdiction will require the creation of a "legal fiction." In spite of the reluctance by some scholars to fix basepoints for legal purposes independent of the actual geographical conditions, geographical fictions have occupied a venerable place in legal history. For example, in order for a British court to assert jurisdiction in the eighteenth century, the court articulated that the island of Minorca (an island in the Mediterranean Sea now belonging to Spain) was said to be located within the parish of Mary-le-Bow in the ward of Cheap in the city of London (*Mostyn* v. *Fabrigas*, 1774).

It remains to be seen whether this argument will prove legally defensible as an interpretation of the Convention. A coastal state will need to argue that a chart for purposes of the "limits of the territorial sea" or defining the "breadth of the exclusive zone" can be differentiated from a chart being used for other maritime purposes. While a "chart" is not defined by UNCLOS, the president of the Directing Committee of the International Hydrographic Bureau has opined that states should use the definition of "nautical chart" provided for in Regulation 2(2) of Chapter V of the 1974 Safety of Life at Sea Convention: "a special-purpose map or a specially compiled database from which such map is derived, that is issued officially by or on the authority of a government, authorized Hydrographic Office or other relevant government institution" (Maratos, 2008).

Legal scholars preparing reports for the International Law Association (ILA) have debated whether the provision of charts by coastal states as required by Articles 16, 47, and 75 of the Convention can secure maritime delimitations that have changed or are expected to change due to sea-level rise. There is disagreement among scholars. Judge Jesus from the International Tribunal for the Law of the Sea, writing in his personal capacity as a Cape Verde citizen, noted that "a substantial rise in sea level, whatever the cause, should not entail the loss of States' ocean space and their rights over maritime resources, already recognized by the 1982 Convention and by the community of nations" (Jesus, 2003: 602). The ILA convened an "International Committee on Baselines under the International Law of the Sea" to identify existing law on "the normal baseline" and determine if there needed to be "further clarification or development of that law" in light of changes such as sea-level rise. The Baselines Committee concluded its work in 2012 by rejecting the possibility of freezing baselines. As the committee concluded: "it follows that if the legal baseline changes with human-induced expansions of the actual low-water line to seaward, then it must also change with contractions of the actual-low water line to the landward" (ILA, 2012: 422). The committee observed that even if coastal states have the discretion to choose points from which it will measure its low-water line, those lines must have some grounding in existing geographical conditions. The committee opined that "Coastal States may protect and preserve territory through physical reinforcement, but not through the legal fiction of a charted line that is unrepresentative of the actual low-water line" (ILA, 2012: 424).

The ILA convened a second committee to evaluate the impact of sea-level rise on land and maritime boundaries. In November 2012, the "Committee on International Law and Sea Level Rise" was tasked with understanding "the implications under international law" of territorial loss due to sea-level rise and "to develop proposals for the progressive development of international law in relation to the possible loss of all or of parts of state territory and maritime zones due to sea-level rise" (ILA, 2016: 1). The second committee diverged from the baselines committee regarding the possibility of a "legal fiction" to protect state interests in existing maritime zones and maritime delimitations. In its 2016 report, the sea-level rise committee contemplated the possibility of either freezing baselines or freezing maritime zone outer limits to create maps that states could rely upon for delineating maritime zones and that would not unfairly dispossess sea-level-vulnerable states of maritime entitlements (ILA, 2016: 14–16) In 2018, delegates approved an International Law Association resolution in support of this position on the basis of the need to maintain international peace and security. (ILA, 2018).

With a rise in sea level changing the physical integrity of island features, an additional complication arises with defining archipelagic baselines for a nation constituted of a group of islands. Under Article 47 of the Convention, states are permitted to make claims over an area defined by the "outermost points of the outermost islands and drying reefs of the archipelago." Climate change may affect what geographical features can be identified as "drying reefs" and islands (UNCLOS, 1982: Arts. 6 and 121). According to Article 121, an island must be "above water at high tide" and must be capable of sustaining "human habitation or economic life." While the South China Sea arbitration involved a contentious dispute and the decision by the Permanent Court of Arbitration does not have precedential effect, the tribunal offered a number of observations about what qualifies as an island. Specifically, an island must have permanent sources of freshwater and have soil resources capable of supporting agricultural production; otherwise, it would be simply a rock. The sources of island freshwater can include groundwater as well as rainfall harvesting (South China Sea Arbitration, 2016: para. 615).

The arbitration tribunal in this one case took the unusual decision of finding that while the record appeared to indicate that some of the features in question were capable of "enabling the survival of small groups of people" (even though the capacity was "distinctly limited") that the tribunal would also be investigating the "historical evidence of human habitation and economic life" (South China Sea Arbitration, 2016: para. 616). This is an unusual analytic move because Article 121(3) only requires an evaluation of whether a feature "cannot sustain human habitation or economic life" on its own. The Convention's negotiators agreed to the use of the disjunctive "or," suggesting that a party attempting to assert a claim that a feature is an island must either prove the ability of the feature to "sustain human habitation" or to "sustain economic life" independent of an export economy. In the South China Sea arbitration case, the tribunal appeared to read Article 121(3) as conjunctive. Even though the tribunal was reluctantly willing to accept that some of the features in question were capable of sustaining human habitation, there was on the record insufficient historic evidence that the features had sustained permanent human habitation and were able to sustain economic life (South China Sea Arbitration, 2016: paras. 624–25). "Temporary extraction" exercises for purposes of exporting resources were not enough to constitute "economic life."

In the context of climate change, the tribunal's analysis of what constitutes an island might cut more than one way for an archipelagic state or a single island state such as Niue. Impacts of climate change on islands are likely to include sea-level rise for at least some islands. While it is unclear that currently inhabited islands will be fully submerged under the current projections of sea-level rise, the tribunal's decision raises some future questions to be answered regarding jurisdictional assertions of an island state.

The coastal or archipelagic state retains a certain degree of flexibility for the state in defining its jurisdictional reach because Article 16(2) of the Convention requires the coastal state to deposit a copy of the "officially recognized" chart with the secretary-general of the United Nations. Presumably, archipelagic states may deposit with the secretary-general charts with identified basepoints that are, at the time of the chart submission, above high tide. Given the lack of clarity about what impact sea-level rise may have on marine entitlements, it would be in the best interest of states to make submissions of charts to the United Nations defining basepoints for purposes of maritime delimitations. Even where there may be disputed boundaries, the unilateral submission of charts would put other states on notice of an individual state's understanding of its jurisdictional reach. Fifty-five states have already made such submissions for purposes of defining the territorial sea. (See the appendix to this chapter for a list of states that have filed marine charts defining a territorial sea, archipelagic baseline, or an exclusive economic zone.) With so many states having submitted charts defining the territorial sea and/or the EEZ on the basis that the charts accurately reflect the state's understanding of its UNCLOS-defined maritime entitlements, customary international law may be crystallizing such that the charts can be regarded as evidence of state practice regarding basepoints and baselines. Given the increasing amount of practice in this area, it is reasonable to conclude that the charts are intended to secure and "freeze" baselines. If states could have anticipated in the 1970s the extent of sea-level rise when the Convention was being negotiated, they would never have concluded or ratified a treaty that would have led to a future substantial loss of rights from climate-related oceanic changes.

Rethinking Maritime Oil and Gas Exploitation

While the three mitigation and adaptation strategies described earlier offer opportunities for parties to the Convention to address various climate impacts, the Convention text also includes resource development law to encourage the growth of major GHG producers. Oil and gas production contribute approximately 6.4 percent of global GHGs (*The Guardian*, 2011). The products of these industries, when burned by other industries, contribute even more to climate change. Because energy demand is expected to increase by 50 percent by 2030, offshore resources are anticipated to provide up to a quarter of that energy (41 billion tons) (World Ocean Review, 2010: 142) (see Chapter 24). In 2007, offshore oil met 37 percent of annual oil production needs (World Ocean Review, 2010: 143). The Convention encourages continued development, with states having the right to exploit marine nonliving resources that include gas and oil, on their continental shelves (UNCLOS, 1982: Arts. 76–77). Notably, the right to extract is conditioned upon not infringing certain rights of other states. Specifically, Article 78(2) of the Convention provides that "[t]he exercise of the rights of the coastal State over the continental shelf must not infringe or result in any unjustifiable interference with navigation and other rights and freedoms of other States as provided for in this Convention."

Depending on the extent of ocean acidification that results from the deposition of carbon dioxide into ocean waters from fossil fuels extracted from the ocean, it might be argued by some states that coastal state exploitation of offshore oil and gas resources is interfering with other Convention rights such as the freedom to fish. While the regime of the continental shelf has been embraced by coastal states and many states have submitted requests to the Commission on the Limits of the Continental Shelf for expert recommendations about the extent of their shelves, it appears that this part of the Convention may be maladaptive in terms of climate change impacts. For all additional fossil fuel extraction, some amount of carbon is likely to enter the atmosphere that will contribute to atmospheric warming and acidification. Very few states have been willing to restrict offshore development for purposes of reducing climate impacts. Even for states that have been willing to make changes, such as the United States in December 2016 withdrawing offshore lands for oil exploitation, subsequent political administrations may abrogate these decisions, as has proven true in the United States.

Conclusion

Even though the terms "global warming" and "climate change" are not mentioned in the Convention, the text that was negotiated may be interpreted to further bolster states' strategies for climate mitigation and adaptation in at least three ways. First, greenhouse gases are pollutants under the Convention, and states have obligations to prevent, reduce, and control marine pollutants or, in other words, mitigate GHG production. Second, states must manage fisheries to maintain or restore stocks that take into consideration "relevant environmental ... factors" that would include migratory shifts based on warming and acidifying oceans. And third, states may endeavor to secure maritime baselines in spite of sea-level changes by submitting maritime charts to the United Nations as required under the Convention. While states may undertake some of these strategies on their own, success will depend on a high level of commitment to cooperating with other states, including not just parties to the Convention but also states that accept the Convention as reflecting customary international law. Achieving this high level of cooperation will require unprecedented levels of investment in ocean management.

The Convention offers the beginning of a "blueprint" for states to coordinate resources to work together to restore stocks and allocate catches based on emerging environmental factors. That said, all efforts to foster cooperation will be for naught if states simply continue to exploit and utilize carbon-intensive fuels. The right to exploit fossil fuels on states' continental shelves must therefore be questioned where it undermines other Convention rights.

It may be hoped that UNCLOS will be a "constitution for the oceans" that guides states in meeting global ocean challenges presented by climate change, ranging from sea-level rise to acidification. However, the Convention as drafted is replete with contradictions, particularly when a state weighs its responsibilities against its rights. On the one hand, the Convention identifies responsibilities for marine pollution reduction and fisheries management in light of changing environmental conditions. On the other hand, it provides a rights-based framework for oil and gas exploitation. States must decide in the long term whether to sacrifice living marine resources in order to continue exploiting in the short term nonliving marine resources. This will require future negotiations among states to design international rules that, if followed, might avert an impending ocean crisis.

References

Broecker, W. (1975). Climate change: Are we on the brink of a pronounced global warming? *Science*, **189**, 460–3.

Caron, D. (1990). When law makes climate change worse: Rethinking the law of baselines in light of a rising sea level. *Ecology Law Quarterly*, **17**, 621–53.

Cheung, W. W., Sarmiento, J. L., Dunne, J., et al. (2013) Shrinking of fish exacerbates impacts of global ocean changes on marine ecosystems. *Nature Climate Change*, **3**(3), 254–8.

Church, J. A., Clark, P. U., Cazenave, A., et al. (2013) Sea level change. In *Climate Change 2013: The Physical Science Basis. Contribution of Working Group I to the Fifth Assessment Report of the Intergovernmental Panel on Climate Change*. Cambridge, UK: Cambridge University Press.

Coral Triangle Initiative, Regional Plan of Action. (2009). Available at www .coraltriangleinitiative.org/library/cti-regional-plan-action.

EU Directorate General for Internal Policies. (2015). *Fisheries Management and the Arctic in the Context of Climate Change*. Available at www.europarl.europa.eu/ RegData/etudes/STUD/2015/563380/IPOL_STU(2015)563380_EN.pdf.

Fairall, C. W., Bradley, E. F., Godfrey, et al. (1996). The cool skin and the warm layer effects on sea surface temperatures. *Journal of Geophysical Research*, **101**, 1295–1308.

Griffin, P. (2017). *Carbon Majors Report 2017*. Available at https://b8f65cb373b1 b7b15feb-c70d8ead6ced550b4d987d7c03fcdd1d.ssl.cf3.rackcdn.com/cms/reports/ documents/000/002/327/original/Carbon-Majors-Report-2017.pdf.

The Guardian. (2011). Which industries and activities emit the most carbon?, April 28. Available at https://www.theguardian.com/environment/2011/apr/28/industries-sectors-carbon-emissions.

ICAO (Convention on International Civil Aviation). (1944). 15 U.N.T.S. 295.

ILA (International Law Association). (2012). Report of the Seventh-Fifth Conference held in Sofia, Baselines under the International Law of the Sea: Committee Report.

ILA (International Law Association). (2016). *International Law and Sea Level Rise Committee Report*, Johannesburg Conference, August 7–11.

ILA (International Law Association). (2018). Resolution 5/2018 Committee on International Law and Sea Level Rise.

IPCC "A" (Intergovernmental Panel on Climate Change), Porter, J. R., Xie, L, et al. (2014). Food security and food production systems. In *Climate Change 2014: Impacts, Adaptation, and Vulnerability. Part A: Global and Sectoral Aspects. Contribution of Working Group II to the Fifth Assessment Report of the Intergovernmental Panel on Climate Change.* Available at www.ipcc.ch/pdf/assessment-report/ar5/wg2/WGIIAR5-Chap7_FINAL.pdf.

IPCC "B", Kovats, R. S., Valentini, R., et al. (2014). Europe. In *Climate Change 2014: Impacts, Adaptation, and Vulnerability. Part B: Regional Aspects. Contribution of Working Group II to the Fifth Assessment Report of the Intergovernmental Panel on Climate Change.* Available at www.ipcc.ch/pdf/assessment-report/ar5/wg2/WGIIAR5-Chap23_FINAL.pdf.

Jesus, J. (2003). Rocks, new-born islands, sea level rise and maritime space. In J. A. Frowein, K. Scharioth, I. Winkelmann and R. Wolfrum., eds., *Negotiating for Peace- Liber Amicorum Tono Eitel*, Berlin: Springer, 529–42.

Lam, V. W., Cheung, W. W., and Sumaila, U. R. (2016) Marine capture fisheries in the Arctic: Winners or losers under climate change and ocean acidification? *Fish and Fisheries*, **17**(2), 335–57.

Maratos, A. (2008). President of the Directing Committee of the International Hydrographic Bureau, "Nautical Charts and UNCLOS." Available at www.hydro-international.com/content/article/nautical-charts-and-unclos.

McIlgorm, A., Hanna, S., Knapp, G., et al. (2010). How will climate change alter fishery governance?: Insights from seven international case studies. *Marine Policy*, **34**(1), 170–7.

Mostyn *v.* Fabrigas 1 Cowp. 161, 98 Eng. Rep. 1021 (K.B. 1774) (Mansfield, L.J.).

Resolution LP.4(8) on the Amendment to the London Protocol to Regulate the Placement of Matter for Ocean Fertilization and other Marine Geoengineering Activities (adopted Oct. 18, 2013), IMO Doc. LC/35/15 Annex 4.

Revelle, R. and Suess, H. (1957). Carbon dioxide exchange between atmosphere and ocean and the question of an increase of atmospheric CO2 during the past decades, *Tellus IX*, **9**(1), 18–27.

Segel, A., De Young, C., and Soto, D. (2016). Food and Agriculture Organization, *Climate Change Implications for Fisheries and Aquaculture: Summary of the Findings of the Intergovernmental Panel on Climate Change Fifth Assessment Report*, FIAP CA1122. Available at www.fao.org/3/a-i5707e.pdf.

South China Sea Arbitration, Philippines v. *China, Award*, PCA Case No 2013-19, ICGJ 495 (PCA 2016), July 12, 2016.

UNCLOS (United Nations Convention on the Law of the Sea). (1982). 21 ILM 1261.

UNFCCC (United Nations Framework Convention on Climate Change). (1992). 1771 UNTS 107.

Viljhalmsson, H. (2007). Impact of changes in natural conditions on ocean resources. In M. Nordquist, R. J. Long and T. H. Heidar., eds., *Law, Science, and Ocean Management*. The Netherlands: Brill Publishers, pp. 225–70.

Williamson, P., Wallace, D. W., Law, C. S., et al. (2012) Ocean fertilization for geoengineering: A review of effectiveness, environmental impacts and emerging governance. *Process Safety and Environmental Protection*, **90**(6), 475–88.

World Ocean Review. (2010). *World Ocean Review-1*. Available at http://world oceanreview.com/wp-content/downloads/wor1/WOR1_english.pdf.

Appendix

Public Marine Chart Submissions under the United Nations Convention on the Law of the Sea Publicizing Territorial Seas, Archipelagic Regions, and Exclusive Economic Zones

Marine chart submissions from 78 states are significant for the formation of customary international law regarding state understandings of their practice of designating basepoints and baselines for purposes of asserting maritime jurisdiction. In terms of state practice, the following list suggests that customary international law may be crystallizing, where states understand that filing their marine charts under UNCLOS fixes their basepoints and baselines for certain purposes. Fifty-five states have submitted charts defining either baselines for the measurement of territorial seas or outer limits of territorial seas. Thirty-nine states have declared the extent of their EEZ jurisdiction on marine charts.

Table A20.1. *Marine chart submissions*

State	Year and Chart Submissions	State	Year and Chart Submissions
Algeria	2018 – EEZ	Lebanon	2011 – EEZ
Argentina	1996 – territorial sea and EEZ	Lithuania	2006 – territorial sea, contiguous zone and EEZ
Australia	2000 – territorial sea	Madagascar	2002 – territorial sea
Bahamas	2008 – archipelagic baselines	Marshall Islands	2016 – territorial sea, EEZ and archipelagic baselines
Bangladesh	2016 – territorial sea	Mauritius	2008 – territorial sea and archipelagic baselines
Belgium	2014 – territorial sea, EEZ	Micronesia	2018 – EEZ
Brazil	2015 – territorial sea, EEZ	Myanmar	2008 – territorial sea
Chile	2000 – territorial sea, the contiguous zone, EEZ	Nauru	1999 – territorial sea and EEZ
China	1996 – territorial sea, 2004 – territorial sea and EEZ with Vietnam 2012 – territorial sea Diaoyu Islands	Netherlands	2002 – territorial sea
Comoros	2011 – archipelagic baselines	New Zealand	2006 – territorial sea and EEZ
Congo (Republic of)	2018 – territorial sea, EEZ	Nicaragua	2013 – territorial sea

Table A20.1. (*cont.*)

State	Year and Chart Submissions	State	Year and Chart Submissions
Cook Islands	2014 – EEZ	Niue	2014 – territorial sea and EEZ
Costa Rica	1997 – EEZ	Norway	1996 – EEZ and territorial sea
			2003–2005 – territorial sea for mainland Norway, Svalbard, Jan Mayen, and Bouvet Island
Cote D'Ivoire	2016 – territorial sea, contiguous zone and EEZ	Pakistan	1999 – territorial sea
Croatia	2005 – EEZ as an "Ecological and Fisheries Protection Zone"	Palau	2008 – EEZ
Cuba	2009 – EEZ	Papua New Guinea	2002 – archipelagic baselines
Cyprus	1996 – territorial sea	Philippines	2009 – archipelagic baselines
	2004 – EEZ		
Ecuador	2012 – territorial sea and EEZ with Colombia	Qatar	2017 – territorial sea
	2017 – EEZ with Costa Rica		
Equatorial Guinea	1999 – territorial sea and EEZ	Romania	1997 – territorial sea
Fiji	2015 – territorial sea and archipelagic baseline, and EEZ	Russia	2013 – EEZ in Sea of Okhotsk
Finland	1997 – territorial sea and EEZ	St. Vincent and the Grenadines	2014 – territorial sea and archipelagic baselines
France	2009–2013 – EEZ and territorial sea for French Islands	Samoa	2018 – territorial sea
	2013 – EEZ in Mediterranean Sea		
	2015 – territorial sea for Antarctic territories		
	2018 – territorial sea for French Antilles		
Gabon	1999 – territorial sea	Sao Tome and Principe	1998 – archipelagic baselines, outer limits of territorial sea, contiguous zone, and EEZ
Germany	1995 – territorial sea and EEZ in Baltic Sea and North Sea	Saudi Arabia	2010 – territorial sea
Ghana	2018 – territorial sea and EEZ	Seychelles	2009 – territorial sea and archipelagic baselines
Grenada	2009 – territorial sea and archipelagic baselines	Solomon Islands	2018 – EEZ
Guyana	2015 – territorial sea	Somalia	2014 – EEZ
Honduras	2000 – territorial sea	Spain	2000 – EEZ as outer reach of Fisheries Protection Zone in the Mediterranean Sea
			2018 – EEZ
India	2010 – territorial sea	Sudan	2017 – territorial sea

Table A20.1. (*cont.*)

State	Year and Chart Submissions	State	Year and Chart Submissions
Indonesia	2009 – archipelagic baselines	Suriname	2017 – territorial sea
Iraq	2011– territorial sea	Tanzania	2013 – territorial sea
Ireland	2006 – EEZ	Trinidad and Tobago	2004 – territorial sea and archipelagic baselines
Italy	1996 – territorial sea	Tunisia	1998 – territorial sea
Jamaica	1996 – archipelagic baselines	Tuvalu	2013 – territorial sea, archipelagic baselines, and EEZ
Korea	2017 – Territorial Sea	United Arab Emirates	2009 – territorial sea
Japan	2008 – territorial sea	United Kingdom	2004 – EEZ for "British Indian Ocean Territory" 2014 – EEZ
Kenya	2006 – territorial sea and EEZ	Uruguay	1999 – territorial sea and EEZ
Kiribati	2015 – territorial sea, EEZ and archipelagic baselines	Vanuatu	2010 – territorial sea and archipelagic baselines 2018 – EEZ
Korea	2017– territorial sea	Vietnam	2004 – territorial sea and EEZ in Gulf of Tonkin
Kuwait	2016 – territorial sea	Yemen	2015 – territorial sea
Latvia	2011 – territorial sea		

Marine chart submissions to fulfill UNCLOS obligations under Articles 16(2) (territorial sea), 47(9) (archipelagic), and 75(2) (EEZ) can be found at http://www.un.org/Depts/los/ LEGISLATIONANDTREATIES/depositpublicity.htm

Part VI
Policies for Ocean Governance

21

The Plastic–Climate Nexus
Linking Science, Policy, and Justice

PETER STOETT AND JOANNA VINCE

Introduction

Among the plethora of marine ecosystem challenges of this century, two have received increasing attention in recent years. One of these challenges, the primary subject of this book, is climate change. Another challenge is that of plastic. Marine plastic pollution can be subdivided into three basic categories: (1) *macroplastics*, including debris such as fishing nets, large pieces of Styrofoam, and parcels that have been lost or discarded from cargo ships; (2) *microplastics*, which comprise particles under 5 mm in diameter that remain when plastic objects, including plastic nurdles (which are used in various production processes), enter the sea and phytodegrade; and (3) *nanoplastics*, the end state of microplastic degradation, which are invisible to the naked eye and, because they are so small (1,000 times smaller than an algal cell)), are more likely than microplastics to pass through biological membranes. All of these categories of plastics are regrettable intrusions on natural ecosystems, but the latter are particularly worrisome for those concerned with environmental change.

In this chapter, we argue that all types of plastic pollution are also troublesome irritants for those concerned specifically with climate governance. The links between plastic debris, which is reshaping ocean ecology, and the broader threat of climate change are rarely explored. Our analysis suggests that there are several interlinked variables and relevant ethical and policy-related considerations. In particular, the climate change and the microplastics narratives – discovery, scientific advances, technological and policy innovations – need to come together if we are to make solid progress in response to both of these wicked problems.

We develop three primary ideas in this chapter. First, there are clear links between the oceans and climate change agenda and the marine debris agenda. Microplastics especially are contributing to the climate change problem and vice versa (Royer et al., 2018). Second, there are intriguing governance parallels between the two problems, such as existential threat responses, shared but differentiated responsibilities, and common heritage issues. The most pronounced intersection is at the level of ecological justice. Third, we argue that climate change governance should take the microplastics plague into account. Similarly, those dealing with plastic debris have no choice but to take climate change into serious consideration.

In the remainder of this chapter we discuss why plastic debris should be a topic of grave concern to the climate change mitigation and adaptation community, and we show why it is a vital issue for environmental governance and ecological justice. We explore similarities between these two mega-issues from policy and justice perspectives, looking especially at human impacts and concerns about shared governance. We examine international legal principles that can be applied to both issues, including common heritage, shared but differentiated responsibilities, and the precautionary principle. Finally, we take a pragmatic look at what can be done to integrate these twin issues as part of wider efforts to achieve effective ocean governance amidst climate change.

The Emergence of Plastic Pollution

Synthetic polymers can be manufactured from fossil fuels or biomass. The complete biodegradation of plastic occurs when none of the original polymer remains, a result of microbial action breaking plastic down into carbon dioxide, methane, and water. Most plastic in circulation today is a petroleum and natural gas by-product. It does not biodegrade fully. Instead, it photodegrades into tiny particles that can both release toxic chemicals and absorb persistent organic pollutants and hydrophobic chemicals. Easily mistaken for food, this material then enters the marine food chain (Provencher, 2014). It has even been found in organisms living in the deep sea (Taylor et al., 2016). Plastic has also been found in the most remote ocean areas and identified in every type of marine habitat (Ivar do Sul and Costa, 2014). For example, a French team studying plankton found remarkably high plastic levels in the Southern Ocean in 2012. Remote Arctic sea ice has been found to contain high concentrations of microplastics, and the extent to which melting ice will release various anthropogenic particulates back into the ocean is not yet fully known (Obbard et al., 2014). Plastic debris acts as a transport system for invasive alien species and diminishes sunlight and oxygen levels, contributing to dead zones.

The positive correlation between loss of marine biodiversity and reductions in marine ecosystem services has been known for quite some time (Worm et al., 2006). However, although plastic pollution has been observed for decades, awareness of microplastics is relatively new, dating back to the late 1990s. There was solid scientific evidence that plastic debris, even at the micro-level, was an emerging problem in the Atlantic Ocean in the early 1970s (Carpenter et al., 1972; Colton, Knapp, and Burns, 1974), but it was not until the 1990s and early 2000s that a flood of research emerged finding microplastic abundance in the oceans and, still later, in freshwater systems. An alarming European Commission report suggested that approximately 10 percent of the global plastic manufactured each year, about 265 million tons, ends up in the oceans or other water systems (EC, 2011). In some areas, densities have been recorded at 100,000 particles per square meter of water (Wright, Thompson, and Galloway, 2013).

That the entire life cycle of plastic is vast and global should not surprise us. An industry producing over 260 million tons of product annually, and using roughly 8 percent of global oil production in the process, leaves a tremendous ecological footprint (Redclift, 1996; Thompson, 2009; Clapp, 2012). The world's shorelines are the most obvious repository of plastic debris, from fishing material to degraded plastic bottles. The discovery of five so-called "garbage patches" in the oceanic gyres, first in the Pacific and then elsewhere, rang loud alarm bells (Moore et al., 2001, Kaiser, 2010, Titmus and Hyrenbach, 2011). The five gyres serve as conductor belts for masses of immeasurable subsurface microplastic. Early pelagic studies even indicated a higher abundance of microplastics than zooplankton (the building block of the marine food chain) in some areas (Moore et al., 2001). This is especially troubling, given the important role zooplankton play in regulating climate. Plastic debris also has a direct impact on wildlife health and contributes to positive feedback loops: for example, plastic bags look very similar to jellyfish in the marine environment, leading sea turtles to ingest the plastic and suffocate or starve as a consequence. The reduction in the number of sea turtles, in turn, exacerbates the growing problem of invasive jellyfish abundance (see Schuyler et al., 2014; Wilcox et al., 2015). One author sums up the wildlife impact of plastic marine debris as "entanglement, ingestion, smothering, hangers-on, hitch-hiking and alien invasions" (Gregory, 2009; see also Desforges et al., 2014). There is no evidence that even biodegradable plastics break down once consumed by mammals such as sea turtles or seabirds (Muller et al., 2012).

Plastic debris originates from many sources. Nurdles used as preproduction pellets escape the product cycle at various stages and end up in oceans (Ogata et al., 2009; Hammer et al., 2012). Microbeads used in cosmetic cleansers and other personal products, including toothpaste, are flushed into water systems and are too small to be stopped by wastewater filtration systems. Three-quarters or more of litter in the ocean comes from land-based sources (Derraik, 2002; Hardesty et al., 2014; Jambeck et al., 2015). Consequently, microplastic pollution, whether through nurdles and pellets, microfibers in clothing released through washing processes, or personal cosmetic products, needs to be managed before it reaches the ocean. Regulation of the use of microbeads in cosmetic products has been timely; however, enforcement and implementation are slow. Resilient plastic bags are ubiquitous across continents and have joined bottle caps, six-pack rings, straws, cigarette butts, and other plastic products in waterways considered pristine before the 1970s. Increasing evidence suggests polyester, released through the washing of clothes, is one of the most abundant microplastics found on shorelines (Browne et al., 2011). Even the synthetic rubber in car tires results in microplastic debris.

As societies move toward an aquaculture-based global protein regime, many observers have concluded that it is already a major source of plastic pollution (Austin, 2014; Mathalon and Hill, 2014). One study found that aquaculture was a major contributor of marine litter off the coasts of Chile, South Korea, Japan, and other areas (Hinojosa and Thiel, 2009). Another area of future concern will be the rise of 3D printing, "using plastic 'ink' [which] will guarantee expanded use of polymeric feedstocks" (Moore, 2015: vii). Even the recycling industry, which is vital for reducing plastic pollution, comes with its own pollution and climate change issues, although exciting ideas for secondary uses of plastics, such as the construction of roads and sidewalks in the Netherlands, are appearing (Valentine, 2015).

Meanwhile, new evidence suggests that the deep sea is harboring much more plastic than was previously imagined. A 2014 study found that microplastic fibers were "up to four orders of magnitude more abundant (per unit volume) in deep sea sediments from the Atlantic Ocean, Mediterranean Sea and Indian Ocean than in contaminated sea-surface waters" (Woodall et al., 2014; see also Van Cauwenberghe et al., 2013). Extricating this plastic is not feasible. And scientists are now investigating "life in the plastisphere," where entirely new microbial communities are evolving (Zettler, Mincer, and Amaral-Zettler, 2013).

For both plastic pollution and climate change, technological solutions exhibit serious limitations. We believe that they should not form the basis of an action strategy in the future. Robotic or captained pelagic vacuum cleaners (such as the crowd-sourced SeaVax) or trailing suction hopper dredging ships (such as the *Queen of the Netherlands*) can no doubt make a substantive impact. However, it would take a virtual navy of such ships to affect significant change, given the mass of plastic in our oceans, lakes, and rivers. That said, this is not a reason to consider plastic retrieval to be futile, and governments with exclusive economic zones surely have an obligation, when possible, to undertake surface cleanup programs. Given their difficult location in areas beyond national jurisdiction, a multilateral fund such as that suggested by the Global Oceans Commission could be utilized to finance retrieval operations in the oceanic gyres. The Ocean Cleanup Project proposed conducting retrieval operations in the North Pacific Garbage Patch, while another study indicates that more beneficial outcomes would be attained by focusing retrieval areas off the coast of East Asia, from where a majority of plastic litter now originates. Such operations could include capture closer to coasts to prevent marine litter being swept into the open ocean. This would reduce the amount of plastic that degrades into microplastic, and it could also reduce the amount of microplastic and plankton overlap (Sherman and van Sebille, 2016).

As with climate change, larger-scale geoengineering designs are more problematic. Stimulating phytoplankton presence and activity by increasing nutrient availability in areas swamped by plastic waste is one such proposal which would, theoretically, also increase carbon dioxide (CO_2) uptake (see Chapter 26). Unfortunately, introducing iron or nitrogen into areas with especially high plastic density may cause as many problems as it solves. Stimulating plastic-bound bacteria to increase their nitrogen-fixing capacity is another idea that seems rather farfetched to most scientists. In general, stimulating phytoplankton can lead to eutrophication and "dead zones" with low oxygen content in which anaerobic bacteria thrive; their respiration produces nitrous oxide, a greenhouse gas about 265 times more potent than CO_2. Biodegradable plastics, meanwhile, do not offer a solution to the aquatic plastics problem. Most biodegradable plastic products do not decompose unless a temperature of 50°C is reached, which is (to put it mildly) highly undesirable in any aquatic context.

Microplastic Pollution and Climate Change

Climate-Related Impacts of Plastic Pollution in Oceans

The production of plastic products and the deluge of aquatic plastics is contributing to problems associated with climate change. The plastics industry itself has been estimated to produce a 10 percent contribution to total "global warming potential" if its relative contribution to other material groups is included (STAP, 2011). This is probably a low estimate that does not include the climate implications of the global recycling industry. Beyond this, microplastics and nanoplastics can be ingested by zooplankton and microzooplankton (Cole et al., 2013), which negatively affects algal feeding, potentially increasing algal blooms and disrupting zooplankton growth and fecundity. Phytoplankton use sunlight to bind water and carbon dioxide, creating oxygen and organic matter for other organisms to eat. If organisms are ingesting microplastic, or their permeable membranes are absorbing nanoplastic, they are ingesting less organic carbon-based matter, and thus the carbon cycling function of ocean ecosystems may be compromised. Research in this area is just beginning. Membrane permeability is a core concern when we link nanoplastic pollution with the climate-regulating functions of the oceans (see Koelmans et al., 2015). This is admittedly speculative at this juncture, but the implications could be severe for the global carbon cycle because oceans act as sinks for at least half of anthropocentric carbon emissions. This is a potential "tipping point" or positive feedback loop that we cannot afford to ignore.

In addition, there is emerging evidence that plastic ingestion by coral reefs adds to the threat climate change and ocean acidification (see Chapter 22) pose to these precious biodiverse resources (Hall et al., 2015). In general, increased water temperatures and acidification can accelerate the breakdown of macroplastic into micro- and nanoplastic. Increased carbon levels in the oceans are already contributing to ocean acidification and other problems, but microzooplankton and zooplankton form the basis of the marine food chain, and their collapse would spell biological disaster. Ocean currents may shift due to climate change, delivering plastic to (increasingly rare) untainted areas, and sea-level rise will claim more litter from shorelines with tidal activity and extreme weather events such as hurricanes and cyclones. A combination of shifting temperature zones and the use of plastic debris as vectors will exacerbate the problem of invasive species, including that of travelling microbes (Derraik, 2002). Warming waters will also release plastic and related toxic material (and potential pathogens) captured in marine ice (CBD, 2012). In short, climate change and heavy microplastic pollution are working in tandem to lower marine ecosystem resilience.

Plastic Pollution and Climate Justice

Where the plastic–climate nexus is most clear, however, is in the normative sphere. Climate change and plastic pollution are both social justice issues where deleterious impacts are suffered by those who have typically contributed the least to the problems and are least able to escape them. The overlap between plastic pollution and climate change is just beginning to receive serious scientific investigation, but the political similarities are already quite striking. For example, in both cases common but differentiated responsibilities are involved. This is well known in the case of greenhouse gas emissions, where a minority of countries are effectively responsible for the majority of pollution. In the plastic context, according to one estimate, "20 countries, out of a total of 192 coastlines, are responsible for 83% of the plastic debris put into the world's oceans" (Tibbetts, 2015: A91). Another study concluded that reducing waste by 50 percent in the top 20 countries would result in a nearly 40 percent decline in inputs of plastics to the oceans (Leonard, 2015). More pollution (carbon, methane, plastic) now comes from the Asian region than others, but historically both industrial emissions and plastic were largely European and American innovations. It would be a constructive exercise to begin discussing the historical and contemporary plastic footprint of countries and even of individuals.

Most visibly, perhaps, small-island states are as disproportionately affected by aquatic microplastics as they are by extreme weather events and sea-level rise (see the chapters in Part II). They have contributed relatively little to these problems (see Costa and Barletta, 2015). This was acknowledged in the landmark 2011 Honolulu Commitment, whereby parties "recognized the need to address the special requirements of developing countries, in particular the Least Developed Countries and Small Island Developing States, and their need for financial and technical assistance, technology transfer, training and scientific cooperation to enhance their ability to prevent, reduce and manage marine debris as well as to implement this commitment and the Honolulu Strategy" (see later). Even with financial and technological assistance, implementing such policies can be difficult, as has been the case in the Pacific (see Chasek, 2010; Vince et al., 2017). These environmental justice concerns permeate both the plastic and climate change adaptation discourses for good reason.

Both climate change and microplastic pollution are severe threats to human health (on climate see Machalaba, Romanelli, and Stoett, 2017; on plastics see van Cauwenberghe and Janssen, 2014). Similarly, both climate and plastic pollution threaten the very existence of ecosystems on which humans and especially riparian and coastal communities are dependent. This makes both issues subject to the heritage-of-mankind concept, which asserts common responsibilities for conserving resources for future generations. The scientific uncertainty about the long-term impact of micro- and nanoplastic abundance on climate change – not only climate change impacts in the oceans but also on a global scale (because of the oceans' regulatory role in climate) – gives rise to the relevance of the precautionary principle in this regard. According to this principle, avoiding pollution is required to avoid potentially harmful consequences. Intergenerational justice is also a significant concept in both contexts, because lack of action today will certainly exacerbate hardship in the future.

Shared Governance of Climate Change and Marine Plastic Pollution

The governance arrangements involved in managing both climate change and marine plastic pollution are multifaceted and complex. Climate change has been labelled a "wicked" and "super wicked problem" (Lazarus, 2008; Levin et al., 2012), and marine plastic pollution is also categorized in a similar vein (Hastings and Potts, 2013). As with climate change policy, the complexity with regulatory approaches lies with the transboundary and cross-jurisdictional nature of the problem. The vast quantities of marine plastic waste can be localized, or it can travel via ocean currents to the most remote parts of the globe (STAP, 2011), highlighting the difficulties of attributing responsibility to any state, government, or organization for its removal. Microplastics make this even more challenging. The sources are so varied and widespread that technological advances to filter plastic waste from the ocean surface are too small in scale to have an impact.

Regulation is only one approach to managing macroplastic and microplastic marine debris. Economic/market and community-based efforts are also an integral part of the management, prevention, and mitigation of plastic pollution. Similar to climate change, this mixture of governance approaches is developed and implemented at the local, national, regional, and global levels, all at varying scales and degrees of success (Vince and Hardesty, 2016). Importantly, regulating, ceasing, or altering land-sourced microplastic pollution is the responsibility of national governments, and each jurisdiction will have different methods for doing this. In the United States in 2015, the California Assembly voted to prohibit the sale of microbeads in personal care products from 2020. California's actions with regard to marine debris are significant, and already the impact of the "California effect" (Fredriksson et al., 2002; Perkins and Neumayer, 2012) is being seen elsewhere. Other jurisdictions around the world are following suit by introducing microbead bans (see Rochman et al., 2015). However, by the time they are fully implemented, tons of microplastics will have entered the ocean. In Australia, microbead legislation has not been enacted, and state and federal governments have only agreed to voluntary approaches. Rwanda's outright ban on plastic bags and France's recent plastic utensil bans are examples, but resistance to such bans is also fierce in the United States and elsewhere.

Bottom-up, community governance solutions to marine pollution cannot be underestimated. Many of the positive changes to reducing plastic marine pollution have occurred through community action (Ghostnets Australia, 2015; Vince and Hardesty, 2016). Education (Derriak 2002; Duckett and Repaci, 2015) and the use of "citizen science" (Science Communication Unit, 2013; Hidalgo-Ruz and Thiel, 2015; Jambeck and Johnsen, 2015) increase engagement with the community and change the behavior of those involved with the use and removal of plastics from the marine environment. Local communities and nonstate actors are using social license to bring about change where regulation is lagging (Vince and Hardesty, 2016). While the larger nongovernmental organizations (NGOs), such as Greenpeace, Worldwide Fund for Nature, and the International Union for Conservation of Nature, usually take a holistic approach to the problem of plastic debris, other NGOs are issue specific. *Beat the Microbead* is an organization that distributes data on which products do or do not have microbeads, and it is accessible through their website or smartphone app (Beat the Microbead, 2015). Consequently, the consumer has the power through social license to drive changes to corporate policies and products (Morrison, 2014; Cullen-Knox, 2016; Vince and Haward, 2017).

The polarizing effects of politics surrounding plastic pollution are particularly noticeable in the microbead debate. The legislative changes and the voluntarily driven consumer demand have had a number of positive impacts on industries and the use of microbeads. Legislative loopholes remain for those companies that may not necessarily want to change from plastic to natural alternatives. Major retailers Coles and Woolworths in Australia have introduced bans across their stores as a response to consumer demand (Browne, 2016). Johnson & Johnson, and Procter & Gamble have all agreed to remove gradually polyethylene microbeads from their products (Abrams, 2015). However, it is unclear whether the microplastics will be replaced with bioplastics or natural alternatives. Corporations are varied in their responses to this issue and in the general use and life cycle of plastics. Changes to their corporate social responsibility (CSR) policies may have an impact. However, in the meantime the wider community and consumer demand exercised through social license will be a driver of change within the market.

According to a report by the secretariat of the Convention on Biological Diversity and the Scientific and Technical Advisory Panel (GEF, 2012), "many companies now see packaging and plastics sustainability as part of broader corporate social responsibility, and negative brand image is becoming a major driving force which is being harnessed in the interests of improving packaging materials and technologies." In 2011, industry plastics associations came together and developed a global Declaration for Solutions on Marine Litter. Sixty-nine plastics organizations from 35 countries have signed this Declaration (Marine Litter Solutions, 2017). Market and economic governance approaches encourage industry to engage in corporate social responsibility and actively participate in reducing or removing plastic marine debris from the environment. UNEP's *Guidelines on the Use of Market-Based Instruments to Address the Problem of Marine Litter* (2009) identify market-based instruments as taxes, charges, fees, fines, penalties, liability and compensation schemes, subsidies and incentives, and tradable permit schemes. There are several basic principles behind these instruments: the polluter pays principle, the user/beneficiary pays principle, and the principle of full cost recovery. While market-based instruments have been used successfully to reduce macroplastic pollution (for example, Cho, 2009; Hardesty, 2014), little progress has occurred with their application in the management of microplastics, just as their ultimate impact on climate change mitigation remains contested.

Globally, the need to address the plastic problem is increasingly recognized, with discussions on marine plastic pollution occurring at international assemblies such as the World Ocean Summit and G7 (2015) and G20 meetings. Calls for a new legally binding international agreement have been made by a number of scholars (Stoett, 2010; Gold et al., 2013; Chen 2015; Vince and Hardesty, 2016; Raubenheimer and McIlgorm, 2017), demonstrating the large gap in international law for specifically dealing with land-based plastic marine pollution. The United Nations Law of the Sea Convention (UNCLOS) Part XII (Articles 192–237) is dedicated to the protection and preservation of the marine environment (see Chapter 2) and requires states to take all measures "that are necessary to prevent, reduce and control pollution of the marine environment from any source, using for this purpose the best practicable means at their disposal and in accordance with their capabilities, and they shall endeavor to harmonize their policies in this connection" (art. 194). Although UNCLOS sets out the responsibilities of states and necessary measures they need to undertake to minimize pollution, there are doubts that UNCLOS on its own can solve the plastics problem. While UNCLOS recognizes that there are six different sources of marine pollution, including land-based pollution, it does not go into detail about the type of pollutants and technical rules (Palassis, 2011). States are directed to adopt their own laws and regulations dealing with marine pollution and to work with relevant international organizations.

Soft law dominates global efforts to deal with plastic marine debris. While soft law is nonbinding, there is a strong expectation that it will be adhered to (Birnie and Boyle, 1992), and it maintains "a considerable practical significance" (Lyster, 1985). Joyner (2000) argues that soft law facilitates the process that gives rise to customary law and that it encourages compliance. With regard to plastic marine debris, soft law is an essential part of a nascent regime on ocean plastic. There are many examples of this, including:

- The UN Conference on the Environment and Development's Rio Declaration encouraged integrated, precautionary, and anticipatory marine environmental protection (United Nations, 1992). It set out an approach to addressing damaging impacts from air, land, and water; recycling; sewage treatment; and the prevention, reduction, and control of ship-sourced pollution. Plastic marine debris can certainly be added to this list.
- The Conference of the Parties to the CBD (COP CBD) and the Scientific and Technical Advisory Panel of the Global Environment Facility released a report on the impacts of marine debris on marine and coastal biodiversity in 2012. A decision to address the impacts of marine debris on marine and coastal biodiversity was adopted through Decision XI/18 at the 11th Meeting of COP CBD.
- The United Nations Environment Programme (UNEP) released several guidelines for dealing with marine pollution: *Guidelines on Survey and Monitoring of Marine*

Litter (2009), *Guidelines on the Use of Market-Based and Economic Instruments* (2009), and *Marine Litter: A Global Challenge* (2009). The latter report has a number of recommendations for the 13 participating Regional Seas Programs, including, *inter alia,* the development of a Regional Action Plan or strategy to deal with marine pollution; mitigation delivered globally but coordinated at the regional level and implemented at the national level; National Plans of Action that draw on existing legislation; and the coordination of UN organizations working on the marine litter problem.

- The *Honolulu Strategy* was formed with UNEP collaboration in 2012 as a global strategy to reduce marine debris.
- The UNEP Global Partnership on Marine Litter (GPML) was announced in 2012. It is a coordinating forum for stakeholders at all levels working on marine debris prevention and management.
- In 2014 the Global Ocean Commission Report recognized that the majority of marine debris is land sourced and called for a more coordinated effort by governments, the private sector, and civil society to stop plastic pollution before it enters the ocean.
- The parties of the United Nations Sustainable Development Summit (2015) agreed upon the goal to *Conserve and Sustainably Use the Oceans, Seas and Marine Resources*, including to "[b]y 2025, prevent and significantly reduce marine pollution of all kinds, in particular from land-based activities, including marine debris and nutrient pollution."
- The G7 group released an *Action Plan to Combat Marine Litter* in June 2015, which included land-based and sea-based priorities to reduce marine debris.

The diffuse nature of the origin of plastic pollution in the oceans, and the abstract nature of the oceans crisis generally – it is far removed from most citizens' daily lives, or at least their daily consciousness – might make plastic pollution a difficult subject for proponents of global governance to sell to governments. The new discovery of microplastics in freshwater systems, which have an immediate impact on water supplies, fisheries, and other variables essential to human health, will prompt more policy action. The abstract nature of the problem is reshaped by the increased proximity to citizens and politicians.

Recommendations

It would betray any serious effort to integrate ocean governance issues to omit the plastic pollution issue. Its links to climate change demand an even more precise response. Toward this end, we have a number of recommendations:

a) The Honolulu Strategy should be reinvigorated by the adoption of a global accord to curb the purposeful and incidental dispersion of plastic waste into the marine environment (Gold et al., 2013; Chen, 2015; Raubenheimer, 2017). There is a global ethical imperative to establish such an accord, but it will be a long-term project demanding multilevel, adaptive governance. The accord should refer to the plastic–climate nexus to give it greater legitimacy and expand its scope.

b) The potential links between climate change and plastics pollution should receive ample research funding, as well as political consideration. Governments should integrate pollution limitation agendas with their climate change mitigation and adaptation plans. The Transboundary Waters Assessment Program (n.d., 2014) can be helpful in this regard as well.

c) Linked to the earlier point, marine plastic pollution should be seen as a justice issue. Plastic pollution is a fundamental challenge to the UN Sustainable Development Goals' "development-with-dignity agenda." Indeed, climate justice and the risks of plastic abundance in waterways are inseparable concerns and thus should be linked by activists accordingly.

d) More research is needed on the behavioral economics of marine litter (Wyles, 2014). Incentive structures need to be constructed for both governmental policy making and for consumer and producer behavior to reduce plastic consumption, as well as greenhouse gas emissions.

e) The Global Oceans Commission's suggestion to establish a Global Marine Responsibility Fund to help finance related retrieval and research efforts should be explored. Heavy investment in technological advancement for plastics retrieval is critical and should be partially funded by the plastics and fossil fuel industries, part of a broader regime of extended producer responsibility. This can be linked with climate initiatives where possible, and such efforts should take climate impact into consideration.

f) Though the extent of plastic in the oceans is a major problem, and industries such as aquaculture must be regulated, the main problem for future generations is the increase in land-based sources of plastic pollution, including microfibers released into water supplies with clothes washing and through other invisible forms of pollution. Tackling this issue demands fiscal and taxation strategies to reduce waste, updates in water treatment facilities, and harsh penalties for serious violations of dumping regulations. Many of these regulatory measures can also be applied to climate policy and will face similar problems regarding transparency, surveillance, and other monitoring challenges.

g) Governments and NGOs must continue to encourage the active involvement of "citizen science" in both climate and pollution monitoring. We can learn from previous positive experiences with invasive species identification and wastewater management.

h) Clear indicators are necessary for tracking progress. For example, the Convention for the Protection of the Marine Environment of the North-East Atlantic employs three indicators of marine plastic: seabed litter, beach litter, and ingestion by Northern fulmar seabirds. Joint Group of Experts on the Scientific Aspects of Marine Environmental Protection and others are working on operational indicators. Such measurements should be integrated with climate models and assessments.

Conclusion

The aquatic plastic pollution problem can be linked to climate change. Emerging science suggests that ocean acidification, the potential breakdown of the ocean ecosystem food chain, rises in sea levels, invasive species, and plastic marine debris are interlinked variables. Thus, there is a global imperative, based on the four most widely accepted tenets of international environmental law – the precautionary principle, common but differentiated responsibilities, intergenerational equity, and the common heritage of mankind – to prevent further macroplastic, microplastic, and nanoplastic waste from entering rivers, lakes, and oceans. Since those who contribute least to both problems are the most heavily affected, environmental injustice is a common theme.

Climate governance efforts need to take the plastic plague seriously because plastic debris can disrupt the food chain, and by implication the ocean ecosystem's ability to regulate climate. This adds further fuel to arguments that serious tipping points on climate are approaching. Climate governance models assume that ecosystem resilience is a positive factor, necessary for the success of both long- and short-term efforts to mitigate and adapt to climate change. Indeed, the ecosystem-based adaptation approach hinges on this assumption. Plastic debris is a major threat to this resilience and could render many climate governance efforts fruitless if it is not addressed.

Converging international legal principles, actors, and governance mechanisms are involved in the plastic–climate nexus. Plastic production contributes directly to greenhouse gas emissions, and changes to the product cycle can have a positive impact on mitigation efforts. The shipping (see Chapter 23), oil and natural gas (Chapter 24), and tourism industries (Chapter 3) are all key sectors in both marine debris and climate change mitigation. Responses such as beach maintenance and coastal ecosystem conservation will involve many government agencies and clients. The evolving legal framework for plastic waste reduction may provide helpful models for the climate regime, and vice versa: the Paris Agreement, with its limited and self-defined legal obligations, may be the best model for plastic pollution reduction. Though it is also a limited device, UNCLOS will undoubtedly play an organizational role in both of these policy objectives, and it could place emphasis on their continued convergence.

There is no doubt that ocean governance needs to grapple with both climate change and marine plastic pollution as two of the major issues of our time. There is merit in considering the interlinkages between the two issues. It will be valuable to consider common, and perhaps even linked, policy responses to them.

References

Abrams, R. (2015). Fighting pollution from microbeads used in soaps and creams. *The New York Times*, 22 May.

Austin, B. (2014). The aquaculture industry. *Microbiology Today*, **41**(4), 166–9. Available at www.microbiologysociety.org/publications/microbiology-today/past-issues.cfm/publication/water/article/7A993C5D-AE1E-4428-B8F391A1E9D BEFED.

Beat the Microbead. (2015). Available at www.beatthemicrobead.org/en/

Birnie, P. and Boyle, A. (1992). *International Law and the Environment*. Oxford: Clarendon Press

Browne, M. A, Crump, P., Niven, S. J., et al. (2011) Accumulation of microplastic on shorelines worldwide: Sources and sinks. *Environment, Science, and Technology*, **45**, 9175–9.

Browne, R. (2016). Coles and Woolworths ban products containing plastic microbeads, *The Sydney Morning Herald*, January 8. Available at www.smh.com.au/environment/coles-and-woolworths-ban-products-containing-microbeads-20160107-gm1mwm.html.

Carpenter, E. J., Anderson, S. J., Harvey, G. R., Miklas, H. P., and Peck, B. B. (1972). Polystyrene spherules in coastal waters. *Science*, **178**(4062), 749–50.

CBD (Convention on Biological Diversity). (2012). *Impacts of Marine Debris on Biodiversity. CBD Technical Series No. 67*. Montreal: Secretariat of the Convention on Biological Diversity.

Chasek, P. (2010) *Confronting Environmental Treaty Implementation Challenges in the Pacific Islands*. Honolulu: EastWest Center.

Chen, C.-L. (2015). Regulation and management of marine litter. In M. Bergmann, L. Gutow, and M. Klages, eds., *Marine Anthropogenic Litter*. New York: Springer, 395–428.

Cho, D.-O. (2009). The incentive program for fishermen to collect marine debris in Korea. *Marine Pollution Bulletin*, **58**, 415–7.

Clapp, J. (2012). The global reach of plastic waste. In S. Foote, ed., *Histories of the Dustheap*. Cambridge, MA: MIT Press, 199–225.

Cole, M., Lindeque, P., Fileman, E., et al. (2013). Microplastic ingestion by zooplankton. *Environmental Science & Technology*, **47**(12), 6646–55. doi:10.1021/es400663f

Colton, J., Knapp, F., and Burns, B. (1974). Plastic particles in surface waters of the northwestern Atlantic. *Science*, **185**(4150), 491–7.

Costa, M. F. and Barletta, M. (2015). Microplastics in coastal and marine environments of the western tropical and sub-tropical Atlantic Ocean. *Environmental Science: Processes and Impacts*, **17**(11), 1868–79. doi:10.1039/c5em00158g

Cullen-Knox, C., Eccleston, R., Haward, M., Lester, L., and Vince, J. (2016). Contemporary challenges in environmental governance: the rise of social licence. *Environmental Policy and Governance*, **27**(1), 3–13.

Derraik, J. (2002). The pollution of the marine environment by plastic debris: A review. *Marine Pollution Bulletin*, **44**(9), 842–52.

Derraik, J. G. B., (2002). The pollution of the marine environment by plastic debris: a review. Mar. Pollut. Bull. **44**, 842–52.

Desforges, J.P.W., Galbraith, M., Dangerfield, N., and Ross, P. S. (2014). Widespread distribution of microplastic in sub-surface seawater in the northeast Pacific Ocean. *Marine Pollution Bulletin*, **79**(1–2), 94–9.

Duckett, P. E. and Repaci, V. (2015). Marine plastic pollution: Using community science to address a global problem. *Marine and Freshwater Research*, **66**, 665–73.

European Commission (EC). (2011). *Plastic Waste: Ecological and Human Health Impacts, In-depth Report*. Brussels: EC Directorate-General, Environment.

Fredriksson, P. G. and Millimet, D. L. (2002). Is there a "California effect" in US environmental policymaking? *Regional Science and Urban Economics*, **32**(6), 737–64.

GEF – Secretariat of the Convention on Biological Diversity and the Scientific and Technical Advisory Panel. (2012). Impacts of marine debris on biodiversity: Current status and potential solutions. *Montreal Technical Series No. 67*, 61.

GhostNets Australia. (2015). Available at www.ghostnets.com.au/.

Gold, M., Mika, K., Horowitz, C., Herzog, M., and Leitner, L. (2013). Stemming the tide of plastic marine litter: A global action agenda. Pritzker Policy Brief 5.

Gregory, M. (2009). Environmental implications of plastic debris in marine settings – Entanglement, ingestion, smothering, hangers-on, hitch-hiking and alien invasions. *Philosophical Transactions of the Royal Society B: Biological Sciences*, **364**(1526), 2013–25.

Group of Seven (G7). (2015). *Leaders' Declaration*. Berlin, Germany: Schloss Elmau.

Hall, N. M., Berry, K. L. E., Rintoul, L., and Hoogenboom, M. O. (2015). Microplastic ingestion by scleractinian corals. *Marine Biology*, 162(3), 725–32. doi:10.1007/s00227-015-2619-7

Hammer, J., Kraak, M. H., and Parsons, J. R. (2012). Plastics in the marine environment: The dark side of a modern gift. In D. M. Whitacre, M. Fernanda, and F. A. Gunther, eds., *Reviews of Environmental Contamination and Toxicology*. New York: Springer, 1–44.

Hardesty, B., Wilcox, C., Lawson, T., Lansdell, M., and van der Velde, T. (2014). Understanding the effects of marine debris on wildlife. *A Final Report to Earthwatch Australia*. CSIRO. Available at http://apo.org.au/system/files/41318/apo-nid41318-150746.pdf.

Hastings, E. and Potts, T. (2013). Marine litter: Progress in developing an integrated policy approach in Scotland. *Marine Pollution*, **42**, 49–55.

Hidalgo-Ruz, V. and Thiel, M. (2015). The contribution of citizen scientists to the monitoring of marine litter. In M. Bergmann, L. Gutow, and M. Klages, eds., *Marine Anthropogenic Litter*. New York: Springer, 429–45.

Hinojosa, I. A. and Thiel, M. (2009). Floating marine debris in fjords, gulfs and channels of southern Chile. *Marine Pollution Bulletin*, **58**(3), 341–50. doi:10.1016/j.marpolbul.2008.10.020

Ivar do Sul, J. A. and Costa, M. F. (2014). The present and future of microplastic pollution in the marine environment. *Environmental Pollution*, **185**, 352–64. doi:10.1016/j.envpol.2013.10.036

Jambeck, J. R., Geyer, R., Wilcox, C., et al. (2015). Plastic waste inputs from land into the ocean. *Science*, **347**(6223), 768–71.

Jambeck, J. R. and Johnsen, K. (2015). Citizen-based litter and marine debris data collection and mapping. *Computing in Science & Engineering*, **17**, 20–26.

Joyner, C. C. (2000). The international ocean regime at the new millennium: A survey of the contemporary legal order. *Ocean & Coastal Management*, **43**(2), 163–203.

Kaiser, J. (2010). The dirt on ocean garbage patches. *Science*, **328**, 1506.

Koelmans, A., Besseling, E., and Shim, W. J. (2015). Nanoplastics in the aquatic environment. Critical review. *Marine Anthropogenic Litter*, Cham: Springer International Publishing, 325–40. Available at http:// link.springer.com/10.1007/978-3-319-16510-3_12.

Lazarus, R. J. (2008). Super wicked problems and climate change: Restraining the present to liberate the future. *Cornell Law Review*, **94**, 1153.

Leonard, G. (2015). Trashing the ocean: New study provides first estimate of how much plastic flows into the ocean. *Ocean Conservancy*. Available at http://blog.oceanconservancy.org/2015/02/13/trashing-the-ocean-new-study-provides-first-estimate-of-how-much-plastic-flows-into-the-ocean/.

Levin, K., Cashore, B., Bernstein, S., and Auld, G. (2012). Overcoming the tragedy of super wicked problems: Constraining our future selves to ameliorate global climate change. *Policy Sciences*, **45**(2), 123–52.

Lyster, S. (1985). *International Wildlife Law: An Analysis of International Treaties Concerned with the Conservation of Wildlife*. Cambridge, UK: Grotius Publications.

Machalaba, C., Romanelli, C., and Stoett, P. (2017). Global environmental change and emerging infectious diseases: Macrolevel drivers and policy responses. In M. Bouzid, ed., *Examining The Role of Environmental Change on Emerging Infectious Diseases and Pandemics*. Hershey, PA: IGI Global, 24–67.

Marine Litter Solutions. (2017). *Joint Declaration*. Retrieved from: www.marinelitter solutions.com/who-we-are/joint-declaration.aspx

Mathalon, A. and Hill, P. (2014). Microplastic fibers in the intertidal ecosystem surrounding Halifax Harbor, Nova Scotia. *Marine Pollution Bulletin*, **81**(1), 69–79. doi:10.1016/j.marpolbul.2014.02.018

Moore, C. (2015). Foreword. In M. Bergmann, L. Gutow, and M. Klages, eds., *Marine Anthropogenic Litter*. New York: Springer.

Moore, C., Moore, S., Leecaster, M., and Weisberg, S. (2001). A comparison of plastic and plankton in the North Pacific Central Gyre. *Marine Pollution Bulletin*, **42**, 1297–1300.

Morrison, J. (2014) *The Social License. How to Keep Your Organization Legitimate*. Basingstoke: Palgrave Macmillan.

Muller, C., Townsend, K., and Matschullat, J. (2012). Experimental degradation of polymer shopping bags in the gastrointestinal fluids of sea turtles. *Science of the Total Environment*, **416**, 464–7.

Obbard, R. W., Sadri, S., Wong, Y. Q., et al. (2014). Global warming releases microplastic legacy frozen in Arctic Sea ice. *Earth's Future*, **2**, 315–20.

Ogata, Y., Takada, H., Mizukawa, K., et al. (2009). International pellet watch: Global monitoring of persistent organic pollutants (POPs) in coastal waters. 1. Initial phase data on PCBs, DDTs, and HCHs. *Marine Pollution Bulletin*, **58**(10), 1437–46.

Palassis, S. (2011). Marine pollution and environmental law. In R. Baird and D. Rothwell, eds., *Australian Coastal and Marine Law*, Annandale NSW: The Federation Press, 228–63.

Perkins, R. and Neumayer, E. (2012) Does the "California effect" operate across borders? Trading and investing-up in automobile emission standards. *Journal of European Public Policy*, **19**, 217–37.

Provencher, J., Bond, A. L., and Mallory, M. L. (2014). Marine birds and plastic debris in Canada: A national synthesis, and a way forward. *Environmental Reviews*, **23**(1), 1–13.

Raubenheimer, K. and McIlgorm, A. (2017). Is the Montreal Protocol a model that can help solve the global marine plastic debris problem? *Marine Policy*, **81**, 322–9.

Redclift, M. (1996). *Wasted: Counting the Costs of Global Consumption*, London: Earthscan.

Rochman, C. M., Kross, S. M., Armstrong, J. B., et al. (2015). Scientific evidence supports a ban on microbeads. *Environmental Science & Technology*, **49**(18), 10759–61. doi: 10.1021/acs.est.5b03909.

Royer, S.-J., Ferrón, S., Wilson, S. T., and Karl, D. M. (2018). Production of methane and ethylene from plastic in the environment. *PloS One*, **13**(8), e0200574.

Schuyler, Q. A., Wilcox, C., Townsend, K., Hardesty, B. D., and Marshall, N. J. (2014). Mistaken identity? Visual similarities of marine debris to natural prey items of sea turtles. *BMC Ecology*, **14**(1), 14.

Science Communication Unit. (2013). *Science for Environment Policy Indepth report: Environmental Citizen Science*. Report produced for the European Commission DG Environment. Bristol: University of the West of England. Available at: http://ec.europa.eu/science-environment-policy

Sherman, P. and van Sebille, E. (2016). Modeling marine surface microplastic transport to assess optimal removal locations. *Environmental Research Letters*, **11**(1), 014006. doi:10.1088/1748-9326/11/1/014006

STAP. (2011). Marine debris as a global environmental problem: Introducing a solutions based framework focused on plastic. *A STAP Information Document*. Washington, DC: Global Environment Facility. Available at www.thegef.org/gef/sites/thegef.org/files/publication/STAP%20MarineDebris%20-%20website.pdf.

Stoett, P. (2010). Framing bioinvasion: Biodiversity, climate change, security, trade, and global governance. *Global Governance: A Review of Multilateralism and International Organizations*, **16**(1), 103–120.

Taylor, M. L., Gwinnett, C., Robinson, L. F., and Woodall, L. C. (2016). Plastic microfibre ingestion by deep-sea organisms. *Scientific Reports*, 6. Available at www.nature.com/articles/srep33997.pdf

The Honolulu Commitment. (2011). Available at www.unep.org/pdf/PressReleases/Honolulu_Commitment-FINAL.pdf

Thompson, R. C., Swan, S. H., Moore, C. J., and Vom Saal, F. S. (2009). Our plastic age. *Philosophical Transactions of the Royal Society B: Biological Sciences*, **364**, 373–6.

Tibbetts, J. H. (2015). Managing marine plastic pollution: Policy initiatives to address wayward waste. *Environmental Health Perspectives*, **123**(4), A90–93. doi:10.1289/ehp.123-A90

Titmus, A. J. and Hyrenbach, K. D. (2011). Habitat associations of floating debris and marine birds in the North East Pacific Ocean at coarse and mesospatial scales. *Marine Pollution Bulletin*, **62**(11), 2496–506.

Transboundary Waters Assessment Programme (TWAP). (n.d.). Available at www.geftwap.org/twap-project.

Transboundary Waters Assessment Programme (TWAP). (2014). Assessing transboundary aquatic ecosystems globally through the UNEP-GEF Transboundary Waters Assessment Programme (TWAP): Insights for freshwater ecosystems. [PowerPoint slides]. Available at www.unep.org/delc/Portals/119/ForumBasin Organization/twap-bilay.pdf.

UNEP *Guidelines On Survey And Monitoring Of Marine Litter* (2009), UNEP, Nairobi.

UNEP, *Marine Litter a Global Challenge* (2009), UNEP, Nairobi.

UNEP Regional Seas Programme. (2009). Marine Litter: A Global Challenge. Nairobi, Kenya: United Nations Environment Program.

United Nations. (1992). Rio Declaration on Environment and Development. Available at http://www.un.org/documents/ga/conf151/aconf15126-1annex1.htm/

Valentine, K. (2015). Netherlands company introduces plastic roads that are more durable, climate friendly than asphalt. *Think Progress*. Available at http://thinkprogress.org/climate/2015/07/22/3682552/plastic-roads-netherlands/.

Van Cauwenberghe, L. and Janssen, C. R. (2014). Microplastics in bivalves cultured for human consumption. *Environmental Pollution*, **193**(6), 5–70.

Van Cauwenberghe, L., Vanreusel A., Mees J., and Janssen, C. R. (2013). Microplastic pollution in deep-sea sediments. *Environmental Pollution* **182**, 495–9.

Vince, J. and Hardesty, B. D. (2016). Plastic pollution challenges in marine and coastal environments: From local to global governance. *Restoration Ecology*, **25**(1), 123–8.

Vince, J., Brierley, E., Stevenson, S., and Dunstan, P. (2017). Ocean governance in the South Pacific region: Progress and plans for action. *Marine Policy*, **79**, 40–45.

Vince, J. and Haward, M. (2017). Hybrid governance of aquaculture: Opportunities and challenges. *Journal of Environmental Management*, **201**, 138–44.

Wilcox, C., Mallos, N., Leonard, G., Rodriguez, A., and Hardesty, B. (2015). Estimating the consequences of marine litter on seabirds, turtles and marine mammals using expert elicitation. *Marine Policy*. **65**, 107–14.

Woodall, L. C., Sanchez-Vidal, A., Canals, M., et al. (2014). The deep sea is a major sink for microplastic debris. *Royal Society Open Source*, December. Available at http://rsos.royalsocietypublishing.org/content/1/4/140317.

Worm, B., Barbier, E. B., Beaumont, N., et al. (2006). Impacts of biodiversity loss on ocean ecosystem services. *Science*, **314**(5800), 787–90.

Wright, S. L., Thompson, R. C., and Galloway, T. S. (2013). The physical impacts of microplastics on marine organisms: A review. *Environmental Pollution*, **178**, 483–92.

Wyles, K., Pahl, S., and Thompson, R. (2014). Perceived risks and benefits of recreational visits to the marine environment: Integrating impacts on the environment and impacts on the visitor. *Ocean and Coastal Management*, **88**, 53–63.

Zettler, E., Mincer, T., and Amaral-Zettler, L. (2013). Life in the plastisphere: Microbial communities on plastic marine debris. *Environmental Science and Technology*, **47**(13), 7137–46. doi:10.1021/es401288x

22

Financing Emission Reductions
Forest Mechanisms as a Model for Coral Reefs

KAMLESHAN PILLAY, YANASIVAN KISTEN, ALBERTUS SMIT,
AND DAVID GLASSOM

Introduction

Even though the 21st Conference of the Parties (COP) to the United Nations Framework Convention on Climate Change (UNFCCC), held in Paris in 2015, yielded a positive outcome in the form of a legally binding climate agreement – the Paris Agreement – there is uncertainty regarding the implementation of Nationally Determined Contributions (NDCs) of individual countries. Despite this uncertainty, there is a likelihood that mitigation of global greenhouse gas emissions will be achieved to some degree through the management and augmentation of natural carbon stocks (Hurteau et al., 2008; Edwards et al., 2010). In this respect, Reduced Emissions from Deforestation and forest Degradation (REDD) and, more recently REDD+ mechanisms, are fund-based schemes introduced by the UNFCCC to protect the declining forest biome (Lederer, 2011). Their aim is to reduce carbon emissions by protecting forests. REDD+ has mainly operated between Nordic and European Union countries, on one hand, and developing countries with a high coverage of forest biome, on the other.

The inclusion of forest protection and REDD+ in a new climate agreement is seen by many climate scientists as being essential due to the high carbon sequestration and storage capacity of forests (mostly tropical forests) (Kelly, 2010). Aside from functioning as carbon reserves, forests are recognized as one of the biomes that possess the highest biodiversity per unit area (Turner et al., 2007; Venter and Koh, 2012). The protection of these "biodiversity hotspots," as they are commonly referred to by the scientific community, are not only essential in the climate change context but also in a social context (Groom and Palmer, 2012) because 800 million people live within or in close proximity to forests (Chomitz et al., 2006).

In addition to the preservation of forests, there are other biomes that contribute to the storage and sequestration of carbon, supplement livelihoods, and serve as biological hotspots (Siikamäki et al., 2013). Coral reefs are a case in point. These environments possess some of the largest amounts of biodiversity per unit area on Earth (Kushlan, 2009). Coral reefs consist of calcifying organisms that are able to sequester carbon for their growth. Coral reefs also provide habitats for a diverse array of organisms, which contributes to the organic carbon stocks within complex food webs.

Initially, coral reef ecosystems were considered to be carbon sinks (Ware et al., 1992). However, Laffoley and Grimsditch (2009) have shown from ocean chemistry that they are in fact sources of carbon, contributing approximately 50 Tg C per year. Despite this fact, it is possible that coral reef protection activities could still qualify as offset credits, because the protection of these environments could ensure consistent levels of carbon, rather than transforming them into a stronger source of carbon.

The inclusion of "blue forest" coastal ecosystems, such as mangrove forests, seagrass, and salt marshes under carbon offset schemes has recently been explored in the academic literature (Lau, 2013; Locatelli et al., 2014; Siikamäki et al., 2013; Ahmed and Glaser, 2016). Mangrove forests are particularly well suited to the generation of carbon offsets due to their status as carbon sinks (Locatelli et al., 2014). This chapter considers whether coral reefs could similarly be classified as "blue carbon," specifically under "payment for ecosystem services" (PES) schemes such as REDD+. Using REDD+ as a model, we identify problems with using coral reefs as carbon offsets. We outline some of the structural and procedural changes that would need to be present in a hypothetical coral reef protection mechanism (CRPM).

Coral Reefs and Climate Change

The locations of coral reefs are mostly dependent on the temperature range of seawater, with most coral reefs existing at the low to mid-latitudes (although they sometimes extend to higher latitudes). Most coral reefs occur in the South Pacific, with 43.98 percent of reefs in the jurisdiction of Indonesia, Australia, and the Philippines (UNEP, 2001) (see Chapters 6 and 7). Despite occurring in relatively nutrient-poor areas, coral reefs are highly productive due to symbiotic relationships between animals and primary producers (Coffroth et al., 2006). Coral reefs are of vital ecological importance, aiding in the resilience of coastlines (Kushlan, 2009) while providing raw materials for consumption and for medicines (Moberg and Folke, 1999). It is estimated that approximately 500 million people depend on resources from coral reefs (NOAA, 2014).

Climate change poses extreme risks to the survival of coral reefs because they are highly sensitive to warm temperatures (Baker et al., 2008). Warmer temperatures reduce the capacity of seawater to hold inorganic carbon, which results in less carbon being bio-available to coral species (Guinotte and Fabry, 2008). This also contributes to ocean acidification, which in turn leads to reef dissolution. Warmer temperatures of ocean waters also adversely affect symbiotic zooxanthellae with devastating consequences for the corals themselves (Lesser, 1997). Lastly, other concerns include improper tourism practices, dangerous (dynamite) and highly exploitative fishing techniques, overfishing, and sand mining (Coral Reef Alliance, 2013).

Coral reefs are highly productive and biodiverse marine ecosystems usually found in warm, clear, and nutrient-poor shallow-water coastal areas (Baker et al., 2008; Kushlan, 2009). The successes of coral reefs are largely due to the symbiotic relationship between reef-building cnidarian coral polyps and unicellular algae known as zooxanthellae (Muscatine and Portner, 1977; Coffroth et al., 2006). This symbiotic relationship allows for high primary productivity in a low-nutrient environment because nutrients are tightly coupled within the system: Corals provide inorganic nutrients via excretion, which is absorbed and fixed by zooxanthellae (Muscatine and Portner, 1977). Zooxanthella then provide materials to the hosts, which greatly aid in metabolic functions and allows corals to build massive reef structures that serve as habitats, food sources, and nursery areas for many species (Muscatine and Portner, 1977; Moberg and Folke, 1999).

Coral reefs have been seen as a carbon source due to the calcification process. However, current models may not be accurate, and future human activities, such as eutrophication, may switch coral reefs from a source of carbon to a sink for it (Andersson and Mackenzie, 2004; Borges et al., 2005). Coral reefs allow for high productivity and fixing of biological carbon in animals that move between ecosystems and may be important for fisheries (Moberg and Folke, 1999; Baker et al., 2008).

Some reef-building corals are highly sensitive to their environment and may become bleached following stress events, as seen with the 1998 El Niño warming event (Baker et al., 2008). Coral bleaching is the "whitening of corals due to a loss of symbiotic algae and/or their pigments" (Glynn, 1984; Brown, 1997; Douglas, 2003). Coral bleaching is a general response to a range of stressors, including "extreme temperatures, high irradiance, high heavy metal and pathogenic microorganism concentrations amongst others" (Douglas, 2003). However, large-scale bleaching is caused primarily by increases in water temperature beyond the thermal tolerance of coral colonies (Baker et al., 2008). Zooxanthellae are expelled because of the damaging oxidants they produce under stress, depriving the corals of the energetic benefits provided (Lesser, 1997; Douglas, 2003). This results in mass coral mortality and subsequent degradation of the reef and its ecosystem (Douglas, 2003; Baker et al., 2008). The degradation of the ecosystem results in massive losses in biodiversity and productivity, and thus its carbon storage potential.

With mean temperatures and extreme temperature events expected to increase in the future, coral bleaching episodes are also predicted to become more frequent (Glynn, 1993; Hoegh-Guldberg, 1999; Hughes et al., 2003; Donner et al., 2005; Hoegh-Guldberg et al., 2007). The extent of damage due to warming events is a function of depth, reef topography, water flow and circulation, area coverage and community structure (Obura, 2005; Ateweberhan et al., 2013), and the duration of the stress. Several factors may result in a coral reef system being either resilient or susceptible to change: depth allows for more refugia to exist during warming events; thus, there is an increased chance of survival for corals (Bongaerts et al., 2010). Furthermore, certain growth form structures and different coral species may also be more resistant to bleaching (Nyström, 2006). The genotype of the coral symbionts also influences susceptibility; for example, slow-growing massive species have been shown to be resistant to warming events, but not on a repeated basis, whereas faster-growing branching corals have shown a high scope for adaptation (Guest et al., 2012). Early life stages of corals are also at a higher risk, which lowers the scope for recovery following warming events (Kurihara, 2008). Lastly, the ability of coral reefs to recover once they have been bleached varies: coral reefs with resistant community structures and absent local drivers of bleaching can experience faster recovery rates (Sheppard et al., 2008).

Ocean acidification is the reduction in seawater pH as a result of higher bicarbonate ion concentrations: the increased dissolved carbon dioxide (CO_2) concentration shifts the inorganic carbon system balance in seawater toward bicarbonate, a weak acid, which results in a reduced carbonate and aragonite saturation in seawater (Ateweberhan et al., 2013). Ocean acidification causes these compounds to be less bioavailable, and therefore calcification is compromised, leading to slower reef accretion rates and ultimately the dissolution of reefs (Doney et al., 2009; Andersson and Mackenzie, 2011; Albright et al., 2016). Ocean acidification also results in effects on living colonies, such as weaker skeletons or slow linear growth rates. Other environmental variables have interaction effects with ocean acidification (Kroeker et al., 2010). Calcification is a biological process, responsible for creating large calcium carbonate skeletons over time, which results in reef building, and therefore the process does have a thermal optimum (Silverman et al., 2007). Even though some research has stated calcification increases with rising temperatures, this only occurs in a narrow temperature range and up to a thermal maximum (24 to 27°C). A further increase in sea surface temperature (SST) causes coral stress and death (Clausen and Roth, 1975; Coles and Jokiel, 1978; Marshall and Clode, 2004). Lower pH also disrupts other biological activities, such as photosynthesis of the algal symbiont (Gao and Zheng, 2010), respiration (Feely et al., 2010), and nitrogen fixation (Vitousek et al., 2002). Ocean acidification may harm the reproductive cycles of coral and thereby the overall survival of the species. Corals in their early life stages (juveniles and larvae) are particularly vulnerable to bleaching (Kurihara, 2008).

Infectious coral diseases have increased in severity and frequency in recent times (Hoegh-Guldberg et al., 2007), and a vast amount of coral reef degradation has been attributed to this increase (Ateweberhan et al., 2013). Although it is unclear whether causality can be established between raised SST and the presence of infectious diseases, there is a consensus that poor environmental conditions, such as increased sedimentation, turbidity, excess nutrients (N, P, and Fe), and enhanced algal growth (Jyothi and Awasthi, 2013), can result in corals being more susceptible to death. Some studies have established that coral diseases, such as black band (Kuta and Richardson, 2002), white pox (Paterson et al., 2002), dark spots, and yellow band disease (Gil-Agudelo et al., 2004), have increased in frequency with an increased SST. An investigation by Rosenberg and Ben Haim (2002) suggested that physiological processes in corals are compromised at high SST. This susceptibility is accentuated as pathogens may become active with increased SST. Increased susceptibility of corals increases the rate of coral reef disease transfer, thereby affecting the entire community structure (Bruno et al., 2007).

Besides bleaching, declining calcification, and disease, corals are possibly affected by various other anthropogenic-driven changes. For example, sea-level rise may overwhelm reefs, and marine pollution may affect various physiological processes (Baker et al., 2008). Coupled with overfishing and eutrophication, which often perturb coral reefs already under thermal stress, the shift in community structure is toward algal dominance, and this further reduces biodiversity and resilience (Baker et al., 2008; Porzio et al., 2011). Nonclimatic stressors acting on coral reefs have made these systems extremely vulnerable and have pushed them near to their tipping points; the continuation of CO_2 emissions that drive climatic trends toward more frequent extreme temperature events and higher mean seawater temperatures will have long-term impacts on coral reef structure (Scheffer and Carpenter, 2003).

Lessons From REDD+

The REDD+ mechanism has suffered from five major flaws associated with (1) definitional issues, (2) reference levels, (3) permanence, (4) leakage, and (5) structure (Madeira, 2008; IIED, 2009; Harvey et al., 2010). Definitional issues refer to insight in a policy context as to what exactly constitutes a forest. Defining what a forest is of paramount importance, as it determines policy structure, allocation of resources, finance allowances, and inevitably whether the scheme is successful or not (Sasaki and Putz, 2009). In REDD+, there is no differentiation between natural forests and plantations, and this is of concern, as plantations have severe secondary impacts on the environment. Competition between monoculture and indigenous species, compromised groundwater supplies, loss of soil quality, and land degradation are just some of the consequences when replacing natural forests with artificial forests (plantations) (Bremer and Farley, 2010).

The major question regarding definitional issues within a CRPM is whether artificial reefs should count as natural reefs. Artificial reefs made from rubble or concrete may result in coral growth, but these "ecosystems" may not be considered under a CRPM, as they may fall short of delivering carbon sequestration *and* biodiversity benefits. Definitional issues are important in this context, as the replacement of other coastal habitats (lagoons, estuaries, mangroves) in favor of artificial coral reefs to gain remuneration could be a potential issue. Artificial coral reefs may also result in toxicity (Gómez-Buckley and Haroun, 1994), damage to ecosystems by concentrating fish in certain areas, and overfishing (Smith et al., 2015). If toxicity can be reduced from artificial coral reefs, then this could allow for an increase in biodiversity overall, which could be seen as a co-benefit.

Reference levels are the baseline to which emission reductions need to be compared; but there is still conjecture as to whether baselines should be determined on a country-by-country level (Busch et al., 2011). Reference levels are also important, as they determine which countries can participate in the scheme: countries with coral reefs with already high levels of protection would not benefit as much as countries with poor environmental regulation of coral reefs (Herold et al., 2012). Furthermore, the distribution of coral reefs is not even. Coral reefs exist mainly in tropical regions and midlatitudes and are confined to warm waters with average maximum temperatures of 30°C (Schoepf et al., 2015); therefore, coral reefs are restricted to fewer areas, with possibly limited participation from landlocked states (UNEP, 2001; Coral Reef Alliance, 2013). This is an important difference between REDD+ and a potential CRPM. Even though forests are also not present in all states, if plantations are included in a formalized mechanism under the UNFCCC, then REDD+ can potentially work in every country. In terms of coral reefs, even if artificial reefs are included in a CRPM, they still cannot be constructed in all coastal waters due to the requirement of warm waters.

Carbon leakage in the context of REDD+ refers to losses of carbon from other activities as a result of REDD+ forest protection. For example, reduced fuel wood availability for local communities living near protected forests may result in higher usage of paraffin and other fuels, which may increase carbon emissions. Leakage in a coral reef context is also important. As mentioned earlier, hundreds of millions of people are reliant on coral reefs (NOAA, 2014). The dependency of livelihoods on coral reefs requires that leakage be prevented by ensuring that the needs of communities do not favor environmental practices that cause emissions. The issue of leakage is very closely entwined to the matter of permanence. Poverty alleviation and rural livelihoods being dependent on protected coral reefs could be extremely problematic if a scheme is not implemented for a long enough period and lacks sufficient financial compensation to sustain leakage and proper monitoring, reporting, and verification (MRV) (Venter and Koh, 2012). Indeed, MRV is a significant barrier to the implementation of a CRPM, with the development of baselines being especially difficult.

Fund Based or Market Based?

A flaw of REDD+ that must be sufficiently addressed if a CRPM is to be adopted is that of structure. The issues of leakage and permanence can only be addressed in a coral reef context if adequate financing is available. The structure of a CRPM being a fund-based, as the REDD+ scheme currently is, or a market-based mechanism, such as the Clean Development Mechanism (CDM), will determine how much financing is available. Regardless of whether a mechanism is fund based or market based, it needs to be able to generate a sufficient amount of finance to be able to prevent carbon leakage over the entire project life cycle while encouraging co-benefits. There are two streams of thought in this respect: if a CRPM is implemented under the UNFCCC, then it is possible that financing could be made available; otherwise, finance would need to be generated from the mechanism itself.

If a CRPM is adopted as voluntary mechanism, as with REDD+, then it is questionable whether it could be a fund-based mechanism as opposed to a market-based mechanism. Even though the REDD+ mechanism has been able to foster cooperation between developed and developing nations, it has failed to collaborate with as many least developed countries (LDCs) as it was initially envisioned to do. Furthermore, with the recession of 2009 and subsequent financial slowdown, it is questionable whether fund-based mechanisms can offer long-term financing. This is quite a serious concern in the REDD+ framework, as a breakdown of a REDD+ project, through a lack of financing, could lead to improper practices (e.g., overharvesting and slash-and-burn agriculture) that results in the instantaneous loss of carbon, thus counteracting the entire REDD+ scheme (Palmer, 2011). In a coral reef context, the concerns are similar. Although the destruction of coral reefs for ornamental harvesting, for example, does not result in immediate carbon emissions, it does result in an overall decline in resilience that, over time, can reduce the capacity of a coral reef ecosystem to sequester carbon.

A market-based mechanism can offer a solution to long-term finance availability. According to Linacre et al. (2011), carbon markets account for approximately US $142 billion. REDD+ as a market-based mechanism would be able to generate approximately US$7 billion per year, which could be reinvested back into the scheme to promote co-benefits. It was also estimated that a purely fund-based mechanism needed between US$2 and US$28 million a year (Lederer, 2011). Estimations could differ appreciably between the protection of coral reefs and forests. Furthermore, a market-based solution could be entirely ineffective if the issue of a low carbon credit price is not addressed. Low prices of carbon credits are a result of the absence of regulatory policies to drive their demand. A mixed system, with a market- and fund-based mechanism, could allow for more long-term financing to be available and thereby increase the resilience of the mechanism with regard to financial flows while still relying on committed funds from cooperating countries. A concern of a potentially created CRPM is participation. According to Venter and Koh (2012), only 0.54 percent of projects within the market-based CDM system were afforestation and reforestation projects (the CDM does not recognize projects that avoid destruction and deforestation, so there is a potential that this percentage could be higher).

Co-benefits of Coral Protection

Some scholars have questioned the primary objective of REDD+. Is it a purely carbon-offsetting mechanism, or can it contribute to emission reductions while also promoting biodiversity protection, poverty alleviation, and preservation of cultural practices associated with the protection of forests (Visseren-Hamakers et al., 2012)? Just as this is an important matter in the context of forest protection, so, too, is it a concern regarding coral reef protection. With livelihoods being dependent on coral reefs, particularly in Africa, Asia, and Latin America, it is imperative that a CRPM address and incorporate the needs of local populations. This could be done by facilitating employment of local people who would be marginalized by the implementation of a CRPM. Job creation can also be beneficial, as it can prevent carbon leakage. The approach of integrating the needs of local communities will differ from country to country.

Besides job creation for local communities, it is also vital that people who are indirectly dependent on the coral reefs understand how these ecosystems play a role in maintaining their livelihoods. This will promote greater buy-in and involvement in the decision-making processes by local communities, thus enhancing the effectiveness of reef protection efforts. If a CRPM scheme is able to foster more integration with local communities, then it will likely be more successful in the long term. Transparency and accountability need to be incorporated into a CRPM. Payment for ecosystem services schemes like REDD+ implemented at the national level have failed in the past due to the revolt by local communities as benefits failed to filter down to people who were in fact sacrificing their livelihoods.

According to Lawlor et al. (2010), there are four essential requirements for integrating local communities into a REDD+. These are local assistance and community involvement, adaptive management in response to rapid changes, transparency and accountability of management (particularly with regard to how funds are disseminated), and discussion forums where locals and management can be heard. Similar requirements are envisioned for a CRPM.

As REDD+ mostly involves partnerships between developing countries that have indigenous populations who live within forests, there is a concern that ill-defined land rights could cause implementation problems (Corbera et al., 2011). Conversely, coral reefs fall within the exclusive economic zones of countries, thereby allowing states to possess permanent sovereignty over their resources (see Chapters 2 and 18). However, community involvement is still important as it can determine the success or failure of a CRPM. Ultimately, local communities through malpractice and unsustainable practices have the ability to disrupt a well-functioning CRPM, thereby reducing its effectiveness. Moreover, if a CRPM can provide local employment and deliver emission reductions, then it could deliver on all pillars of sustainability, something the REDD+ mechanism has struggled to do.

The ability of REDD+ to deliver the co-benefit of biodiversity protection is perhaps the most contentious issue inhibiting its implementation. As noted previously, in many REDD+ schemes, there is no distinction between plantations and natural forests. This is problematic, as REDD+ schemes may include plantations that discount indigenous species, outcompeting them and causing other secondary environmental impacts. Therefore, REDD+ may deliver in terms of emission reductions as fast-growing plantations may be able to sequester appreciably more carbon than natural forests. However, it could simultaneously fail to promote the co-benefit of biodiversity protection.

Coral reef destruction and the removal of coral reduces overall biodiversity and the stability and resilience of the ecosystem. It is therefore necessary that a CRPM incorporate biodiversity protection as a co-benefit. Marine protected areas (see Chapter 11) have been shown to increase the line fishery yields in surrounding waters – the so-called "spillover" effect (Alós et al., 2015). It has been suggested by Venter and Koh (2012) that linking biodiversity credits to a REDD+ scheme (either fund or market based) can provide finance for biodiversity protection, thereby integrating and promoting co-benefits into the mechanism. There is a major difference between coral reefs and forests (following the current REDD+ definition): a functioning coral reef inherently relies on a high level of biodiversity; a lack of this compromises the entire ecosystem functioning and leads to reduced resilience to change. Conversely, forests in a REDD+ context could be plantations and thereby do not rely on biodiversity as a prerequisite to deliver emission reductions. Analogously, artificial reefs are structures that allow for reef building and can be described as the substratum onto which corals may attach themselves (Carr and Hixon, 1997). Even though they may induce negative impacts such as toxicity and overfishing, they inherently promote increased biodiversity and productivity. Therefore, biodiversity protection is a concern within a CRPM, but if a CRPM is implemented correctly, it will inherently promote increased biodiversity.

Conclusion

The protection, management, and augmentation of natural carbon stocks are likely to play a crucial role in determining how countries meet their mitigation objectives. The REDD+ mechanism is an established scheme that allows for the protection of forest environments in developing countries. Forests are well noted for their carbon sequestration and storage capacity, high biodiversity, and role in supporting livelihoods. Carbon credits yielded from REDD+ schemes can also be used by states to meet their emission goals. In this chapter we have explored how a Coral Reef Protection Mechanism might be designed using the REDD+ scheme as a framework. Issues with REDD+ as a payment-for-ecosystem services mechanism have been well documented by academics. These include the financing structure, delivering of co-benefits, definitional discrepancies, and the matter of permanence. These issues could be applied to a theoretical PES scheme to provide policy makers with knowledge and insight into the functioning of an actual CRPM.

Some of the key findings of our research include the following: First, despite coral reefs being sources of carbon, these environments could yield carbon credits if their protection results in the ecosystem becoming a lesser source of carbon or is transformed into a sink for carbon. Second, as is the case with REDD+, what is defined as a "coral reef" is critical to establishing which protection activities are included under a CRPM. Artificial reefs may qualify under a CRPM if they are able to deliver carbon sequestration and biodiversity benefits. However, issues of toxicity, concentration of fish, and overfishing could deter the inclusion of artificial reefs under a CRPM. Third, reference levels may be difficult to establish, as the distribution of coral reefs is not uniform. Coral reefs are restricted to environments with specific conditions. Furthermore, a CRPM may have limited country participation, as coral reefs are only located in the jurisdiction of certain coastal countries. Fourth, it is critical that a CRPM be able to deliver carbon sequestration and co-benefits such that carbon leakage is prevented. A CRPM must address issues of poverty alleviation and rural livelihoods.

Perhaps the most critical element of a CRPM is that of the financial structure of the scheme and whether it is fund or market based. A fund-based CRPM may be difficult to maintain through donor financing, whereas a market-based solution may be better for the long-term financial sustainability of the scheme. Irrespective of which financial structure is selected, a CRPM must be able to generate a sufficient amount of financing to be able to prevent carbon leakage over the entire project life cycle while encouraging co-benefits.

Given the difficulty of assessing the transition of a coral reef from a carbon source to a carbon sink, coral reef protection or rehabilitation activities are more aligned to marine and biodiversity offset schemes (Bos et al., 2014). Issues related to co-benefits within REDD+ are still useful in the development of a PES for coral reefs. The question of a fund-based versus a marked-based instrument is still applicable, with the exception that a market-based instrument would need to trade biodiversity credits. Regardless of the type of CRPM that might be chosen, this case shows how oceans and climate governance can be linked. This can be done effectively if lessons from existing climate governance mechanisms are applied when crafting new mechanisms for ocean governance.

References

Ahmed, N. and Glaser, M. (2016). Coastal aquaculture, mangrove deforestation and blue carbon emissions: Is REDD+ a solution? *Marine Policy*, **66**, 58–66.

Albright, R., Caldeira, L., Hosfelt, J., et al. (2016). Reversal of ocean acidification enhances net coral reef calcification. *Nature*, **531**, 362–5.

Alós, J., Puiggrós, A., Diaz-Gil, C., et al. (2015). Empirical evidence for species-specific export of Fish Naïveté from a no-take marine protected area in a coastal recreational hook and line fishery. *PLOS One*, **10**, 1–16.

Andersson A. J. and Mackenzie F. T. (2004). Shallow-water oceans: A source or sink of atmospheric CO_2? *Frontiers in Ecology and the Environment*, **2**, 348–53.

Andersson, A. J. and Mackenzie, F.T. (2011). Ocean acidification: Setting the record straight. *Biogeosciences Discussions*, **8**, 6161–90.

Ateweberhan, M., Feary, D. A., Keshavmurthy, S., et al. (2013). Climate change impacts on coral reefs: Synergies with local effects, possibilities for acclimation, and management implications. *Marine Pollution Bulletin*, **74**, 526–39.

Baker, A. C., Glynn, P. W., and Riegl, B. (2008). Climate change and coral reef bleaching: An ecological assessment of long-term impacts, recovery trends and future outlook. *Estuarine, Coastal and Shelf Science*, **80**, 435–71.

Bongaerts, P., Ridgway, T., Sampayo, E., and Hoegh-Guldberg, O. (2010). Assessing the "deep reef refugia" hypothesis: Focus on Caribbean reefs. *Coral Reefs*, **29**, 309–27.

Borges, A., Delille, B., and Frankignoulle, M. (2005). Budgeting sinks and sources of CO_2 in the coastal ocean: Diversity of ecosystems counts. *Geophysical Research Letters*, **32**, 1–4.

Bos, M., Pressey, R. L., and Stoeckl, N. (2014). Effective marine offsets for the Great Barrier Reef World Heritage Area. *Environmental Science and Policy*, **42**, 1–15.

Bremer, L. L. and Farley, K. A. (2010). Does plantation forestry restore biodiversity or create green deserts? A synthesis of the effects of land-use transitions on plant species richness. *Biodiversity and Conservation*, **19**, 3893–915.

Brown, B.E. (1997). Coral bleaching: Causes and consequences. *Coral Reefs*, **16**, 129–38.

Bruno, J. F., Selig, E. R., Casey, K. S., et al. (2007). Thermal stress and coral cover as drivers of coral disease outbreaks. *PLoS Biology*, **5**, e124.

Busch, J., Godoy, F., Turner, W. R., and Harvey, C. A. (2011). Biodiversity co-benefits of reducing emissions from deforestation under alternative reference levels and levels of finance. *Conservation Letters*, **4**, 101–15.

Carr, M. H. and Hixon, M.A. (1997). Artificial reefs: The importance of comparisons with natural reefs. *Fisheries*, **22**, 28–33.

Chomitz, K. M., Buys, P., De Luca, G., Thomas, T. S., and Wertz-Kanounnikoff, S. (2006). *At Loggerheads? Agricultural Expansion, Poverty Reduction and Environment in the Tropical Forests*. World Bank Policy Research Report. Washington, DC: World Bank.

Clausen, C. D. and Roth, A. A. (1975). Effect of temperature and temperature adaptation on calcification rate in the hermatypic coral Pocillopora damicornis. *Marine Biology*, **33**, 93–100.

Coffroth M. A., Lewis C. F., Santos, S. R., and Weaver, J. L. (2006). Environmental populations of symbiotic dinoflagellates in the genus Symbiodinium can initiate symbioses with reef cnidarians. *Current Biology*, **16**, R985–87.

Coles, S. L. and Jokiel, P. L. (1978). Synergistic effects of temperature, salinity and light on hermatypic coral Montipora verrucosai. *Marine Biology*, **49**, 187–95.

Coral Reef Alliance. (2013). *Coral Reefs and Global Climate Change*. Available at www.coral.org/resources/issue_briefs_b.

Corbera, E., Estrada, M., May, P., Navarro, G., and Pacheco, P. (2011). Rights to land, forests and carbon in REDD+: Insights from Mexico, Brazil and Costa Rica. *Forests*, **2**, 301–42.

Doney, S. C., Fabry, V. J., Feely, R. A., and Kleypas, J. A. (2009). Ocean acidification: The other CO_2 problem. *Annual Review of Marine Science*, **1**, 169–92.

Donner, S. D., Skirving, W. J., Little, C. M., Oppenheimer, M., and Hoegh-Guldberg, O. (2005). Global assessment of coral bleaching and required rates of adaptation under climate change. *Global Change Biology*, **11**, 2251–65.

Douglas, A. (2003). Coral bleaching – how and why? *Marine Pollution Bulletin*, **46**, 385–92.

Edwards, D. P., Fisher, B., and Boyd, E. (2010). Protecting degraded rainforests: Enhancement of forest carbon stocks under REDD+. *Conservation Letters*, **3**, 313–6.

Feely, R. A., Alin, S. R., Newton, J., et al. (2010). The combined effects of ocean acidification, mixing, and respiration on pH and carbonate saturation in an urbanised estuary. *Estuarine, Coastal and Shelf Science*, **88**, 442–9.

Gao, K. and Zheng, Y. (2010). Combined effects of ocean acidification and solar UV radiation on photosynthesis, growth, pigmentation and calcification of the coralline alga Corallina sessilis (Rhodophyta). *Global Change Biology*, **16**, 2388–98.

Gil-Agudelo, D. L., Smith, G. W., Garzón-Ferreira, J., Weil, E., and Peterson, D. (2004). Dark spots disease and yellow band disease, two poorly known coral diseases with high incidence in Caribbean reefs. In E. Rosenberg and Y. Loya, eds., *Coral Health and Disease*. Berlin: Springer, 337–50.

Glynn, P. W. (1984). Widespread coral mortality and the 1982/83 El Niño warming event. *Environmental Conservation*, **11**, 133–46.

Glynn, P. W. (1993). Coral reef bleaching: Ecological perspectives. *Coral Reefs*, **12**, 1–17.

Gómez-Buckley, M. C. and Haroun, R. J. (1994). Artificial reefs in the Spanish coastal zone. *Bulletin of Marine Science*, **55**, 1021–8.

Groom, B. and Palmer, C. (2012). REDD+ and rural livelihoods. *Biological Conservation*, **154**, 42–52.

Guest, J. R., Baird, A. H., Maynard, J. A., et al. (2012). Contrasting patterns of coral bleaching susceptibility in 2010 suggest an adaptive response to thermal stress. *PLoS ONE*, **7**, e33353.

Guinotte, J. M. and Fabry, V. J. (2008). Ocean acidification and its potential effects on marine ecosystems. *Annals of the New York Academy of Sciences*, **1134**, 320–42.

Harvey, C. A., Dickson, B., and Kormos, C. (2010). Opportunities for achieving biodiversity conservation through REDD, *Conservation Letters*, **3**, 53–61.

Herold, M., Angelsen, A., Verchot, L.V., Wijaya, A., and Ainembabazi, J. H. (2012). A stepwise framework for developing REDD+ reference levels. In A. Angelsen, M. Brockhaus, W. D. Sunderlin, and V. Verchot, eds., *Analysing REDD+: Challenges and Choices*. Bogor: Center for International Forestry Research, 279–301.

Hoegh-Guldberg, O. (1999). Coral bleaching, climate change and the future of the world's coral reefs. *Marine and Freshwater Research*, **50**, 839–66.

Hoegh-Guldberg, O., Mumby, P. J., Hooten, A. J., et al. (2007). Coral reefs under rapid climate change and ocean acidification. *Science*, **318**, 1737–42.

Hughes, T. P., Baird, A. H., Bellwood, D. R., et al. (2003). Climate change, human impacts and the resilience of coral reefs. *Science*, **301**, 929–33.

Hurteau, M. D., Koch, G. W., and Hungate, B. A. (2008). Carbon protection and fire risk reduction: Toward a full accounting of forest carbon offsets. *Frontiers in Ecology and the Environment*, **6**, 493–8.

IIED. (2009). *REDD: Protecting Climate Forests and Livelihoods*. Available at www.iied.org/redd-protecting-climate-forests-livelihoods.

Jyoti, J. and Awasthi, M. (2013). Factors influencing algal growth. In S. Kumar and S. K. Tyagi, eds., *Recent Advances in Bioenergy Research VII*. Kapurthala: Sardar Swaran Singh National Institute of Renewable Energy, India, 315–23.

Kelly, D. J. (2010). The case for social safeguards in post-2012 agreement on REDD. *Law, Environment and Development Journal*, **6**, 61.

Kroeker, K. J., Kordas, R. L., Crim, R. N., and Singh, G.G., (2010). Meta-analysis reveals negative yet variable effects of ocean acidification on marine organisms. *Ecology Letters*, **13**, 1419–34.

Kurihara, H. (2008). Effects of CO_2-driven ocean acidification on the early developmental stages of invertebrates. *Marine Ecology Progress Series*, **373**, 275–284.

Kushlan, K. (2009). Coral reefs: The failure to regulate at the international level. *Environs: Environmental Law and Policy*, **33**, 317–42.

Kuta, K. and Richardson, L. (2002). Ecological aspects of black band disease of corals: relationships between disease incidence and environmental factors. *Coral Reefs*, **21**, 393–8.

Laffoley, D. and Grimsditch, G., (2009). The management of natural coastal carbon sinks: Coral reefs. Gland, Switzerland, IUCN Report, 8.

Lau, W. W. Y., (2013). Beyond carbon: Conceptualizing payments for ecosystem services in blue forests on carbon and other marine and coastal ecosystem services. *Ocean and Coastal Management*, **83**, 5–14.

Lawlor, K., Weinthal, E., and Olander, L. (2010). Institutions and policies to protect rural livelihoods in REDD+ regimes. *Global Environmental Politics*, **10**, 1–11.

Lederer, M. (2011). From CDM to REDD+—What do we know for setting up effective and legitimate carbon governance? *Ecological Economics*, **70**, 1900–7.

Lesser, M. P. (1997). Oxidative stress causes coral bleaching during exposure to elevated temperatures. *Coral Reefs*, **16**, 187–192.

Linacre, N., Kossoy, A., and Guigon, P. (2011). *State and Trends of the Carbon Market 2011*. Washington, DC: World Bank.

Locatelli, T., Binet, T., Kairo, J. G., et al. (2014). Turning the tide: How blue carbon and payments for ecosystem services (PES) might help save mangrove forests. *AMBIO*, **43**, 981–95.

Madeira, E. C. M., (2008). *Policies to Reduce Emissions from Deforestation and Degradation (REDD) in Developing Countries: An Examination of the Issues Facing the Incorporation of REDD into Market-Based Climate Policies*. Washington, DC: Resource for the Future.

Marshall, A. T. and Clode, P. (2004). Calcification rate and the effect of temperature in a zooxanthellate and azooxanthellate scleractinian reef coral. *Coral Reefs*, **23**, 218–24.

Moberg, F. and Folke, C., (1999). Ecological goods and services of coral reef ecosystems. *Ecological Economics*, **29**, 215–33.

Muscatine, L. and Porter, J. W. (1977). Reef corals: Mutualistic symbioses adapted to nutrient-poor environments. *Bioscience*, **27**, 454–60.

NOAA. (2014). *Coral Reefs: An Important Part of Our Future*. Available at www .noaa.gov/features/economic_0708/coralreefs.html.

Nyström, M. (2006). Redundancy and response diversity of functional groups: Implications for the resilience of coral reefs. *AMBIO: A Journal of the Human Environment*, **35**, 30–35.

Obura, D. O. (2005). Resilience and climate change: Lessons from coral reefs and bleaching in the Western Indian Ocean. *Estuarine, Coastal and Shelf Science*, **63**, 353–72.

Palmer, C., (2011). Property rights and liability for deforestation under REDD+: Implications for "permanence" in policy design. *Ecological Economics*, **70**, 571–76.

Patterson, K. L., Porter, J. W., Ritchie, K. B., et al. (2002). The etiology of white pox, a lethal disease of the Caribbean elkhorn coral, Acropora palmata. *Proceedings of the National Academy of Sciences of the United States of America*, **99**, 8725–30.

Porzio, L., Buia, M. C., and Hall-Spencer, J.M. (2011). Effects of ocean acidification on macroalgal communities. *Journal of Experimental Marine Biology and Ecology*, **400**, 278–87.

Rosenberg, E. and Ben Haim, Y. (2002). Microbial diseases of corals and global warming. *Environmental Microbiology*, **4**, 318–26.

Sasaki, N. and Putz, F. E. (2009). Critical need for new definitions of "forest" and "forest degradation" in global climate change agreements. *Conservation Letters*, **2**, 226–32.

Scheffer, M. and Carpenter, S. R. (2003). Catastrophic regime shifts in ecosystems: Linking theory to observation. *Trends in Ecology and Evolution*, **18**, 648–56.

Schoepf, V., Stat, M., Falter, J. L., and McCulloch, M. T., (2015). Limits to the thermal tolerance of corals adapted to the highly fluctuating, naturally extreme temperature environment. *Nature Scientific Reports*, **5**, Article Number: 17639.

Sheppard, C. R. C., Harris, A., and Sheppard, A. L. S. (2008). Archipelago-wide coral recovery patterns since 1998 in the Chagos Archipelago, central Indian Ocean. *Marine Ecology Progress Series*, **362**, 109–17.

Siikamäki, J., Sanchirico, J. N., Jardine, S., McLaughlin, D., and Morris, D., (2013). Blue carbon: Coastal ecosystems, their storage, and potential for reducing emissions. *Environment: Science and Policy for Sustainable Development*, **55**, 14–29.

Silverman, J., Lazar, B., and Erez, J. (2007). Effect of aragonite saturation, temperature, and nutrients on the community calcification rate of a coral reef. *Journal of Geophysical Research: Oceans*, **112**(C5).

Smith, J. A., Lowry, M. B., and Suthers, I. M. (2015). Fish attraction to artificial reefs not always harmful: A simulation study. *Ecology and Evolution*, **5**, 4590–602.

Turner, W. R., Brandon, K., Brooks, T. M., et al. (2007). Global conservation of biodiversity and ecosystem services. *BioScience*, **57**, 868–73.

UNEP. (2001). *UNEP-WCMC World Atlas of Coral Reefs: The most Detailed Assessment Ever of the Status and Distribution of the World's Coral Reefs.* Available at http://coral.unep.ch/atlaspr.htm.

Venter, O. and Koh, L. P. (2012). Reducing emissions from deforestation and forest degradation (REDD+): Game changer or just another quick fix? *Annals of the New York Academy of Sciences*, **1249**, 137–50.

Visseren-Hamakers, I. J., McDermott, C., Vijge, M. J., and Cashore, B. (2012). Trade-offs, co-benefits and safeguards: Current debates on the breadth of REDD+. *Current Opinion in Environmental Sustainability*, **4**, 646–53.

Vitousek, P. M., Cassman, K., Cleveland, C., et al. (2002). Towards an ecological understanding of biological nitrogen fixation. *Biogeochemistry*, **57**, 1–45.

Ware, J. R., Smith, S. V., and Reaka-Kudla, M. L. (1992). Coral reefs: Sources or sinks of atmospheric CO_2? *Coral Reefs*, **11**, 127–30.

23

Capturing a Moving Target
Decarbonizing Shipping through Informational Governance

JUDITH VAN LEEUWEN

Introduction

International shipping is important for the global economy, as approximately 90 percent of all traded goods are transported by ships (Lister et al., 2015). As a consequence of increasing economic growth and international trade, the demand for seaborne transport, and the resulting greenhouse gas (GHG) emissions, keep rising. Between 2007 and 2012 international shipping contributed 3.1 percent of annual global carbon dioxide (CO_2) emissions and 2.8 percent of all GHG emissions (International Maritime Organization, 2014). Projected growth in GHG emissions from shipping range between 50 and 250 percent by 2050, depending on future economic growth and energy developments (International Maritime Organization, 2014). GHG emissions from shipping were exempted from the 1992 UN Framework Convention on Climate Change and its 1997 Kyoto Protocol, as well as the 2015 Paris Agreement (Martinez Romera, 2016; Scott et al., 2017). The International Maritime Organization (IMO), an intergovernmental, specialized UN agency with a responsibility for the safety, security, and pollution prevention of ships (see Chapter 2), was asked to develop regulations for shipping instead. The first IMO regulations, focusing on increasing the energy efficiency of ships through technological and operational measures, were adopted in 2011.

Despite its importance for the global economy, relatively little social scientific analysis has been done of the system that governs shipping's environmental impacts (Lister et al., 2015). Some scholarly attention has been given to the functioning and limits of the IMO and its Marine Environmental Protection Committee (MEPC), which plays a vital role in governing safety and environmental aspects from shipping (Tan, 2006; Van Leeuwen, 2010; Karim, 2015). Since 1958, the IMO has adopted over 53 conventions related to maritime safety, environmental risks, and liability and compensation for maritime claims (Lister, 2015). Moreover, strong support exists for IMO and its conventions, and efforts to find regulatory solutions for new issues are commonplace within the IMO (Ringbom, 1999). Maritime governance is predominantly a state-based endeavor in which adopting environmental standards through international law plays a key role (cf Van Leeuwen, 2010, 2015). This formal regulation largely depends on the capacity of state actors to cooperate and enforce regulation of the shipping sector. Scholarly attention has gone to the emerging of private initiatives in greening shipping (Wuisan et al., 2012; Lister, 2015; Scott et al., 2017) and the development of corporate social responsibility in the sector (Yliskylä-Peuralahti and Gritsenko, 2014; Lund-Thomsen et al., 2016; van Leeuwen and van Koppen, 2016).

Only fairly recently, informational tools have been developed through collaboration between governments, businesses, and civil organizations (Lister, 2015), as well as within IMO. These tools focus on disclosing environmental information to businesses and other actors with the aim of integrating environmental considerations in daily business transactions and management. Using the concept of informational governance (Mol, 2006; Toonen and Mol, 2016), this chapter reflects on this new development of actors using such informational tools to increase the environmental sustainability of the maritime sector. This chapter therefore reviews the current use of, as well as the potential of, such informational tools in governing GHG emissions from shipping.

In the next section, the informational governance perspective will be introduced. After that, an introduction is given into the climate change impacts of shipping and the characteristics of the shipping industry. An explanation of how the IMO currently regulates GHG emissions from shipping, including the informational tools developed by the IMO, is given in the next section. The emergence of voluntary informational tools in the shipping industry is then discussed, while the use and extent to which informational governance is living up to its potential in maritime governance is discussed in the sixth section. The chapter ends with some concluding remarks.

Informational Governance

Since the 1990s, a scholarly debate has emerged over the implications of the shift from government to governance in terms of environmental governance (Pierre and Peters, 2000; Van Kersbergen and Van Waarden, 2004; Van Leeuwen, 2010). Decision-making authority and processes have seen a shift from hierarchical decision making within the context of the nation-state to multiactor arrangements for decision making at local, national, regional, and international scales (Van Leeuwen, 2010). As a consequence of individualization and globalization processes, nation-state capacity is increasingly under pressure (Pierre and Peters, 2000). While intergovernmental decision making is one of the ways in which states seek to regain control over environmental issues, nonstate actors have also been active in complementing governmental regulation of environmental issues.

These nonstate efforts are analyzed through the conceptual lenses of nonstate market-driven governance (Cashore, 2002), private governance (Falkner, 2003; Pattberg, 2005), and public–private partnerships (Glasbergen et al., 2007). Empirically, such studies typically focus on certification and labelling systems such as the Forest and Marine Stewardship Councils (Pattberg, 2005; Gulbrandsen, 2009; Bush et al., 2013). A core characteristic of such systems, however, is the gathering and disclosure of environmental information to facilitate integrating environmental sustainability considerations in production and consumption practices of private actors and consumers (Toonen and Mol, 2016). This increasingly important role of information and its disclosure in (nonstate) governance practices is conceptualized under the term "informational governance" (Mol, 2006) or "governance-by-disclosure" (Gupta and Mason, 2014) and is a relatively new topic within sustainability studies (Soma et al., 2016). Other examples of such informational activities are environmental reporting and benchmarking or citizen-led monitoring systems (Mol, 2006). Informational governance takes the network society in the information age (Castells, 2000) as a starting point for discussing the potential transformative role of information, as well as the fact that political battlefields move to information, symbols, and the media (Mol, 2006).

Information and knowledge have always been important for effective decision making. Conventional regulation relies on expert-based knowledge and (natural) science-based monitoring systems to design and enforce regulation. The concept of informational governance, however, refers to the idea that information becomes a transformative resource and thus fundamentally changes processes, institutions, and practices of governance (Mol, 2006; Toonen and Mol, 2016). Informational governance points attention to (new) information systems, such as certification systems and environmental reporting, and how they are used by actors to move toward sustainability. What is more, it argues that it is those actors that can use information's transformative capacity that are at the forefront in steering society. The central actors in informational governance are thus those that have access to, produce, verify, and have control over information (Mol, 2006).

How information systems become transformative and have a governing effect depends on whether information becomes widely available and accessible and whether actors and markets are responsive to this information (Mol, 2015; Toonen and Mol, 2016). Informational governance thus depends on transparency, that is disclosing information, as a key principle. Transparency and information disclosure are often analyzed with a positive connotation, as increased transparency is expected to lead to empowerment of weak actors and to more sustainable outcomes (Gupta and Mason, 2014; Mol, 2015). New information systems and mechanisms that aim for mandatory or voluntary information disclosure stimulate new governance and enforcement practices and dynamics (Toonen and Mol, 2016). Public and private governance initiatives that employ targeted disclosure of information as a way to evaluate and/or steer the behavior of selected actors are core to informational governance (cf Gupta and Mason, 2014; Mol, 2015). Information disclosure alone, however, does not necessarily lead to changes in behavior and practices. Informational governance only works under certain conditions. It has to be well designed; otherwise, it can lead to information overload or to issues of reliability and trust (Toonen and Mol, 2016). In order to live up to its transformative potential, an information system has to generate reliable information which becomes embedded into the decision-making processes of a wide variety of actors (cf Scott et al., 2017). The remainder of this chapter seeks to analyze the extent to which the emergence of informational governance in the shipping industry lives up to its transformative potential.

Shipping and Climate Change

Greenhouse gas emissions from shipping mainly originate from the burning of fossil fuels by engines that are used to propel these ships. The dominant fuel that is used is heavy "bunker" fuel oil. Heavy fuel oil is a residue of petroleum distillation and therefore characterized by much higher sulfur content. Shipping not only contributes to climate change through CO_2 emissions but also through sulfur oxides (SO_x), nitrous oxides (NO_x), particulate matter (PM), and black carbon emissions (see Chapter 15). In 2015, the estimated volume of world seaborne goods, transported by almost 91,000 ships, surpassed 10 billion tons (in weight) for the first time in history (UNCTAD Secretariat, 2016). However, how much heavy fuel oil and other fossil fuels these ships have used to transport their cargo, and the associated GHG emissions, are unknown. So far, the IMO has undertaken three studies to estimate the GHG emissions from international shipping, with the last one assessing the period of 2007 to 2012 (International Maritime Organization, 2014).

In order to reduce GHG emissions from shipping, three main avenues exist: adoption of operational measures, adoption of technical measures to increase energy efficiency (changing a ship's design) or switching to alternative (renewable) forms of energy. Operational measures require little or no investment, but their adoption is limited (Poulsen and Sornn-Friese, 2015). Technological options have the potential to increase the energy efficiency of ships up to 90 percent, but barriers for their adoption still exist (Gilbert et al., 2014). Technological options are usually only cost-efficient for ships that are under construction because existing ships would have to undergo expensive retrofitting. Even though fuel prices are a large part of the operational expenses (Poulsen and Sornn-Friese, 2015) and energy-efficiency measures pay themselves back over a certain period of time, it is in most cases the charterer who benefits from fuel efficiencies and not the ship owner itself (Scott et al., 2017). In addition, a shift to alternative fuels is marked by many uncertainties over the practical implementation and ways to upscale alternative fuel-production infrastructure (Gilbert et al., 2014).

In reducing GHG emissions from international shipping, a global approach is advocated by the shipping industry and maritime states (Hackmann, 2012; Scott et al., 2017). Since environmental regulation is still mainly considered to be an extra cost and burden (Lister et al., 2015; van Leeuwen and van Koppen, 2016), unilateral regulation creates a competitive disadvantage for those affected by it. Affected ship owners might decide to register their flag in another country, avoiding this extra regulatory burden and associated costs. This is possible because of the existence of flag states with open registries that do not require a genuine link between the ship and the state where the ship is registered (e.g., Panama and Liberia). The majority of the world's merchant fleet (in terms of cargo-carrying capacity) is already registered in such open registries (DeSombre, 2006). Unilateral regulation is therefore regarded as counterproductive.

Climate Change Regulation and the IMO's Informational Tools

Because of the desire for uniform regulation, the International Maritime Organization, and its 1973 International Convention for the Prevention of Pollution from Ships, as modified by the Protocol of 1978 (the so-called MARPOL Convention; see Chapter 2), play important roles in regulating GHG emissions from shipping. Annex VI of the MARPOL Convention regulates air emissions from shipping, most notably SO_x, NO_x, PM, volatile organic compounds, and ozone-depleting substances. While the current cap on the sulfur content of bunker fuel is 3.5 percent, this cap will be further reduced to 0.5 percent on January 1, 2020. In so-called emission control areas, more stringent requirements exist, specifically 0.10 percent (from January 1, 2015). Three tiers set limits for NO_x emissions, depending on the age of the ship or whether the ship operates in an emission control area. Black carbon is a relatively new issue and is not currently regulated (see Chapter 15 for a proposal to regulate it).

Alongside these air emissions, energy-efficiency measures were adopted in 2011 as part of Annex VI of the MARPOL Convention, specifically the Energy Efficiency Design Index (EEDI) and the Ship Energy Efficiency Management Plan (SEEMP). In 2016, IMO followed the example of the EU, which adopted a Monitoring, Reporting, and Verification (MRV) Regulation for CO_2 emissions in 2013 by adopting a global data collection system for CO_2 emissions. At the same time, a roadmap was adopted for developing a comprehensive IMO strategy on the reduction of GHG emissions from ships, with measures to be adopted in 2023 (International Maritime Organization, 2016a). While market-based mechanisms have also been subject to debate, progress on their adoption in recent years has come to a standstill (see also van Leeuwen and van Koppen, 2016).

In the remainder of this section, the EEDI and SEEMP and their information-based system of compliance through certification will be discussed first. After that, the global data collection system that will generate data on actual fuel use and emissions, and which can be considered as an informational tool as well, will be discussed.

The EEDI, SEEMP, and IMO's Certification System

The EEDI requires a minimum energy efficiency level per capacity mile for different ship types and size segments (Marine Environment Protection Committee, 2011). As of January 1, 2013, new ships above 400 gross tonnage (the molded volume of all enclosed spaces of the ship) have to meet the reference level for their ship type. These reference levels change every five years to continuously stimulate technical innovation. For the first five years (2015–2019), a 10 percent reduction is required, while a 30 percent reduction is planned for 2025. Compliance is proven through the

International Energy Efficiency Certificate issued by the ship's administration or an authorized classification society. This certificate is valid for the whole life of a ship and should be on-board during port state inspections and audits.

This system of certification, intended to verify and communicate compliance, is used for many standards under MARPOL and other IMO Conventions. Flag states (i.e., the states in which ships are registered; see Chapter 2) do not always have direct access to their ship, nor do they always have the capacity and expertise to do all required inspections to verify compliance with the MARPOL Convention. This is complicated further by the fact that the majority of ships are registered in a handful of flag states with open registries, as noted previously. In addition, the relationship between the country where the ship is registered and where it trades is often weak. That is why the majority of flag states have a contract with a classification society (Vorbach, 2001; Tan, 2006). These classification societies issue the certificates required by the IMO conventions for construction and maintenance of a ship and technical equipment on board.

In addition to inspections by classification societies, ships are inspected by the port state while they are in port. Port states gained the power to inspect and detain incoming ships (irrespective of the flag of the ship) under the MARPOL Convention (Mitchell, 1994), as well as under the UN Convention on the Law of the Sea. Following the Amoco Cadiz oil spill in 1978, France organized a conference on the enforcement of IMO standards by port states. This conference led to the adoption of the Paris Memoranda of Understanding (MoU) on Port State Control in 1984 by 14 European states (currently the Paris MoU has 27 members, as well as Canada, and covers the waters of the European coastal states and the North Atlantic basin). Other regions have followed suit, and eight other MoUs exist. Generally, such an MoU includes the agreement to inspect 25 percent of ships visiting the port against standards set by the IMO and the International Labor Organization. After inspection, a port state is allowed to request the rectification of a deficiency, detain, or ban a ship. The outcome of inspections is usually registered in a shared database and can be used to develop performance lists of individual ships, classification societies, and flag states.

The second energy-efficiency measure is the SEEMP, which can be considered an informational tool as well (Marine Environment Protection Committee, 2012). Having a SEEMP has been mandatory for all ships above 400 gross tonnage since January 1, 2013. A SEEMP is based on a continuous cycle of planning, implementing, monitoring, and self-evaluation to improve a ship's energy management and thus also its energy efficiency. The SEEMP may also be combined with the safety, environmental, or energy management system of a company as long as a specific SEEMP is developed for each ship. Potential efficiency measures are, among others, fuel-efficient operations (e.g., improved voyage planning and weather routing), optimized ship or cargo handling (e.g., optimum ballast conditions), hull maintenance, propulsion system (maintenance), waste heat recovery, and improved use of fleet capacity. An Energy Efficiency Operational Indicator (EEOI), which provides insight into the energy use of a ship during a certain period of time, such as the amount of CO_2 emitted per unit of transport work, may also be used as part of the

SEEMP. The International Energy Efficiency Certificate is proof of compliance with this regulation. In sum, the SEEMP puts forward the requirement that companies develop an information system on the specificities of the ship and its use in relation to energy efficiency potential. Based on this information, companies can identify and select measures to increase the energy efficiency of a ship.

The Global CO₂ Data Collection System

Reacting to the slow progress of the IMO, and the anticipated limited effectiveness of the EEDI and SEEMP to achieve an actual overall reduction of CO_2 emissions from shipping (Gilbert et al., 2014; Bows-Larkin, 2015), the EU adopted a strategy to integrate CO_2 emissions from shipping into its climate change policies in June 2013 (European Commission, 2013). The first step of the strategy was taken with the adoption of the Regulation on Monitoring, Reporting, and Verification of CO_2 emissions from maritime transport in 2015. From January 1, 2018, onwards, ships over 5,000 gross tonnage calling at EU ports are required to collect and publish verified annual data on their CO_2 emissions, including the amount of CO_2 emitted during journeys from and to EU ports and within EU ports, as well as other parameters, such as distances travelled, time at sea, and cargo carried. Based on this data, the average energy efficiency of a ship can be determined. These ships must also carry a document of compliance from an accredited MRV verifier. In addition to providing more insight into the performance of individual ships, this MRV regulation is expected to lead to a 2 percent reduction in annual CO_2 emissions from participating ships (van Leeuwen and van Koppen, 2016).

In October 2016, the IMO followed the EU with the adoption of a similar global data collection system for CO_2 emissions from large ships of 5,000 gross tonnage or more. The ships that fall under this system account for approximately 85 percent of CO_2 emissions from international shipping (International Maritime Organization, 2016a). The IMO's system differs from the EU's MRV regulation, especially with regard to the level of detail of the reported data. At the end of each year, a ship is required to report this data to the flag state. After determining whether the data meets IMO's requirements, the flag state issues a statement of compliance. The flag state in turn transfers this data to the IMO Ship Fuel Oil Consumption Database. Based on this database, the IMO secretariat will produce an annual report, which summarizes the data that has been collected, for the IMO's MEPC.

Voluntary Informational Tools for Reducing GHG Emissions From Shipping

In addition to IMO regulations and informational tools aimed at reducing GHG emissions from shipping, multiple voluntary informational tools have emerged focusing on improving the environmental performance, including the energy efficiency, of ships. Table 23.1 shows an overview and characteristics of voluntary informational tools that will be discussed in this section. These voluntary informational tools aim to

Table 23.1. *Characteristics of informational tools in maritime governance*

	Green Award	Environmental Ship Index	Clean Shipping Index	Clean Cargo Working Group	Shipping efficiency.org
Scope	Quality, safety, and environment	Air emissions	Air and water emissions	Air and water emissions and separately for CO_2	CO_2 emissions
Shipping segment	Seagoing and inland shipping	All	All	Container ships	All
Supplier of information	Ship owner	Ship owner	Ship owner	Ship owner	IHS Maritime database, ship owners and operators, classification societies, yards and engine manufacturers
Verification of data	Third-party verification	Self-declaration, subject to checks and obvious errors	Third-party verification	Self-declaration, methods of CCWG have been audited	By ship owners, rating method has been audited
User-target groups	Ports and marine service providers	Ports and marine service providers	Cargo owners	Cargo owners	Charterers, financial institutions, and ports
Accessibility of information	Public	Public	Members	Members	Public

disclose information to ports and cargo owners, who both engage in business transactions with ship owners (see also Lister, 2015; Lister et al., 2015; Scott et al., 2017).

Voluntary Informational Tools for Ports

A set of informational tools has been developed for seaports. Some ports have developed their own individual award and economic incentive programs whereby ships that reduce air and water emissions are able to get reductions in port fees. Such individual programs exist, for example, in Singapore, Vancouver, and San Francisco (Svensson and Andersson, 2012), as well as in 15 ports in Europe (COGEA, 2017). Similarly, such a system also exists globally through the Green Award. The Green Award has a comprehensive approach as it incorporates quality, safety, and environmental standards. It uses a combination of office audits and surveys to certify participating ships. A certificate is valid for three years. A list of ships that are certified is publicly available on the Green Award website (Green Award, 2017). Having a certificate gives a range of advantages for the ship owner, including tools to deal with air emissions, shorter port visits, reduced port fees, and projection of a better image. In 2014–2015, 225 seagoing ships, 550 inland ships, and 60 incentive providers (i.e., ports and maritime services companies) were certified (Green Award, 2015).

While the Green Award is comprehensive in scope, the Environmental Ship Index developed by the World Ports Climate Initiative only focuses on air emissions of NO_x, SO_x, and CO_2. Specifically, the index evaluates the amount of NO_x and SO_x that is emitted by a ship, while for CO_2 it uses the Energy Efficiency Operational Indicator of the IMO. While developed for ports, it can also be used by cargo owners to reward and select ships with lower air emissions (Environmental Ship Index, n.d.). The data used for the index are based on self-declaration and are randomly checked for inconsistencies and obvious mistakes. As of mid-2017, the total number of ships with a valid ESI score is 6553, and their scores are publicly available on the website of the Environmental Ship Index (Environmental Ship Index, 2017).

Voluntary Informational Tools for Charterers and Cargo Owners

In addition to informational tools for ports, informational tools exist that facilitate cargo owners to select ships to transport their cargo, based on a ship's environmental profile. For example, the Clean Shipping Index was launched in 2010 by the Swedish Clean Shipping Project. Based on information provided by the ship owner, ships are scored on air emissions (NO_x, SO_x, PM, and CO_2) as well as the discharge of chemicals and waste (water). Verification is ensured by a third party (e.g., a classification society). Only members are able to access the database, which shows the index score of different ships. In addition to an index, the Clean Shipping Project functions as a network to exchange best practices and information (Wuisan et al., 2012).

The Clean Cargo Working Group (CCWG) of the Business for Social Responsibility uses a similar approach, and also focuses on creating a dialogue between cargo and ship owners about the environmental footprint of container ships. The CCWG represents over 85 percent of the container shipping market, as well as some multinational cargo owners like Nike and Ikea (Clean Cargo Working Group, 2017). Besides being a network and facilitating dialogue, the CCWG has also developed a Performance Metrics Tool. The aspects involved include CO_2, SO_x, NO_x, waste/water/ chemicals management, EMS, and transparency (Svensson and Andersson, 2012). For CO_2 emissions, it developed a CO_2 Emission Accounting Methodology, which calculates grams of CO_2 per container per kilometer based on actual fuel consumed, actual distance travelled, and the maximum capacity of the vessel (Business for Social Responsibility, 2017). Environmental data have been gathered for 3,300 container ships, and are only available for members of the working group. Audits are conducted to ensure a limited level of assurance (Business for Social Responsibility, 2017).

Shippingefficiency.org is a final example and has developed a GHG rating, similar to those of electrical appliances, ranging from A (most energy efficient) to G (least energy efficient). It has rated over 76,000 ships, and these ratings are available on the website. It uses a plethora of information sources – such as the IHS Maritime database, ship owners and operators, classification societies, yards, and engine manufacturers – to rate a ship. It reverses the burden of proof to the ship owner to check the data used and indicate the accuracy of the data, although the GHG rating method itself has been audited.

Capturing a Moving Target through Informational Governance?

To what extent has information become a transformative source in reducing GHG emissions from shipping? As argued earlier, data reliability and availability are important prerequisites for informational governance. While reliability is sometimes difficult to assess, it is important to know what kind of information becomes available and where this information comes from. Furthermore, the disclosure of this information in itself does not necessarily lead to a transformative force. More important is that the data becomes available for actors that subsequently use it in their policy and business decisions and as such become empowered in creating incentives to reduce GHG emissions. This section will first discuss the increased data availability of GHG emissions in shipping before looking at the extent to which different actors actively use this information.

Increasing Data Availability and Transparency?

As a result of the informational tools in maritime governance, available information about the environmental profile of individual ships is growing. However, the scope of the voluntary and mandatory informational tools differ (see Table 23.1). In two of the voluntary indexes, multiple environmental aspects are taken into account, whereas two others (i.e., the CCWG and shippingefficiency.org) have a tool that focuses solely on CO_2 emissions, and the Environmental Ship Index focuses on a broader set of air emissions. It should also be noted that only the CCWG works with actual CO_2 emissions based on fuel use and other parameters, whereas the others use the EEDI or other ship design information to account for CO_2 emissions in their index. The request for actual GHG emission data and information is expected to come more forcefully through the mandatory data reporting systems of the EU and IMO. It remains to be seen whether these data will be incorporated into the voluntary informational tools and in turn increase the accuracy of these tools.

In addition, only for three of the five voluntary informational mechanisms is the information on the environmental characteristics of a ship publicly available. Interestingly, the informational mechanism that targets cargo owners has a membership-based disclosure of information (i.e., Clean Shipping Index and the Clean Cargo Working Group), although it should be noted that the CCWG publishes aggregated emission reports on their website. The mandatory EU informational tool will be publicly available, while the one from IMO is only intended for the MEPC. What is more, the data supplied to the MEPC will be anonymized aggregated data, meaning that it is not possible to trace back the energy efficiency of individual ships. However, the SEEMP of every ship has to be updated to document the methodologies that will be used for collecting the required data and reporting that data to the flag administration (Dufour, 2016). Availability of the SEEMP to other actors is, however, very limited as well. It also remains to be seen whether companies will start publishing environmental information in annual or environmental reports, something that is not common practice yet (Rahim et al., 2016).

To assess the reliability of the data, it is relevant to look at who supplies the data. In all but one case (i.e., shippingefficiency.org), it is the ship owners who supply these data, which means that the submitter has vested interest in these data (Scott et al., 2017). The verification of data by an external party therefore is important. Two of the voluntary informational tools (Green Award and Clean Shipping Index) use third-party verification in their procedures. Furthermore, the data system of the EU requires third-party verification of data that is submitted to the EU. The IMO ensures reliability by requiring flag states to check the submitted data against the requirements (and to issue a statement of compliance if this is the case). The reliability of the three voluntary informational tools that use self-diagnoses will increase automatically when they start using data from the EU MRV or global data collection system, as these are already based on verified data. Next to the question of who supplies the data, there are also concerns about the reliability of data, because calculating fuel consumed is very difficult (although the EEDI and SEEMP will contribute to establishing a framework for that) (Rahim et al., 2016).

Empowerment of Actors?

For information systems and tools to become transformative, it is necessary that actors use the information in decision-making processes. It is therefore relevant to know to whom information becomes available and how this information (might) affect the actor in its policy or management decisions. The target group for which information is disclosed through the various informational tools discussed in this chapter differs substantially. The main actors are ports and marine service providers, charterers and cargo owners, ship owners, the MEPC, and the public.

Ports use information acquired through the Green Award and the Environmental Ship Index to reward more environmental ships. However, they do not necessarily use it to attract more ships and increase their market potential in terms of handling more cargo. Also for the ship owners, reduced port fees do not drive the uptake of environmental measures, as such fees are a minor part of the operational costs, and documentation requirements are demanding while few ports participate in these schemes (Lister et al., 2015). While ports know more about which ship is green, they do not create additional incentives to become more green. In addition, the use of the Environmental Ship Index remains limited to a few active and forerunning ports (Fenton, 2017), and the same can be said for the Green Award.

Competition in the shipping industry focuses on market access to cargo owners. The transformative power of the Clean Shipping Index and CCWG lies in the increasing importance of environmental consideration in business transactions. Cargo owners that use environmental information in selecting ships to transport cargo create a competitive advantage for more environmentally friendly ships. Overall, however, for most cargo owners, price and reliability remain the essential factors considered in chartering a ship (Lister et al., 2015). The segment in which informational governance is creating a sustainability effect is in container shipping (Lister et al., 2015), with 85 percent of the container industry being part of the CCWG. This development is driven by big-brand cargo owners, who themselves are advancing their environmental management efforts to include the transport of their cargo (Lister, 2015). Similarly, the motivation for ship owners to participate is limited to regulation or economic considerations – in this case, pressure by cargo owners (Van Leeuwen, 2016). Ship owners lack an internal environmental management strategy that goes beyond regulatory compliance, and subsequently refrain from seeking to expand market access through improved environmental performance and reputation (van Leeuwen and van Koppen, 2016). In general, ship owners have difficulty with regarding these informational tools as a legitimate and accepted governance arrangement, as they distort the level playing field (Wuisan et al., 2012; Lister, 2015).

Ship owners play a role in providing the information in most of the informational tools discussed in this chapter. However, they are also offered a tool to use this information for internal purposes through the SEEMP. The SEEMP should, similar to the more general safety and environmental management systems, allow ship owners to develop a management system for the use and organization of information for the purpose of identifying and selecting energy-efficiency measures. However, the SEEMP regulation does not require actual improvements in energy efficiency of a ship (Johnson et al., 2013). Since the SEEMP is a management-based standard, the certificate that verifies compliance is proof of the existence of the SEEMP, not of the energy-efficiency outcomes of the SEEMP (Scott et al., 2017). The actual transformative power will very much depend on internal management procedures of the ship owner and on how such a SEEMP is implemented and used to improve energy efficiency (Scott et al., 2017). Some are indeed pessimistic about its actual effectiveness (Johnson et al., 2013). In the projections of the IMO, only 1 of the 16 scenarios, which already take the implementation of the EEDI and SEEMP into account, projects a decrease of absolute CO_2 emissions (International Maritime Organization, 2014). The fact that most ship owners do not have the ambition to do more than what is required by formal regulation (van Leeuwen and van Koppen, 2016) and do not directly benefit from reduced fuel costs make such a pessimistic scenario likely.

IMO's global data system will greatly improve the evidence for the state of GHG emissions from the sector. This information is intended for decision-making processes within the MEPC and IMO, and allows "a decision to be made on whether any further measures are needed to enhance energy efficiency and address greenhouse gas emissions from international shipping. If so, proposed policy options would then be considered" (International Maritime Organization, 2016b). However, it remains to be seen to what extent this evidence will be able to contribute to resolving the recurring debate between proponents of the common, but differentiated, responsibility principle that dominates the climate regime (i.e., the developing countries, some of which are powerful flag states) and those favoring the application of the no more favorable treatment principle, which requires port states to enforce IMO standards uniformly to all ships in their ports irrespective of the flag state of the ship (i.e., the developed countries). This debate has been one of the reasons why progress on developing IMO standards on GHG emissions has been slow.

Finally, the public (including environmental NGOs) gets more access to energy efficiency and GHG emission data as well. It can already get insight into a ship's environmental performance through the voluntary informational tools, but the EU MRV regulation will add a more comprehensive and aggregated analysis of GHG emissions from shipping. However, it can be questioned whether this information will actively be used, because shipping is not a very visible economic activity (Lister, 2015; Toonen and Mol, 2016), and only a handful of environmental NGOs work actively on shipping. As a consequence, public awareness on shipping's potential environmental impact remains low, and it is uncertain whether this will change in the near future.

Conclusion

Despite the importance of international shipping for the global economy and its contribution to climate change, efforts to curtail GHG emissions from shipping are only just being implemented. Part of the relatively new regulations on GHG emissions are the use of informational tools, both within the IMO and by nonstate actors. This chapter discussed the use and (potential) contribution of informational tools to regulating shipping's GHG emissions. As a result of the EU MRV regulation in particular (and to some extent the IMO's global data collection system), more information will become available on the actual energy efficiency of individual ships. In addition, through the voluntary informational tools, the set of actors that will use such information is growing from ship owners, flag states, and the IMO, to business actors like ports and cargo owners and the public. Along with this change, pressure to become more energy efficient will no longer only come from formal regulation but also from market pressures of ports and cargo owners requesting insight into the energy efficiency of ships.

However, this is not actual practice yet. The transformative power of information to improve the energy efficiency of ships is potentially there but is not displayed yet. Even though ports and cargo owners have informational tools to consider environmental performance of ships in their daily operations at their disposal, the use of these informational tools is limited to frontrunners, and especially targets container ships. In addition, information on actual GHG emissions and energy efficiency is not part of these indexes yet (as they use the EEDI or similar tools to estimate the potential energy efficiency rather than real emission data).

Whether this will change in the near future depends on a number of developments. First, the EU MRV regulation and IMO's CO_2 global data collection system should be implemented. Although how the IMO and the EU will subsequently use the increased data in their decision making remains to be seen, as within the IMO the implementation of the principle of common but differentiated responsibility is still subject to debate. Second, the voluntary informational tools discussed earlier should make more use of these data becoming available. What is more, the use of the indices should become more widespread. But what in the end might be most important to achieving more energy efficiency is the way in which ship owners integrate GHG emissions and energy efficiency information in their energy efficiency management and invest in reducing GHG emissions. Only then will environmental information become a transformative power. It is thus not only the design of informational tools that matters but also the wider institutional and market setting and barriers within which a sector operates that matters for the effectiveness in which information is transformative for achieving sustainability.

With the expected growth in shipping, it is not likely that current energy efficiency regulations and informational tools, even when they live up to their potential, will lead to an absolute reduction of CO_2 emissions. For that to happen, a transition toward CO_2-neutral or renewable energy is needed, which at the moment is not at all considered within the IMO. There is thus still a long way to go before decarbonizing shipping becomes a reality.

References

Bows-Larkin, A. (2015). All adrift: Aviation, shipping, and climate change policy. *Climate Policy*, **15**(6), 681–702.

Bush, S. R., Toonen, H., Oosterveer, P., and Mol, A. P. (2013). The "devils triangle" of MSC certification: Balancing credibility, accessibility and continuous improvement. *Marine Policy*, **37**, 288–93.

Business for Social Responsibility. (2017). *Clean Cargo Working Group (CCWG) 2016 Progress Report, Business for Social Responsibility*. Available at www.bsr .org/reports/Clean_Cargo_Progress_Report_2016.pdf.

Cashore, B. (2002). Legitimacy and the privatization of environmental governance: How non–state market–driven (NSMD) governance systems gain rule–making authority. *Governance*, **15**(4), 503–29.

Castells, M. (2000). *The Information Age: Economy, Society and Culture; Volume 1: The Rise of the Network Society*. Malden, Oxford: Blackwell Publishers.

Clean Cargo Working Group. (2017). *Our Accomplishments*. Available at www.bsr .org/en/collaboration/groups/clean-cargo-working-group.

COGEA. (2017). *Study on Differentiated Port Infrastructure Charges to Promote Environmentally Friendly Maritime Transport Activities and Sustainable Transportation*. Brussels: European Commission.

DeSombre, E. R. (2006). *Flagging Standard; Globalization and Environmental, Safety and Labor Regulations at Sea*. Cambridge, MA, London: MIT Press.

Dufour, J. (2016). *EU MRV vs. IMO Fuel Consumption Data Collection System*. Available at www.green4sea.com/eu-mrv-vs-imo-fuel-consumption-data-collection-system.

Environmental Ship Index. (n.d.). *Introduction*. Available at www.environmentalship index.org/Public/Home.

Environmental Ship Index. (2017). *List of Participating Ships*. Available at www .environmentalshipindex.org/Public/Ships.

European Commission. (2013). Integrating maritime transport emissions in the EU's greenhouse gas reduction policies. *COM(2013) 479 Final*. Brussels: European Commission.

Falkner, R. (2003). Private environmental governance and international relations: Exploring the links. *Global Environmental Politics*, **3**(2), 72–87.

Fenton, P. (2017). The role of port cities and transnational municipal networks in efforts to reduce greenhouse gas emissions on land and at sea from shipping–An assessment of the World Ports Climate Initiative. *Marine Policy*, **75**, 271–7.

Gilbert, P., Bows-Larkin, A., Mander, S., and Walsh, C. (2014). Technologies for the high seas: Meeting the climate challenge. *Carbon Management*, **5**(4), 447–61.

Glasbergen, P., Biermann, F., and Mol, A. P. J. (2007). *Partnerships, Governance and Sustainable Development; Reflections on Theory and Practice*. Cheltenham: Edward Elgar.

Green Award. (2015). *Green Award Foundation Annual Report 2014–2015*. Available at http://annualreport2014-2015.greenaward.org/index.php/greenaward/?page=3.

Green Award. (2017). *All Seagoing Ships*. Available at www.greenaward.org/greenaward/ 13-all-seagoing-ships.html.

Gulbrandsen, L. H. (2009). The emergence and effectiveness of the Marine Stewardship Council. *Marine Policy*, **33**(4), 654–60.

Gupta, A. and Mason, M. (2014). *Transparency in Global Environmental Governance: Critical perspectives*. Cambridge, MA, London: MIT Press.

Hackmann, B. (2012). Analysis of the governance architecture to regulate GHG emissions from international shipping. *International Environmental Agreements: Politics, Law and Economics*, **12**(1), 85–103.

International Maritime Organization. (2014). *Third IMO Greenhouse Gas Study 2014*. London: International Maritime Organization.

International Maritime Organization. (2016a). *New Requirements for International Shipping as UN Body Continues to Address Greenhouse Gas Emissions*.

Available at www.imo.org/en/MediaCentre/PressBriefings/Pages/28-MEPC-data-collection--.aspx.

International Maritime Organization. (2016b). *IMO Takes Further Action on Climate Change.* Available at www.imo.org/en/MediaCentre/PressBriefings/Pages/11-data-collection-.aspx.

Johnson, H., Johansson, M., Andersson, K., and Södahl, B. (2013). Will the ship energy efficiency management plan reduce CO_2 emissions? A comparison with ISO 50001 and the ISM code. *Maritime Policy & Management*, **40**(2), 177–90.

Karim, M. S. (2015). *Prevention of Pollution of the Marine Environment from Vessels: The Potential and Limits of the International Maritime Organisation.* Chem, Heidelberg, New York, Dordrecht, London: Springer.

Lister, J. (2015). Green shipping: Governing sustainable maritime transport. *Global Policy* **6**(2), 118–29.

Lister, J., Poulsen, R. T., and Ponte, S. (2015). Orchestrating transnational environmental governance in maritime shipping. *Global Environmental Change*, **34**, 185–95.

Lund-Thomsen, P., Poulsen, R. T., and Ackrill, R. (2016). Corporate social responsibility in the international shipping industry: State-of-the-art, current challenges and future directions. *The Journal of Sustainable Mobility*, **3**(2), 3–13.

Marine Environment Protection Committee. (2011). *Resolution MEPC.203(62); Amendments to the Annex of the Protocol of 1997 to amend the International Convention for the Prevention of Pollution from Ships, 1973, as modified by the Protocol of 1978 relating thereto.* London: International Maritime Organization.

Marine Environment Protection Committee. (2012). *Resolution MEPC.213(63); 2012 Guidelines for the Development of a Ship Energy Efficiency Management Plan (SEEMP).* London: International Maritime Organization.

Martinez Romera, B. (2016). The Paris agreement and the regulation of international bunker fuels. *Review of European, Comparative & International Environmental Law*, **25**(2), 215–27.

Mitchell, R. B. (1994). *Intentional Oil Pollution at Sea; Environmental Policy and Treaty Compliance.* Cambridge, MA: MIT Press.

Mol, A. P. (2006). Environmental governance in the Information Age: The emergence of informational governance. *Environment and Planning C: Government and Policy*, **24**(4), 497–514.

Mol, A. P. (2015). Transparency and value chain sustainability. *Journal of Cleaner Production*, **107**, 154–61.

Pattberg, P. (2005). What role for private rule-making in global environmental governance? Analysing the Forest Stewardship Council (FSC). *International Environmental Agreements: Politics, Law and Economics*, **5**(2), 175–89.

Pierre, J. and Peters, B. G. (2000). *Governance, Politics and the State.* London: MacMillan Press.

Poulsen, R. T. and Sornn-Friese H. (2015). Achieving energy efficient ship operations under third party management: How do ship management models influence energy efficiency? *Research in Transportation Business & Management*, **17**, 41–52.

Rahim, M. M., Islam, M. T., and Kuruppu, S. (2016). Regulating global shipping corporations' accountability for reducing greenhouse gas emissions in the seas. *Marine Policy*, **69**, 159–70.

Ringbom, H. (1999). Preventing pollution from ships – Reflections on the 'adequacy' of existing rules. *Review of European Community & International Environmental Law*, **8**(1), 21–8.

Scott, J., Smith, T., Rehmatulla, N., and Milligan, B. (2017). The promise and limits of private standards in reducing greenhouse gas emissions from shipping. *Journal of Environmental Law*, **29**(2), 231–62.

Soma, K., Termeer, C. J., and Opdam, P. (2016). Informational governance–A systematic literature review of governance for sustainability in the Information Age. *Environmental Science & Policy*, **56**, 89–99.

Svensson, E. and Andersson, K. (2012). *Inventory and Evaluation of Environmental Performance Indices for Shipping*. Gothenburg: Chalmers University of Technology.

Tan, A.K.-J. (2006). *Vessel-Source Marine Pollution; The Law and Politics of International Regulation*. Cambridge, UK: Cambridge University Press.

Toonen, H. M. and Mol, A. P. J. (2016). Governing the marine environment through information: Fisheries, shipping, and tourism. In B. H. MacDonald, S. S. Soomai, E. M. De Sante, and P. G. Wells, eds., *Science, Information and Policy interface for Effective Coastal and Ocean Management*. Boca Raton, FL: CRC Press (Taylor & Francis Group), 125–51.

UNCTAD Secretariat. (2016). *Review of Maritime Transport*. New York and Geneva: UNCTAD.

Van Kersbergen, K. and Van Waarden F. (2004). 'Governance' as a bridge between disciplines: Cross-disciplinary inspiration regarding shifts in governance and problems of governability, accountability and legitimacy. *European Journal of Political Research*, **43**(2), 143–71.

Van Leeuwen, J. (2010). *Who Greens the Waves? Changing Authority in the Environmental Governance of Shipping and Offshore Oil and Gas Production*. Wageningen: Wageningen Academic Publishers.

Van Leeuwen, J. (2015). The regionalization of maritime governance: Towards a polycentric governance system for sustainable shipping in the European Union. *Ocean & Coastal Management*, **117**, 23–31.

Van Leeuwen, J. (2016). Governing the Artic in the era of the Anthropocene: Does corporate authority matter in Arctic shipping governance? In P. Pattberg and F. Zelli, eds., *Environmental Politics and Governance in the Anthropocene; Institutions and Legitimacy in a Complex World*. London, New York: Routledge, 127–45.

Van Leeuwen, J. and van Koppen, C. (2016). Moving sustainable shipping forward. *Journal of Sustainable Mobility*, **3**(2), 42–66.

Vorbach, J. E. (2001). The vital role of non-flag state actors in the pursuit of safer shipping. *Ocean Development & International Law*, **32**(1), 27–42.

Wuisan, L., van Leeuwen, J., and van Koppen, C. S. A. (2012). Greening international shipping through private governance: A case study of the Clean Shipping Project. *Marine Policy*, **36**(1), 165–73.

Yliskylä-Peuralahti, J. and Gritsenko, D. (2014). Binding rules or voluntary actions? A conceptual framework for CSR in shipping. *WMU Journal of Maritime Affairs*, **13**(2), 251–68.

24

Energy from the Sea
Challenges and Opportunities

CHRISTINA I. REICHERT AND JOHN VIRDIN

Introduction

Since the first offshore oil production well in the United States was drilled off Summerland, California, in 1896, offshore oil and gas production has flourished, reaching 30 percent of global crude oil output – approximately 30 million barrels per day – by 2015 (National Commission on the British Petroleum Deepwater Horizon Oil Spill and Offshore Drilling, 2011; Energy Information Administration, 2016). Perhaps unsurprisingly, the ocean was a major source of energy in the twentieth century. Now in the twenty-first century, even as offshore oil and gas extraction continues, use of renewable ocean energy has begun, reaching 0.5 gigawatts (GW) of commercial generation capacity operating in 2016, with another 1.7 GW under construction (World Energy Council, 2016). The ocean's importance as a source of energy is expected to continue, if not grow, in coming decades. The mix of sources will likely change, with more electricity provided by renewable sources. The ocean has a perhaps underappreciated role to play as the world grapples with how to decarbonize and meet the ambitious targets of the 2015 Paris Agreement on climate change (Paris Agreement, 2016).

Although offshore oil and gas extraction is expected to continue in the coming decades, the worldwide potential for the oceans to generate new *renewable* sources of energy is enormous. Can the oceans feasibly play a role in global decarbonization of energy? The Deep Decarbonization Pathways Project (2015) analyzed how much renewable ocean energy is required for the world to meet global objectives to limit climate change. It found that doing this would require global installation of 29 GW of offshore wind by 2020, 100 GW by 2030, 268 GW by 2040, and 616 GW by 2050. Other technologies for harvesting renewable ocean energy may become commercially viable, for example, wave, current, and ocean thermal energy conversion (OTEC).

In addition to achieving the emissions reductions targets of the Paris Agreement, growth in renewable energy production from the ocean would contribute to meeting the global Sustainable Development Goals (SDGs), agreed to by the international community at the 2015 meeting of the United Nations General Assembly (UN General Assembly, Res. 70/1). Specifically, increased renewable ocean energy would aid in reaching the target to increase economic benefits to small-island developing states (see Chapters 8 and 18) from sustainable use of ocean resources as part of SDG 14's aim to conserve and sustainably use the ocean (Mauritius Prime Minister's Office, 2013). Renewable energy from the sea would also help to meet the target to increase substantially the share of renewable energy in the global mix as part of SDG 11 on affordable and clean energy, and it would contribute to the climate action described in SDG 13.

If the ocean can play a role in global decarbonization, what will the bridge from fossil fuels to renewables look like? In this chapter, we synthesize some of the current information and projections available to answer this question. We take stock of the potential for the ocean to help governments meet the Paris Agreement targets and relevant SDGs. We analyze the challenges and opportunities for the ocean to provide more renewable energy as part of the global mix, and we discuss the likely prospects for more traditional offshore fossil fuel extraction over the same time. We point to several options for public policies that could help to address the challenges, and capture the opportunities, of renewable ocean energy.

Traditional Ocean Energy Sources: Offshore Oil and Gas Drilling

The offshore oil and gas sector has traditionally provided the majority of the ocean's contribution to the global energy mix. Based on the assumption that approximately 32 percent of global oil and gas activities were offshore (APEC, 2014), the Organisation for Economic Co-operation and Development (OECD) estimates that the global value added of offshore oil and gas exploration and production in 2010 was USD 504 billion. At that time, around 207 floating oil and gas platforms and more than 9,000 fixed offshore platforms were operating, mainly concentrated in the biggest offshore oil- and gas-rich sedimentary basins, such as the North Sea, Mediterranean Sea, Arab-Persian Gulf, West and East Africa, North and South America, India, the North and South China Sea, and West Australia (OECD, 2016). Based on estimates of world offshore crude oil production by physiographical location and region published by the International Energy Agency (IEA), shallow-water production accounted for 93 percent of total output (around USD 468 billion value added), compared to 7 percent for deep-water production (around USD 35 billion value added) (IEA, 2012).

Any discussion of the ocean's potential contribution to meeting the Paris Agreement targets and decarbonization must acknowledge that carbon sources of ocean energy will continue to grow. Currently, some 37 percent of the world's proven oil reserves are thought to exist in the oceans, with around one-third of these in deep water (IEA, 2012). While the world moves toward a greener electricity system, offshore oil and gas may play a bridging role in the transition toward a greener ocean energy system.

In the coming years, independent of what happens on land, oil production from the sea floor is expected to grow much more slowly than offshore gas production.

For example, while the IEA (2014b) projects that offshore oil and gas will grow at about 3.5 percent per year up to 2030, total offshore crude oil production is expected to see a significant increase, whereas offshore gas production is expected to have strong growth in both shallow and deep water. The OECD explains that the persistence of low oil and gas prices will affect these projections, particularly given the high costs and environmental effects of ultra-deep-water exploration and production (OECD, 2016). This change reflects a shift in oil production from the sea floor to deeper waters, even while gas production from the sea floor continues apace in other available waters.

The Future of Ocean Energy: Renewables

As the OECD (2016) notes, the global ocean contains a massive source of renewable energy waiting to be harnessed. New sources include tides, waves, currents, osmosis, and OTEC, with commercial interest in these sources growing significantly. According to the Ocean Energy Systems Implementing Agreement (OES, 2012), by 2050 the global ocean could provide 337 GW of wave and tidal energy, and possibly as much again from OTEC, generating up to 1.2 million jobs in industries widely distributed along supply chains. Already in Europe, for example, renewable ocean energy supply chains include the manufacture of tidal turbines, hydro-turbines and steel spare parts, wave power plants and generators, and wave-power attenuators (OECD, 2016). This potential has been widely recognized in recent decades, leading to the development and launch of many large prototypes. However, to date obstacles remain in transitioning from the prototype phase to widespread commercialization of renewable ocean energy production.

Offshore Wind

Among the various sectors of renewable ocean energy, offshore wind has been unique in its trends and conditions, progressing from the first small pilot project to a nascent industry with the potential for significant further growth. From a baseline of global installed capacity producing over 7 GW in 2010, there is potential to reach on the order of 40 to 60 GW by 2020, with further potential for growth by an order of magnitude by 2050 to give wind energy a sizeable global market share (IEA 2014b; OECD, 2016). Although there is a broad range of projections, the most optimistic suggest that some 400 GW of offshore wind capacity could be installed by 2030, and approximately 900 GW by 2050, if the industry can drive down costs across all elements of the supply chain to become cost-effective compared to alternative energy sources (OECD, 2016). The bulk of the deployment of offshore wind is expected in the European Union (EU) (likely to be some half of the global market) and China (roughly a quarter of the market), India, and the United States (with some 20 percent of the market) (OECD, 2016).

The International Renewable Energy Agency (IRENA) estimates that offshore wind energy has even greater location and generation potential than onshore wind energy (IRENA, 2016). Wind speeds at sea are higher than on land, and fewer obstacles contribute to turbulence. Consequently, offshore wind energy is generally more efficient than onshore wind energy. Offshore wind energy faces proportionally higher grid connection and construction costs than onshore wind energy, given the distance from the turbines to shore, and the technology must be able to withstand severe marine environments. Additional costs come from assessing and maintaining sites at sea. Operations also require marine transportation for access to turbines.

Offshore wind turbines mirror those used for onshore wind generation, although offshore turbines must include additional measures to prevent corrosion and withstand more strenuous weather conditions (BOEM, n.d. b). Offshore turbines are typically larger than turbines deployed onshore, with nominal capacities typically from 2 to 6 megawatts (MW) and tower heights greater than 200 feet. The engineering and design of offshore wind facilities depend on site-specific conditions, including water depth, wind speed, and the geology of the seabed. Offshore wind developers are continuing to push the envelope to make floating foundations commercially available by 2020 so that offshore areas with mid-depth water conditions (30 to 50 meters) can be used as lower-cost alternatives to fixed-bottom foundations (IRENA, 2016; Patterson, 2017). Offshore wind farms at depths of less than 50 to 60 meters can implement fixed-bottom foundations, which are commercially deployed. Nevertheless, development at greater depths may only be efficient with floating platforms. Further research and development are necessary to make floating platforms more cost-efficient and to move them beyond the demonstration phase (IRENA, 2016).

Opportunities exist to make offshore wind energy more cost-effective. Grid-connection transmission lines can be shorter for offshore wind than for onshore wind, so there is greater potential for cost reductions. Additionally, the use of high-voltage direct current connections could reduce transmission losses, permitting turbines to be placed farther offshore. An interoperator maintenance concept that uses a joint fleet and logistics infrastructure could lower costs. Moreover, opportunities for technological innovation exist in the areas of energy storage and smart-grid information technology systems that control power output through communications between a remote wind plant and a centralized control center (IRENA, 2016).

Average capital costs for offshore wind facilities increased in the mid- to late-2000s due to deeper water sites, longer distances to shore, siting complexity, and higher contingency reserves, but costs appear to be falling in recent years. A study by Navigant Consulting (2014) found average capital costs of USD 5,187 per kilowatt (kW) for 2013 offshore wind installations for which data were available, a decrease of approximately USD 200 per kW compared to 2012. Cost varies significantly depending on the size of the facility, site conditions, and technology costs at the time of construction. For example, the cost of the 400-MW Anholt wind farm in Denmark exceeded USD 1.5 billion, but the cost for Denmark's Middelgrunden wind farm was roughly USD 50 million (4C Offshore, 2016a, 2016b). In the United

States, the Block Island Wind Farm in Rhode Island cost approximately USD 300 million, and the project developer predicts that the wind farm will reduce the average rate payer's electricity cost on Block Island by 40 percent (Deepwater Wind, 2016; Schlossberg, 2016). The estimated cost of Hawaii's 30-year wind energy plan is USD 1.9 billion, or roughly USD 5 million per MW of offshore wind capacity, due in part to the lack of sufficient manufacturing or harbor facilities (Harball, 2015).

Wave Energy Converters

Wave energy converters (WECs) use the kinetic energy of waves to generate electricity. There are over 100 pilot projects and demonstrations worldwide for different WECs, but only a small number are near commercialization. The projects closest to commercialization are undergoing smaller-scale open water tests throughout the duration of their life cycle, testing the device under a variety of stressors in all seasons. Individual devices within these stages range in capacity from 18 kW to 1 MW, with capacity factors – average power generated divided by the rated peak power – being between 30 and 50 percent (IRENA, 2014a).

Because wave regimes vary so widely across different geographical regions, a variety of wave energy technologies exists. According to the US Department of Interior's Bureau of Ocean Energy Management (n.d.), there are four general WEC technology categories: (1) terminator devices placed perpendicular to the waves to capture the wave power; (2) attenuators that lie perpendicular to the waves and create energy as the waves move sections of the devices up and down; (3) point absorbers that include a floating structure and utilize the rise and fall of the ocean at a single point to generate electricity; and (4) overtopping devices where incoming waves fill reservoirs and then the water is released through hydro-turbines. Due to this variety of technologies, the industry suffers from a lack of cohesion and limited supply chains for necessary components. Steps to overcome these challenges include deployment of arrays and development of more dedicated infrastructure, such as ports and transmission grids, to support future installation, operation, and management. Wave energy could also reduce infrastructure costs by seeking opportunities to develop synergistically with other offshore industries, such as aquaculture or wind energy (IRENA, 2014a).

Wave energy is most abundant in areas where winds are strongest, especially between 40 and 60 degrees latitude around the globe (Behrens et al., 2012). There are other strong currents and trade winds outside of this range that result in very stable wave sources. Waves tend to be stronger in deeper waters, but from a transmission and cost perspective, it is cheaper and more efficient to have power closer to shore and thus closer to where the demand is located. Thus, islands formed by volcanoes can be ideal locations for this technology. These islands – for example, Mauritius – usually have narrow, steep continental shelves, leading to deep ocean waters near shore and thus higher energy potential at lower costs.

A study by IRENA (2014a) estimates the levelized cost of electricity (LCOE) for wave energy to fall between USD 0.13 and USD 0.70 per kW hour (kWh) for a 10-MW deployment. The agency also projects 2030 LCOE to decrease by around USD 0.13 to 0.25 per kWh if deployment can reach 2 GW worldwide, which would give a considerable scope for economies of scale. *The New York Times* reported that Carnegie Wind Energy's WEC devices currently operate at a LCOE of USD 0.40 per kWh (Yee, 2015). This same report projected that large wave farms of 100 MW would reduce this cost to USD 0.12 to 0.15 per kWh.

Tidal Energy Technology

Similar to wave energy, tidal energy technologies are still in the developmental phase. Currently, there are two general categories of tidal energy technologies. The first uses potential energy created by the different water heights during high and low tides. The tidal range technology relies on a barrage with gates that open and close to control water flow. The facilities may generate power during either high tide or low tide. Operators may close the gates when the tide is highest and then release water through the turbines, or close the gates when the tide reaches its lowest level and generate power when water flows from the sea into the reservoir through the turbines. There is a tidal technology option that allows two-way power generation via both ebb and flood tides. Tidal range technologies generally require at least a five-meter difference between water levels at high and low tides (IRENA, 2014c).

The second category of technology is tidal current, also known as tidal stream, which uses the kinetic energy of the tides to create electricity. Tidal current itself can be broken into three different categories. There are horizontal axis or vertical axis cross-flow turbines, which may resemble underwater wind turbines that rotate slowly. The majority of existing tidal current projects, as well as research and development investments, are currently pursuing horizontal axis turbines. Another tidal current technology is a reciprocating device that has blades called hydrofoils. As the tides flow on either side of the hydrofoils, they move up and down to drive a rotating shaft or hydraulic system that ultimately produces electricity. The final tidal current technology consists of any other designs that do not fall into the previous two categories. This includes rotating screw-like devices as well as tidal kites. A hybrid form of tidal technology that combines both tidal range and tidal current is also under development (IRENA, 2014c).

Construction costs for the La Rance (France) and Sihwa (South Korea) tidal energy projects were USD 340 per kW and USD 117 per kW, respectively (IRENA, 2014c). A key difference in the project costs is the up-front costs due to the need to construct a dam in La Rance and using an already existing dam in Sihwa. IRENA estimates that costs for future projects will decrease with increased deployment, with the LCOE projected at USD 0.32 to 0.57 per kWh. As seen with these two projects, costs for future projects will greatly depend on location and existing infrastructure.

Deep Ocean Applications

Deep ocean water applications (DOWA) include both OTEC and seawater air conditioning (SWAC), which tend to be complementary technologies. OTEC uses the temperature differences between the upper and the deeper layers of the ocean to generate electricity and works best where the temperature difference between surface and deep seawater is about 20°C. In areas with close access to deep cold seawater and a flat coastal shelf, transmission and construction costs may be lowered. This technology has existed since the 1930s, and there are at least 11 pilot and demonstration OTEC projects worldwide, although at-sea OTEC plants are primarily in the pilot stage (IRENA, 2014b).

Three kinds of OTEC systems exist: closed cycle, open cycle, and hybrid. Closed-cycle systems use liquids with low boiling points to rotate a turbine and generate electricity. Warm surface seawater vaporizes the low boiling point fluid to rotate a turbine, and cold deep seawater condenses the vapor back into liquid that is recycled through the system. Open-cycle systems boil warm surface seawater to drive a low-pressure turbine, and cold deep seawater condenses the steam back into liquid. Hybrid systems flash-evaporate warm surface seawater, which vaporizes a low boiling point liquid to drive a turbine to produce electricity (IRENA, 2014b).

With all three of these systems, spent cold deep seawater can flow directly into a cooling system for use in air conditioning, known as a SWAC. More than seven institutions or cities worldwide use SWAC, ranging from 50 tons (building scale) to 58,000 tons (city scale). SWAC can provide energy savings from 80 to 90 percent compared to conventional air conditioning systems. Moreover, SWAC technologies have a short economic payback period that can be anywhere from three to seven years (IRENA, 2014b). OTEC has a 90 to 95 percent capacity factor, meaning that this technology can create nonintermittent, baseload power using ocean energy. OTEC also has multiple outputs, such as SWAC, aquaculture purposes, and desalinization of seawater, which make the plants more economical. The technological challenges with this technology, however, are an energy loss of 20 to 30 percent from pumping and its very low heat engine efficiency, with a maximum of 7 percent (IRENA, 2014b).

Although, like other ocean energy resources, OTEC tends to have high capital costs – a 105-kW OTEC plant costs an estimated USD 5 million to build, with the heat generator accounting for one-third of that cost – experts predict that scaling up the capacity of OTEC plants will significantly decrease costs (Hodgkins, 2015; Deep Sea Water Research Institute, 2016). For example, researchers at IRENA (2014b) predict that the LCOE for a 100-MW OTEC plant would approach USD 0.07 to 0.19 per kWh. Studies also suggest that government bonds with interest rates of around 4.2 percent over 20 years in lieu of private loans with interest rates of 8 percent over 15 years could reduce the LCOE of OTEC power to around USD 0.10 for a 5- to 10-MW plant and around USD 0.03 for a 100-MW plant (Vega, 2010, 2012).

Algae Biogas

Scientists have used different feedstocks, including algae, to create biogas that can be turned directly into biofuels, such as ethanol and biodiesel. Algae can grow in a range of aquatic environments, produce biogas very rapidly, grow on land not used for traditional agriculture, and minimize land use compared with other biofuel feedstocks. Some researchers say algae could be 10 or even 100 times more productive than traditional bioenergy feedstocks (DOE, n.d.; NREL, n.d.). Scientists are currently studying how to reduce the cost and increase the productivity of implementing algae for biofuel (Huntley et al., 2015). The main economic challenges for this technology are its ability to compete with existing commodity products with tight margins, primarily crude oil, soy meal, and corn meal.

Researchers are currently testing various algae species for use in biogas grown through two main technologies: raceway ponds and photobioreactors. Each technology struggles with strain isolation, nutrient sourcing and utilization, production management, harvesting, co-product development, fuel extraction refining, and residual biomass utilization (Hannon et al., 2010). Duke University's Marine Algae Industrialization Consortium (MAGIC) project combines biofuel with high-value bio-products, such as industrial chemicals, bio-based polymers, and proteins (Duke University, 2015). MAGIC researchers reviewed ten different algae studies, finding that capital costs for algae biogas can range from an estimated USD 45 to 95 million and operating costs can range from an estimated USD 2 to 20 per day (Gerber et al., 2016). The researchers expect promising economic competitiveness by obtaining biofuel as well as nutraceuticals, specialty pigments, and food (Johnson, 2016). The US National Aeronautics and Space Administration's (NASA) Offshore Membrane Enclosure for Growing Algae (OMEGA) project sought to grow algae, clean wastewater, capture carbon dioxide, and produce biofuel without competing with agriculture for water, fertilizer, or land (NASA, n.d.). The OMEGA study estimated that 204,320 floating photobioreactors would have capital costs of about USD 547 million and would require USD 100 million per year in revenue (Trent, 2012; Danan, 2013).

Government Policies to Increase Renewable Ocean Energy

For offshore renewable energy technologies to substantially help the world reach its climate goals, public policies that support these technologies will be needed. This section outlines potential policies related to government permitting and marine spatial planning, research and development, and project funding and development, drawing examples largely (but not exclusively) from the United States.

Government Permitting Programs and Marine Spatial Planning

For developed renewable ocean energy technologies like offshore wind, one obstacle to commercialization is complex, time-intensive, and costly government permitting processes created to protect natural resources, commercial and recreational fisheries (see chapters in Part III), and marine areas with special significance due to their conservation, recreational, ecological, historical, scientific, cultural, archaeological, educational, and aesthetic qualities. An example of how these processes may affect offshore wind projects in the United States comes from the 2001 "Cape Wind" 468-MW project off the coast of Massachusetts.

Developers of the Cape Wind project obtained the required permits, but after Congress shifted authority over offshore wind projects to the US Department of Interior in 2005, the developers had to start over (Gerrard, 2017). By 2011, the project had completed most of the permitting processes but then faced lawsuits from wealthy property owners on Cape Cod (Kennedy, 2005; Seelye, 2013). After spending more than USD 70 million fighting (and largely winning) regulatory and legal battles, in 2015, the utilities that had signed power purchase agreements to buy the power output from the Cape Wind project backed out of their contracts (Powell, 2012; Susskind and Cook, 2015). In July 2016, the US Court of Appeals for the District of Columbia Circuit largely ruled for the Cape Wind developers, but held that the US Bureau of Ocean Energy Management was required to supplement its environmental review of the project (under the National Environmental Policy Act) and the US Fish and Wildlife Service had erred when evaluating the project's effect on threatened and endangered species (under the Endangered Species Act) (*Public Employees for Environmental Responsibility v. Hopper*, 827 F.3d 1077 [D.D.C. 2016]).

In the United States, the US Secretary of the Interior developed "Smart from the Start" to help reduce permitting timelines from expected 7 to 10 years to half that or less (DOE, 2011; Gerrard, 2017). This program sought to restructure the regulatory approval process. It was intended to expedite leasing of wind energy areas that were identified by an interagency process so as to designate areas with fewer user conflicts and resource issues (BOEM, n.d. a, n.d. b). The program was also intended to create a parallel expedited track for building offshore transmission infrastructure to support projects in the 11 currently identified wind energy areas.

The only successful offshore wind energy project in the United States – off the coast of Block Island, Rhode Island – benefited greatly from marine spatial planning as a tool to simplify the permitting process (Burger, 2011; Susskind and Cook, 2015; Boehnert, 2016; Gerrard, 2017). After the state of Rhode Island determined that it could expand its renewable energy production offshore, it began studying its coasts' marine ecology, climate, cultural and historical resources, fisheries, tourism, and recreation (Gerrard, 2017). The state created the Rhode Island Ocean Special Area Management Plan (Ocean SAMP) in 2010 for zoning both state (within 3 miles of shore) and federal (between 3 and 200 miles of shore) waters (R.I. Coastal Resource Management Council, 2010). This plan was incorporated into the state's coastal zone management plan, which provides US states the authority to review activities occurring in federal waters off their coasts under the Coastal Zone Management Act (16 U.S.C. §§ 1451–66). Under the plan, Rhode Island identified a three-mile area southeast of Block Island as an ideal location for a wind turbine demonstration site

(RI Coastal Ress. Mgmt. Council, 2010). After issuing a request for proposals for a five-turbine, 20-MW demonstration wind farm off the coast of Block Island, the state of Rhode Island received and accepted an application from Deepwater Wind. Not only did the Rhode Island Ocean SAMP ensure that the state considered the competing interests of its coastal resources, it also meant that the state had already conducted many of the required studies for state and federal permits (Gerrard, 2017). Thus, when Deepwater Wind began obtaining the required federal and state permits, it was able to complete that process within two years, substantially less than Cape Winds' 16-year process that is still ongoing.

Marine spatial planning has been used to balance stakeholders' competing interests in particular parts of coastal regions. Policy makers can use this mechanism to accomplish important goals that could push offshore renewable energy forward. For example, Scotland used marine spatial planning to locate offshore areas ideal for wind, wave, and tidal technology implementation, with the goal of establishing a coherent strategic vision as well as objectives and policies to further the achievement of sustainable development of offshore renewable technologies (Offshore Wind Industry Group, 2013; Scottish Government, 2015). Throughout the process of developing a marine spatial plan, policy makers must research both the energy and other resources available across their coastal zone, thereby creating a collection of the information required for the environmental reviews and permitting process for offshore energy projects. They must engage stakeholders who may otherwise object to the energy project, lessening the opposition to offshore energy projects so that the outcome is more similar to Block Island than Cape Wind in the United States. Doing this can ensure that the public is involved early in the process, which is vital for long-lived policy solutions.

Research and Development Support

One of the most prevalent, and most vital, roles that governments play in developing and implementing offshore renewable energy is supporting research and funding emerging technologies as a public good. The main obstacle facing many renewable ocean energy technologies is the cost of developing and implementing them at this early stage of growth. As discussed earlier, aside from offshore wind, most of the renewable ocean energy technologies are primarily in the demonstration stage, and government funding for research and development could be instrumental for their future expansion. Coordinated public and private research efforts to establish innovation hubs and testing facilities provided a key driver for existing projects, such as Hawaii's deployment of multiple ocean energy technologies. Parallels can be drawn between the benefits of government funding for more developed renewable energy technologies as well as traditional electricity generation technologies.

There are examples worldwide where governments have provided funding for the research and development of renewable ocean energy technologies. In the United

States, Hawaii received USD 9.7 million in grants toward wave energy projects (DOE, 2014, 2015). In 2008, the Scottish government launched the Saltire Prize Challenge for the development of a wave or tidal energy device that generates at least 10 GW hours (GWh) over a continuous two-year period using only the power of the sea (Scottish Government, 2013). The Scottish government also implemented a National Renewables Infrastructure Fund to help with project development (Offshore Wind Scotland, 2017). The New Zealand government enacted the Energy Efficiency and Conservation Act of 2000, which led to a National Energy Efficiency and Conservation Strategy in September 2001. The law established the Energy Efficiency and Conservation Authority, which is the primary government agency that provides information, advice, public awareness, research into renewables, and incentives (such as grant schemes) to promote renewable energy (IEA, 2014a). The Danish government has supported long-term research and development, premium tariffs, and ambitious national targets since the 1980s in an effort to bolster domestic renewable technologies (Danish Energy Agency, 2012). In each of these examples, public policies to fund and support research and development for offshore renewable technologies played a key role in moving the industry forward.

Project Funding and Incentives

A similar area ripe for additional public policy support is project funding and incentives once technologies are ready for implementation. This support could fall into categories, including tax incentives, renewable energy credits (RECs), feed-in tariffs, and requests for proposals (RFPs) with specific buying mandates (Griset, 2011).

Tax incentives could be similar to the US Production Tax Credit, which provides owners of qualifying renewable-energy projects with an indexed tax credit of 2.1 cents per KWh of qualified electricity produced (Goodward and Gonzalez, 2010). This incentive has been known as "the primary federal incentive for wind energy" and could serve the same purpose for renewable ocean energy (Griset, 2011).

In some US states, governments have created renewable portfolio standards (RPSs) that require load-serving entities to source a specific portion of their energy from qualified renewable energy resources, represented as RECs (Griset, 2011). Load-serving entities can build their own renewable projects or purchase RECs from other entities to meet the RPS. Policy makers could leverage RPSs to help develop off-shore renewable energy. This has been done in Maryland, where the legislature amended their existing RPS – which requires 25 percent of electricity consumed in the state to come from renewable sources by 2020 – to include offshore wind, creating a carve-out for offshore wind not to exceed 2.5 percent of total retail electricity sales (Maryland Offshore Wind Energy Act, 2013). This requirement launched the planning process for offshore wind development off the coast of Maryland (Hicks, 2017).

Feed-in tariffs can target a specific type of generation resource, thereby providing long-term contracts with fixed prices for offshore renewable projects. The Canadian province of Ontario, for example, passed a feed-in tariff in 2009 that would offer off-shore wind developers long-term contracts at fixed prices – USD 0.16 per KWh, escalating with inflation as measured by a consumer price index (Griset, 2011).

Similarly, RFPs require utilities to purchase a specified amount of energy through long-term contracts, and these could include offshore renewable energy. For example, Massachusetts issued an RFP for offshore wind projects, requiring that electricity distribution companies enter into cost-effective, long-term contracts for offshore wind energy generation equal to approximately 1,600 MW of aggregate capacity no later than mid-2027 (Massachusetts General Laws, ch. 25A, § 11F). Rhode Island lawmakers assisted the Block Island project by amending legislation establishing criteria for long-term power purchase agreements and by helping to facilitate regulatory approval for the undersea transmission cable necessary to connect the project to the island's electricity grid (Gerrard, 2017).

Conclusion

New renewable technologies offer opportunities for using the oceans to provide energy in a world experiencing climate change.

To shorten the timeline needed to utilize responsibly and inclusively utilize emerging technologies for renewable ocean energy, policy makers should consider the experiences and policies outlined earlier. They should aim to learn from other projects across the globe and from prior innovations in more traditional energy technologies. Policy makers can implement marine spatial planning as a mechanism for pushing renewable ocean energy forward. Throughout that planning processes, policy makers can collect the information required for environmental reviews and the permitting process, engage stakeholders who may otherwise object to the energy project, and involve the public early in the process.

The cost of developing and implementing renewable ocean-energy technologies can be a roadblock. Government funding for research and development, as well as project funding and incentives, will be an essential element for expanding renewable ocean energy. Coordinated public and private research efforts to establish innovation hubs and testing facilities can help developers overcome obstacles. Government assistance can also support renewable ocean energy projects once they are installed at large scales, as suggest by the US experience with tax incentives, RECs, feed-in tariffs, and RFPs with specific buying mandates.

In combination, such policy approaches can bolster the ability of the oceans to assist the world in mitigating and adapting to climate change.

Acknowledgments

The authors thank Hallie Cramer, Cassidee Kido, Brandon Morrison, and Dhara Patel for the research they conducted with support from the Energy Initiative as part of the Bass Connections Course, "History and Future of Ocean Energy" at Duke University. This research benefited greatly from collaboration with Manta Nowbuth, dean of the Faculty of Ocean Studies at the University of Mauritius; Silvia Romero Martinez, Senior Energy Specialist; and Yuvan Beejadur, Natural Resource Management Specialist, both at the World Bank. The authors are also grateful for the excellent comments from Eric A. Bilsky, Assistant General Counsel of Oceana, Inc., and Stephen E. Roady, Professor of the Practice of Law at Duke University.

References

4C Offshore. (2016a). *Anholt Offshore Wind Farm.* Available at www.4coffshore .com/windfarms/anholt-denmark-dk13.html.

4C Offshore. (2016b). *Middelgrunden Offshore Wind Farm.* Available at www .4coffshore.com/windfarms/middelgrunden-denmark-dk08.html.

APEC. (2014). *Ocean and Fisheries Working Group, Asia-Pacific Economic Cooperation, Marine Sustainable Development Report.* Available at http:// publications.apec.org/publication-etail.php?pub_id=1552.

Behrens, S., Griffin, D., Hayward, J., et al. (2012). *Ocean Renewable Energy: 2015– 2050: An Analysis of Ocean Energy in Australia.* Available at www.marine .csiro.au/~griffin/articles/Ocean-renewable-energy-2015-2050.pdf.

BOEM. (n.d. a). *Ocean Wave Energy.* Available at www.boem.gov/Renewable-Energy-Program/Renewable-Energy-Guide/Ocean-Wave-Energy.aspx.

BOEM. (n.d. b). *Offshore Wind Energy.* Available at www.boem.gov/Renewable-Energy-Program/Renewable-Energy-Guide/Offshore-Wind-Energy.aspx.

Boehnert, J. M. (2016). A new blueprint for coastal zone management. *Natural Resources & Environment,* **30**(4), 52.

Bureau of Oceans and International Environmental and Scientific Affairs (OES). (2012). *An International Vision for Ocean Energy.* Available at www.ocean-energy-systems.org/documents/81542_oes_vision_brochure_2011.pdf.

Burger, M. (2011). Consistency conflicts and federalism choice: Marine spatial planning beyond the states' territorial seas, *Environmental Law Review,* **41**, 10602.

Coastal Zone Management Act. 16 United States Code §§ 1451–66.

Danan, N., (2013). NASA OMEGA Project: The ocean as a platform for biofuel, *PlanetOS,* June 21. Available at https://planetos.com/blog/nasa-omega-project-the-ocean-as-a-platform-for-biofuel.

Danish Energy Agency. (2012). *Energy Policy in Denmark.* Available at www.ens .dk/sites/ens.dk/files/dokumenter/publikationer/downloads/energy_policy_in_ denmark_-_web.pdf.

Deep Decarbonization Pathways Project. (2015). *Pathways to Deep Decarbonization.* Available at http://deepdecarbonization.org/wp-content/uploads/2016/03/DDPP_ 2015_REPORT.pdf.

Deep Sea Water Research Institute. (2016). *Kume Guide.* Available at http:// kumeguide.com/Industry/DeepSeaWater/ResearchInstitute.

Deepwater Wind. (2016). *Block Island Wind Farm Now Fully Financed.* Press Release of Deepwater Wind, July 16. Available at http://dwwind.com/press/ block-island-wind-farm-now-fully-financed.

Department of Energy (DOE). (n.d.). *Algal Biofuels.* Available at http://energy.gov/ eere/bioenergy/algal-biofuels.

Department of Energy (DOE). (2011). *A National Offshore Wind Strategy: Creating an Offshore Wind Energy Industry in the United States.* Available at www .energy.gov/sites/prod/files/2013/12/f5/national_offshore_wind_strategy.pdf.

Department of Energy (DOE). (2014). *Energy Department Announces $10 Million For Wave \ Energy Demonstration at Navy's Hawaii Test Site.* Press Release of US Department of Energy, April 28. Available at http://energy.gov/eere/ articles/energy-department-announces-10-million-wave-energy-demonstration- navy-s-hawaii-test.

Department of Energy (DOE). (2015). *Innovative Wave Power Device Starts Producing Clean Power in Hawaii.* Press Release of US Department of Energy, July 6. Available at http://energy.gov/eere/articles/innovative-wave-power- device-starts-producing-clean-power-hawaii.

Duke University. (2015). *Marine Algae Industrialization Consortium (Magic) Awarded $5.2m DOE Project.* Press Release of Duke University, July 9. Available at http://oceanography.ml.duke.edu/johnson/2015/07/09/arine-algae- industrialization-consortium-magic-awarded-5-3m-doe-project.

Energy Information Administration (EIA). (2016). Offshore production nearly 30% of global crude oil output in 2015. *Today in Energy,* October 25. Available at www.eia.gov/todayinenergy/detail.php?id=28492.

Gerber, L. N., Tester, J. W., Beal, C. M., Huntley, M. E., and Sills, D. L. (2016). Target cultivation and financing parameters for sustainable production of fuel and feed from microalgae. *Environmental Science & Technology.* **50**(7), 3333–41. Available at http://pubs.acs.org/doi/pdf/10.1021/acs.est.5b05381.

Gerrard, M. B. (2017), Legal pathways for a massive increase in utility-scale renew- able generation capacity. *Environmental Law Review,* **47**, 10591.

Goodward, J. and Gonzalez, M. (2010). *Bottom Line on Renewable Energy Tax Credits,* World Resources Institute. Available at www.wri.org/publication/bottom- line-renewable-energy-tax-credits.

Griset, T. J. (2011). Harnessing the ocean's power: Opportunities in renewable ocean energy resources. *Ocean & Coastal Law Journal,* **16**, 395.

Hannon, M., Gimpel, J., Tran, M., Rasala, B., and Mayfield, S. (2010). Biofuels from algae: Challenges and potential. *Biofuels,* **1**(5), 763–84. Available at www .ncbi.nlm.nih.gov/pmc/articles/PMC3152439/pdf/nihms269384.pdf.

Harball, E. (2015). Utility-scale offshore wind proposal floated for Hawaii, *ClimateWire,* March 23. Available at www.eenews.net/climatewire/stories/ 1060015507.

Hicks, J. (2017). *Maryland regulators greenlight two major offshore wind projects. Washington Post,* May 12. Available at www.washingtonpost.com/local/md-politics/ maryland-regulators-greenlight-two-major-offshore-wind-projects/2017/05/12/ 38f045c8-3712-11e7-b373-418f6849a004_story.html?utm_term=.1aaf85ddd6df.

Hodgkins, K. (2015). Hawaii's new OTEC power plant harvests energy stored in warm ocean water. *Digital Trends,* August. Available at www.digitaltrends .com/cool-tech/hawaii-ocean-thermal-energy-conversion.

Huntley, M. E., Johnson, Z. I., Brown, S. L., et al. (2015). Demonstrated large-scale production of marine microalgae for fuels and feed. *Algal Research,* **10**, 249.

International Energy Agency (IEA). (2012). *World Energy Outlook 2012.* Available at http://dx.doi.org/10.1787/weo-2012-en.

International Energy Agency (IEA). (2014a). *Marine Energy Deployment Fund.* Available at www.iea.org/policiesandmeasures/pams/newzealand/name-24171- en.php.

International Energy Agency (IEA). (2014b). *World Energy Outlook 2014*. Available at http://dx.doi.org/10.1787/weo-2014-en.

International Renewable Energy Agency (IRENA). (2014a). *Ocean Thermal Energy Conversion Technology Brief*. Available at www.irena.org/DocumentDownloads/ Publications/Ocean_Thermal_Energy_V4_web.pdf.

International Renewable Energy Agency (IRENA). (2014b). *Wave Energy Technology Brief*. Available at www.irena.org/documentdownloads/publications/ wave-energy_v4_web.pdf.

International Renewable Energy Agency (IRENA). (2014c). *Tidal Energy Technology Brief*. Available at www.irena.org/documentdownloads/publications/ tidal_energy_v4_web.pdf.

International Renewable Energy Agency (IRENA). (2016). *Wind Power Technology Brief* Available at www.irena.org/DocumentDownloads/Publications/IRENA-ETSAP_Tech_Brief_Wind_Power_E07.pdf.

Johnson, Z. I. (2016). Interviewed by C. Reichert (March 3).

Kennedy, R. F. Jr. (2005). An ill wind off Cape Cod, *New York Times*, December 16.

Mauritius Prime Minister's Office. (2013). *The Ocean Economy: A Roadmap for Mauritius*. Available at www.oceaneconomy.mu/PDF/Brochure.pdf.

Maryland Offshore Wind Energy Act of 2013.

Massachusetts General Laws, chapter 25A, § 11F (2015).

National Aeronautical Science Administration (NASA). (n.d.). *OMEGA Project 2009–2012*. Available at www.nasa.gov/centers/ames/research/OMEGA/#.V1y_ZFUrLrc.

National Commission on the British Petroleum Deepwater Horizon Oil Spill and Offshore Drilling. (2011). *Deep Water: The Gulf Oil Disaster and The Future of Offshore Drilling*. Available at www.gpo.gov/fdsys/pkg/GPO-OILCOMMISSION/ pdf/GPO-OILCOMMISSION.pdf.

Navigant Consulting for U.S. Department of Energy. (2014). *Offshore Wind Market and Economic Analysis: 2014 Annual Market Assessment*. Available at http:// energy.gov/sites/prod/files/2014/09/f18/2014%20Navigant%20Offshore%20Wind %20Market%20%26%20Economic%20Analysis.pdf.

National Renewable Energy Laboratory (NREL). (n.d.). *Biofuels Basics.* Available at http://www.nrel.gov/workingwithus/re-biofuels.html.

Organisation for Economic Co-operation and Development (OECD). (2016). *The Ocean Economy in 2030*. Available at http://dx.doi.org/10.1787/9789264251724-en.

Offshore Wind Industry Group. (2013). *Scotland's Offshore Wind Route Map: Developing Scotland's Offshore Wind Industry to 2020 and Beyond*. Available at www.gov.scot/Resource/0041/00413483.pdf.

Offshore Wind Scotland. (2017). Available at www.offshorewindscotland.org.uk.

Paris Agreement. (2016). Annex 1, United Nations Doc. FCCC/CP/2015/10 (Jan. 29, 2016).

Patterson, B. (2017). *Know how to build a floating turbine? A guide is being drafted*, *Energy & Environmental News*, September 25. Available at www.eenews.net/ climatewire/2017/09/25/stories/1060061549.

Powell, T. H. (2012). Revisiting federalism concerns in the offshore wind energy industry in light of continued local opposition to the Cape Wind Project. *Boston University Law Review*, **92**, 2023.

Public Employees for Environmental Responsibility v. *Hopper*, 827 F.3d 1077 (D.D.C. 2016).

Rhode Island Coastal Resources Management Council. (2010). *Rhode Island Ocean SAMP*. Available at http://sos.ri.gov/documents/archives/regdocs/released/pdf/CRMC/6208.pdf.

Schlossberg, T. (2016). America's first offshore wind farm spins to life, *New York Times*, December 14. Available at www.nytimes.com/2016/12/14/science/wind-power-block-island.html?mcubz=1.

Scottish Government. (2013). *Marine Scotland, Wave Energy in Scottish Waters: Initial Plan Framework*. Available at www.gov.scot/resource/0042/00423949.pdf.

Scottish Government. (2015). *Scotland's National Marine Plan: 11. Offshore Wind and Marine Renewable Energy*. Available at www.gov.scot/Publications/2015/03/6517/12.

Seelye, K. Q. (2013). Koch brother wages 12-year fight over wind farm, *New York Times*, October 22. Available at www.nytimes.com/2013/10/23/us/koch-brother-wages-12-year-fight-over-wind-farm.html.

Susskind, L. and Cook, R. (2015). The cost of contentiousness: A status report on offshore wind in the eastern United States. *Vermont Environmental Law Journal*, **33**, 204.

Trent, J. and NASA Ames Research Center. (2012). *OMEGA (Offshore Membrane Enclosures for Growing Algae) – A Feasibility Study for Wastewater To Biofuels*. Available at www.energy.ca.gov/2013publications/CEC-500-2013-143/CEC-500-2013-143.pdf.

UN General Assembly. (2015). Resolution 70/1 (October 21, 2015).

Vega, L. (2010). Economics of ocean thermal energy conversion (OTEC): An update, *Offshore Technology Conference*. Available at http://hinmrec.hnei.hawaii.edu/wp-content/uploads/2010/01/OTEC-Economics-2010.pdf.

Vega, L. (2012). Ocean thermal energy conversion. *Encyclopedia of Sustainability Science and Technology*, **7**, 296–328. Available at http://hinmrec.hnei.hawaii.edu/wpcontent/uploads/2010/01/OTEC-Summary-Aug-2012.pdf.

World Energy Council. (2016). *World Energy Resources*. Available at www.worldenergy.org/wp-content/uploads/2016/10/World-Energy-Resources-Full-report-2016.10.03.pdf.

Yee, A. (2015). Catching waves and turning them into electricity, *New York Times*, April 22. Available at www.nytimes.com/2015/04/23/business/energy-environment/catching-waves-and-turning-them-into-electricity.html.

25

Climate Change and Navies

Bracing for the Impacts

KAPIL NARULA

Introduction

Climate change is one of the greatest challenges of the present time (G20, 2015). In the words of Thomas F. Stocker, co-chairman of the Intergovernmental Panel on Climate Change (IPCC), climate change "threatens our planet, our only home" (Gillis, 2013). The challenge of climate change is unique. The process has complex dynamics, and humans have incomplete and imperfect knowledge about the working of the climate. The global climate is a nonstationary system with a large number of variables, which adds to the complexity of modelling and forecasting climate change. While the effect of climate change is universal, it is nonhomogeneous and differs in scope and severity. On the one hand, there are cascading nonlinear impacts, which increase the threats. On the other hand, there are significant time lags, and the large time constant of the system poses a challenge for observing the results of mitigation action. As the anthropogenic sources of climate change are spread over various sectors of the economy, there are a large number of actors, which increases the complexity of climate governance. Countries tend to act in their self-interest, and their differing capacity and capability add another dimension to the challenge of governing climate change.

This chapter undertakes an assessment of these complicated threats from climate change to national security, focusing particularly on security in the maritime realm. Naval forces will inevitably be forced to respond to the impacts of climate change on oceans. Adapting to climate change to ensure optimal naval preparedness is increasingly becoming the need of the hour. Acknowledging climate change as a threat to naval readiness is the foremost step, which would trigger follow-on policy responses from naval planners. A few measures to counter the impact of climate change on navies are proposed in this chapter. The case of the US Navy is briefly presented to outline some efforts in this direction. The chapter concludes that climate change presents a clear and present danger to the effectiveness of naval forces. Consequently, it is important that appropriate actions are taken by the navies across the world to brace for the impacts of climate change.

Threats of Climate Change to National Security

Climate change has wide-ranging economic, social, and environmental impacts, which transcend geographical and political boundaries and have different implications for different countries. In the United States, a high-level Military Advisory Board, comprising retired three- and four-star flag officers from the Army, Navy, Air Force, and Marine Corps, was convened in 2007 by the Center of Naval Analyses (CNA), Washington, D.C., to study pressing issues which threatened America's national security. The study concluded that "climate change acts as a threat multiplier for instability in some of the most volatile regions of the world" and "projected climate change will add to tensions even in stable regions of the world" (CNA Military Advisory Board, 2007). Due to its implications for US national security, the report recommended that "the consequences of climate change should be fully integrated into national security and national defense strategies." These findings were reinforced by another Military Advisory Board, which reconfirmed that climate change impinges on national security (CNA Military Advisory Board, 2014).

The report of the Defense Science Board Task Force on *Trends and Implications of Climate Change on National and International Security* noted that climate change will only grow in concern for the United States and its security interests. The report offered guidance to the Department of Defense (DoD) on how to become a leader in mitigating and adapting to the growing effects of climate change (Office of the Under Secretary of Defense for Acquisition, Technology and Logistics, 2011). The Pentagon also acknowledged that climate change is an "urgent and growing threat to our national security" and identified the most serious and likely climate-related security risks for the US defense forces. The report, titled *National Security Implications of Climate-Related Risks and a Changing Climate*, examined the ways in which the combatant commands are integrating mitigation of climate-related risks into their planning processes and the resources required for an effective response (US Department of Defense, 2015).

The impact of climate change on defense preparedness has been acknowledged at the highest level in the United States. The Quadrennial Defense Review (QDR) is a legislatively mandated review of the US DoD priorities that is undertaken every four years. It assesses the threats and rebalances DoD's strategies and force levels for fulfilling US national security needs by identifying military capabilities that could contribute to overcoming those threats (US Department of Defense, 2017). QDR 2010 first acknowledged the threat of climate change on military operations and stated that "climate change and energy will play significant roles in the future security environment," and the effect of climate change and its complex interplay with other stressors "may spark or exacerbate future conflicts" (US Department of Defense, 2010). QDR 2014 re-emphasized the impact of climate change on both military missions and installations. The document noted that "[t]he impacts of climate change may increase the frequency, scale, and complexity of future missions, including defense support to civil authorities, while at the same time undermining the capacity of our domestic installations to support training activities." It also noted that "in many areas, the projected impacts of climate change will be more than threat multipliers; they will serve as catalysts for instability and conflict" (US Department of Defense, 2014b).

The implications of climate change on military operations have also been acknowledged by the United Kingdom (UK Ministry of Defense, 2012), Australia (Barrie et al., 2015), the European Union (Council of the European Union, 2013), NATO (NATO Parliamentary Assembly, 2015), the Organization for Security and Co-operation in Europe (Maas et al., 2010), and other organizations (King, 2014). Furthermore, in the run-up to the Paris climate change conference in 2015, an international conference on the "Implications of Climate Change for Defense" was convened where authorities and defense experts from 35 countries, the UN, the African Union, and the EU discussed the consequence of climate change on the armed forces. In his statement to the conference, UN Secretary-General Ban Ki-moon warned that "climate change is a threat multiplier and can both exacerbate conflicts within and between States, as well as threaten international peace and security" (United Nations Secretary-General, 2015).

The fifth Assessment Report (AR5) of the IPCC has analyzed the widespread impacts of climate change. These ranged from recurring flooding, drought, temperature extremes, frequent and more severe extreme weather events, sea-level rise, and so forth (IPCC, 2014a). Apart from having direct impacts, events triggered by climate change exacerbate threats to food, water, and energy security, and they have been observed to affect the society at micro and macro levels. Natural calamities triggered by climate change disrupt supply and distribution chains and create resource shortages, triggering social unrest and thereby endangering national security. The armed forces of a country have the responsibility of strengthening internal security and hence need to be prepared to respond as well as withstand the impacts of climate change. Table 25.1 identifies various events that may occur due to climate change and lists their likely impact on military forces.

Likely Impacts on Naval Forces

For the purpose of this chapter, naval forces include national navies, coast guards, and marine corps. These services perform various roles, such as providing maritime security, maritime law enforcement, search and rescue (SAR), humanitarian assistance/disaster relief (HA/DR), and environmental protection in the maritime domain. While there are specialized agencies for responding to disasters on land, very often it is observed that at sea these responsibilities spill over to naval forces. Specific impacts of climate change on naval forces are described next.

Reduced Naval Preparedness

The impact of climate change undermines military readiness and limits the ability of naval forces to perform their functions. Climate change–related extreme events

Table 25.1. *Climate change events and their likely direct impact on military forces*

Event	Likely Direct Impact on Military Forces
Regional Warming	None in the short term
Heat Extremes	Larger cooling requirement for equipment, changes in equipment design, constraints on continuous operational exploitation of equipment, possible excess personnel mortality due to heat-related stress
Precipitation (intraseasonal variability, extremes)	Risk of flooding, inundation, damage to infrastructure
Drought/decrease in water availability	Loss of operational/training time; additional effort to mitigate and overcome water scarcity
Lower groundwater recharging	Impact on forward areas and temporary facilities
Sea-level rise (SLR)	Risk of flooding of coastal installations, permanent loss of land
Tropical cyclones	Damage to infrastructure, inundation of land, large effort in evacuation and relocation
Flooding (sea-level rise and storm surges)	Economic damages, loss of capacity to undertake operations

lead to a breakdown in communication and disrupt naval supply chains. The capability to transport personnel and material is often affected, which compromises the ability to undertake operations in a timely and effective manner. This was observed in October 2014 when Cyclone Hudhud hit the city of Vishakhapatnam and caused massive damages to infrastructure in the Indian navy's largest base on the eastern seaboard. It was estimated that the loss from damage to the infrastructure in the Eastern Naval Command was around 300 million USD (Bureau Report, 2014). While there was no damage to the ships and submarines, the cyclone caused extensive damage to the infrastructure in the naval dockyard and community areas. Communication and power lines were damaged, and availability of essential commodities like water, petrol, and electricity was disrupted for several days, resulting in a complete breakdown in the functioning of the naval base. A similar loss of life and ability to undertake operations was observed in 2013 when the super cyclone Haiyan devastated the Philippine navy's Tacloban naval base (First Post World, 2013). It was reported that ships refused to enter the harbor for many hours after the winds had subsided, delaying the rescue efforts (Singh, 2015).

Climate change can also affect supply chains for naval marine forces that are undertaking operations in faraway lands. Droughts, flooding, and extreme precipitation in areas where naval forces are involved in warfighting missions can lead to severe stress on water and food supply lines, affecting the mission directly.

Risks to Coastal Naval Infrastructure

Sea-level rise (SLR) presents a major threat to coastal naval infrastructure. AR5 predicts that global mean sea level will continue to rise during the twenty-first century. The report forecasts that by the end of the twenty-first century, it is very likely that sea level will rise in more than about 95 percent of the ocean area, and about 70 percent of the coastlines worldwide would experience a sea level change within ±20 percent of the global mean. While the projected rates of SLR vary across studies and depend on the concentration of CO_2 in the atmosphere, IPCC forecasts that the global average sea level could rise by nearly 1 meter by 2100, relative to a 1986–2005 baseline in the high CO_2 concentration case (IPCC, 2014b). An important point to note is that this is the global average and the SLR would be different across regions and would threaten coastal naval infrastructure nonuniformly.

The consequence of SLR on coastal naval installations is a function of local topology, tidal amplitudes, extremities of storm surges, and effectiveness of flood prevention infrastructure. Nevertheless, irrespective of the location of the base around the world, it is certain that the naval establishments would be more vulnerable and would face additional risks to infrastructure. While the increase in SLR and its effect along the coast would be gradual, its impact would be amplified during extreme precipitation events, cyclones, and storm surges. It is very likely that events related to climate change would disrupt operations in the short to medium term and would have large economic costs. As the frequency and the strength of cyclones and storms are forecasted to increase in the coming years, failure to build a resilient and climate-proof naval infrastructure in the face of the threat of climate change could affect the ability of the navy to undertake naval missions and to mobilize adequate force levels.

Increased Risk to Personnel

Increasing intensity of storms presents a risk to the life of personnel. It is reported that in October 2016, US military cargo planes evacuated about 700 family members of troops and other staff from the naval base in Guantánamo Bay, Florida, ahead of Hurricane Matthew (Rosenberg, 2016). A more recent example of the increased vulnerability of naval personnel to extreme weather events is the evacuation of over 5,000 personnel from Key West, a Naval air station in Florida, in anticipation of the disastrous impact of Hurricane Irma, a Category 5 storm that hit the east coast of the United States in early September 2017 (Press TV, 2017). Mandatory orders were issued to evacuate active-duty service members, civilians, and their families from the station, leaving behind a skeleton crew of approximately 50 to 60 personnel for manning essential functions at the installation. It was reported by CNN that aircraft, including fixed-wing planes and helicopters, were likely to be moved inland from Jacksonville and Mayport, Florida, and submarines were also preparing to evacuate from naval submarine base Kings Bay in Georgia (CNN, 2017).

Indirect and Uncertain Fallouts

Climate change may lead to severe stress on the availability of food and water in vulnerable societies. Changes in precipitation and temperature may affect crop yields, and it is postulated that the extreme drought in large parts of Syria from 2006 to 2011 was exacerbated by climate change. This drought, accompanied by poor governance by the incumbent Assad regime, led to internal displacement of around two million people and is cited as one of the reasons that contributed to the civil war (Kelley et al., 2015). The internal strife and fighting in Syria led to an unprecedented inflow of refugees into Europe from land as well as maritime routes. The European migrant crisis has seen the inflow of more than 2.5 million asylum seekers in 2015 and 2016 from Syria, Eritrea, Afghanistan, Kosovo, Nigeria, Iraq, Senegal, Gambia, and Mali and has been a major cause of concern to the EU (Bogdan and Fratzke, 2015). Small-island states like Kiribati and the Maldives also face an existential threat from SLR apart from the threat of cyclones, which could trigger a deluge of climate refugees in the future.

In order to contain large-scale migration across the maritime domain and to prevent the loss of life at sea due to sinking of unseaworthy boats used by migrants, the EU set up a "Task Force Mediterranean" in 2013. Apart from augmenting border surveillance, the task force has contributed to enhancing maritime domain awareness, as well as to fighting against trafficking, smuggling, and organized crime (European Commission, 2013). This increased maritime patrolling has now transformed into a full-time commitment of naval assets. North Atlantic Treaty Organization (NATO) maritime forces have also joined the efforts of the national governments of Turkey and Greece and the EU's FRONTEX agency to conduct reconnaissance, monitoring, and surveillance of the maritime domain in response to the growing migration from the sea (North Atlantic Treaty Organization, 2016).

Weak maritime law enforcement led to illegal fishing and toxic waste dumping at sea, thereby lowering fish catch in coastal waters off Africa. These have, in part, been identified as causes that contributed to the rise of Somali fishermen embracing piracy in the Gulf of Aden and the Gulf of Guinea, the Indian Ocean, and off the coast of the Horn of Africa between 2005 and 2015 (Zalmanowicz, 2014). This led to the permanent international presence of warships from Japan, China, India, Australia, and other countries for escorting the national fleet of merchant ships transiting through the region.

Piracy is also linked to terrorist funding (Zlutnick, 2009). The Combined Maritime Forces (CMF), which is a multinational naval partnership of 32 member nations, has been formed to promote maritime security. The CMF is composed of three task forces: CTF 150: Maritime security and counter-terrorism, CTF 151: Counter piracy, and CTF 152: Arabian Gulf security and cooperation (Combined Maritime Forces, 2017). Its focus areas are defeating terrorism, preventing piracy, encouraging regional cooperation, and promoting a safe maritime environment apart from responding to environmental and humanitarian crises.

These naval mobilizations have a huge financial as well as an operational cost for countries and strain military resources, apart from diverting the attention of the navies from national security missions. Such continued operations also have strategic implications and lead to militarization of the region. The setting up of a permanent Chinese naval base in Djibouti to support the People's Liberation Army (PLA) Navy presence in the area is a recent example of how poor maritime governance and uncertain fallouts can lead to situations that can fundamentally alter the geopolitics of a region.

The Changing Role of Naval Forces in Response to Climate Change

Apart from lowering military readiness and increased risks to life and infrastructure, climate change could lead to changes in naval operations and strategy. The location of climate change–induced disasters cannot be predicted with accuracy, which introduces a large uncertainty. It is therefore paramount that the naval forces, irrespective of their location, insulate themselves against the impact of climate change and are well prepared to undertake operations at a short notice. These operations, which are spread over a large spectrum, vary according to the capability of the naval forces of the respective country.

Change in Role Leading to Increased Strain on Resources

As a part of a country's armed forces, the navy performs various roles such as offensive, defensive, strategic, diplomatic, and benign (Indian Navy, 2017). The multidimensional capability of the naval forces at sea, below the sea surface, and in the airspace above sea enables it to undertake SAR operations in the maritime domain. Apart from this, naval forces are often the first responders to provide HA/DR, as they can be quickly mobilized and have a large outreach as well as endurance to undertake such tasks. The mobility of naval assets such as ships, landing crafts, and boats, coupled with reliable communications and a disciplined work force, help in restoring law and order and in providing immediate relief to the disaster-afflicted areas. International maritime forces have also cooperated in tandem with civilian or UN-sponsored agencies to provide disaster relief by distributing food, water, and relief material, especially in island states. With a forecasted increase in the frequency as well as the severity of climate-related disasters, there is a strong possibility that naval forces across the world would be increasingly called upon to help civilian authorities in undertaking relief operations. This additional role is likely to limit their ability to respond to military contingencies and would lead to an increased strain on resources.

Strategic Changes

Climate change is a "threat multiplier" and may increase instability in strategically significant regions of the world. Threats to water and food security, increases in glacial melt and flooding, rainfall variability, coastal inundation, and depleting fish stocks in areas where maritime boundaries have not been delineated can act as stressors leading to increased tensions and instability in maritime regions. This can increase the likelihood of naval forces being drawn into skirmishes at sea due to the increased presence of naval assets and face-offs in the affected areas.

Climate change has a visible impact on reducing sea ice in the Arctic, and the region has emerged as an area of increasing strategic significance. Melting sea ice has led to the possibility of opening up year-round sea routes in the Arctic Ocean, apart from increasing the feasibility of commercial exploitation of maritime and seabed resources. This has led to increased geostrategic concerns, especially among the Arctic member countries. A regular commercial sea route in the Arctic would open up the possibility of accidents in the harsh environment, and it is likely that SAR missions in the Arctic would have to be supported by naval ships in the Arctic Ocean (see Chapter 14).

The Russian navy is increasing its influence in the Arctic and is building capabilities to support operations in the region. Russia launched the world's biggest nuclear-powered icebreaker, *Arktika*, in mid-2016, which is the first in a series of three ships (Hambling, 2016). In 2016, it also launched the *Ilya Muromets,* an icebreaking ship that will exclusively be used by the Russian navy and is expected to join the Northern Fleet in 2017 (Koreneva and Rotenberg, 2016). The Arctic region figures prominently in Russia's military doctrine, and it has plans to refurbish its Arctic air force. Construction of two corvettes, which are designed for use in the Arctic region and are capable of acting as ice breakers, is also in the pipeline.

The US Navy expects the Arctic "to remain a low threat security environment where nations resolve differences peacefully" and has a roadmap for the Arctic region. It anticipates its role as a supporter of US Coast Guard (USCG) operations and to provide SAR and HA/DR assistance. The US Coast Guard currently has three icebreakers and is planning to add more in order to close the "icebreaker gap" with the Russians in the Arctic (Schwartz and Stavridis, 2016). The Canadian navy is also planning to maintain a more sizeable presence in the Arctic waters. The presence of naval assets is likely to increase tensions in the polar region, which would then demand suitable changes in the naval strategies of stakeholder countries.

Multifunctionality and Force Structure

Considering the geographical scale and the severity of climate disasters, naval forces are likely to be frequently involved in relief operations. Thus, there will be a requirement for ships to perform multiple roles, such as SAR, large-scale evacuation, humanitarian assistance, and disaster relief to support these relief missions. Ships would therefore need to be built in a way that allows flexibility in undertaking different operations with minor modifications. Specialized hospital ships may have to be constructed, manned, and operated on a semi-permanent or permanent basis by larger navies to support relief operations. In the long run, specific capabilities may need to be added to the existing fleet to perform the earlier noted functions. This may alter the structure of the naval forces. Additionally, personnel need to be trained for undertaking missions such as large-scale evacuation; providing relief, medical assistance and distribution of aid; and the like. While such missions, when conducted in faraway lands, would add to the power projection capability and will lead to an increase in the soft power of the country, they will also rob the naval forces of critical manpower as well as material resources.

Acknowledging Climate Change as a Threat to Naval Readiness

Notwithstanding the scientific evidence that the IPCC presented in its fifth Assessment Report (AR5), the world still has a large number of climate change deniers. Despite the increasing evidence of the disastrous impacts of climate change, the science of climate change continues to be questioned. This is a grave challenge, as it restricts policy action and prevents allocation of resources to build climate resilience. But is there a strong case for taking action against increased threats from climate change for naval forces? Almost all countries in the world supported the Paris climate agreement. Hence, it is implied that they unanimously agree with the threats posed by climate change, the need to take action for adapting to climate change, and the proposed measures to mitigate climate change. While this acknowledgment provides proof that countries take climate change seriously, it is hard to see why this thinking has not percolated in military circles. This question therefore needs careful examination, as there is not enough evidence to show that countries are considering the impact of climate change on naval strategy and operations.

It is acknowledged that there is a large uncertainty in the range of possible impacts of climate change, as well as the location and the type of climate change–induced disaster. The location and nature of future disasters cannot be predicted in the medium to long term. However, it is important to note that this uncertainty does not lower the impact of climate change or the associated risk. On the contrary, it increases the threat perception. Further, as the severity and the scale of impact cannot be forecasted, the time to respond to short-term warnings is inadequate, leaving the naval assets unprotected to the vagaries of nature.

Convincing the naval leadership, both civilian and military, to take action against climate change is one of the foremost challenges today. One of the main arguments that is presented for not taking action is that we may not have complete information about the next climate change disaster. Second, the high cost of climate-proofing naval stations leads to the problem of "asymmetric risks vs costs." While the dilemma is acknowledged, the question that needs to be answered is: Do we have "sufficient" information to build a case for naval forces to act against climate change? The answer here is subjective and depends on the perception and the understanding of the military leadership to the threats that climate change poses to naval readiness. A deeper understanding of the long-term, multidecadal commitment, which is required to mitigate and adapt to climate change, is also essential.

Notwithstanding the earlier comments, it is evident that climate change poses a risk to naval preparedness and affects naval operations. Therefore, efforts are needed to integrate the risks from climate change in military as well as naval strategy. The first step in this direction would be to accept that the threat of climate change for navies is real. An acknowledgment by the top naval leadership in uniform, as well as their civilian counterparts in ministries of defense, would go a long way in preparing the ground for further action. The next step would be to identify the vulnerabilities and to assess the impact of climate change on naval installations, assets, operations, and effectiveness. As a part of this vulnerability assessment, the threats to naval infra-structure could be evaluated by using models and mapping tools for high-risk installa-tions, and a contingency plan to respond to these risks needs to be put in place. As an example, the US DoD had ordered an evaluation of all US defense installations across the world to identify the vulnerability of military bases to climate change. Such an assessment can then become a starting point for in-depth analysis and action.

Mitigation of risks and adaptation form the next layer of action. Protection of coastal bases from SLR and storm surges by construction of floodwalls and addi-tional infrastructure for protection needs to be assessed. Plans for new installations should be re-examined so as to incorporate the impact of SLR on naval installa-tions. For example, placement of generator rooms and electrical accessories in areas that are prone to flooding such as basements and below the water line on jetties should be avoided. The ability of a naval base to recover rapidly when damage occurs needs to be augmented. Backup facilities for electricity, water, communica-tions, firefighting, and pumping out arrangements would have to be installed, or the existing ones need to be weather-proofed.

The last aspect would be to increase the response capacity and resilience of the naval forces by addressing the full spectrum of climate change issues and their pro-jected impacts. Scenario analysis and conducting tabletop exercises to undertake large-scale disaster relief operations with imposed constraints, such as reduced avail-ability of naval ships and personnel, may be undertaken to evaluate the impact of climate change on the effectiveness of missions.

In order to have a structured response and to ensure long-term capacity develop-ment, the aforementioned aspects need to be included in planning and policy docu-ments. While strengthening resilience to climate impacts is critical, naval forces should also contribute to lowering their carbon footprint in the long term. Mitigation efforts such as incorporating energy efficiency in naval operations, energy conservation, and installation of renewable energy plants in naval bases would avoid further buildup of atmospheric CO_2. Greening naval operations will also have co-benefits, such as lowering of energy costs, increasing the range of ships at sea, and reducing the number of refueling stops at sea. When accompanied by improve-ments in resilience, these measures will go a long way in responding to the impact of climate change in the long run.

The last argument to settle in the question of incorporating climate change in naval strategy is the question of "when is the right time?" It is a question of "when" and not "if." Early action equals better preparedness, and the previously mentioned efforts would go a long way in equipping the naval forces with the science, tools, and mechanisms to respond to the impacts of climate change so as to maintain their effectiveness.

The Case of the US Navy

The United States, which has the largest and the most powerful navy in the world, is playing a leadership role in adapting to climate change and in its mitigation. It has undertaken various measures to address climate change in its defense strategy by integrating climate change considerations into its plans and operations (Reinhardt and Toffel, 2017). The Department of the Navy: Energy, Environment and Climate Change is the nodal agency that coordinates the US Navy's and Marine Corps' energy, environmental, and climate change programs. The US Department of Navy (DoN) realized that climate change has many implications for naval operations. It identified various drivers for change, such as the changing Arctic, the potential impact of SLR and changing storm patterns on naval installations and plans, and the increasing role of the Navy in providing humanitarian assistance, among others (Naval Studies Board Division on Engineering and Physical Sciences, 2011). Considering the growing impact of climate change on naval operations, the Chief of Naval Operations (CNO) created Task Force Climate Change (TFCC) in May 2009 to "address the naval implications of a changing Arctic and global environment" and "to make recommendations to Navy leadership regarding policy, investment, action, and to lead public discussion" on climate change (Department of the Navy, 2017).

One of the main impacts of climate change is the melting of sea ice in the Arctic and the increase of maritime activity in the region. This has opened up a new maritime frontier for the US DoN in the region, a development with far-reaching implications. The US Navy's TFCC released the Navy Arctic Roadmap in 2009, which was updated in 2014. The *US Navy Arctic Roadmap, 2014–2030* outlined the Navy's strategic approach for the Arctic Ocean and includes an implementation plan to support the desired end states. The roadmap examines various aspects of the US Navy's perspective of the region as a theater for operations and outlines the ways and means for near-term (until 2020), mid-term (2020 to 2030), and far-term operations (beyond 2030). Specific actions for operating safely in the Arctic include specialized training; additional measures for safe navigation; undertaking Command, Control, Communications, Computers, Intelligence, Surveillance, and Reconnaissance operations; additional infrastructure and facilities; modifications to platforms, weapons, support equipment, and sensors; enhanced maritime domain awareness; and so forth (US Navy Task Force Climate Change, 2014). The document also acknowledges that "changes in the environment must be continuously examined and taken into account."

Over the years, the US defense strategy has referred to climate change as a "threat multiplier." It has acknowledged that a changing climate will affect the ability of the DoD to defend the nation and will have a real impact on the military and the way it executes its missions. With an aim to address the growing concerns of climate change, the DoD first released the "FY 2012 Climate Change Adaptation Roadmap" and its updated version, the "FY 2014 Climate Change Adaptation Roadmap," which identifies various actions and plans to increase the resilience of the US DoD to the impacts of climate change.

Three broad goals were established in the roadmap: identify and assess the effects of climate change on the DoD, integrate climate change considerations across the DoD and manage associated risks, and collaborate with internal and external stakeholders on climate change challenges. The potential effects of climate change on the DoD were identified under four main groups: plans and operations, training and testing, built and natural infrastructure, and acquisition and supply chain. Drawing from the assessment of more than 7,000 bases, installations, and other facilities which was conducted across the world to assess the vulnerability of US bases, the roadmap laid out the way ahead for integrating climate change considerations into the DoD's plans, operations, and training for managing associated risks. The DoD is also considering the impacts of climate change in war games and defense planning scenarios, along with working with the combatant commands to address the impacts of climate change in their areas of responsibility (US Department of Defense, 2014a).

While future actions for adaptation and mitigation by the US naval forces are under a cloud due to the Trump administration's decision to withdraw from the Paris climate change agreement, it is evident that the impact of climate change is becoming more and more visible. Slowing down or no further action against climate change would erode the base that has painstakingly been built up by the US DoD. It is therefore important that the momentum is maintained and adaptation plans are regularly upgraded to account for dynamic changes in the climate.

Conclusion

Climate change is one of the greatest threats facing societies. The impact of climate change on oceans is widespread and is potentially catastrophic for life on Earth. There is sufficient evidence to conclude that climate change exacerbates threats to national security and poses direct and indirect threats to the military. The immediate threats from climate change to naval forces are increased risk to coastal infrastructure from sea-level rise, storm surges and inundation, and risk to personnel. The changing role of the navy in response to climate change has been observed in recent years. It is very likely that naval forces across the world will be called on to undertake humanitarian assistance and disaster relief missions to support civilian agencies, especially in island states. The nature of naval operations may undergo strategic and tactical changes, and the number and type of naval assets may have to be reworked while incorporating multifunctionality into platform design.

As the severity and the scale of climate-related disasters cannot be predicted with certainty, it leaves little reaction time to prepare for remedial action. The widespread threat and the potentially large impact introduce asymmetrical costs, which bring an additional level of complexity. The meagre allocation of resources for climate adaptation adds to the existing challenges. Nevertheless, building resilience and adaptation capacity is essential for maintaining the effectiveness of naval forces. Early action equals better preparedness, and therefore countries that take immediate measures would be better equipped to brace for the impact of climate change. Some specific measures to prepare the naval forces for the impact of climate change have been proposed, and it is up to individual countries to examine them in their specific context. The case of US Navy and the efforts that it has taken to improve its readiness to respond to the impact of climate change is an example of a way ahead for other navies. Navies across the world can be better prepared to brace for the impact of climate change if they acknowledge the threats it poses and take timely action to build resilience and adaptation capacity.

References

Barrie, C., Steffen, W., Pearce A., and Thomas, M. (2015). *Be Prepared: Climate Change, Security and Australia's Defense Force.* Climate Council of Australia Limited.

Bogdan, N. B. and Fratzke, S. (2015). *Europe's Migration Crisis in Context: Why Now and What Next?* Migration Policy Institute. Available at www.climate council.org.au/uploads/fa8b3c7d4c6477720434d6d10897af18.pdf.

Bureau Report. (2014). Hudhud Impact: Navy's ENC suffers ₹2,000-cr loss. *The Hindu Business Line.* Available at www.thehindubusinessline.com/news/national/hudhud-impact-navys-enc-suffers-2000cr-loss/article6512272.ece.

CNA Military Advisory Board. (2007). *National Security and the Threat of Climate Change.* Alexandria, VA: The CNA Corporation.

CNA Military Advisory Board. (2014). *National Security and the Accelerating Risks of Climate Change.* Alexandria, VA: CNA Corporation.

CNN. (2017). US Navy to evacuate 5,000 as military preps for Hurricane Irma. September 7. *CNN Politics.* Available at http://edition.cnn.com/2017/09/05/politics/key-west-navy-evacuation-hurricane-irma/index.html.

Combined Maritime Forces. (2017). *About Combined Maritime Forces (CMF).* Available at https://combinedmaritimeforces.com/about/.

Council of the European Union. (2013). *Council Conclusions on EU Climate Diplomacy. Foreign Affairs Council Meeting Luxembourg.* June 24. Available at https://climateandsecurity.org/2013/07/03/european-attention-to-the-security-implications-of-climate-change/#more-3323.

Department of the Navy. (2017). *Climate Change.* Available at http://greenfleet.dodlive.mil/climate-change/.

European Commission. (2013). *Communication from The Commission to The European Parliament and The Council on the Work of the Task Force Mediterranean.* 4.12.2013 COM (2013) 869. Brussels: European Commission. Available at https://ec.europa.eu/home-affairs/sites/homeaffairs/files/what-is-new/news/news/docs/20131204_communication_on_the_work_of_the_task_force_mediterranean_en.pdf.

First Post World. (2013). Typhoon Haiyan: Tacloban worst hit, Philippines pleads for aid, First *Post World,* November 11. Available at www.firstpost.com/world/typhoon-haiyan-tacloban-worst-hit-philippines-pleads-for-aid-1223257.html.

G20. (2015). *G20 Leaders' Communiqué Antalya Summit*, November 15-16, 2015. Available at http://g20.org.tr/g20-leaders-commenced-the-antalya-summit/.

Gilis, J. (2013). U.N. Climate Panel endorses ceiling on global emissions, *New York Times*, September 27. Available at www.nytimes.com/2013/09/28/science/global-climate-change-report.html.

Hambling, D. (2016). Russia built a big, bad nuclear-powered icebreaker to win the Arctic. *Popular Mechanics*. Available at www.popularmechanics.com/military/navy-ships/a21484/russia-nuclear-powered-icebreaker/

Indian Navy. (2017). *Role of Navy*. Available at www.indiannavy.nic.in/content/role-navy.

IPCC. (2014a). Summary for policymakers. In C. B. Field, V. R. Barros, D. J. Dokken, et al., eds., *Climate Change 2014: Impacts, Adaptation, and Vulnerability. Part A: Global and Sectoral Aspects. Contribution of Working Group II to the Fifth Assessment Report of the Intergovernmental Panel on Climate Change*. Cambridge, UK: Cambridge University Press, 1–32.

IPCC. (2014b). Climate Change 2014: Synthesis Report. Contribution of Working Groups I, II and III to the Fifth Assessment Report of the Intergovernmental Panel on Climate Change [Core Writing Team, R. K. Pachauri and L. A. Meyer (eds.)]. Geneva, Switzerland: IPCC.

Kelley, C., Mohtadi, S., Cane, M., Seager, R., and Kushnir, Y. (2015). Climate change in the Fertile Crescent and implications of the recent Syrian drought. *PNAS*, 112(11), 3241–6.

King, W. C. (2014). *Climate Change: Implications for Defense – Key Findings from the Intergovernmental Panel on Climate Change Fifth Assessment Report*. European Climate Foundation, Global Military Advisory Council on Climate Change, Institute for Environmental Security, and the University of Cambridge's Institute for Sustainability Leadership.

Koreneva, M. and Rotenberg, O. (2016*). Russia unveils new navy icebreaker in Arctic military focus. Yahoo News*. Available at http://gmaccc.org/wp-content/uploads/2014/06/AR5_Summary_Defence.pdf.

Maas, A., Briggs, C., Cheterian, V., et al. (2010). *Shifting Bases, Shifting Perils: A Scoping Study on Security Implications of Climate Change in the OSCE Region and Beyond*. Berlin: Adelphi Research gemeinnützige GmbH.

NATO Parliamentary Assembly. (2015). *Resolution 427 on Climate Change and International Security*. Available at climateobserver.org/wp-content/uploads/2015/10/193-stc-15-e-rev.-1-bis-climate-change-and-international-security.pdf.

Naval Studies Board Division on Engineering and Physical Sciences. (2011). *National Security Implications of Climate Change for U.S. Naval Forces*. Washington, DC: The National Academies Press.

North Atlantic Treaty Organization. (2016). *Assistance for the Refugee and Migrant Crisis in the Aegean Sea*. June 27. Available at www.nato.int/cps/en/natohq/topics_128746.htm.

Office of the Under Secretary of Defense for Acquisition, Technology and Logistics (2011). *Report of The Defense Science Board Task Force on Trends and Implications of Climate Change for National and International Security*. Washington, DC: Defense Science Board.

Press TV. (2017). US Navy evacuating 5,000 as it prepares for Hurricane Irma, September 6. *Popular Military*. Available at http://popularmilitary.com/us-navy-evacuating-5000-prepares-hurricane-irma/.

Reinhardt, F. L. and Toffel, M. W. (2017). *Managing climate change: Lessons from the U.S. Navy. Harvard Business Review.* Available at https://hbr.org/2017/07/managing-climate-change.

Rosenberg, C. (2016). Guantánamo Navy base begins evacuating 700 parents and kids, plus pets, ahead of Hurricane Matthew. *Miami Herald.* October 1. Available at www.miamiherald.com/news/nation-world/world/americas/guantanamo/article105393586.html.

Schwartz, N. A. and Stavridis, J. G. (2016). Russia unveils new navy icebreaker in Arctic military focus. *The Wall Street Journal.* Available at www.wsj.com/articles/a-quick-fix-for-the-u-s-icebreaker-gap-1454542242.

Singh, A. (2015). Climate change and maritime security in the Indian Ocean region. *Journal of Defence Studies*, **9**(1), 63–82.

UK Ministry of Defense. (2012). *Defense Infrastructure Organization Estate and Sustainable Development.* Available at www.gov.uk/guidance/defence-infrastructure-organisation-estate-and-sustainable-development#mod-climate-change-strategy.

United Nations Secretary-General. (2015). *Secretary-General's Message to International Conference on the Implications of Climate Change for Defense.* Available at www.un.org/sg/en/content/sg/statement/2015-10-14/secretary-generals-message-international-conference-implications.

US Department of Defense. (2010). *Quadrennial Defense Review Report 2010.* Available at https://dod.defense.gov/Portals/1/features/defenseReviews/QDR/QDR_as_of_29JAN10_1600.pdf.

US Department of Defense. (2014a). *2014 Climate Change Adaptation Roadmap.* Available at www.acq.osd.mil/eie/downloads/CCARprint_wForward_e.pdf.

US Department of Defense. (2014b). *Quadrennial Defense Review Report 2014.* Available at http://archive.defense.gov/pubs/2014_quadrennial_defense_review.pdf.

US Department of Defense. (2015). *National Security Implications of Climate-Related Risks and a Changing Climate.* 8-6475571. Available at https://fas.org/man/eprint/dod-climate.pdf.

US Department of Defense (2017). *Quadrennial Defense Review.* Available at www.defense.gov/News/Special-Reports/QDR/.

US Navy Task Force Climate Change. (2014). *The United States Navy Arctic Roadmap for 2014 to 2030.* Available at www.navy.mil/docs/USN_arctic_roadmap.pdf.

Zalmanowicz, B. (2014). The rise and fall (?) of Somali pirates. *IHLS.* Available at https://i-hls.com/archives/27056.

Zlutnick, D. (2009). Sea bandits: Poverty, business and the rise of Somali piracy. *The Nor Easter.* Issue no. 4. Available at http://noreaster.neanarchist.net/issue4/article4.html.

26

Geoengineering at Sea
Ocean Fertilization as a Policy Option

JOHN T. OLIVER AND STEVEN M. TUCKER

Introduction

Human-induced climate change has emerged as one of the greatest challenges the world has ever faced. The most recent reports of the Intergovernmental Panel on Climate Change document global vulnerabilities and discuss the need to make huge investments and fundamental changes to mitigate adverse future impacts (IPCC, 2013, 2014). These reports make several brief mentions of "geoengineering," specifically carbon dioxide removal from the atmosphere, as a possible strategy for mitigating climate change (see, e.g., IPCC, 2014, Box 3.3, 89, 119, 123). However, they do not discuss the potential geoengineering strategy of *ocean fertilization* to enhance phytoplankton populations, thereby increasing carbon uptake in the world's oceans. The 2014 US National Climate Assessment includes a brief mention of ocean fertilization as a potential remedial measure (see US Global Change Research Program, 2014: 651).

While engineering approaches to enhance environmental resilience must be scientifically established as environmentally safe and cost-effective, it is time to decide which remedial approaches to the challenges of climate change humankind should pursue in earnest. Ocean fertilization should be given full consideration as one of these potential approaches. In our view, ocean fertilization could be employed in conjunction with other global efforts to reduce greenhouse gases in the atmosphere, such as expanding the use of renewable energy sources; protecting and expanding forests, kelp beds, and other carbon sinks; and employing additional mitigation and adaptation measures to enhance climate resilience and restore global environmental equilibrium.

This chapter seeks to raise awareness of the potential of ocean fertilization as a major part of the solution to climate change and declining ocean productivity. The concept of ocean fertilization warrants further scientific research, investment, and, if determined to be environmentally safe and cost-effective, large-scale implementation in suitable ocean regions through appropriate international governance.

The Vital Role of Phytoplankton in the Carbon Cycle

Scientists have established a very close relationship between global temperatures, sea levels, atmospheric carbon dioxide, and ocean acidification over the past 400,000 years and longer. During most of that period, the level of carbon dioxide in the atmosphere hovered around 280 parts per million (ppm). Since the beginning of the Industrial Revolution, however, it has increased rapidly, now exceeding 400 ppm (NOAA, 2017). This increase is unprecedented in Earth's geologic history. There is a nearly universal scientific consensus that the only credible explanation for such a dramatic rise is the burning of fossil fuels in increasingly prodigious quantities (IPCC, 2013, 2014). The world's ocean waters have absorbed much of this increase (Abraham, 2017).

Phytoplankton, such as diatoms, play a vital role in nature at the bottom of the oceanic food chain. These tiny plants produce half of the world's oxygen supply (Roach, 2004). However, there is growing evidence that global levels of phytoplankton in the ocean have declined markedly over the past century, reducing the ability of the ocean to remove carbon dioxide from the atmosphere and to supply oxygen, in the process undermining the ocean-based food chain and aggravating ocean acidification (Boyce et al., 2010; Marshall et al., 2017). Massive harvesting efforts and the reduction in the abundance of fish, sea birds, and marine mammals from the global ocean ecosystem in recent decades detract from the role marine life plays in the ocean's nutrients cycle. As a result, the exchange of nutrients between the ocean depths and sea surface, fertilizing phytoplankton and powering the ocean as a natural carbon sink, is increasingly impeded. If marine mammals, birds, and other such sea animals were to make a sustained global comeback, this vital natural carbon sink would be enhanced (MacKenzie, 2011).

Given adequate sunlight, water, space, and nutrients, phytoplankton populations in surface waters will grow and reproduce almost exponentially until a constraint, such as nutrient availability, is reached (NASA, 2010). During this process, phytoplankton consume carbon dioxide and other nutrients, mature, reproduce, and create oxygen (NASA, 2010). This may seem to be a modest incremental contribution, but the oxygen these microscopic marine plants produce in the aggregate is critical: ocean-based photosynthesis creates half of the Earth's atmospheric oxygen (Roach, 2004). Periods of enhanced growth, reproduction, and proliferation of phytoplankton are called "blooms" (NASA, 2010). Although a bloom can last for weeks or months, the life span of individual phytoplankton is only a few days (NASA, 2010). Individual organisms are either consumed by predators or die and gradually sink through the water column. If their microscopic remains sink to depths that are undisturbed by surface weather, currents, and other factors, the carbon they contain will likely be sequestered for eons. England's famous White Cliffs of Dover are composed of massive numbers of phytoplankton skeletons that were deposited on the ocean floor during the Cretaceous Period around 100 million years ago (Witty, 2011; American Geophysical Union, 2016). These and other deposits elsewhere in the world contain trillions of tons of sequestered carbon.

As the primary producer, phytoplankton also provide direct and indirect nutrition for organisms up through the entire food chain. They provide the critical link between the sun's energy and the ocean's biologic components. While grazing by zooplankton, shellfish, and other higher-level predators is a significant factor in limiting the proliferation of phytoplankton communities and potential carbon sequestration, it necessarily leads to increased localized oceanic productivity. Several of the ocean's massive "biological deserts" offer ample sunlight and suitable water characteristics, yet generally contain little phytoplankton or other biological activity (Kerr, 2008). As phytoplankton form the base of the entire marine food web, waters that are nutrient limited and unable to support phytoplankton communities also provide little support for all other kinds of oceanic life.

Natural Ocean Fertilization

Nature has carried out large-scale ocean fertilization for many millions of years. Scientists have long recognized that the most productive areas of the world's oceans are those near the coasts and certain upwelling areas in shallows or along the continental shelf breaks that receive nutrients through natural processes on a consistent basis (Brink, 2004; Sigman and Hain, 2012). For example, wind blowing across certain land areas, such as the Sahara and Gobi Deserts, carries aloft nutrient-rich dust that settles on and fertilizes the northeast Atlantic, the Mediterranean Sea, the Sea of Japan, and other highly productive ocean regions. Powerful ocean currents, such as the Gulf Stream, Humboldt Current, and North Pacific Current, encounter shallow waters and bring vast amounts of nutrients from the deep ocean up into surface waters. These phenomena catalyze rich plankton blooms and support the remarkable biological productivity found on the Grand Banks, along the coasts of Chile and Peru, and in the central Bering Sea. Rivers, tidal activity, and winds distribute iron and other nutrients to most coastal waters. Large whales descend thousands of feet to consume squid and other creatures in the deep ocean and then defecate in surface waters. While doing so, these cetaceans help fertilize phytoplankton in the ocean (Roman et al., 2014). Other marine fish, birds, and mammals contribute to successful ocean fertilization through their feeding, diving, and other life processes.

Episodically, and far more violently, volcanic activity occasionally sends iron-rich dust over vast portions of the ocean, often with dramatic results. Volcanic eruptions can trigger temporary and large-scale increases in biological activity in ocean space, beginning with plankton blooms. In the short term, these blooms can remove huge quantities of carbon dioxide and produce a temporary, major enhancement of the oceanic food chain in the affected region. When Mount Pinatubo erupted on the Philippine island of Luzon in 1991, for example, strong upper-level winds distributed massive amounts of volcanic dust containing iron and other nutrients across the central and southern Pacific Ocean, South China Sea, and even eastern parts of the Indian Ocean. This phenomenon produced three measurable outcomes: (1) large plankton blooms over nutrient-poor ocean areas downwind of the eruption, especially in the South China Sea (Langmann, 2014); (2) a reduction in the rate of increase of atmospheric carbon dioxide over the following two years (Wright, 2003); and (3) a 0.5°C reduction in the Earth's temperature over this same period (Self et al., 1999).

On August 7, 2008, Kasatochi, a volcano in Alaska's Aleutian Island chain, erupted just as a storm system was passing through the area. The volcano sent massive amounts of ash and gases into the atmosphere, which traveled thousands of miles downwind, spreading iron-rich dust across a vast area of the Gulf of Alaska and the northeast Pacific Ocean during the late summer. The timing of the volcanic eruption was critical to the phytoplankton bloom. Both the storm and the summer sunlight contributed to this massive bloom (Choi, 2010). The level of photosynthesis and concentrations of chlorophyll in the ocean increased dramatically (Hamme et al., 2010; Langmann, 2014). Although scientists have not been able to establish a causal connection with certainty, this dramatic increase in biological productivity in the Gulf of Alaska may well have set the stage for a huge run of western Canadian salmon two years later (Olgun et al., 2013). Based on decades of declining salmon runs, fisheries experts had predicted that 1.7 million sockeye and other salmon would return to spawn in British Columbia's Fraser River in 2010. In actuality, fishery experts recorded the return of over 34 million fish that season (Parsons and Whitney, 2012). Interestingly, the previous record run for Fraser River salmon took place in 1958, two years after a volcanic eruption had occurred on the Kamchatka Peninsula in eastern Siberia (Jones, 2010).

Human-Engineered Ocean Fertilization

Partially based on these real-life experiences, a native Canadian tribe paid $2.5 million to the Haida Salmon Restoration Corporation in 2012 to spread 100 tons of iron-based fertilizer in the northeastern Pacific Ocean region, several hundred miles west of Haida Gwaii, in western British Columbia, where the salmon from their rivers matured (Tollefson, 2012). This fertilization effort produced an unprecedented phytoplankton bloom in that portion of the ocean (Batten and Gower, 2014). As record levels of salmon returned to their home rivers to spawn, some observers called the effort a "phenomenal success" (Revkin, 2014). Indeed, the 2013 salmon run defied all expectations, more than quadrupling previous years' yields from 50 million to 226 million fish (Zubrin, 2014). Despite these remarkable results, this intervention was of concern to many, largely because it was conducted without prior governmental review and approval or adequate scientific vetting. Even so, the Haida Salmon Restoration Corporation and the Haida Gwaii native community leaders are pushing ahead with developing and applying the concept of iron nutrient fertilization to increase ocean productivity for their communities well into the future (Haida Salmon Restoration, 2017).

Several companies and scientific institutes have been researching the potential of human-engineered ocean fertilization for safely promoting plankton growth in those extensive parts of the ocean that do not receive adequate nutrients on a consistent basis. These companies include Planktos (Planktos Ecosystems) and Green Sea Venture (Green Sea Venture, n.d.), both registered in the United States, as well as the Australian-based Ocean Nourishment Corporation (Ocean Nourishment) and the Canadian-based Haida Salmon Restoration Corporation (Haida Salmon Restoration). Each of these, as well as scientific research institutions such as the Moss Landing Marine Laboratories in California (Moss Landing), Woods Hole Oceanographic Institution in Massachusetts (Woods Hole, 2017), British National Oceanography Centre (National Oceanography Centre), and German Alfred Wegener Institute for Polar and Marine Research (Alfred Wegener Institut), have been investigating the science underlying ocean fertilization. All of these companies, institutions, and nongovernmental organizations (NGOs), as well as other scientists and researchers, are looking for ways to systematically and safely promote plankton growth in the ocean's biological deserts. For example, on its website, the management of Ocean Nourishment notes:

It is now widely accepted that to reach acceptable climate stabilization targets it will be necessary to both reduce emissions and remove legacy carbon from the atmosphere. Ocean Nourishment is a technology that removes legacy carbon from the atmosphere and sequesters this into the largest carbon sink on the planet – the deep ocean.

Human-engineered ocean iron fertilization is a rather recent concept in the context of climate science. The late oceanographer John H. Martin developed an hypothesis that the low level of phytoplankton in many ocean regions was caused by a persistent lack of adequate iron and other essential nutrients (NASA, n.d.). Martin speculated that increased iron concentrations, the phytoplankton growth they stimulated in the ocean, and the resulting large-scale removal of greenhouse gases from the atmosphere could help explain past ice ages. Conversely, he suggested that the lack of nutrients explains the extensive biological deserts present in large parts of the ocean today. Martin researched the role of iron as a phytoplankton micronutrient and posited that using iron fertilization to increase phytoplankton abundance and primary productivity in the ocean would increase biological demand for carbon dioxide and act as a sink for excess quantities of the substance (NASA, n.d.). Researchers have taken the experiments that Martin proposed into the open ocean and largely confirmed the validity of his theories.

International, scientifically endorsed, open-ocean studies have examined the effects of fertilization, using iron and other nutrients, on phytoplankton and the ocean environment. For example, during field research conducted in 2004 in the Southern Ocean as part of the European Iron Fertilization Experiment (EIFEX) scientists fertilized an area approximately 14 kilometers in diameter with seven tons of iron sulfate (Biello, 2012; Smetacek et al., 2012). The scientists followed the development of the ensuing large planktonic bloom for five weeks and found that the addition of iron resulted in a massive growth of phytoplankton from the surface down to about 100 meters (Helmholz Association, 2012). At the end of the study, the researchers discovered that more than half of the bloom's plankton diatoms died and sank rapidly, reaching depths of over 1,000 meters (Helmholz Association, 2012). Natural processes removed most of the dead diatoms and their captured carbon, fixed by photosynthesis within their shells during the time of active growth in surface waters, into the ocean depths (Marshall, 2012). Based on nature's systematic and episodic ocean fertilization over eons, it is likely that further rigorous scientific research will establish that ocean fertilization, carried out in nutrient-deficient ocean waters during appropriate seasons on scales similar to that of natural processes, can promote phytoplankton activity in a manner that is both effective and safe for the environment. Moreover, based on the EIFEX results, it is likely that a good portion of the carbon fixed in the phytoplankton will be sequestered on the ocean floor for eons.

A 2012 article in *Scientific American* estimated that ocean fertilization might be able to sequester as much as 1 billion metric tons of carbon dioxide per year, compared with annual human emissions of more than 8 billion metric tons and rising (Biello, 2012). Victor Smetacek of the Alfred Wegener Institute Helmholz Centre for Polar and Marine Research in Germany and his collaborators are even more optimistic, estimating that widespread ocean fertilization could remove between 0.5 and 1 gigaton of carbon from the atmosphere annually (Smetacek et al., 2012). This quantity equates to 25 to 33 percent of the carbon added annually from all sources (CBS News, 2012).

Iron fertilizer is very inexpensive and has the potential to be extremely cost-effective. Minute amounts of iron added to the surface waters of iron-depleted "biological deserts" during seasons in which there is adequate sunlight can trigger remarkably vibrant phytoplankton blooms (Perry, 2012). Current estimates of the amount of iron and other nutrients required to sequester significant amounts (gigatons) of carbon dioxide range widely (Powell, 2007). Moreover, introduction of this amount of nutrients would induce growth that could help restore the plankton deficiency worldwide (Powell, 2007). Of course, the cost of distributing vast amounts of iron-rich fertilizer to appropriate parts of the ocean could reach billions of dollars each year. However, even this amount would not be high when compared to the enormous costs of other efforts to mitigate or adapt to the consequences of climate change, declining phytoplankton and fish stocks, and ocean acidification. Whether assessed from an ecosystem-services perspective or in the context of a suitable carbon market, it is likely this fertilization process, even if done on an ongoing basis in distant parts of the ocean, could become cost-effective.

Concerns about Ocean Fertilization

Despite the tremendous potential of ocean fertilization, much of the international scientific and policy community appears reluctant to consider the concept, let alone foster its further research, development, and implementation. In 2015, the editors of *Scientific American* expressed concern about the "environmental skittishness" that opposed even the most rigorous scientific research into geoengineering proposals to deal with climate change, including ocean fertilization (The Editors, 2015). They argued: "Civilization may depend on such geo-engineering methods as the planet keeps warming. We need tests to get them right – and stop people from doing them wrong" (The Editors, 2015). An early report of the Intergovernmental Panel on

Climate Change briefly discussed "ocean fertilization" as one of the "apparently promising techniques," but it continued: "These options tend to be speculative and many of their environmental side effects have yet to be assessed; detailed cost estimates have not been published; and they are without a clear institutional framework for implementation" (IPCC, 2007, para. 11.2.2). Other organizations have also expressed concerns. For example, in June 2007, the States Party to the London Dumping Convention (see Chapter 2 of this volume) issued a formal statement of concern, noting that "the potential for large scale ocean iron fertilization to have negative impacts on the marine environment and human health" (IMO, 2007).

Some critics are concerned that ocean fertilization could create harmful algal blooms, hypoxia, toxins, and other adverse conditions. Lisa Speer of the Natural Resources Defense Council argues, "At best, [ocean fertilization] technology appears to be speculative in terms of its ability to permanently sequester carbon. At worst, it could turn out to be a disastrous experiment that could have major impacts on the ocean ecosystems that billions of people rely on for food, employment, and for life itself" (quoted in Harris, 2012). Other experts have questioned the long-term feasibility of the concept to remove large amounts of CO_2 from the atmosphere (Zeebe and Archer, 2005). Because of this often-fervent opposition, there have been no scientific research cruises to test the efficacy and environmental safety of ocean fertilization since 2009, and we do not know of any planned for the foreseeable future (Biello, 2012).

As a general rule, however, negative effects of phytoplankton blooms are only observed in coastal waters – as "red tides" – where systems already rich in biologic activity can be degraded by hypoxia or trophic impacts and excessive bioaccumulation. Most species of ocean-based phytoplankton are either harmless or beneficial (NASA, 2010). Properly employed, ocean fertilization would increase phytoplankton growth only in regions of the deep ocean, far out to sea, where iron and other nutrient deficiencies would otherwise be the primary limiting factor for marine productivity. The largest ocean fertilization research projects under active consideration are just a small fraction of the size of most natural current- or wind-fed and volcanic-induced blooms. Natural blooms in the wake of major dust storms have been studied since the beginning of the twentieth century, and no deep-water die-offs or other adverse consequences have been reported (Boyd et al., 2007). Moreover, if a particular experiment, or even large-scale employment of ocean fertilization, demonstrated adverse consequences in a particular ecosystem, fertilization efforts in that area could stop, with lessons learned collected and analyzed. As the introduced nutrients were consumed, natural conditions would reassert themselves.

The Way Forward

International organizations and experts, private companies and academia, and responsible government agencies should be at the forefront of responsible research into ocean fertilization. They should allocate appropriate levels of funding and commitment to the research effort. Moreover, these entities should ensure that any research proposal that involves significant intervention with environmental processes proceeds in a manner wholly consistent with the precautionary principle: if an action or policy raises concern of a risk of causing harm to the environment or to humans, in the absence of scientific consensus that the action or policy is not harmful, the burden of proof that it is not harmful falls on the proponents of the initiative (Hatzell-Nichols, 2017). All responsible proponents of ocean fertilization agree that there must be rigorous research to prove that large-scale ocean fertilization can be implemented without significant adverse environmental and human impacts, and that it should be done in a way that is cost-effective.

In addition to removing of vast quantities of carbon dioxide, a fertilization effort of this scope could significantly increase ocean productivity. Fisheries generate worldwide revenues of hundreds of billions of dollars each year and are important to meet the growing protein needs of the increasing global population, particularly in the developing world (The World Bank, 2017) (see chapters in Part III of this volume). Given these critically important global interests, even a modest increase in ocean productivity would more than pay for an extensive and ongoing fertilization initiative. Additionally, nutrient fertilization and the attendant removal of carbon dioxide from Earth's ocean waters could help mitigate ocean acidification and thereby help to restore the far healthier historic levels of marine phytoplankton.

Wise maritime governance would be critical if and after ocean fertilization were shown to be environmentally safe and cost-effective. It would not be too difficult or expensive for vessels engaged in international trade to carry distribution systems for iron and other nutrients while transiting through the ocean's biological deserts. Assuming that the safety and efficacy of ocean fertilization were proven to the complete satisfaction of the international community, the International Maritime Organization (IMO) could endorse and regulate the practice. The IMO has long required that ships engaged in international trade follow strict design, construction, and inspection rules; carry a wide variety of navigational, communications, environmental, lifesaving, and other equipment; and comply with standard protocols for training, certification, and underway operations (see Chapters 2 and 23 of this volume). After appropriate consideration, the IMO could prescribe measures and adopt standards for installing ocean fertilization systems, much as they have long prescribed navigational and safety equipment. Whenever a vessel equipped with such a system were sailing through a portion of the ocean in the proper season (during which the phytoplankton in that ecosystem would benefit significantly from the introduction of appropriate nutrients), GPS-linked computers could automatically trigger the ship's installed system to dispense appropriate amounts of nutrients. Potentially hundreds of the largest commercial vessels could carry and operate such fertilizing equipment as part of their routine international voyages. Ports of call could maintain supplies of iron sulfate and other nutrients available for restocking, as they have always done to replenish fuel bunkers, food stores, and other supplies. In this way, many thousands of tons of nutrients could be dispensed in nutrient-deficient ocean regions during appropriate seasons of the year at little incremental cost.

A complementary approach would be for individual coastal states or local or regional fishing organizations to pay to fertilize areas of the ocean to increase the health and quantity of the commercial fish stocks maturing in those waters. Just as farmers pay to fertilize their arable fields to increase crop yields and vitality as part of prudent land stewardship, those who rely on fishing might find it cost-effective to fertilize ocean space to increase their long-term success, as did the native peoples of Haida Gwaii in the eastern Pacific in 2012. The concept of paying taxes or fees to improve yield might be particularly appropriate given the public-trust character of marine ecosystems upon which commercial fisherman have long relied (Lam, 2012). An internationally approved and funded carbon offset program could help encourage the establishment and ongoing successful operation of such ocean fertilization efforts.

Even if the results of ocean fertilization activities met the hopes and expectations of its most fervent proponents, of course, the approach would constitute only one part of the major mitigation efforts needed to blunt and eventually reverse the adverse effects of climate change. The world must continue to take the necessary steps to significantly reduce the level of greenhouse gases discharged into the atmosphere by converting to renewable sources of energy; protecting and expanding jungles and woodlands; promoting kelp beds and mangrove swamps; improving fishing practices; nurturing fish, birds, and marine mammals; and otherwise acting as better stewards of Earth and its ocean ecosystems. Mankind must also enhance ocean productivity and conservation efforts while combating the many sources of destructive pollution affecting ocean waters.

Conclusion

Although proponents of ocean fertilization must proceed with caution, rely on the best available science and practices, and apply the precautionary principle, the potentially extraordinary benefits of ocean fertilization are simply too great to ignore. As David Biello, the environment and energy editor at *Scientific American*, observed in an article on ocean fertilization (Biello, 2014):

The ocean is no longer a vast, unknowable wilderness, whose mysterious gods must be placated before it can be crossed. Instead, it's become the first viable arena for large-scale manipulation of the planetary environment. A land animal has come to tame the heaving, alien world of the sea and, though doing so can make us uncomfortable, in the end it might undo a great deal of the damage we have already done.

Farmers have been preoccupied with husbanding and enhancing the fertility of the soil in their farmlands, fields, and rice paddies for thousands of years. Indeed, had the global community not embraced and matured the technology of fertilizing arable land, the world would not now be able to feed its current and growing population of 7.5 billion people (US Department of Agriculture, 2014; Farming First, 2017). Of course, poor farming techniques, including overfertilization, have led to serious environmental problems. Just as the world has developed prudent soil fertilization methods and the best scientific research steadily improves on past practices, humankind should now seriously undertake research to determine if fertilizing large portions of the ocean, which are now biological deserts, can provide a major carbon sink for the benefit of everyone while also improving the health and productivity of the ocean. States, international organizations, and scientific associations should press for additional scientific studies and work to set prudent, forward-looking governance parameters for such research to go forward. Ocean fertilization may well be one important measure to realize more effective ocean governance and stewardship amidst the many challenges of climate change.

Acknowledgment

The authors acknowledge and appreciate the editorial assistance that Lieutenant Commander Molly K. Waters provided in condensing a longer version of this chapter.

References

Abraham, J. (2017). Scientists study ocean absorption of human carbon pollution, *The Guardian*. Available at www.theguardian.com/environment/climate-consensus-97-per-cent/2017/feb/16/scientists-study-ocean-absorption-of-human-carbon-pollution.

Alfred Wegener Institut Hemholtz Sentrum für Polar und Meeresforschung (n.d.). Available at www.awi.de/en.html.

American Geophysical Union. (2016). Giant algal bloom sheds light on formation of the White Cliffs of Dover. *Joint Release*, September 15. Available at https://news.agu.org/press-release/giant-algal-bloom-sheds-light-on-formation-of-white-cliffs-of-dover/.

Batten, S. and Gower, J. (2014). Did the iron fertilization near Haida Gwaii in 2012 affect the pelagic lower trophic level ecosystem? *Journal of Plankton Research*, 36(4), 925–32. Available at http://plankt.oxfordjournals.org/content/36/4/925.ful.

Biello, D. (2012). Controversial spewed iron experiment succeeds as carbon sink. *Scientific American*. Available at www.scientificamerican.com/article/fertilizing-ocean-with-iron-sequesters-carbondioxide/.

Biello, D. (2014). Engineering the ocean: Once you know what plankton can do, you'll understand why fertilising the ocean with iron is not such a crazy idea. *Aeon*. Available at http://aeon.co/magazine/nature-and-cosmos/can-tiny-plankton-help-reverse-climate-change/.

Boyce, D. G., Lewis, M. R., and Worm, B. (2010). Global phytoplankton decline over the past century. *Nature*, **466**, 591–6. Available at www.nature.com/nature/journal/v466/n7306/full/nature09268.html?foxtrotcallback=true.

Boyd, P. W., Jickells, T., Law, C. S., et al. (2007). Mesoscale iron enrichment experiments 1993–2005: Synthesis and future directions. *Science*, 315(5812), 612–17. Available at http://science.sciencemag.org/content/315/5812/612.

Brink, K. H. (2004). The grass is greener in the coastal ocean. *Oceanus Magazine*, **42**(1). Available at www.whoi.edu/oceanus/viewArticle.do?id=2503.

CBS News. (2012). Rogue iron dump in ocean stirs controversy. *Livescience.com*, October 19. Available at www.cbsnews.com/news/rogue-iron-dump-in-ocean-stirs-controversy/.

Choi, C. A. (2010). Volcano blows hole in global warming fight, *NBC News*. Available at www.nbcnews.com/id/39632361/ns/technology_and_science-science/t/volcano-blows-hole-global-warming-fight/#.WkO5DHkUl7g.

Farming First. (2017). *Soil Health: The Key to Boosting Productivity, Sustainability, and Resilience*. Available at https://farmingfirst.org/2017/06/soil-health-the-key-to-boosting-productivity-sustainability-and-resilience/.

Green Sea Venture, Inc. (n.d.). Available at www.greenseaventure.com/.

Haida Salmon Restoration Corporation. (2017). Available at http://haidasalmon restoration.com.

Harris, R. (2012) Can adding iron to oceans slow global warming. *NPR*, July 18. Available at http://www.npr.org/2012/07/18/156976147/can-adding-iron-to-oceans-slow-global-warming.

Hatzell-Nichols, L. (2017). *A Climate of Risk: Precautionary Principles, Catastrophes, and Climate Change*. New York: Routledge (Taylor & Francis Group).

Helmholz Association. (2012). Researchers publish results of an iron fertilization experiment. *EurekAlert: The Global Source for Science News*. Available at www.eurekalert.org/pub_releases/2012-07/haog-rpr071712.php.

Hamme, R. C., Webley, P. W., Crawford, W. R., et al. (2010). Volcanic ash fuels anomalous plankton bloom in subarctic northeast Pacific. *Geophysical Research Letters*, **37**(19). Available at http://dx.doi.org/10.1029/2010GL044629.

International Maritime Organization (IMO). (2007). *Scientific Groups Cautious over Iron Fertilization of the Oceans to Sequester CO_2*. Available at http://www.imo .org/blast/mainframe.asp?topic_id=1472anddoc_id=8214.

Intergovernmental Panel on Climate Change (IPCC). (2007). *IPCC Fourth Assessment Report: Climate Change 2007*. Available at www.ipcc.ch/publications_ and_data/ar4/wg3/en/ch11s11-2-2.html.

Intergovernmental Panel on Climate Change (IPCC). (2013). *Climate Change 2013: The Physical Science Basis*. Available at www.climatechange2013.org/.

Intergovernmental Panel on Climate Change (IPCC). (2014). *Climate Change 2014: Synthesis Report*. Available at http://ipcc.ch/report/ar5/syr/.

Jones, N. (2010). Sparks fly over theory that volcano caused salmon boom. *Nature*, October 29. Available at www.nature.com/news/2010/101029/full/news.2010.572 .html.

Kerr, R. A. (2008). The ocean's biological deserts are expanding. *Science Magazine*, January 25. Available at http://news.sciencemag.org/2008/01/oceans-biological-deserts-are-expanding.

Lam, M. E. (2012). Of fish and fishermen: Shifting societal baselines to reduce environmental harm in fisheries. *Ecology and Society*, **17**(4). Available at www .ecologyandsociety.org/vol17/iss4/art18/.

Langmann, B. (2014). On the role of climate forcing by volcanic sulfate and volcanic ash. *Advances in Meteorology*. Available at www.hindawi.com/journals/amete/2014/340123/.

MacKenzie, D. (2011). *The Starving Ocean*. Available at www.fisherycrisis.com/.

Marshall, K. N., Kaplan, I. C., Hodgson, E. E., et al. (2017). Risks of ocean acidification in the California Current food web and fisheries: Ecosystem model projections. *Global Change Biology*, **23**(4), 1525–39. Available at https://news-oceanacidification-icc.org/2017/01/13/risks-of-ocean-acidification-in-the-california-current-food-web-and-fisheries-ecosystem-model-projections/.

Marshall, M. (2012). Geo-engineering with iron might work after all. *New Scientist*, July 8. Available at www.newscientist.com/article/mg21528744.100-geo-engineering-with-iron-might-work-after-all.html.

Moss Landing Marine Laboratories. (2017). Available at www.mlml.calstate.edu.

National Atmospheric and Space Administration (NASA). (n.d.). John Martin (1935–1993), *Earth Observatory*. Available at https://earthobservatory.nasa.gov/Features/Martin/martin_4.php and https://earthobservatory.nasa.gov/Features/Martin/martin_5.php.

National Atmospheric and Space Administration (NASA). (2010). What are phytoplankton?, *Earth Observatory*. Available at https://earthobservatory.nasa.gov/Features/Phytoplankton/.

National Oceanic and Atmospheric Administration (NOAA). (2017). *Trends in Atmosphere Carbon Dioxide*, July. Available at www.esrl.noaa.gov/gmd/ccgg/trends/.

National Oceanography Centre, Natural Environment Research Council. (2017). Available at http://noc.ac.uk.

Ocean Nourishment Corporation. (n.d.). Available at www.oceannourishment.com/ocean-solutions.

Olgun, N., Duggen, S., Langmann, B., et al. (2013). Geochemical evidence of oceanic iron fertilization by the Kasatochi volcanic eruption in 2008 and the potential impacts on Pacific sockeye salmon. *Marine Ecology Progress Series*, **488**, 81–8. Available at https://oceanrep.geomar.de/21945/

Parsons, T. R. and Whitney, F. A. (2012). Did volcanic ash from Mt. Kasatochi in 2008 contribute to a phenomenal increase in Fraser River sockeye salmon (*Oncorhynchus nerka*) in 2010? *Fisheries Oceanography*, **21**(5), 374–77. Available at http://dx.doi.org/10.1111/j.1365-2419.2012.00630.x.

Perry, W. (2012). Could fertilizing the ocean reduce global warming, *Live Science*. Available at www.livescience.com/21684-geoengineering-iron-fertilization-climate.html.

Planktos Ecosystems (n.d.). Available at http://planktos.com/.

Powell, H. (2007). Fertilizing the ocean with iron: Should we add iron to the sea to reduce greenhouse gases in the air? *Oceanus*, November 13. Available at www.whoi.edu/oceanus/viewArticle.do?id=34167.

Revkin, A. C. (2014). A fresh look at iron, plankton, carbon, salmon and ocean engineering. *Dot Earth: New York Times Blog*, July 18. Available at http://dotEarth.blogs.nytimes.com/2014/07/18/a-fresh-look-at-iron-plankton-carbon-salmon-and-ocean-engineering/?_php=trueand_type=blogsand_r=0.

Roach, J. (2004). Source of half Earth's oxygen gets little credit. *National Geographic News*, June 4. Available at http://news.nationalgeographic.com/news/2004/06/0607_040607_phytoplankton.html.

Roman, J., Estes, J. A., Morissette, L., et al. (2014). Whales as marine ecosystem engineers. *Frontiers in Ecology and the Environment*, **12**(7), 377–85. Available at www.researchgate.net/publication/263782441_Whales_as_marine_ecosystem_engineers.

Self, S., Zhao, J. X., Holasek, R. E., Torres, R. C., and King, A. J. (1999). The atmospheric impact of the 1991 Mount Pinatubo eruption. *USGS Study*. Available at http://pubs.usgs.gov/pinatubo/self/.

Sigman, D. M. and Hain, M. P. (2012). The biological productivity of the ocean. *Nature Education*. Available at www.nature.com/scitable/knowledge/library/the-biological-productivity-of-the-ocean-70631104.

Smetacek, V., Klaas, C., Strass, V. H., et al. (2012). Deep carbon export from a Southern Ocean iron-fertilized diatom bloom. *Nature*, **487**(7407), 313–19. Available at www.nature.com/articles/nature11229.

The Editors. (2015). Tech to cool down global warming should be tested now. *Scientific American*, January 2. Available at www.scientificamerican.com/article/tech-to-cool-down-global-warming-should-be-tested-now/.

Tollefson, J. (2012). Ocean-fertilization project off Canada sparks furor. *Nature*, October 23. Available at www.nature.com/news/ocean-fertilization-project-off-canada-sparks-furore-1.11631.

US Department of Agriculture, Economic Research Service. (2014). *With Adequate Productivity Growth, Global Agriculture Is Resilient to Future Population and Economic Growth*, December 1. Available at www.ers.usda.gov/amber-waves/2014/december/with-adequate-productivity-growth-global-agriculture-is-resilient-to-future-population-and-economic-growth/.

US Global Change Research Program. (2014). Mitigation – Chapter 27. *National Climate Assessment: Full Report*. Available at http://nca2014.globalchange.gov/report.

Witty, M. (2011). The White Cliffs of Dover are an example of natural carbon sequestration. *ResearchGate*. Available at www.researchgate.net/publication/274661547_The_White_Cliffs_of_Dover_are_an_Example_of_Natural_Carbon_Sequestration.

Woods Hole Oceanographic Institution, (2017). Available at www.whoi.edu/occi.

The World Bank. (2017). *Oceans, Fisheries and Coastal Economies*. Available at www.worldbank.org/en/topic/environment/brief/oceans.

Wright, L. (2003). In aftermath of volcanic eruptions, photosynthesis waxes, carbon dioxide wanes, *Scientific American*, March 28. Available at www.scientificamerican.com/article/in-aftermath-of-volcanic/.

Zeebe, R. E. and Archer, D. (2005). Feasibility of ocean fertilization and its impact on future atmospheric CO2 levels, *Geophysical Research Letters*, May 10. Available at http://onlinelibrary.wiley.com/doi/10.1029/2005GL022449/full.

Zubrin, R. (2014). The salmon are back – Thank human ingenuity, *National Review*, April 22. Available at www.nationalreview.com/article/376258/.

Part VII

Conclusion

27

Ocean Governance Amidst Climate Change
An Essay on the Future

PAUL G. HARRIS

Introduction

Each of the chapters in this volume presents specific insights for the governance of oceanic change. These insights range from analyses of underlying forces and institutions to accounts of real-world experiences on the sea and proposals for how to theorize or implement effective policies. In this concluding chapter, the aim is to highlight some of the lessons that may be learned from the preceding chapters and to contemplate briefly possible futures for governing oceanic change. The future that we can expect to have ought to shape ocean governance – but ocean governance can also shape the future that we will actually have.

Oceanic Change: Governing for the Future

In this volume, we have focused on a number of major issue areas – islands and coasts, marine fisheries, polar seas, governance institutions, international law, and global policies – that will be central to the governance of the oceans as the consequences of climate change become more evident. What are some of the big ideas or lessons that we can take away from the case studies in these issue areas?

The Durability of Sustainable Development

The case studies in Part II on vulnerable islands and coasts highlight the enduring significance and importance of *sustainable* development – economic development that is *environmentally* sustainable and able to address inequities in societies effectively. The communities of islands and coasts are among those that will be adversely affected most immediately and dramatically by climate change. They will experience sea-level rise, higher storm surges, more dramatic weather, and extreme degradation of environments along coastlines and within nearby waters. At the same time, these communities are often economically vulnerable, and sometimes they exist within states where powerful economic forces exacerbate this vulnerability. In spite of climate change, without major changes in social policies, members of these communities may have few options but to exploit increasingly vulnerable seas. Indeed, as oceanic change harms local ecosystems, the incentives to exploit those ecosystems unsustainably will likely increase as people compete for fewer natural resources. This points to the importance of development that helps island and coastal communities to escape poverty so that they can participate fully and beneficially in strategies for adapting to climate change. This will require financial resources, often from abroad. It may also require more than a little reform of policy processes, both for formulation and implementation, to ensure that those resources are used widely for the benefit of people and the environment rather than for a minority of politically well-connected actors.

Management for the Common Good

The case studies in Part III on fisheries and pelagic seas remind us that as oceans change, so, too, will fisheries. Many fish species will likely migrate to new areas, and entire fisheries may disappear as fish migrate from their normal ranges. Simultaneously, some fisheries may blossom, at least temporarily, as species move into new areas. Many fisheries already encompass multiple jurisdictions, including coastal areas, territorial seas, exclusive economic zones, and ocean areas beyond national jurisdiction. Managing fisheries, whether locally, nationally, regionally, or internationally, has already proven to be extraordinarily difficult. Most of the world's fisheries are not being exploited sustainably, and many of them have collapsed. If fisheries are to remain important sources of food in the future, fisheries governance will have to be far more effective in coping with all of the traditional challenges combined with the effects of oceanic change. Effective governance of fisheries amidst climate change will have to take a long-term view, in the process managing the varying interests of stakeholders that may have perverse incentives to continue to overexploit fishing grounds even as the seas lose their fecundity. Effective governance will require more coordination and cooperation across jurisdictions; this will demand much more compromise and sharing of resources than exists today, especially beyond national jurisdictions. The past has demonstrated the difficulty of doing this when oceanic conditions were relatively stable. Certainly, achieving success in managing fisheries in the future will put the governance of oceanic change to the test.

Extreme Change, Extreme Choices

A question for climate governance is fundamentally whether it is acceptable to tolerate, let alone encourage, behaviors that make the problem worse. The case studies in Part IV, which focus on changing polar seas, suggest that the policy directions of many governments are wrongheaded – even as the world shifts into high gear to combat climate change, many policies are actually encouraging it. Scientists are increasingly describing what can reasonably be characterized as horrifying changes in Earth's polar seas, ranging from incredible temperature increases and extreme loss of sea ice in the Arctic Ocean to new concerns that the Antarctic ice cap is far more vulnerable to global warming than previously thought, and thus may add greatly to sea-level rise in this century and beyond. A response to these changes could be to aggressively step up efforts to mitigate climate-changing pollution and to find creative means for governing changing polar oceans. However, the most visible responses to environmental changes in the Arctic and Antarctic have been to look for ways of economically exploiting the changes in those areas – to find ways of making money from them – even at the expense of the environment. In the Arctic, this has involved plans to increase shipping traffic and to open up new areas to the extraction of minerals generally and fossil fuels in particular. To do the latter will obviously make the problem of climate change worse: burning the oil and gas extracted from the Arctic will contribute to the global warming that is devastating the Arctic's ecosystems (to say nothing of the effects of Artic warming, such as changes in weather in Europe and North America, felt in other regions). Similarly, in both the Arctic and Antarctic, governments are looking for ways to expand fisheries – to go after fish that are moving from lower latitudes into these regions and to exploit wider areas as they become more accessible. The question for effective governance of oceanic change is whether these inclinations to exploit polar seas for economic gain can be overcome. Instead of looking for ways to profit from climate change in polar areas, what may be needed is a more extreme approach: hands off. After all, limiting environmental harm and coping with environmental change are hallmarks of good environmental governance, not least amidst climate change.

Adapting Regimes to Climate Change

The chapters in Part V on institutions and law for ocean governance critically examine multilateral international approaches to ocean governance amidst climate change. An important lesson that comes from those chapters is that it is vital for the climate and ocean regimes – the collection of agreements, institutions, and laws for collective action in these areas – to be recrafted so that they are much more aligned than at present. Global environmental governance has generally been characterized historically by relatively isolated attempts to deal with specific issues: an ozone regime to manage stratospheric ozone, an endangered-species regime to manage endangered species, a climate regime to manage climate change, an oceans regime to manage oceans, and so forth. This is not to say that diplomats and international policy makers are not aware of the connections among environmental issues; normally, they do understand the overlaps. In part, issues have been isolated to bring a measure of relative simplicity to international and global policy challenges that are monumentally complex in both technical and political terms. What is more, new efforts are underway to coordinate global environmental regimes. In the case of oceanic change, efforts are now beginning to bring significant rationality to the overlapping objectives and policies of the climate and ocean regimes (and, indeed, other regimes related to climate). This will have to continue, and indeed the pace and scale of these coordination efforts will have to increase greatly, if extant regimes for climate and oceans are to be effective sources of governance for oceanic change in the future.

Mitigation and Adaptation

The chapters in Part VI on policies for ocean governance remind us that the oceans play two overlapping parts in the climate change story: they have a central role in the degree and scale of climate change, and in turn they have an unfortunate role in suffering from the consequences of climate change. Effective policies for oceanic change will contribute to managing the oceans in ways that help to mitigate climate change and enable people and communities to adapt to the climate changes to come. At sea, this will mean protecting marine ecosystems as much as possible for the future, both for the sake of those ecosystems per se but also for people and communities that will rely upon them for sustenance and vitality. Governance of oceanic change will require continually improving scientific understanding of the linkages between ocean mitigation and ocean adaptation. It will need willingness among policy makers at all levels to act on scientific understanding in accordance with the precautionary principle: to act even if uncertainty remains, because the dangers of inaction are far too great. Realizing both mitigation and adaptation will require new resources: new institutional mechanisms that coordinate more actors in more issue areas; new economic principles that are not based on endless growth in material consumption (notably of fossil fuels); and new attitudes regarding human, national, and global security that finally put the environment – including Earth's oceans and seas – at the top of political agendas within and among states. The oceans can continue to buffer many of the effects of climate change as humanity picks up the pace of trying to limit its effects in the future. That will require governance that does not rely on miracles to happen, that goes beyond business as usual, and that pushes absolutely everyone to do all that we can to prevent and cope with climate change.

Climate Futures: Of Miracles, Business as Usual, and Doing What We Can

What will the governance of oceanic change look like in the future? A range of possibilities exist, some more or less likely than others.

Miracles

The least likely future is the miraculous one – the one that those who, like President Donald Trump, seem to believe – that climate change is of little (or no) concern. For them, global warming is perceived to be, at most, normal, mild, and hardly worth changing lifestyles to address. Alas, such beliefs do not comport with reality; they are fanciful fictions of a stable environment that is passing into history. What is more, such conceptions do not account for what is happening to Earth's oceans and seas: dramatic warming, unprecedented melting of sea ice, nearly back-to-back coral bleaching events, and transoceanic acidification. One is tempted to hope for a miracle that will magically make all of these impacts, and so many others, go away, or for these impacts to be a figment of the collective imagination. One wishes that the climate skeptics' and climate deniers' accusations – that scientists are strangely conspiring to hoodwink the world into believing climate change – were true. Alas, the future will not countenance such hopes. Climate change, and its corollaries in oceanic change, are becoming all too real. Yet as noted in Chapter 1, the belief in miracles may not go away anytime soon. Trump and other climate deniers, such as those in the Australian government, will use their power to act on their mythical beliefs. This means far less robust policies and action to mitigate greenhouse gas pollution, and indeed it means more policies to encourage industries and societies to remain heavily reliant on fossil fuels. Effective governance of oceanic change will require confronting assiduously such policies based on myth (or greed disguised as myth) until they are replaced more or less universally by policies informed by the realities of climate change. This will include promotion of oceanic policies that aim to adapt to the impacts of climate change for the benefit of marine ecosystems and nearby human communities. It will mean discouraging oceanic policies that unsustainably exploit the changes brought on by climate change – and especially discouraging policies, such as drilling for oil and gas in the Arctic Ocean, that will exacerbate both climate change and oceanic change.

Business as Usual

The most likely future is a variation on what might be called "business as usual light." A light version of business as usual involves industries, societies, and individuals continuing to premise their lives on continued growth in consumption, albeit increasingly of less environmentally harmful "stuff." But such "green consumption" will not have the necessary effect if more people join the global consuming classes, which seems destined to happen despite the impacts of climate change. Business as usual in this sense means moving toward more renewable forms of energy because that is what is sustainable for both economies and ecology. It means using technologies to reduce the human impact on the environment. Indeed, the world is now in the early stages of a transition away from fossil fuels. This is good news for humanity, good news for the global environment, and good news for the oceans. But this transition is happening *much too slowly*. Fossil fuels remain by far the most important source of energy globally. Even to cut their use by half in the next decade or two, which would be a miracle in itself, would not be enough to realize the objectives of the United Nations Framework Convention on Climate Change – to "prevent dangerous anthropogenic interference with the climate system" (see Chapter 1). Global warming would be mitigated, but it would continue nonetheless. Oceanic change might be mitigated, but it would continue nonetheless. Truly effective governance of oceanic change is not compatible with even an environmentally lighter-touch form of business as usual. The cornucopia of modernity, even one dominated by "green" consumption, can at best move the world two steps forward – only to have it move one step back. A lighter version of business as usual is better than a traditional version; it is better than nothing. But is still bad for the oceans and the planet more generally. We will have to do much more than that.

Doing What We Can

What is the alternative to business as usual in the future? One has to be realistic about such a question. Any careful analysis of global affairs generally and ocean affairs specifically points to fundamental weakness in how the world is governed. Many of these weaknesses can be addressed; indeed, they are being addressed in myriad ways by more enlightened governments, nongovernmental organizations, and other actors. The future depends on doing this far more in the near future and beyond. We ought not expect miracles, but we can expect governments and other actors to do *all that they can* to govern oceans effectively as climate change takes a firmer grip on the environment. Indeed, we can and ought to expect everyone, including individuals, to do all that *we* can. This means making choices in favor of the Earth's climate and ocean systems when those choices present themselves, as they do every day for nearly everyone who is not preoccupied with survival. Such an approach to climate change and ocean governance is not very exciting or innovative. It is really quite mundane. But that is partly the point: if it is so mundane yet so valuable to our collective well-being, it is a kind of wonder that we have not been doing it already. Given our collective dependence on the oceans, it is a wonder that they are not right at the top of the global agenda. That, too, will have to change.

Conclusion

Climate change is here to stay; it will grow worse, and the great expanses of the oceans will be changed forever. To be sure, the issues examined in this volume are hardly the only ones that are important for climate change and ocean governance. That said, they are among those that cannot be ignored: climate change will be felt directly and harshly among islands, coasts, fisheries, and polar seas, and dealing with the causes and impacts of oceanic change effectively will depend vitally on whether institutions, laws, and policies are able to respond. Looked at collectively, the chapters in this volume paint a picture of what is needed to govern oceanic change. It will require that we not waste more time hoping for miracles. It will require that we move far beyond business as usual. It will require quite dramatic changes in collective and individual behaviors to reduce drastically humanity's impact on the oceans. The reasonable and rational approach to governance is to formulate and implement policies that do everything that is possible to mitigate and to cope with oceanic change, given the myriad variables that affect environmental governance, not least of which human and ecological needs, technological capabilities, financial resources, and, of course, politics. By doing what is possible within these constraints, the global community might at least muddle along to minimize oceanic change and simultaneously adapt to many of the difficulties it has already started to bring. This is not a very optimistic way to end, but perhaps it is a realistic one. What we can say for certain is that climate change makes ocean governance more vital than ever. The sooner everyone realizes this, accepts it, and acts upon it, the more livable the future is likely to be.

Index